Probability & Statistics

About the Author

Athanasios Papoulis was educated at the Polytechnic University of Athens and at the University of Pennsylvania. He started teaching in 1948 at the University of Pennsylvania, and in 1952 he joined the faculty of the then Polytechnic Institute of Brooklyn. He has also taught at Union College, U.C.L.A., Stanford, and at the TH Darmstadt in Germany.

A major component of his work is academic research. He has consulted with many companies including Burrough's, United Technologies, and IBM, and published extensively in engineering and mathematics, concentrating on fundamental concepts of general interest. In recognition of his contributions, he received the distinguished alumnus award from the University of Pennsylvania in 1973, and, recently, the Humboldt award given to American scientists for internationally recognized achievements.

Professor Papoulis is primarily an educator. He has taught thousands of students and lectured in hundreds of schools. In his teaching, he stresses clarity, simplicity, and economy. His approach, reflected in his articles and books, has been received favorably throughout the world. All of his books have international editions and translations. In Japan alone six of his major texts have been translated. His book *Probability, Random Variables, and Stochastic Processes* has been the standard text for a quarter of a century. In 1980, it was chosen by the Institute of Scientific Information as a citation classic.

Every year, the IEEE, an international organization of electrical engineers, selects one of its members as the outstanding educator. In 1984, this prestigious award was given to Athanasios Papoulis with the following citation:

> *For inspirational leadership in teaching through thought-provoking lectures, research, and creative textbooks.*

PROBABILITY & STATISTICS

Athanasios Papoulis
Polytechnic University

PRENTICE HALL
Englewood Cliffs, NJ 07632

Library of Congress Cataloging-in-Publication Data

PAPOULIS, ATHANASIOS
Probability & statistics/Athanasios Papoulis.
p. cm.
Bibliography: p.
Includes index.
ISBN 0-13-711698-5
1. Probabilities. 2. Mathematical statistics. I. Title.
II. Title: Probability and statistics.
QA273.P197 1990
519.2--dc20
89-33219
CIP

Editorial/production supervision: *Colleen Brosnan*
Interior design: *Lorraine Mullaney*
Cover design: *Lorraine Mullaney*
Manufacturing buyer: *Mary Noonan*

Cover art:
"*Pensive Athena,* Goddess of Wisdom," 460-450 BC, Acropolis.

Printed in the United States of America
10 9 8 7 6 5 4 3 2 1

ISBN 0-13-711698-5

Prentice-Hall International (UK) Limited, *London*
Prentice-Hall of Australia Pty. Limited, *Sydney*
Prentice-Hall Canada Inc., *Toronto*
Prentice-Hall Hispanoamericana, S.A., *Mexico*
Prentice-Hall of India Private Limited, *New Delhi*
Prentice-Hall of Japan, Inc., *Tokyo*
Simon & Schuster Asia Pte. Ltd., *Singapore*
Editora Prentice-Hall do Brasil, Ltda., *Rio de Janeiro*

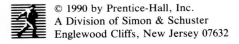

Contents

Preface

Probability is a difficult subject. A major reason is uncertainty about its meaning and skepticism about its value in the solution of real problems. Unlike other scientific disciplines, probability is associated with randomness, chance, even ignorance, and its results are interpreted not as objective scientific facts, but as subjective expressions of our state of knowledge. In this book, I attempt to convince the skeptical reader that probability is no different from any other scientific theory: All concepts are precisely defined within an abstract model, and all results follow logically from the axioms. It is true that the practical consequences of the theory are only inductive inferences that cannot be accepted as logical certainties; however, this is characteristic not only of statistical statements, but of all scientific conclusions.

The subject is developed as a mathematical discipline; however, mathematical subtleties are avoided and proofs of difficult theorems are merely sketched or, in some cases, omitted. The applications are selected not only because of their practical value, but also because they contribute to the mastery of the theory. The book concentrates on basic topics. It also includes a simplified treatment of a number of advanced ideas.

In the preparation of the manuscript, I made a special effort to clarify the meaning of all concepts, to simplify the derivations of most results, and to unify apparently unrelated concepts. For this purpose, I reexamined the conventional approach to each topic, departing in many cases from traditional methods and interpretations. A few illustrations follow:

In the first chapter, the various definitions of probability are analyzed and the need for a clear distinction between concepts and reality is stressed. These ideas are used in Chapter 8 to explain the difference between probability and statistics, to clarify the controversy surrounding Bayesian statistics, and to develop the dual meaning of random numbers. In Chapter 11, a comprehensive treatment of the method of least square is presented, showing the connection between deterministic curve fitting, parameter estimation, and prediction. The last chapter is devoted to entropy, a topic rarely discussed in books on statistics. This important concept is defined as a number associated to a partition of a probability space and is used to solve a number of ill-posed problems in statistical estimation. The empirical interpretation of entropy and the rationale for the method of maximum entropy are related to repeated trials and typical sequences.

The book is written primarily for upper division students of science and engineering. The first part is suitable for a one-semester junior course in probability. No prior knowledge of probability is required. All concepts are developed slowly from first principles, and they are illustrated with many examples. The first three chapters involve mostly only high school mathematics; however, a certain mathematical maturity is assumed. The level of sophistication increases in subsequent chapters. Parts I and II can be covered in a two-semester senior/graduate course in probability and statistics.

This work is based on notes written during my stay in Germany as a recipient of the Humboldt award. I wish to express my appreciation to the Alexander von Humboldt Foundation and to my hosts Dr. Eberhard Hänsler and Dr. Peter Hagedorn of the TH Darnstadt for giving me the opportunity to develop these notes in an ideal environment.

Athanasios Papoulis

PART ONE

PROBABILITY

1 The Meaning of Probability

Most scientific concepts have a precise meaning corresponding, more or less exactly, to physical quantities. In contrast, probability is often viewed as a vague concept associated with randomness, uncertainty, or even ignorance. This is a misconception that must be overcome in any serious study of the subject. In this chapter, we argue that the theory of probability, like any other scientific discipline, is an exact science, and all its conclusions follow logically from basic principles. The theoretical results must, of course, correspond in a reasonable sense to the real world; however, a clear distinction must always be made between theoretical results and empirical statements.

1-1 Introduction

The theory of probability deals mainly with *averages of mass phenomena* occurring sequentially or simultaneously: games of chance, polling, insurance, heredity, quality control, statistical mechanics, queuing theory, noise. It has been observed that in these and other fields, certain averages approach a constant value as the number of observations increases, and this value remains the same if the averages are evaluated over any subsequence selected prior to the observations. In a coin experiment, for example, the ratio of heads to tosses approaches 0.5 or some other constant, and the same ratio is obtained if one considers, say, every fourth toss. The purpose of the theory is to describe and predict such averages in terms of probabilities of events. The probability of an event $\mathscr{A}$ is a number $P(\mathscr{A})$ assigned to $\mathscr{A}$. This number is central in the theory and applications of probability; its significance is the main topic of this chapter. As a measure of averages, $P(\mathscr{A})$ is interpreted as follows:

If an experiment is performed n times and the event $\mathscr{A}$ occurs $n_{\mathscr{A}}$ times, then almost certainly the relative frequency $n_{\mathscr{A}}/n$ of the occurrence of $\mathscr{A}$ is close to $P(\mathscr{A})$

$$P(\mathscr{A}) \simeq \frac{n_{\mathscr{A}}}{n} \tag{1-1}$$

provided that n is sufficiently large. This will be called the *empirical* or *relative frequency* interpretation of probability.

Equation (1-1) is only a heuristic relationship because the terms *almost certainly, close,* and *sufficiently large* have no precise meaning. The relative frequency interpretation cannot therefore be used to define $P(\mathscr{A})$ as a theoretical concept. It can, however, be used to estimate $P(\mathscr{A})$ in terms of the observed $n_{\mathscr{A}}$ and to predict $n_{\mathscr{A}}$ if $P(\mathscr{A})$ is known. For example, if 1,000 voters are polled and 451 respond Republican, then the probability $P(\mathscr{A})$ that a voter is Republican is about .45. With $P(\mathscr{A})$ so estimated, we predict that in the next election, 45% of the people will vote Republican.

The relative frequency interpretation of probability is objective in the sense that it can be tested experimentally. Suppose, for example, that we wish to test whether a coin is fair, that is, whether the probability of heads equals .5. To do so, we toss it 1,000 times. If the number of heads is "about" 500, we conclude that the coin is indeed fair (the precise meaning of this conclusion will be clarified later).

Probability also has another interpretation. It is used as a measure of our state of knowledge or belief that something is or is not true. For example, based on evidence presented, we conclude with probability .6 that a defendant is guilty. This interpretation is subjective. Another juror, having access to the same evidence, might conclude with probability .95 (beyond any reasonable doubt) that the defendant is guilty.

We note, finally, that in applications involving predictions, both interpretations might be relevant. Consider the following weather forecast: "The probability that it will rain tomorrow in New York is .6." In this forecast, the number .6 is derived from past records, and it expresses the relative frequency of rain in New York under similar conditions. This number, however, has no relevance to tomorrow's weather. Tomorrow it will either rain or not rain. The forecast expresses merely the state of the forecaster's knowledge, and it helps us decide whether we should carry an umbrella.

Concepts and Reality

Students are often skeptical about the validity of probabilistic statements. They have been taught that the universe evolves according to physical laws that specify exactly its future (determinism) and that probabilistic descriptions are used only for "random" or "chance" phenomena, the initial conditions of which are unknown. This deep-rooted skepticism about the "truth" of probabilistic results can be overcome only by a proper interpretation of the meaning of probability. We shall attempt to show that, like any other scientific discipline, probability is an exact science and that all conclusions follow logically from the axioms. It is, of course, true that the correspondence between theoretical results and the real world is imprecise; however, this is characteristic not only of probabilistic conclusions but of all scientific statements.

In a probabilistic investigation the following steps must be clearly distinguished (Fig. 1.1).

Step 1 *(Physical).* We determine, by a process that is not and cannot be made exact, the probabilities $P(\mathscr{A}_i)$ of various physical events $\mathscr{A}_i$.

Step 2 *(Conceptual).* We assume that the numbers $P(\mathscr{A}_i)$ satisfy certain axioms, and by deductive logic we determine the probabilities $P(\mathscr{B}_i)$ of other events $\mathscr{B}_i$.

Step 3 *(Physical).* We make physical predictions concerning the events $\mathscr{B}_i$ based on the numbers $P(\mathscr{B}_i)$ so obtained.

In steps 1 and 3 we deal with the *real world,* and all statements are inexact. In step 2 we replace the real world with an abstract *model,* that is, with a *mental construct* in which all concepts are precise and all conclusions follow from the axioms by deductive logic. In the context of the resulting theory, the probability of an event $\mathscr{A}$ is a number $P(\mathscr{A})$ satisfying the axioms; its "physical meaning" is not an issue.

Figure 1.1

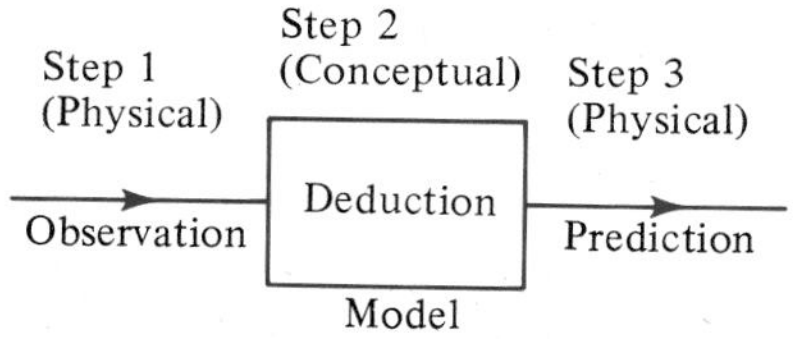

We should stress that model formation is used not only in the study of "random" phenomena but in all scientific investigations. The resulting theories are, of course, of no value unless they help us solve real problems. We must assign specific, if only approximate, numerical values to physical quantities, and we must give physical meaning to all theoretical conclusions. The link, however, between theory (model) and applications (reality) is always inexact and must be separated from the purely deductive development of the theory. Let us examine two illustrations of model formulation from other fields.

Geometry Points and lines, as interpreted in theoretical geometry, are not real objects. They are mental constructs having, by assumption, certain properties specified in terms of the *axioms*. The axioms are chosen to correspond in some sense to the properties of real points and lines. For example, the axiom "one and only one line passes through two points" is in reasonable agreement with our perception of the corresponding property of a real line.

Starting from the axioms, we derive, by pure reasoning, other properties that we call *theorems*. The theorems are then used to draw various useful conclusions about real geometric objects. For example, we prove that the sum of the angles of a conceptual triangle equals 180°, and we use this theorem to conclude that the sum of the angles of a real triangle equals approximately 180°.

Circuit Theory In circuit theory, a resistor is by definition a two-terminal device with the property that its voltage V is proportional to the current I. The proportionality constant

$$R = \frac{V}{I} \tag{1-2}$$

is the resistance of the device.

This is, of course, only an idealized model of a real resistor and (1-2) is an axiom (Kirchoff's law). A real resistor is a complicated device without clearly defined terminals, and a relationship of the form in (1-2) can be claimed only as a convenient idealization valid with a variety of qualifications and subject to unknown "errors." Nevertheless, in the development of the theory, all these uncertainties are ignored. A real resistor is replaced by a mental concept, and a theory is developed based on (1-2). It would not be useful, we must agree, if at each stage of the theoretical development we were concerned with the "true" meaning of R.

Returning to statistics, we note that, in the context of an abstract model (step 2), the probability $P(\mathscr{A})$ is interpreted as a number $P(\mathscr{A})$ that satisfies various axioms but is otherwise arbitrary. In the applications of the theory to real problems, however (steps 1 and 3), the number $P(\mathscr{A})$ must be given a physical interpretation. We shall establish the link between model and reality using three interpretations of probability: relative frequency, classical, and subjective. We introduce here the first two in the context of the die experiment.

Example 1.1

We wish to find the probability of the event $\mathcal{A}$ = {even} in the single-die experiment.

In the relative frequency interpretation, we rely on (1-1): We roll the die n times, and we set $P(\mathcal{A}) = n_{\mathcal{A}}/n$ where $n_{\mathcal{A}}$ is the number of times the event $\mathcal{A}$ occurs. This interpretation can be used for any die, fair or not. In the classical interpretation, we assume that the six faces of the die are *equally likely;* that is, they have the same probability of showing (this is the "fair die" assumption). Since the event {even} occurs if one of the three outcomes f_2 or f_4 or f_6 shows, we conclude that $P\{\text{even}\} = 3/6$. This conclusion seems logical and is generally used in most games of chance. As we shall see, however, the equally likely condition on which it is based is not a simple consequence of the fact that the die is symmetrical. It is accepted because, in the long history of rolling dice, it was observed that the relative frequency of each face equals 1/6. ■

Illustrations

We give next several examples of simple experiments, starting with a brief explanation of the empirical meaning of the terms *trials, outcomes,* and *events.*

A *trial* is the single performance of an experiment. Experimental *outcomes* are various observable characteristics that are of interest in the performance of the experiment. An *event* is a set (collection) of outcomes. The *certain event* is the set $\mathcal{S}$ consisting of all outcomes. An *elementary event* is an event consisting of a single outcome. At a single trial, we *observe* one and only one outcome. An event occurs at a trial if it contains the observed outcome. The certain event occurs at every trial because it contains every outcome.

Consider, for example, the die experiment. The outcomes of this experiment are the six faces $f_1, \ldots, f_6$; the event {even} consists of the three outcomes f_2, f_4, and f_6; the certain event $\mathcal{S}$ consists of all six outcomes. A trial is the roll of the die once. Suppose that at a particular trial, f_2 shows. In this case we observe the outcome f_2; however, many events occur, namely, the certain event, the event {even}, the elementary event $\{f_2\}$, and 29 other events!

Example 1.2

The coin experiment has two outcomes: heads (h) and tails (t). The event heads = $\{h\}$ consists of the single outcome h. To find the probability $P\{h\}$ of heads, we toss the coin n = 1,000 times, and we observe that heads shows n_h = 508 times. From this we conclude that $P\{h\} \simeq .51$ (step 1). This leads to the expectation that in future trials, about 51% of the tosses will show heads (step 3).

One might argue that this is only an approximation. Since a coin is symmetrical (the "equally likely" assumption), the probability of heads is .5. Had we therefore kept tossing, the ratio n_h/n would have approached .5. This is generally true; however, it is based on our long experience with coins, and it holds only for the limited class of fair coins. ■

Example 1.3

We wish to find the probability that a newborn child is a girl. In this experiment, we have again two outcomes: boy (b) and girl (g). We observe that among the 1,000 recently born children, 489 are girls. From this we conclude that $P\{g\} \simeq .49$. This leads to the expectation that about 49% of the children born under similar circumstances will be girls.

Here again, there are only two outcomes, and we have no reason to believe that they are not equally likely. We should expect, therefore, that the correct value for $P\{g\}$ is .5. However, as extensive records show, this expectation is not necessarily correct. ■

Example 1.4

A poll taken for the purpose of determining Republican (r) or Democratic (d) party affiliation specifies an experiment consisting of the two outcomes r and d. A trial is the polling of one person. It was found that among 1,000 voters questioned, 382 were Republican. From this it follows that the probability $P\{r\}$ that a voter is Republican is about .38, and it leads to the expectation that in the next election, about 38% of the people will vote Republican.

In this case, it is obvious that the equally likely condition cannot be used to determine $P\{r\}$. ■

Example 1.5

Daily highway accidents specify an experiment. A trial of this experiment is the determination of the accidents in a day. An outcome is the number k of accidents. In principle, k can take any value; hence, the experiment has infinitely many outcomes, namely all integers from 0 to ∞. The event $\mathscr{A} = \{k = 3\}$ consists of the single outcome $k = 3$. The event $\mathscr{B} = \{k \le 3\}$ consists of the four outcomes $k = 0, 1, 2,$ and 3.

We kept a record of the number of accidents in 1,000 days. Here are the number n_k of days on which k accidents occurred:

k	0	1	2	3	4	5	6	7	8	9	10	>10
n_k	13	80	144	200	194	155	120	75	14	4	1	0

From the table and (1-1) it follows with $n_{\mathscr{A}} = n_3 = 200$, $n_{\mathscr{B}} = n_0 + n_1 + n_2 + n_3 = 437$, and $n = 1{,}000$; hence

$$P(\mathscr{A}) = P\{k = 3\} = .2 \qquad P(\mathscr{B}) = P\{k \le 3\} \simeq .437$$

■

Example 1.6

We monitor all telephone calls originating from a station between 9:00 and 10:00 A.M. We thus have an experiment, the outcomes of which are all time instances between 9:00 and 10:00. A single trial is a particular call, and an outcome is the time of the call. The experiment therefore has infinitely many outcomes. We observe that among the last 1,000 calls, 248 occurred between 9:00 and 9:15. From this we conclude that the probability of the event $\mathscr{A}$ = {the call occurs between 9:00 and 9:15} equals $P(\mathscr{A}) \simeq .25$. We expect, therefore, that among all future calls occurring between 9:00 and 10:00 A.M., 25% will occur between 9:00 and 9:15.

Example 1.7

The age t at death of a person specifies an experiment with infinitely many outcomes. We wish to find the probability of the event $\mathscr{A}$ = {death occurs before 60}. To do so, we record the ages at death of 1,000 persons, and we observe that 682 of them are less than 60. From this we conclude that $P(\mathscr{A}) \simeq 682/1000 \simeq .68$. We should expect, therefore, that 68% of future deaths will occur before the age of 60.

Regularity and Randomness We note that to make predictions about future averages based on past averages, we must impose the following restrictions on the underlying experiment.

A. Its trials must be performed under "essentially similar" conditions (regularity).

B. The ratio $n_{\mathscr{A}}/n$ must be "essentially" the same for any subsequence of trials selected prior to the observation (randomness).

These conditions are heuristic and cannot easily be tested. As we illustrate next, the difficulties vary from experiment to experiment. In the coin experiment, both conditions can be readily accepted. In the birth experiment, A is acceptable but B might be questioned: If we consider only the subsequence of births of twins where the firstborn is a boy, we might find a different average. In the polling experiment, both conditions might be challenged: If the voting does not take place soon after the polling, the voters might change their preference. If the polled voters are not "typical," for example, if they are taken from an affluent community, the averages might change.

1-2 The Four Interpretations of Probability

The term *probability* has four interpretations:

1. Axiomatic definition (model concept)
2. Relative frequency (empirical)
3. Classical (equally likely)
4. Subjective (measure of belief)

In this book, we shall use only the axiomatic definition as the basis of the theory (step 2). The other three interpretations will be used in the determination of probabilistic data of real experiments (step 1) and in the applications of the theoretical results to real experiments (step 3). We should note that the last three interpretations have also been used as definitions in the theoretical development of probability; as we shall see, such definitions can be challenged.

Axiomatic

In the axiomatic development of the theory of probability, we start with a *probability space*. This is a set $\mathcal{S}$ of abstract objects (elements) called outcomes. The set $\mathcal{S}$ and its subsets are called events. The probability of an event $\mathcal{A}$ is by definition a number $P(\mathcal{A})$ assigned to $\mathcal{A}$. This number satisfies the following three axioms but it is otherwise arbitrary.

I. $P(\mathcal{A})$ is a nonnegative number:

$$P(\mathcal{A}) \geq 0 \tag{1-3}$$

II. The probability of the event $\mathcal{S}$ (certain event) equals 1:

$$P(\mathcal{S}) = 1 \tag{1-4}$$

III. If two events $\mathcal{A}$ and $\mathcal{B}$ have no common elements, the probability of the event $\mathcal{A} \cup \mathcal{B}$ consisting of the outcomes that are in $\mathcal{A}$ or $\mathcal{B}$ equals the sum of their probabilities:

$$P(\mathcal{A} \cup \mathcal{B}) = P(\mathcal{A}) + P(\mathcal{B}) \tag{1-5}$$

The resulting theory is useful in the determination of averages of mass phenomena only if the axioms are consistent with the relative frequency of probability, equation (1-1). This means that if in (1-3), (1-4), and (1-5) we replace all probabilities by the corresponding ratios $n_{\mathcal{A}}/n$, the resulting equations remain approximately true. We maintain that they do.

Clearly, $n_{\mathcal{A}} \geq 0$; furthermore, $n_{\mathcal{S}} = n$ because the certain event occurs at every trial. Hence,

$$P(\mathcal{A}) \simeq \frac{n_{\mathcal{A}}}{n} \geq 0 \qquad P(\mathcal{S}) \simeq \frac{n_{\mathcal{S}}}{n} = 1$$

in agreement with axioms I and II. To show the consistency of axiom III with (1-1), we observe that if the events $\mathcal{A}$ and $\mathcal{B}$ have no common elements and at a specific trial the event $\mathcal{A}$ occurs, the event $\mathcal{B}$ does not occur. And since the event $\mathcal{A} \cup \mathcal{B}$ occurs when either $\mathcal{A}$ or $\mathcal{B}$ occurs, we conclude that $n_{\mathcal{A}\cup\mathcal{B}} = n_{\mathcal{A}} + n_{\mathcal{B}}$. Hence,

$$P(\mathcal{A} \cup \mathcal{B}) \simeq \frac{n_{\mathcal{A}\cup\mathcal{B}}}{n} = \frac{n_{\mathcal{A}}}{n} + \frac{n_{\mathcal{B}}}{n} \simeq P(\mathcal{A}) + P(\mathcal{B})$$

in agreement with (1-5).

Model Formation We comment next on the connection between an abstract space $\mathcal{S}$ (model) and the underlying real experiment. The first step in model formation is the correspondence between elements of $\mathcal{S}$ and experimental outcomes. In Section 1-1 we assumed routinely that the outcomes of an experiment are readily identified. This, however, is not always the case.

The actual outcomes of a real experiment can involve a large number of observable characteristics. In the formation of the model, we select from all these characteristics the ones that are of interest in our investigation. We demonstrate with two examples.

Example 1.8

Consider the possible models of the die experiment as interpreted by the three players X, Y, and Z.

X says that the outcomes of this experiment are the six faces of the die, forming the space $\mathcal{S} = \{f_1, \ldots, f_6\}$. In this space, the event {even} consists of the three outcomes f_2, f_4, and f_6.

Y wants to bet on *even* or *odd* only. He argues, therefore, that the experiment has only the two outcomes *even* and *odd*, forming the space $\mathcal{S} = \{\text{even, odd}\}$. In this space, {even} is an elementary event consisting of the single outcome *even*.

Z bets that the die will rest on the left side of the table and f_1 will show. He maintains, therefore, that the experiment has infinitely many outcomes consisting of the six faces of the die and the coordinates of its center. The event {even} consists not of one or of three outcomes but of infinitely many. ■

Example 1.9

In a polling experiment, a trial is the selection of a person. The person might be Republican or Democrat, male or female, black or white, smoker or nonsmoker, and so on. Thus the observable outcomes are a myriad of characteristics. In Example 1.4 we considered as outcomes the characteristics "Republican" and "Democrat" because we were interested only in party affiliation. We would have four outcomes if we considered also the sex of the selected persons, eight outcomes if we included their color, and so on. ■

Thus the outcomes of a probabilistic model are precisely defined objects corresponding not to the myriad of observable characteristics of the underlying real experiment but only to those characteristics that are of interest in the investigation.

Note The axiomatic approach to probability is relatively recent* (Kolmogoroff, 1933); however, the axioms and the formal results had been used earlier. Kolmogoroff's contribution is the interpretation of probability as an abstract concept and the development of the theory as a precise mathematical discipline based on measure theory.

Relative Frequency

The relative frequency interpretation (1-1) of probability states that if in n trials an event $\mathcal{A}$ occurs $n_{\mathcal{A}}$ times, its probability $P(\mathcal{A})$ is approximately $n_{\mathcal{A}}/n$:

$$P(\mathcal{A}) \simeq \frac{n_{\mathcal{A}}}{n} \tag{1-6}$$

* A. Kolmogoroff, "Grundbegriffe der Wahrscheinlichkeits Rechnung," *Ergeb Math und ihrer Grenzg.* Vol. 2, 1933.

provided that n is sufficiently large and the ratio $n_{\mathscr{A}}/n$ is nearly constant as n increases.

This interpretation is fundamental in the study of averages, establishing the link between the model parameter $P(\mathscr{A})$, however it is defined, and the empirical ratio $n_{\mathscr{A}}/n$. In our investigation, we shall use (1-6) to assign probabilities to the events of real experiments. As a reminder of the connection between concepts and reality, we shall give a *relative frequency interpretation* of various axioms, definitions, and theorems based on (1-6).

The relative frequency cannot be used as the definition of $P(\mathscr{A})$ because (1-6) is an approximation. The approximation improves, however, as n increases. One might wonder, therefore, whether we can define $P(\mathscr{A})$ as a limit:

$$P(\mathscr{A}) = \lim_{n\to\infty} \frac{n_{\mathscr{A}}}{n} \tag{1-7}$$

We cannot, of course, do so if n and $n_{\mathscr{A}}$ are experimentally determined numbers because in any real experiment, the number n of trials, although it might be large, it is always finite. To give meaning to the limit, we must interpret (1-7) as an *assumption* used to define $P(\mathscr{A})$ as a theoretical concept. This approach was introduced by Von Mises* early in the century as the foundation of a new theory based on (1-7). At that time the prevailing point of view was still the classical, and his work offered a welcome alternative to the concept of probability defined independently of any observation. It removed from this concept its metaphysical implications, and it demonstrated that the classical definition works in real problems only because it makes implicit use of relative frequencies based on our long experience. However, the use of (1-7) as the basis for a deductive theory has not enjoyed wide acceptance. It has generally been recognized that Kolmogoroff's approach is superior.

Classical

Until recently, probability was defined in terms of the classical interpretation. As we shall see, this definition is restrictive and cannot form the basis of a deductive theory. It is, however, important in assigning probabilities to the events of experiments that exhibit geometric or other symmetries.

The classical definition states that if an experiment consists of N outcomes and $N_{\mathscr{A}}$ of these outcomes are "favorable" to an event $\mathscr{A}$ (i.e., they are elements of $\mathscr{A}$), then

$$P(\mathscr{A}) = \frac{N_{\mathscr{A}}}{N} \tag{1-8}$$

* Richard Von Mises, *Probability, Statistics, and Truth,* English edition, H. Geiringer, ed. (London: G. Allen and Unwin Ltd., 1957).

In words, the probability of an event $\mathscr{A}$ equals the ratio of the number of outcomes $N_{\mathscr{A}}$ favorable to $\mathscr{A}$ to the total number N of outcomes.

This definition, as it stands, is ambiguous because, as we have noted, the outcomes of an experiment can be interpreted in several ways. We shall demonstrate the ambiguity and the need for improving the definition with an example.

Example 1.10

We roll two dice and wish to find the probability p that the sum of the faces that show equals 7. We shall analyze this problem in terms of the following models.

(a) We consider as experimental outcomes the $N = 11$ possible sums 2, 3, . . . , 12. Of these, only the outcome 7 is favorable to the event $\mathscr{A} = \{7\}$, hence $N_{\mathscr{A}} = 1$. If we use (1-8) to determine p, we must conclude that $p = 1/11$.

(b) We count as outcomes the $N = 21$ pairs 1-1, 1-2, 1-3, . . . , 6-6, not distinguishing between the first and the second die. The favorable outcomes are now $N_{\mathscr{A}} = 3$, namely the pairs 1-6, 2-5, and 3-4. Again using (1-8), we must conclude that $p = 3/21$.

(c) We count as outcomes the $N = 36$ pairs distinguishing between the first and the second die. The favorable outcomes are now the $N_{\mathscr{A}} = 6$ pairs 1-6, 6-1, 2-5, 5-2, 3-4, 4-3, and (1-8) yields $p = 6/36$.

We thus have three different solutions for the same problem. Which is correct? One might argue that the third is correct because the "true" number of outcomes is 36. Actually *all three models can be used to describe the die experiment.* The third leads to the correct solution because its outcomes are "equally likely." For the other two models, we cannot determine p from (1-8). ■

Example 1.10 leads to the following refinement of (1-8): The probability of an event $\mathscr{A}$ equals the ratio of the number of outcomes $N_{\mathscr{A}}$ favorable to $\mathscr{A}$ to the total number N of outcomes, provided that all outcomes are equally likely.

As we shall see, this refinement does not eliminate the problems inherent in the classical definition. We comment next on the various objections to the classical definition as the foundation of a precise theory and on its value in the determination of probabilistic data and as a working hypothesis.

Note Be aware of the difference between the numbers n and $n_{\mathscr{A}}$ in (1-1) and the numbers N and $N_{\mathscr{A}}$ in (1-8). In the former, n is the total number of *trials* (repetitions) of the experiment and $n_{\mathscr{A}}$ is the number of successes of the event $\mathscr{A}$. In the latter, N is the total number of *outcomes* of the experiment and $N_{\mathscr{A}}$ is the number of outcomes that are favorable to $\mathscr{A}$ (are elements of $\mathscr{A}$).

CRITICISMS

1. The term *equally likely* used in the refined version of (1-8) can mean only that the outcomes are *equally probable*. No other interpretation consistent with the equation is possible. Thus the definition is *circular;* the concept to be defined is used in the definition. This often leads to ambiguities about the correct choice of N and, in fact, about the validity of (1-8).
2. It appears that (1-8) is a logical necessity that does not depend on experience: "A die is symmetrical; therefore, the probability that 5 will show equals 1/6." However, this is not so. We accept certain alternatives as equally likely because of our collective experience. The probability that 5 will show equals 1/6 not only because the die exhibits geometric symmetries but also because it was observed in the long history of rolling dice that 5 showed in about 1/6 of the trials.

In the next example, the equally likely condition appears logical but is not in agreement with observation.

Example 1.11

We wish to find the probability that a newborn baby is a girl. It is generally assumed that $p = 1/2$ because the outcomes *boy* and *girl* are "obviously" equally likely. However, this conclusion cannot be reached as a logical necessity unrelated to observation. In the first place, it is only an approximation. Furthermore, without access to long records, we would not know that the boy-girl alternatives are equally likely regardless of the sex history of the baby's family, the season or place of its birth, or other factors. It is only after long accumulation of records that such factors become irrelevant and the two alternatives are accepted as approximately equally likely. ■

3. The classical definition can be used only in a limited class of problems. In the die experiment, for example, (1-8) holds only if the die is fair, that is, if its six outcomes have the same probability. If it is loaded and the probability of 5 equals, say, .193, there is no direct way of deriving this probability from the equation.

   The problem is more difficult in applications involving infinitely many outcomes. In such cases, we introduce as a measure of the number of outcomes length, area, or volume. This makes reliance on (1-8) questionable, and, as the following classic example suggests, it leads to ambiguous solutions.

Example 1.12

Given a circle C of radius r, we select "at random" a chord AB. Find the probability p of the event $\mathcal{A}$ = {the length l of the chord is larger than the length $r\sqrt{3}$ of the side of the inscribed equilateral triangle}.

We shall show that this problem can be given at least three reasonable solutions.

First Solution The center M of the chord can be any point in the interior of the circle C. The point is favorable to the event $\mathscr{A}$ if it is in the interior of the circle C_1 of radius $r/2$ (Fig. 1.2a). Thus in this interpretation of "randomness," the experiment consists of all points on the circle C. Using the area of a region as a measure of the points in that region, we conclude that the measure of the total number of outcomes is the area πr^2 of the circle C and the measure of the outcomes favorable to the event $\mathscr{A}$ equals the area $\pi r^2/4$ of the circle C_1. This yields

$$p = \frac{\pi r^2/4}{\pi r^2} = \frac{1}{4}$$

Second Solution We now assume that the end A of the chord AB is fixed. This reduces the number of possibilities but has no effect on the value of p because the number of favorable outcomes is reduced proportionately. We can thus consider as experimental outcomes all points on the circumference of the circle. Since $l > r\sqrt{3}$ if B is on the 120° arc DBE of Fig. 1.2b, the number of outcomes favorable to A are all points on that arc. Using the length of the arcs as measure of the outcomes, we conclude that the measure of the total number of outcomes is the length $2\pi r$ of the

Figure 1.2

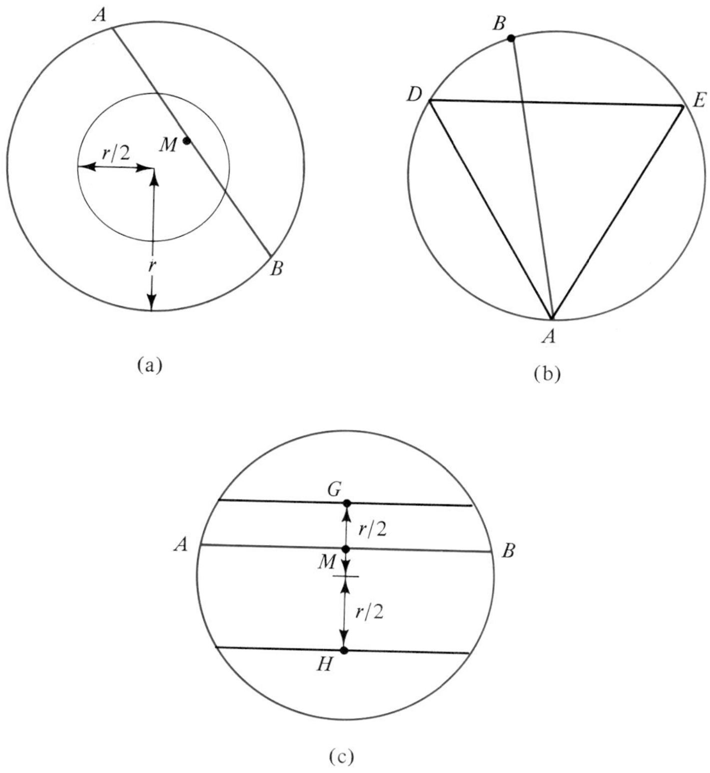

circle and the measure of the favorable outcomes is the length $2\pi r/3$ of the arc DBE. Hence,

$$p = \frac{2\pi r/3}{2\pi r} = \frac{1}{3}$$

Third Solution We assume that the direction of AB is perpendicular to the line FK of Fig. 1.2c. As before, this assumption has no effect on the value of p. Clearly, $l > r\sqrt{3}$ if the center M of the chord AB is between the points G and H. Thus the outcomes of the experiment are all points on the line FK and the favorable outcomes are the points on the segment GH. Using the lengths r and $r/2$ of these segments as measures of the outcomes, we conclude that

$$p = \frac{r/2}{r} = \frac{1}{2}$$

This example, known as *Bertrand paradox*, demonstrates the possible ambiguities associated with the classical definition, the meaning of the terms *possible* and *favorable,* and the need for a clear specification of all experimental outcomes. ■

USES OF THE CLASSICAL DEFINITION

1. In many experiments, the assumption that there are N equally likely alternatives is well established from long experience. In such cases, (1-8) is accepted as self-evident; for example, "If a ball is selected at random from a box containing m black and n white balls, the probability p that the ball is white equals $m/(m + n)$," and "if a telephone call occurs at random in the time interval $(0, T)$, the probability p that it occurs in the interval $(0, t)$ equals t/T." Such conclusions are valid; however, their validity depends on the meaning of the word *random*. The conclusion of the call example that $p = t/T$ is not a consequence of the randomness of the call. Randomness in this case is equivalent to the assumption that $p = t/T$, and it follows from past records of telephone calls.
2. The specification of a probabilistic model is based on the probabilities of various events of the underlying physical experiment. In a number of applications, it is not possible to determine these probabilities empirically by repeating the experiment. In such cases, we use the classical definition as a working hypothesis; we assume that certain alternatives are equally likely, and we determine the unknown probabilities from (1-8). The hypothesis is accepted if its theoretical consequences agree with experience; otherwise, it is rejected. This approach has important applications in statistical mechanics (see Example 2.25).
3. The classical definition can be used as the basis of a deductive theory if (1-8) is accepted not as a method of determining the probability of real events but as an *assumption*. As we show in the next chapter, a deductive theory based on (1-8) is only a special case of the axiomatic approach to probability, involving only experiments in which all elementary events have the same probabil-

ity. We should note, however, that whereas the axiomatic development is based on the three axioms of probability, a theory based on (1-8) requires no axioms. The reason is that if we assume that all probabilities satisfy the equally likely condition, all axioms become simple theorems. Indeed, axioms I and II are obvious. To prove (1-5), we observe that if the events $\mathcal{A}$ and $\mathcal{B}$ consist of $N_{\mathcal{A}}$ and $N_{\mathcal{B}}$ outcomes, respectively, and they are mutually exclusive, their union $\mathcal{A} \cup \mathcal{B}$ consists of $N_{\mathcal{A}} + N_{\mathcal{B}}$ outcomes. And since all probabilities satisfy (1-8), we conclude that

$$P(\mathcal{A} \cup \mathcal{B}) = \frac{N_{\mathcal{A}\cup\mathcal{B}}}{N} = \frac{N_{\mathcal{A}}}{N} + \frac{N_{\mathcal{B}}}{N} = P(\mathcal{A}) + P(\mathcal{B}) \qquad (1\text{-}9)$$

Subjective

In the subjective interpretation of probability, the number $P(\mathcal{A})$ assigned to a statement $\mathcal{A}$ is a measure of our state of knowledge or belief concerning the truth of $\mathcal{A}$. The underlying theory can be generally accepted as a form of "inductive" reasoning developed "deductively." We shall not discuss this interpretation or its effectiveness in decisions based on inductive inference. We note only that the three axioms on which the theory is based are in reasonable agreement with our understanding of the properties of inductive inference. In our development, the subjective interpretation of probability will be considered only in the context of Bayesian estimation (Section 8-2). There we discuss the use of the subjective interpretation in problems involving averages, and we comment on the resulting controversies between "subjectivists" and "objectivists."

Here we discuss a special case of the subjective interpretation of $P(\mathcal{A})$ involving total prior ignorance, and we show that in this case it is formally equivalent to the classical definition.

PRINCIPLE OF INSUFFICIENT REASON This principle states that if an experiment has N alternatives (outcomes) ζ_i and we have no knowledge about their occurrence, we *must* assign to all alternatives the same probability. This yields

$$P\{\zeta_1\} = \cdot \cdot \cdot = P\{\zeta_N\} = \frac{1}{N} \qquad (1\text{-}10)$$

In the die experiment, we must assume that $P\{f_i\} = 1/6$. In the polling experiment, we must assume that $P\{r\} = P\{d\} = 1/2$.

Note that (1-10) is equivalent to (1-8). However, the classical definition, on which Equation (1-8) is based, is conceptually different from the principle of insufficient reason. In the classical definition, we *know,* from symmetry considerations or from past experience, that the N outcomes are equally likely. Furthermore, this conclusion is objective and is not subject to change. The principle of insufficient reason, by contrast, is a consequence of our total ignorance about the experiment. Furthermore, it leads to conclusions that are subjective and subject to change in the face of any evidence.

Concluding Remarks

In this book we present a theory based on the axiomatic definition of probability. The theory is developed deductively, and all conclusions follow logically from the axioms. In the context of the theory, the question "what is probability?" is not relevant. The relevant question is the correspondence between probabilities and observations. This question is answered in terms of the other three interpretations of probability.

As a motivation, and as a reminder of the connection between concepts and reality, we shall often give an *empirical interpretation* (relative frequency) of the various axioms, definitions, and theorems. This portion of the book is heuristic and does not obey the rules of deductive reasoning on which the theory is based.

We conclude with the observation that all statistical statements concerning future events are inductive and must be interpreted as reasonable approximations. We stress, however, that our inability to make exact predictions is not limited to statistics. It is characteristic of all scientific investigations involving real phenomena, deterministic or random. This suggests that physical theories are not laws of nature, whatever that may mean. They are human inventions (mental constructs) used to describe with economy patterns of real events and to predict, but only approximately, their future behavior. To "prove" that the future will evolve exactly as predicted, we must invoke metaphysical causes.

2 Fundamental Concepts

The material in Chapters 2 and 3 is based on the notion of outcomes, events, and probabilities and requires, for the most part, only high school mathematics. It is self-contained and richly illustrated, and it can be used to solve a large variety of problems. In Chapter 2, we develop the theory of probability as an abstract construct based on axioms. For motivation, we make also frequent reference to the physical interpretation of all theoretical results. This chapter is the foundation of the entire theory.

2-1 Set Theory

Sets are collections of objects. The objects of a set are called *elements*. Thus the set

{apple, boy, pencil}

consists of the three elements *apple, pencil,* and *boy*. The elements of a set are usually placed in braces; the order in which they are written is immaterial. They can be identified by words or by suitable abbreviations. For example, $\{h, t\}$ is a set consisting of the elements h for *heads* and t for *tails*.

Similarly, the six faces of a die form the set

$$\{f_1, f_2, f_3, f_4, f_5, f_6\}$$

In this chapter all sets will be identified by script letters* $\mathscr{A}$, $\mathscr{B}$, $\mathscr{C}$, . . . ; their elements will, in general, be identified by the Greek letter ζ. Thus the expression

$$\mathscr{A} = \{\zeta_1, \zeta_2, \ldots, \zeta_N\} \tag{2-1}$$

will mean that $\mathscr{A}$ is a set consisting of the N elements $\zeta_1, \zeta_2, \ldots, \zeta_N$.

The notation

$$\zeta_i \in \mathscr{A}$$

will mean that ζ_i is an element of the set $\mathscr{A}$ (belongs to the set $\mathscr{A}$); the notation

$$\zeta_i \notin \mathscr{A}$$

will mean that ζ_i is not an element of $\mathscr{A}$. Here is a simple illustration. The set $\mathscr{A} = \{2, 4, 6\}$ consists of the three elements 2, 4, and 6. Thus

$$2 \in \mathscr{A} \qquad 3 \notin \mathscr{A}$$

In this illustration, the elements of the set $\mathscr{A}$ are numbers. Note, however, that numbers are used merely for the purpose of identifying the elements; for the specification of the set, their numerical properties are irrelevant.

The elements of a set might be simple objects, as in the preceding examples, or each might consist of several objects. For example,

$$\mathscr{A} = \{hh, ht, th\}$$

is a set consisting of the three elements hh, ht, and th. Sets of this form will be used in experiments involving repeated trials. In the set $\mathscr{A}$, the elements ht and th are different; however, the set $\{th, hh, ht\}$ equals $\mathscr{A}$.

In the preceding examples, we identified all sets explicitly in terms of their elements. We shall also identify sets in terms of the properties of their elements. For example,

$$\mathscr{A} = \{\text{all integers from 1 to 6}\} \tag{2-2}$$

is the set $\{1, 2, 3, 4, 5, 6\}$; similarly,

$$\mathscr{B} = \{\text{all even integers from 1 to 6}\}$$

is the set $\{2, 4, 6\}$.

Venn Diagrams We shall assume that all elements under consideration belong to a set $\mathscr{S}$ called *space* (or *universe*). For example, if we consider children in a certain school, $\mathscr{S}$ is the set of all children in that school.

The set $\mathscr{S}$ is often represented by a rectangle, and its elements by the points in the rectangle. All other sets under consideration are thus represented by various regions in this rectangle. Such a representation is called a *Venn diagram*. A Venn diagram consists of infinitely many points; however, the set $\mathscr{S}$ that it represents need not have infinitely many elements. The diagram is used merely to represent graphically various set operations.

* In subsequent chapters, we shall use script letters to identify only sets representing events of a probability space.

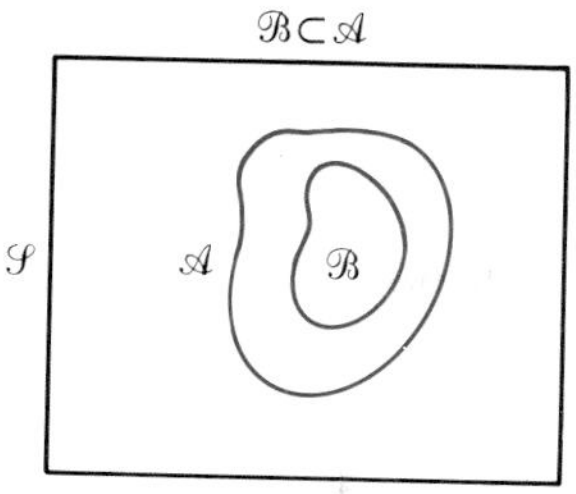

Figure 2.1

Subsets We shall say that a set $\mathcal{B}$ is a subset of a set $\mathcal{A}$ if every element of $\mathcal{B}$ is also an element of $\mathcal{A}$ (Fig. 2.1). The notations

$$\mathcal{B} \subset \mathcal{A} \qquad \mathcal{A} \supset \mathcal{B} \tag{2-3}$$

will mean that the set $\mathcal{B}$ is a subset of the set $\mathcal{A}$. For example, if

$$\mathcal{A} = \{f_1, f_2, f_3\} \qquad \mathcal{B} = \{f_2, f_3\}$$

then $\mathcal{B} \subset \mathcal{A}$. In the Venn diagram representation of (2-3), the set $\mathcal{B}$ is included in the set $\mathcal{A}$.

Equality The notation

$$\mathcal{A} = \mathcal{B}$$

will mean that the sets $\mathcal{A}$ and $\mathcal{B}$ consist of the same elements. To establish the quality of the sets $\mathcal{A}$ and $\mathcal{B}$, we must show that every element of $\mathcal{B}$ is an element of $\mathcal{A}$ and every element of $\mathcal{A}$ is an element of $\mathcal{B}$. In other words,

$$\mathcal{A} = \mathcal{B} \quad \text{iff*} \quad \mathcal{B} \subset \mathcal{A} \quad \text{and} \quad \mathcal{A} \subset \mathcal{B}$$

We shall say that $\mathcal{B}$ is a *proper* subset of $\mathcal{A}$ if $\mathcal{B}$ is a subset of $\mathcal{A}$ but does not equal $\mathcal{A}$. The distinction between a subset and a proper subset will not be made always.

Unions and Intersections Given two sets $\mathcal{A}$ and $\mathcal{B}$, we form a set consisting of all elements that are either in $\mathcal{A}$ or in $\mathcal{B}$ or in both. This set is written in the form

$$\mathcal{A} \cup \mathcal{B} \quad \text{or} \quad \mathcal{A} + \mathcal{B}$$

and it is called the *union* of the sets $\mathcal{A}$ and $\mathcal{B}$ (shaded in Fig. 2.2).

Given two sets $\mathcal{A}$ and $\mathcal{B}$, we form a set consisting of all elements that are in $\mathcal{A}$ and in $\mathcal{B}$. This set is written in the form

$$\mathcal{A} \cap \mathcal{B} \quad \text{or} \quad \mathcal{A}\mathcal{B}$$

and it is called the *intersection* of the sets $\mathcal{A}$ and $\mathcal{B}$ (shaded in Fig. 2.3).

Complement Given a set $\mathcal{A}$, we form a set consisting of all elements of $\mathcal{S}$ that are not in $\mathcal{A}$ (shaded in Fig. 2.4). This set is denoted by $\overline{\mathcal{A}}$ and it is called the *complement* of $\mathcal{A}$.

* *Iff* is an abbreviation for "if and only if."

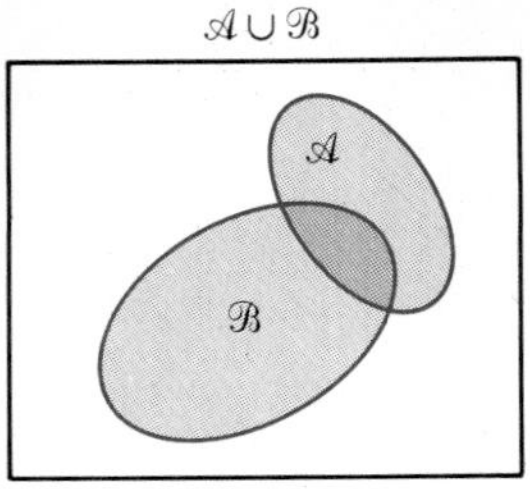

Figure 2.2

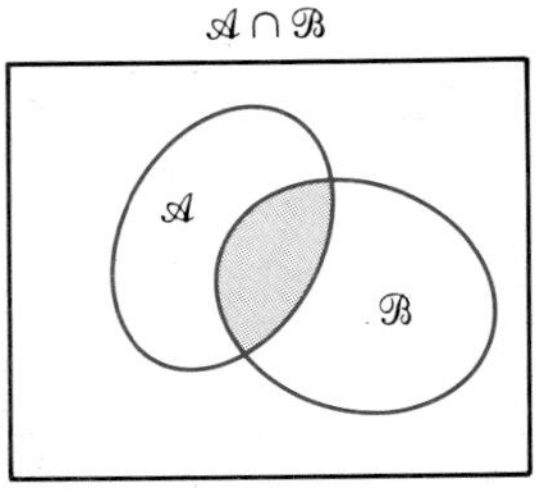

Figure 2.3

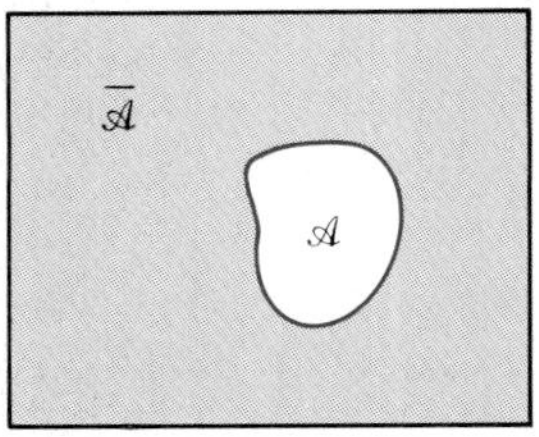

Figure 2.4

Example 2.1

Suppose that $\mathcal{S}$ is the set of all children in a community, $\mathcal{A}$ is the set of children in fifth grade, and $\mathcal{B}$ is the set of all boys. In this case, $\overline{\mathcal{A}}$ is the set of children that are not in fifth grade, and $\overline{\mathcal{B}}$ is the set of all girls. The set $\mathcal{A} \cup \mathcal{B}$ consists of all girls in fifth grade and all the boys in the community. The set $\mathcal{A} \cap \mathcal{B}$ consists of all boys in fifth grade. ■

Disjoint Sets We shall say that the sets $\mathcal{A}$ and $\mathcal{B}$ are *disjoint* if they have no common elements. If two sets are disjoint, their intersection has a meaning only if we agree to define a set without elements. Such a set is denoted by $\varnothing$ and it is called the *empty* (or *null*) set. Thus

$$\mathcal{A} \cap \mathcal{B} = \varnothing$$

iff the sets $\mathcal{A}$ and $\mathcal{B}$ are disjoint.

We note that a set may have finitely many or infinitely many elements. There are two kinds of infinities: countable and noncountable. A set is called

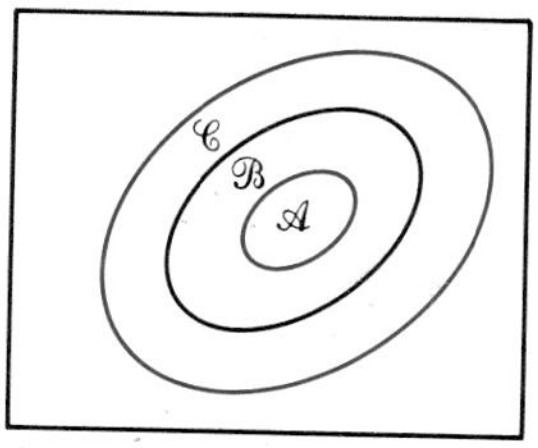

Figure 2.5

countable if its elements can be brought into one-to-one correspondence with the positive integers. For example, the set of even numbers is countable; this is easy to show. The set of rational numbers is countable; the proof is more difficult. The set of all numbers in an interval is noncountable; this is difficult to show.

PROPERTIES The following set properties follow readily from the definitions.

$$\mathscr{A} \cup \mathscr{S} = \mathscr{S} \qquad \mathscr{A} \cap \mathscr{S} = \mathscr{A}$$
$$\mathscr{A} \cup \varnothing = \mathscr{A} \qquad \mathscr{A} \cap \varnothing = \varnothing$$

If $\mathscr{B} \subset \mathscr{A}$, then $\mathscr{A} \cup \mathscr{B} = \mathscr{A}$ and $\mathscr{A} \cap \mathscr{B} = \mathscr{B}$.

Transitive Property If $\mathscr{A} \subset \mathscr{B}$ and $\mathscr{B} \subset \mathscr{C}$, then $\mathscr{A} \subset \mathscr{C}$ (Fig. 2.5).

Commutative Property

$$\mathscr{A} \cup \mathscr{B} = \mathscr{B} \cup \mathscr{A} \qquad \mathscr{A} \cap \mathscr{B} = \mathscr{B} \cap \mathscr{A}$$

Associative Property

$$(\mathscr{A} \cup \mathscr{B}) \cup \mathscr{C} = \mathscr{A} \cup (\mathscr{B} \cup \mathscr{C}) \qquad (\mathscr{A} \cap \mathscr{B}) \cap \mathscr{C} = \mathscr{A} \cap (\mathscr{B} \cap \mathscr{C})$$

From this it follows that we can omit parentheses in the last two operations.

Distributive Law (See Fig. 2.6.)

$$(\mathscr{A} \cup \mathscr{B}) \cap \mathscr{C} = (\mathscr{A} \cap \mathscr{C}) \cup (\mathscr{B} \cap \mathscr{C})$$

Thus set operations are the same as the corresponding arithmetic operations if we replace $\mathscr{A} \cup \mathscr{B}$ and $\mathscr{A} \cap \mathscr{B}$ by $\mathscr{A} + \mathscr{B}$ and $\mathscr{A}\mathscr{B}$, respectively. With

Figure 2.6

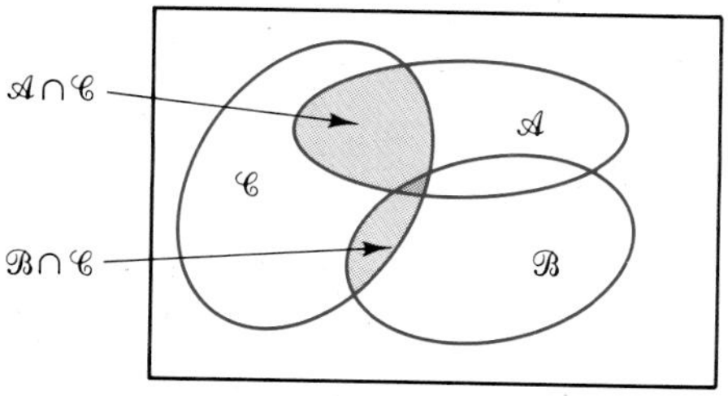

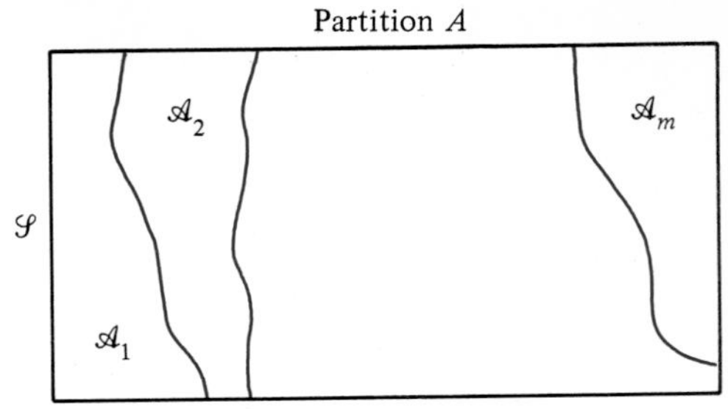

Figure 2.7

the latter notation, the distributive law for sets yields

$$(\mathcal{A} + \mathcal{B})(\mathcal{C} + \mathcal{D}) = \mathcal{A}(\mathcal{C} + \mathcal{D}) + \mathcal{B}(\mathcal{C} + \mathcal{D}) = \mathcal{A}\mathcal{C} + \mathcal{A}\mathcal{D} + \mathcal{B}\mathcal{C} + \mathcal{B}\mathcal{D}$$

Partitions A partition of $\mathcal{S}$ is a collection

$$A = [\mathcal{A}_1, \ldots, \mathcal{A}_m]$$

of subsets $\mathcal{A}_1, \ldots, \mathcal{A}_m$ of $\mathcal{S}$ with the following property (Fig. 2.7): They are disjoint and their union equals $\mathcal{S}$:

$$\mathcal{A}_i \cap \mathcal{A}_j = \varnothing, \quad i \neq j \qquad \mathcal{A}_1 \cup \cdots \cup \mathcal{A}_m = \mathcal{S} \tag{2-4}$$

The set $\mathcal{S}$ has, of course, many partitions. If $\mathcal{A}$ is a set with complement $\overline{\mathcal{A}}$, then

$$\mathcal{A} \cap \overline{\mathcal{A}} = \varnothing \qquad \mathcal{A} \cup \overline{\mathcal{A}} = \mathcal{S}$$

hence, $[\mathcal{A}, \overline{\mathcal{A}}]$ is a partition of $\mathcal{S}$.

Example 2.2

Suppose that $\mathcal{S}$ is the set of all children in a school. If $\mathcal{A}_i$ is the set of all children in the ith grade, then $[\mathcal{A}_1, \ldots, \mathcal{A}_{12}]$ is a partition of $\mathcal{S}$. If $\mathcal{B}$ is the set of all boys and $\mathcal{G} = \overline{\mathcal{B}}$ the set of all girls, then $[\mathcal{B}, \mathcal{G}]$ is also a partition. ∎

Cartesian Product Given two sets $\mathcal{A}$ and $\mathcal{B}$ with elements α_i and β_j, respectively, we form a new set $\mathcal{C}$, the elements of which are all possible pairs $\alpha_i\beta_j$. The set so formed is denoted by

$$\mathcal{C} = \mathcal{A} \times \mathcal{B} \tag{2-5}$$

and it is called the *Cartesian product* of the sets $\mathcal{A}$ and $\mathcal{B}$. Clearly, if $\mathcal{A}$ has m elements and $\mathcal{B}$ has n elements, the set $\mathcal{C}$ so constructed has mn elements.

Example 2.3

If $\mathcal{A}$ = {car, apple, bird} and $\mathcal{B}$ = {heads, tails}, then

$$\mathcal{C} = \mathcal{A} \times \mathcal{B} = \{ch, ct, ah, at, bh, bt\}$$ ∎

The Cartesian product can be defined even if the sets $\mathcal{A}$ and $\mathcal{B}$ are identical.

Example 2.4

If $\mathscr{A} = \{\text{heads, tails}\} = \mathscr{B}$, then

$$\mathscr{C} = \mathscr{A} \times \mathscr{B} = \{hh, ht, th, tt\}$$ ■

Subsets and Combinatorics

Probabilistic statements involving finitely many elements and repeated trials are based on the determination of the number of subsets of $\mathscr{S}$ having specific properties. Let us review the underlying mathematics.

PERMUTATIONS AND COMBINATIONS Given N distinct objects and a number $m \le N$, we select m of these objects and place them in line. The number of configurations so formed is denoted by P_m^N and is called "*permutations* of N objects taken m at a time." In this definition, two permutations are different if they differ by at least one object or by the order of placement.

To clarify this concept, here is a list of all permutations of the $N = 4$ objects a, b, c, d for $m = 1$ and $m = 2$:

$m = 1$

$$a \quad b \quad c \quad d \qquad P_1^4 = 4 \tag{2-6}$$

$m = 2$

$$\begin{matrix} ab & ba & ca & da \\ ac & bc & cb & db \\ ad & bd & cd & dc \end{matrix} \qquad P_2^4 = 12 \tag{2-7}$$

Note that $P_2^4 = 3P_1^4$. This is so because each term in (2-6) generates $4 - 1 = 3$ terms in (2-7). Clearly, 3 is the number of letters remaining after one is selected. This leads to the following generalization.

■ ***Theorem***

$$P_m^N = N(N-1) \cdots (N-m+1) \tag{2-8}$$

■ ***Proof.*** Clearly, $P_1^N = N$; reasoning as in (2-7) we obtain $P_2^N = N(N-1)$. We thus have $N(N-1)$ permutations of the N objects taken 2 at a time. At the end of each permutation so formed, we attach one of the remaining $N - 2$ objects. This yields $(N-2)P_2^N$ permutations of N objects taken 3 at a time. By simple induction, we obtain

$$P_m^N = (N - m + 1)P_{m-1}^N$$

and (2-8) results.

We emphasize that in each permutation, a specific object appears only once, and two permutations are different even if they consist of the same objects in a different order. Thus in (2-7), ab is distinct from ba; the configuration aa does not appear.

Example 2.5

As we see from (2-8),

$$P_2^{10} = 10 \times 9 = 90$$

If the ten objects are the numbers 0, 1, . . . , 9, then P_2^{10} is the total number of two-digit numbers excluding 00, 11, 22, . . . , 99. ■

■ ***Corollary.*** Setting $m = N$ in (2-8), we obtain

$$P_N^N = N(N-1)\cdots 1 = N!$$

This is the number of *permutations* of N objects (the phrase "taken N at a time" is omitted).

Example 2.6

Here are the $3! = 3 \times 2$ permutations of the objects a, b, and c:

abc *acb* *bac* *bca* *cab* *cba* ■

Combinations Given N distinct objects and a number $m \le N$, we select m of these objects in all possible ways. The number of groups so formed is denoted by C_m^N and is called "*combinations* of N objects taken m at a time." Two combinations are different if they differ by at least one object; the order of selection is immaterial.

If we change the order of the objects of a particular combination in all possible ways, we obtain $m!$ permutations of these objects. Since there are C_m^N combinations, we conclude that

$$P_m^N = m!C_m^N \tag{2-9}$$

From this and (2-8) it follows that

$$C_m^N = \frac{N(N-1)\cdots(N-m+1)}{m!}$$

This fraction is denoted by $\binom{N}{m}$. Multiplying numerator and denomination by $(N-m)!$, we obtain

$$C_m^N = \binom{N}{m} = \frac{N!}{m!(N-m)!} \tag{2-10}$$

Note, finally, that

$$C_m^N = C_{N-m}^N$$

This can be established also directly: Each time we take m out of N objects, we leave $N - m$ objects.

Example 2.7

With $N = 4$ and $m = 2$, (2-10) yields

$$C_2^4 = \binom{4}{2} = \frac{4 \times 3}{1 \times 2} = 6$$

Here are the six combinations of the four objects a, b, c, d taken 2 at a time:

ab *ac* *ad* *bc* *bd* *cd* ■

Applications We have N objects forming two groups. Group 1 consists of m identical objects identified by h; group 2 consists of $N - m$ identical objects identified by t. We place these objects in N boxes, one in each box.

We maintain that the number x of ways we can do so equals

$$x = \binom{N}{m} \tag{2-11}$$

▪ ***Proof.*** The m objects of group 1 are placed in m out of the N available boxes and the $N - m$ objects of group 2 in the remaining $N - m$ boxes. This yields (2-11) because there are C_m^N ways of selecting m out of N objects.

Example 2.8

Here are the $C_2^4 = 6$ ways of placing the four objects h, h, t, t in the four boxes B_1, B_2, B_3, B_4:

B_1	h	h	h	t	t	t
B_2	h	t	t	h	h	t
B_3	t	h	t	h	t	h
B_4	t	t	h	t	h	h

$C_2^4 = 6$

■

Example 2.9

This example has applications in statistical mechanics (see Example 2.25). We place m identical balls in $n > m$ boxes. We maintain that the number y of ways that we can do so equals

$$y = \binom{N}{m} \qquad \text{where } N = n + m - 1 \tag{2-12}$$

▪ ***Proof.*** The solution of this problem is rather tricky. We consider the m balls as group 1 of objects and the $n - 1$ interior walls separating the n boxes as group 2. We thus have $N = n + m - 1$ objects, and (2-12) follows from (2-11). In Fig. 2.8, we demonstrate one of the placements of $m = 4$ balls in $n = 7$ boxes. In this case, we have $n - 1 = 6$ interior walls. The resulting sequence of $N = n + m - 1 = 10$ balls and walls is *bwwbbwwwbw*, as shown. ■

Binary Numbers We wish to find the total number z of N-digit binary numbers consisting of m 1s and $N - m$ 0s. Identifying the 1s and 0s as the objects of group 1 and group 2, respectively, we conclude from (2-12) that $z = \binom{N}{m}$.

Figure 2.8

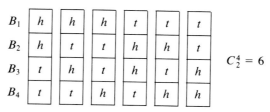

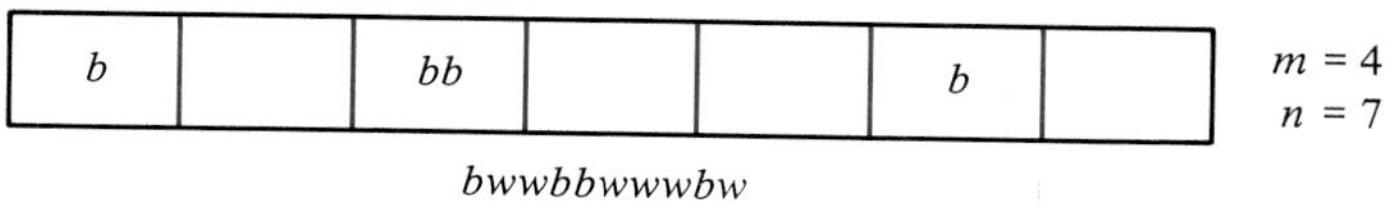

Example 2.10

Here are the $\binom{4}{2} = 6$ four-digit binary numbers consisting of two 1s and two 0s:

1100 1010 1001 0110 0101 0011 ■

Note From the identity (binomial expansion)

$$(a + b)^N = \sum_{m=0}^{N} \binom{N}{m} a^m b^{N-m}$$

it follows with $a = b = 1$ that

$$2^N = \sum_{m=0}^{N} \binom{N}{m} \tag{2-13}$$

Hence, the total number of N-digit binary numbers equals 2^N.

Subsets Consider a set $\mathcal{S}$ consisting of the N elements $\zeta_1, \ldots, \zeta_N$. We maintain that the total number of its subsets, including $\mathcal{S}$ itself and the empty set $\varnothing$, equals 2^N.

■ ***Proof.*** It suffices to show that we can associate to each subset of $\mathcal{S}$ one and only one N-digit binary number [see (2-13)]. Suppose that $\mathcal{A}$ is a subset of $\mathcal{S}$. If $\mathcal{A}$ contains the element ζ_i, we write 1 as the ith binary digit; otherwise, we write 0. We have thus established a one-to-one correspondence between all the N-digit binary numbers and the subsets of $\mathcal{S}$. Note, in particular, that $\mathcal{S}$ corresponds to 11 . . . 1 and $\varnothing$ to 00 . . . 0.

Example 2.11

The set $\mathcal{S} = \{a, b, c, d\}$ has four elements and $2^4 = 16$ subsets:

$\varnothing$ $\{a\}$ $\{b\}$ $\{c\}$ $\{d\}$ $\{a, b\}$ $\{a, c\}$ $\{a, d\}$ $\{b, c\}$
$\{b, d\}$ $\{c, d\}$ $\{a, b, c\}$ $\{a, b, d\}$ $\{a, c, d\}$ $\{b, c, d\}$ $\{a, b, c, d\}$

The corresponding four-digit numbers are as follows:

0000 0001 0010 0100 1000 0011 0101 1001
0110 1010 1100 0111 1011 1101 1110 1111 ■

Generalized Combinations We are given N objects and wish to group them into r classes $A_1, \ldots, A_r$ such that the ith class A_i consists of k_i objects where

$$k_1 + \cdots + k_r = N$$

The total number of such groupings equals

$$C^N_{k_1, \ldots, k_r} = \frac{N!}{k_1!k_2! \cdots k_r!} \tag{2-14}$$

■ ***Proof.*** We select first k_1 out of the N object to form the first class A_1. As we know, there are $\binom{N}{k_1}$ ways to do so. We next select k_2 out of the remaining $N - k_2$ objects to form the second class A_2. There are $\binom{N - k_1}{k_2}$ ways to do so. We then select k_3 out of the remaining $N - k_1 - k_2$ objects to form the third class A_3, and so we continue. After the formation of the A_{r-1} class, there remain $N - k_1 - \cdots - k_{r-1} = k_r$ objects. There is $\binom{k_r}{k_r} = 1$ way of selecting k_r out of the remaining k_r object; hence, there is only one way of forming the last class A_r.

From the foregoing it follows that

$$C^N_{k_1, \ldots, k_r} = \frac{N!}{k_1!(N-k_1)!}\,\frac{(N-k_1)!}{k_2!(N-k_1-k_2)!}\cdots\frac{(k_{r-1}+k_r)!}{k_{r-1}!k_r!}$$

and (2-14) results.

Note that (2-10) is a special case of (2-14) obtained with $r = 2$, $k_1 = k$, $k_2 = N - k$, and $C^N_{k_1,k_2} = C^N_k$

Example 2.12

We wish to determine the number M of bridge hands that we can deal using a deck with $N = 52$ cards.

In a bridge game, there are $r = 4$ hands; each hand consists of 13 cards. With $k_1 = k_2 = k_3 = k_4 = 13$, (2-14) yields

$$M = C^{52}_{13, \ldots, 13} = \frac{52!}{13!13!13!13!} \simeq 5.36 \times 10^{28}$$

■

Let us look at two other interpretations of (2-14).

1. We have r groups of objects. Group 1 consists of k_1 repetitions of a particular object h_1; group 2 consists of k_2 repetitions of another object h_2, and so on. These objects are placed into $N = k_1 + \cdots + k_r$ boxes, one in each box. The total number of ways that we can do so equals $C^N_{k_1, \ldots, k_r}$.
2. Suppose that $\mathcal{S}$ is a set consisting of N elements and that
$$[\mathcal{A}_1, \ldots, \mathcal{A}_r]$$
is a partition of $\mathcal{S}$ formed with the r sets $\mathcal{A}_i$ as in (2-4). If the ith set $\mathcal{A}_i$ consists of k_i elements where k_i are given numbers, the total number of such partitions equals $C^N_{k_1, \ldots, k_r}$.

2-2 Probability Space

In the theory of probability, as in all scientific investigations, it is essential that we describe a physical experiment in terms of a clearly defined model. In this section, we develop the underlying concepts.

■ ***Definition.*** An experimental model is a set $\mathcal{S}$. The elements ζ_i of $\mathcal{S}$ are called *outcomes*. The subsets of $\mathcal{S}$ are called *events*. The empty set $\varnothing$ is called the *impossible event* and the set $\mathcal{S}$ the *certain event*. Two events $\mathcal{A}$

and $\mathcal{B}$ are called *mutually exclusive* if they have no common elements, that is, if $\mathcal{A} \cap \mathcal{B} = \varnothing$. An event $\{\zeta_i\}$ consisting of the single outcome ζ_i is called an *elementary event*. Note the important distinction between the element ζ_i and the event $\{\zeta_i\}$.

If $\mathcal{S}$ has N elements, the number of its subsets equals 2^N. Hence, an experiment with N outcomes has 2^N events (we include the certain event and the impossible event).

Example 2.13

In the single-die experiment, the space $\mathcal{S}$ consists of six outcomes:

$$\mathcal{S} = \{f_1, \ldots, f_6\}$$

It therefore has $2^6 = 64$ events. We list them according to the number of elements in each.

$\binom{6}{0} = 1$ event without elements, namely, the event $\varnothing$.

$\binom{6}{1} = 6$ events with one element each, namely, the elementary events $\{f_1\}, \{f_2\}, \ldots, \{f_6\}$.

$\binom{6}{2} = 15$ events with two elements each, namely, $\{f_1, f_2\}, \{f_1, f_3\}, \ldots, \{f_5, f_6\}$.

$\binom{6}{3} = 20$ events with three elements each, namely, $\{f_1, f_2, f_3\}, \{f_1, f_2, f_4\}, \ldots, \{f_4, f_5, f_6\}$.

$\binom{6}{4} = 15$ events with four elements each, namely, $\{f_1, f_2, f_3, f_4\}, \{f_1, f_2, f_3, f_5\}, \ldots, \{f_3, f_4, f_5, f_6\}$.

$\binom{6}{5} = 6$ events with five elements each, namely, $\{f_1, f_2, f_3, f_4, f_5\}, \ldots, \{f_2, f_3, f_4, f_5, f_6\}$.

$\binom{6}{6} = 1$ event with six elements, namely, the event $\mathcal{S}$.

In this listing, the various events are specified explicitly in terms of their elements as in (2-1). They can, however, be described in terms of the properties of the elements as in (2-2). We now cite various events using both descriptions:

$$\{\text{even}\} = \{f_2, f_4, f_6\} \qquad \{\text{odd}\} = \{f_1, f_3, f_5\}$$
$$\{\text{less than } 4\} = \{f_1, f_2, f_3\} \qquad \{\text{even, less than } 4\} = \{f_2\}$$

Note, finally, that each element of $\mathcal{S}$ belongs to $2^{N-1} = 32$ events. For example, f_1 belongs to the event {odd}, the event {less than 4}, the event $\{f_1, f_2\}$, and 29 other events. ■

Example 2.14

(a) In the single toss of a coin, the space $\mathcal{S}$ consists of two outcomes $\mathcal{S} = \{h, t\}$. It therefore has $2^2 = 4$ events, namely $\varnothing$, $\{h\}$, $\{t\}$, and $\mathcal{S}$.

(b) If the coin is tossed twice, $\mathcal{S}$ consists of four outcomes:

$$\mathcal{S} = \{hh, ht, th, tt\}$$

It therefore has $2^4 = 16$ events. These include the four elementary events $\{hh\}$, $\{ht\}$, $\{th\}$, and $\{tt\}$. The event $\mathcal{H}_1 = \{\text{heads at the first toss}\}$ is not an elementary event. It is an event consisting of the two outcomes hh and ht. Thus $\mathcal{H}_1 = \{hh, ht\}$. Similarly,

$$\mathcal{H}_2 = \{\text{ heads at the second toss}\} = \{hh, th\}$$
$$\mathcal{T}_1 = \{\text{tails at the first toss}\} = \{th, tt\}$$
$$\mathcal{T}_2 = \{\text{tails at the second toss}\} = \{ht, tt\}$$

The intersection of the events $\mathcal{H}_1$ and $\mathcal{H}_2$ is the elementary event $\{hh\}$. Thus

$$\mathcal{H}_1 \cap \mathcal{H}_2 = \{\text{heads at both tosses}\} = \{hh\}$$

Similarly,

$$\mathcal{H}_1 \cap \mathcal{T}_2 = \{\text{heads first, tails second}\} = \{ht\}$$
$$\mathcal{T}_1 \cap \mathcal{T}_2 = \{\text{tails at both tosses}\} = \{tt\}$$ ■

Empirical Interpretation of Outcomes and Events. In the applications of probability to real problems, the underlying experiment is repeated a large number of times, and probabilities are introduced to describe various averages. A single performance of the experiment will be called a *trial*. The repetitions of the experiment form *repeated trials*. In the single-die experiment, a trial is the roll of the die once. Repeated trials are repeated rolls. In the polling experiment, a trial is the selection of a voter from a given population.

At each trial, we observe one and only one *outcome*, whatever we have agreed to consider as the outcome for that experiment. The set of all outcomes is modeled by the certain event $\mathcal{S}$. If the observed outcome ζ is an element of the event $\mathcal{A}$, we say that the event $\mathcal{A}$ *occurred* at that particular trial. Thus at a single trial only one outcome is observed; however, many events occur, namely, all the 2^{N-1} subsets of $\mathcal{S}$ that contain the particular outcome ζ. The remaining 2^{N-1} events do not occur.

Example 2.15

(a) We conduct a poll to determine whether a voter is Republican (r) or Democrat (d). In this case, the experiment consists of two outcomes $\mathcal{S} = \{r, d\}$ and four events:

$$\varnothing, \{r\}, \{d\}, \mathcal{S}$$

(b) We wish to determine also whether the voter is male or female. We now have four outcomes:

$$\mathcal{S} = \{rm, rf, dm, df\}$$

and 16 events. The elementary events are $\{rm\}$, $\{rf\}$, $\{dm\}$, and $\{df\}$. Thus

$$\mathcal{R} = \{\text{Republican}\} = \{rm, rf\} \qquad \mathcal{D} = \{\text{Democrat}\} = \{dm, df\}$$
$$\mathcal{M} = \{\text{male}\} = \{rm, dm\} \qquad \mathcal{F} = \{\text{female}\} = \{rf, df\}$$

$$\mathcal{R} \cap \mathcal{M} = \{\text{Republican, male}\} = \{rm\} \qquad \mathcal{R} \cap \mathcal{F} = \{\text{Republican, female}\} = \{rf\}$$
$$\mathcal{D} \cap \mathcal{M} = \{\text{Democrat, male}\} = \{dm\} \qquad \mathcal{D} \cap \mathcal{F} = \{\text{Democrat, female}\} = \{df\}$$ ■

Note that the impossible event does not occur in any trial. The certain event occurs in every trial. If the events $\mathcal{A}$ and $\mathcal{B}$ are mutually exclusive and

$\mathcal{A}$ occurs at a particular trial, $\mathcal{B}$ does not occur at that trial. If $\mathcal{B} \subset \mathcal{A}$ and $\mathcal{B}$ occurs, $\mathcal{A}$ also occurs. At a particular trial, either the event $\mathcal{A}$ or its complement $\overline{\mathcal{A}}$ occurs. More generally, suppose that $[\mathcal{A}_1, \ldots, \mathcal{A}_m]$ is a partition of $\mathcal{S}$. It follows from Equation (2-4) that at a particular trial, one and only one event of this partition will occur.

Example 2.16

Consider the single-die experiment. The die is rolled and 2 shows. In our terminology, the outcome f_2 is observed. In this case, the event {even}, the event {less than 4}, and 30 other events occur, namely, all subsets of $\mathcal{S}$ that contain the element f_2. ■

Returning to an arbitrary $\mathcal{S}$, we denote by $n_{\mathcal{A}}$ the number of times the event $\mathcal{A}$ occurs in n trials. Clearly, $n_{\mathcal{A}} \le n$, $n_{\mathcal{S}} = n$, and $n_{\varnothing} = 0$. Furthermore, if the events $\mathcal{A}$ and $\mathcal{B}$ are mutually exclusive, then

$$n_{\mathcal{A} \cup \mathcal{B}} = n_{\mathcal{A}} + n_{\mathcal{B}} \tag{2-15}$$

If the events $\mathcal{A}$ and $\mathcal{B}$ have common elements, then

$$n_{\mathcal{A} \cup \mathcal{B}} = n_{\mathcal{A}} + n_{\mathcal{B}} - n_{\mathcal{A} \cap \mathcal{B}} \tag{2-16}$$

Example 2.17

We roll a die 10 times and we observe the sequence

$$f_1 \quad f_5 \quad f_2 \quad f_4 \quad f_6 \quad f_3 \quad f_2 \quad f_4 \quad f_4 \quad f_2$$

In this sequence, the elementary event $\{f_2\}$ occurs 3 times, the event {even} 7 times, and the event {odd} 3 times. Furthermore, if $\mathcal{A}$ = {even} and $\mathcal{B}$ = {less than 5}, then

$$\mathcal{A} \cup \mathcal{B} = \{f_1, f_2, f_3, f_4, f_6\} \qquad \mathcal{A} \cap \mathcal{B} = \{f_2, f_4\}$$

$$n_{\mathcal{A}} = 7 \qquad n_{\mathcal{B}} = 8 \qquad n_{\mathcal{A} \cap \mathcal{B}} = 6 \qquad n_{\mathcal{A} \cup \mathcal{B}} = 9$$

in agreement with (2-16). ■

Note This interpretation of repeated trials is used to relate probabilities to real experiments as in (2-1). If the experiment is performed n times and the event $\mathcal{A}$ occurs $n_{\mathcal{A}}$ times, then $P(\mathcal{A}) \simeq n_{\mathcal{A}}/n$, provided that n is sufficiently large. We should point out, however, that *repeated trials* is also a model concept used to create other models from a given experimental model. Consider the coin experiment. The model of a single toss has only two outcomes. However, if we are interested in establishing averages involving two tosses, our model has four outcomes: *hh*, *ht*, *th*, and *tt*. The model interpretation is thus fundamentally different from the empirical interpretation of repeated tosses. This distinction is subtle but fundamental. It will be discussed further in chapter 3 and will be used throughout the book.

The Axioms

A probabilistic *model* is a set $\mathcal{S}$ the elements of which are experimental *outcomes*. The subsets of $\mathcal{S}$ are events. To complete the specification of the model, we shall assign probabilities to all events.

■ ***Definition.*** The probability of an event $\mathcal{A}$ is a number $P(\mathcal{A})$ assigned to $\mathcal{A}$. This number satisfies the following axioms.

I. It is nonnegative:

$$P(\mathscr{A}) \geq 0 \tag{2-17}$$

II. The probability of the certain event equals 1:

$$P(\mathscr{S}) = 1 \tag{2-18}$$

III. If the events $\mathscr{A}$ and $\mathscr{B}$ are mutually exclusive, then

$$P(\mathscr{A} \cup \mathscr{B}) = P(\mathscr{A}) + P(\mathscr{B}) \tag{2-19}$$

This axiom can be readily generalized. Suppose that the events $\mathscr{A}$, $\mathscr{B}$, and $\mathscr{C}$ are mutually exclusive, that is, that no two of them have common elements. Repeated application of (2-19) yields

$$P(\mathscr{A} \cup \mathscr{B} \cup \mathscr{C}) = P(\mathscr{A}) + P(\mathscr{B}) + P(\mathscr{C})$$

This can be extended to any finite number of terms; we shall assume that it holds also for infinite but countably many terms. Thus if the events $\mathscr{A}_1, \mathscr{A}_2, \ldots$ are mutually exclusive, then

$$P(\mathscr{A}_1 \cup \mathscr{A}_2 \cup \cdots) = P(\mathscr{A}_1) + P(\mathscr{A}_2) + \cdots \tag{2-20}$$

This does not follow from (2-19); it is an additional requirement known as the *axiom of infinite additivity*.

Axiomatic Definition of an Experiment In summary, a model of an experiment is specified in terms of the following concepts:

1. A set $\mathscr{S}$ consisting of the outcomes ζ_i
2. Subsets of $\mathscr{S}$ called events
3. A number $P(\mathscr{A})$ assigned to every event (This number satisfies the listed axioms but is otherwise arbitrary.)

The letter $\mathscr{S}$ will be used to identify not only the certain event but also the entire experiment.

Probability Masses We shall find it convenient to interpret the probability $P(\mathscr{A})$ of an event $\mathscr{A}$ as its probability mass. In Venn diagrams, $\mathscr{S}$ is the entire rectangle, and its mass equals 1. The mass of the region of the diagram representing an event $\mathscr{A}$ equals $P(\mathscr{A})$. This interpretation of $P(\mathscr{A})$ is consistent with the axioms and can be used to facilitate the interpretation of various results.

PROPERTIES In the development of the theory, all results must be *derived* from the axioms. We must not accept any statement merely because it appears intuitively reasonable. Let us look at a few illustrations.

1. The probability of the impossible event is 0:
$$P(\varnothing) = 0 \tag{2-21}$$
■ ***Proof.*** For any $\mathscr{A}$, the events $\mathscr{A}$ and $\varnothing$ are mutually exclusive, hence (axiom III) $P(\mathscr{A} \cup \varnothing) = P(\mathscr{A}) + P(\varnothing)$. Furthermore, $P(\mathscr{A} \cup \varnothing) = P(\mathscr{A})$ because $\mathscr{A} \cup \varnothing = \mathscr{A}$, and (2-21) results.
2. The probability of $\overline{\mathscr{A}}$ equals
$$P(\overline{\mathscr{A}}) = 1 - P(\mathscr{A}) \tag{2-22}$$
■ ***Proof.*** The events $\mathscr{A}$ and $\overline{\mathscr{A}}$ are mutually exclusive, and their union equals $\mathscr{S}$, hence,
$$P(\mathscr{S}) = P(\mathscr{A} \cup \overline{\mathscr{A}}) = P(\mathscr{A}) + P(\overline{\mathscr{A}})$$
and (2-22) follows from (2-18).
3. Since $P(\overline{\mathscr{A}}) \geq 0$ (axiom I), it follows from (2-22) that $P(\mathscr{A}) \leq 1$. Combining with (2-17), we obtain
$$0 \leq P(\mathscr{A}) \leq 1 \tag{2-23}$$
4. If $\mathscr{B} \subset \mathscr{A}$, then
$$P(\mathscr{B}) \leq P(\mathscr{A}) \tag{2-24}$$
■ ***Proof.*** Clearly (see Fig. 2.9),
$$\mathscr{A} = \mathscr{A} \cup \mathscr{B} = \mathscr{B} \cup (\mathscr{A} \cap \overline{\mathscr{B}})$$
Furthermore, the events $\mathscr{B}$ and $\mathscr{A} \cap \overline{\mathscr{B}}$ are mutually exclusive; hence (axiom III)
$$P(\mathscr{A}) = P(\mathscr{B}) + P(\mathscr{A} \cap \overline{\mathscr{B}}) \tag{2-25}$$
And since $P(\mathscr{A} \cap \overline{\mathscr{B}}) \geq 0$ (axiom I), (2-24) follows.
5. For any $\mathscr{A}$ and $\mathscr{B}$,
$$P(\mathscr{A} \cup \mathscr{B}) = P(\mathscr{A}) + P(\mathscr{B}) - P(\mathscr{A} \cap \mathscr{B}) \tag{2-26}$$
■ ***Proof.*** This is an extension of axiom III. To prove it, we shall express the event $\mathscr{A} \cup \mathscr{B}$ as the union of two mutually exclusive

Figure 2.9

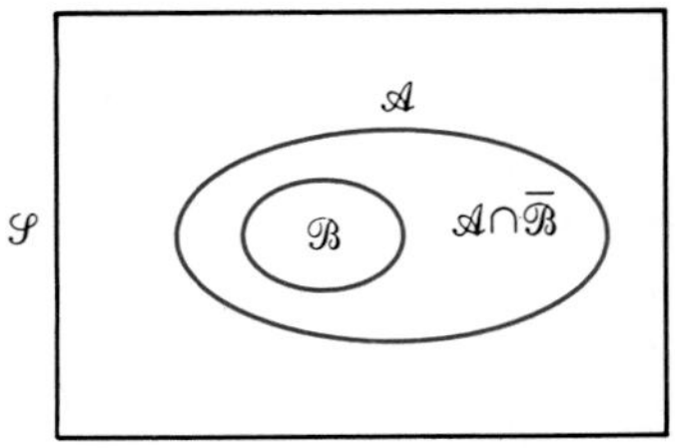

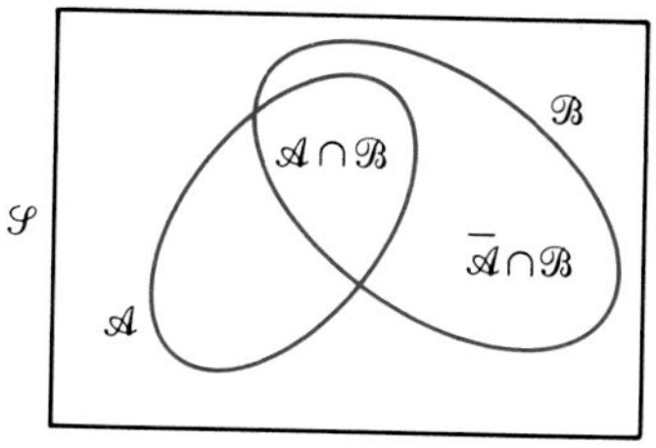

Figure 2.10

events. As we see from Fig. 2.10, $\mathcal{A} \cup \mathcal{B} = \mathcal{A} \cup (\overline{\mathcal{A}} \cap \mathcal{B})$. Hence, as in (2-25),

$$P(\mathcal{A} \cup \mathcal{B}) = P(\mathcal{A}) + P(\overline{\mathcal{A}} \cap B) \tag{2-27}$$

Furthermore,

$$\mathcal{B} = \mathcal{S} \cap \mathcal{B} = (\mathcal{A} \cup \overline{\mathcal{A}}) \cap \mathcal{B} = (\mathcal{A} \cap \mathcal{B}) \cup (\overline{\mathcal{A}} \cap \mathcal{B})$$

(distributive law); and since the events $\mathcal{A} \cap \mathcal{B}$ and $\overline{\mathcal{A}} \cap \mathcal{B}$ are mutually exclusive, we conclude from axiom III that

$$P(\mathcal{B}) = P(\mathcal{A} \cap B) + P(\overline{\mathcal{A}} \cap \mathcal{B}) \tag{2-28}$$

Eliminating $P(\mathcal{A} \cap \mathcal{B})$ from (2-27) and (2-28), we obtain (2-26).

We note that the properties just discussed can be derived simply in terms of probability masses.

Empirical Interpretation. The theory of probability is based on the theorems and on deduction. However, for the results to be useful in the applications, all concepts must be in reasonable agreement with the empirical interpretation $P(\mathcal{A}) \simeq n_{\mathcal{A}}/n$ of $P(\mathcal{A})$. In Section 1-2, we showed that this agreement holds for the three axioms. Let us look at the empirical interpretation of the properties.

1. The impossible event never occurs; hence,

$$n_{\varnothing} = 0 \qquad P(\varnothing) \simeq \frac{n_{\varnothing}}{n} = 0$$

2. As we know, $n_{\mathcal{A}} + n_{\overline{\mathcal{A}}} = n$; hence,

$$P(\overline{\mathcal{A}}) \simeq \frac{n_{\overline{\mathcal{A}}}}{n} = \frac{n - n_{\mathcal{A}}}{n} = 1 - \frac{n_{\mathcal{A}}}{n} \simeq 1 - P(\mathcal{A})$$

3. Clearly, $0 \le n_{\mathcal{A}} \le n$; hence, $0 \le n_{\mathcal{A}}/n \le 1$ in agreement with (2-22).
4. If $\mathcal{B} \subset \mathcal{A}$, then $n_{\mathcal{B}} \le n_{\mathcal{A}}$; hence,

$$P(\mathcal{B}) \simeq \frac{n_{\mathcal{B}}}{n} \le \frac{n_{\mathcal{A}}}{n} \simeq P(\mathcal{A})$$

5. In general [see (2-16)], $n_{\mathcal{A}\cup\mathcal{B}} = n_{\mathcal{A}} + n_{\mathcal{B}} - n_{\mathcal{A}\cap\mathcal{B}}$; hence,

$$P(\mathcal{A} \cup \mathcal{B}) \simeq \frac{n_{\mathcal{A}\cup\mathcal{B}}}{n} = \frac{n_{\mathcal{A}}}{n} + \frac{n_{\mathcal{B}}}{n} - \frac{n_{\mathcal{A}\cap\mathcal{B}}}{n} \simeq P(\mathcal{A}) + P(\mathcal{B}) - P(\mathcal{A} \cap \mathcal{B})$$

Model Specification

An experimental model is specified in terms of the probabilities of all its events. However, because of the axioms, we need not assign probabilities to every event. For example, if we know $P(\mathcal{A})$, we can find $P(\overline{\mathcal{A}})$ from (2-22). We will show that the probabilities of the events of any experiment can be determined in terms of the probabilities of a minimum number of events.

COUNTABLE OUTCOMES Suppose, first, that $\mathcal{S}$ consists of N outcomes $\zeta_1, \ldots, \zeta_N$. In this case, the experiment is specified in terms of the probabilities $p_i = P\{\zeta_i\}$ of the elementary events $\{\zeta_i\}$. Indeed, if $\mathcal{A}$ is an event consisting of the r outcomes $\zeta_{k_1}, \ldots, \zeta_{k_r}$, it can be written as the union of the corresponding elementary events:

$$\mathcal{A} = \{\zeta_{k_1}\} \cup \cdots \cup \{\zeta_{k_r}\} \tag{2-29}$$

hence [see (2-20)],

$$P(\mathcal{A}) = P\{\zeta_{k_1}\} + \cdots + P\{\zeta_{k_r}\} = p_{k_1} + \cdots + p_{k_r} \tag{2-30}$$

Thus the probability of any event $\mathcal{A}$ of $\mathcal{S}$ equals the sum of the probabilities p_i of the elementary events formed with all the elements of $\mathcal{A}$. The numbers p_i are such that

$$p_1 + \cdots + p_N = 1 \qquad p_i \geq 0 \tag{2-31}$$

but otherwise arbitrary.

The foregoing holds also if $\mathcal{S}$ consists of countably many outcomes (see axiom III). It does not hold if the elements of $\mathcal{S}$ are noncountable (points in an interval, for example). In fact, it is not uncommon that the probabilities of all elementary events of a noncountable space equal zero even though $P(\mathcal{S}) = 1$.

Equally Likely Outcomes We shall say that the outcomes of an experiment are equally likely if

$$p_1 = \cdots = p_N = \frac{1}{N} \tag{2-32}$$

In the context of an abstract model, (2-32) is only a special assumption. However, for real experiments, the equally likely assumption covers a large number of applications. In many problems, this assumption is established empirically in terms of observed frequencies or by "reasoning" based on physical "symmetries." This includes games of chance, statistical mechanics, coding, and many other applications.

From (2-30) and (2-32) it follows that if an event $\mathcal{A}$ consists of $N_{\mathcal{A}}$ outcomes, then

$$P(\mathcal{A}) = \frac{N_{\mathcal{A}}}{N} \tag{2-33}$$

This relationship can be phrased as follows: The probability $P(\mathcal{A})$ of an event $\mathcal{A}$ equals the number $N_{\mathcal{A}}$ of elements of $\mathcal{A}$ divided by the total number N of elements. It appears, therefore, that (2-32) is equivalent to the classical definition of probability [see (1-8)]. There is, however, a fundamental difference. In the axiomatic approach to probability, the equally likely condition of (2-32) is an *assumption* used to establish the probabilities of an experimental model. In the classical approach, (2-32) is a *logical conclusion* and is used, in fact, to *define* the probability of $\mathcal{A}$.

Let us look at several illustrations of this important special case.

Example 2.18

(a) We shall say that a coin is *fair* if its outcomes h and t are equally likely, that is, if

$$P\{h\} = P\{t\} = \frac{1}{2}$$

(b) A coin tossed twice generates the space $\mathcal{S} = \{hh, ht, th, tt\}$. If its four outcomes are equally likely (assumption), then

$$P\{hh\} = P\{ht\} = P\{th\} = P\{tt\} = \frac{1}{4} \tag{2-34}$$

In this experiment, the event $\mathcal{H}_1$ = {heads at the first toss} = $\{hh, ht\}$ has two outcomes, hence, $P(\mathcal{H}_1) = 1/2$.

(c) A coin tossed three times generates the space

$$\mathcal{S} = \{hhh, hht, hth, htt, thh, tht, tth, ttt\}$$

Assuming again that all outcomes are equally likely, we conclude that

$$P\{hhh\} = \cdots = P\{ttt\} = \frac{1}{8} \tag{2-35}$$

The event $\mathcal{T}_2$ = {tails at the second toss} = $\{hth, htt, tth, ttt\}$ has four outcomes hence, $P(\mathcal{T}_2) = 1/2$. The event $\mathcal{A}$ = {two heads show} = $\{hhh, hht, thh\}$ has three outcomes, hence, $P(\mathcal{A}) = 3/8$.

We show in Section 3-1 that the equally likely assumptions leading to (2-34) and (2-35) are equivalent to the assumption that the coin is fair *and* the tosses are *independent*. ■

Example 2.19

(a) We shall say that a die is *fair* if its six outcomes f_i are equally likely, that is, if

$$P\{f_1\} = \cdots = P\{f_6\} = \frac{1}{6}$$

The event {even} = $\{f_2, f_4, f_6\}$ has three outcomes, hence, P\{even\} = 3/6.

(b) In the experiment with two dice, we have 36 outcomes f_if_j. If they are equally likely, $\{f_if_j\} = 1/36$. The event $\{11\} = \{f_5f_6, f_6f_5\}$ has two outcomes, hence, $P\{11\} = 2/36$. The event $\{7\} = \{f_1f_6, f_6f_1, f_2f_5, f_5f_2, f_3f_4, f_4f_3\}$ has six outcomes, hence, $P\{7\} = 6/36$. ■

Example 2.20

(a) If a coin is not fair, then

$$P\{h\} = p \qquad P\{t\} = q \qquad p + q = 1 \tag{2-36}$$

For a theoretical investigation, p is an arbitrary number. If the model represents a real coin, p is determined empirically, as in (1-1).

(b) A coin is tossed twice, generating a space with four outcomes. We assign to the elementary events the following probabilities:

$$P\{hh\} = p^2 \qquad P\{ht\} = pq \qquad P\{th\} = qp \qquad P\{tt\} = q^2 \tag{2-37}$$

These probabilities are consistent with (2-30) because

$$p^2 + pq + qp + p^2 = (p + q)^2 = 1$$

The assumption of (2-37) seems artificial. As we show in Section 3-1, it is equivalent to (2-36) and the independence of the two tosses. In preparation, we note the following.

With $\mathcal{H}_1$, $\mathcal{H}_2$, $\mathcal{T}_1$, $\mathcal{T}_2$—the events heads at the first, heads at the second, tails at the first, tails at the second toss, respectively—we have

$$P(\mathcal{H}_1) = P\{hh\} + P\{ht\} = p^2 + pq = p(p + q) = p$$
$$P(\mathcal{H}_2) = P\{hh\} + P\{th\} = p^2 + qp = p(p + q) = p$$
$$P(\mathcal{T}_1) = P\{th\} + P\{tt\} = qp + q^2 = q(p + q) = q$$
$$P(\mathcal{T}_2) = P\{ht\} + P\{tt\} = pq + q^2 = q(p + q) = q$$

The elementary event $\{hh\}$ is the intersection of the events $\mathcal{H}_1$ and $\mathcal{H}_2$; hence, $P(\mathcal{H}_1 \cap \mathcal{H}_2) = P\{hh\}$. Thus

$$\begin{aligned} P(\mathcal{H}_1 \cap \mathcal{H}_2) &= P\{hh\} = p^2 = P(\mathcal{H}_1)P(\mathcal{H}_2) \\ P(\mathcal{H}_1 \cap \mathcal{T}_2) &= P\{ht\} = pq = P(\mathcal{H}_1)P(\mathcal{T}_2) \\ P(\mathcal{T}_1 \cap \mathcal{H}_2) &= P\{th\} = qp = P(\mathcal{T}_2)P(\mathcal{T}_1) \\ P(\mathcal{T}_1 \cap \mathcal{T}_2) &= P\{tt\} = q^2 = P(\mathcal{T}_1)P(\mathcal{T}_2) \end{aligned} \tag{2-38}$$

■

Example 2.21

(a) In Example 2.12(a), $S = \{r, d\}$ and $P\{r\} = p$, $P\{d\} = q$ as in (2-36).

(b) In Example 2.12(b), the space is

$$\mathcal{S} = \{rm, rd, dm, df\}$$

To specify it, we assign probabilities to its elementary events:

$$P\{rm\} = p_1 \qquad P\{rf\} = p_2 \qquad P\{dm\} = p_3 \qquad P\{df\} = p_4$$

where $p_1 + p_2 + p_3 + p_4 = 1$. Thus p_1 is the probability that a person polled is Republican and male. If $\mathcal{S}$ is the model of an actual poll, then $p_1 \simeq n_{rm}/n$ where n_{rm} is the number of male Republicans out of n persons polled.

In this experiment, the event $\mathcal{R}$ = {Republican} consists of two outcomes. Applying (2-31), we obtain

$$P(\mathcal{R}) = P\{rm\} + P\{rf\} = p_1 + p_2$$

Similarly,

$$P(\mathcal{D}) = p_3 + p_4 \qquad P(\mathcal{M}) = p_1 + p_3 \qquad P(\mathcal{F}) = p_2 + p_4$$

■

Equally Likely Events The assumption of equally likely outcomes on which (2-32) is based can be extended to events. Suppose that $\mathcal{A}_1, \ldots, \mathcal{A}_m$ are m events of a partition. We shall say that these events are equally likely if their probabilities are equal. Since the events are mutually exclusive and their union equals $\mathcal{S}$, we conclude as in (2-32) that

$$P(\mathcal{A}_1) = \cdots = P(\mathcal{A}_m) = \frac{1}{m} \tag{2-39}$$

Problems involving equally likely outcomes and events are important in a variety of applications, and their solution is often difficult. However, the difficulties are mainly combinatorial (counting the number of outcomes in an event). Since our primary objective is the clarification of the underlying theory, we shall not dwell on such problems. We shall give only a few illustrations.

Example 2.22

We deal a bridge hand from a well-shuffled deck of cards. Find the probabilities of the following events:

$$\mathscr{A} = \{\text{hand contains 4 aces}\} \qquad \mathscr{B} = \{\text{4 kings}\}$$
$$\mathscr{C} = \{\text{4 aces and 4 kings}\} \qquad \mathscr{D} = \{\text{4 aces or 4 kings}\}$$

The model of this experiment has $\binom{52}{13}$ outcomes, namely, the number of ways that we can take 13 out of 52 objects [see (2-9)]. In the context of the model, the assumption that the deck is well shuffled means that all outcomes are equally likely. In the event $\mathscr{A}$, there are 4 aces, and the remaining 9 cards are taken from the 48 cards that are not aces. Thus the number of outcomes in $\mathscr{A}$ equals $\binom{48}{9}$. This is true also for the event $\mathscr{B}$. Hence,

$$P(\mathscr{A}) = P(\mathscr{B}) = \frac{\binom{48}{9}}{\binom{52}{13}} = .0026$$

The event $\mathscr{C} = \mathscr{A} \cap \mathscr{B}$ contains 4 aces and 4 kings. The remaining 5 cards are taken from the remaining 44 cards that are not aces or kings. Hence,

$$P(\mathscr{C}) = \frac{\binom{44}{5}}{\binom{52}{13}} = 1.7 \times 10^{-6}$$

Finally $\mathscr{D} = \mathscr{A} \cup \mathscr{B}$; hence [see (2-26)],

$$P(\mathscr{D}) = P(\mathscr{A}) + P(\mathscr{B}) - P(\mathscr{C}) = \frac{\left[2\binom{48}{9} - \binom{44}{5}\right]}{\binom{52}{13}} = .0052$$

■

Example 2.23

(a) A box contains 60 red and 40 black balls. A ball is selected at random. Find the probability p_a that the ball is red.

The model of this experiment has 60 + 40 = 100 outcomes. In the context of the model, the random selection is equivalent to the assumption that all outcomes are equally likely. Since there are 60 red balls, the event R = {the select ball is red} has 60 outcomes. Hence, $p_a = P(\mathscr{R}) = 60/100$.

(b) We select 20 balls from the box. Find the probability p_b that 15 of the selected balls are red and 5 black.

In this case, an outcome is the selection of 20 balls. There are $\binom{100}{20}$ ways of selecting 20 out of 100 objects; hence, the experiment has

$\binom{100}{20}$ equally likely outcomes. There are $\binom{60}{15}$ ways of selecting 15 out of the 60 red balls and $\binom{40}{5}$ ways of selecting 5 out of 40 black balls. Hence, there are $\binom{60}{15} \times \binom{40}{5}$ ways of selecting 15 red and 5 black balls. Since all outcomes are equally likely, we conclude that

$$p_b = \frac{\binom{60}{15} \times \binom{40}{5}}{\binom{100}{20}} = .065 \tag{2-40}$$

This can be readily generalized. Suppose that a set contains L red objects and M black objects (two kinds of elements) where $L + M = N$. We select $n \leq N$ of these objects at random. Find the probability p that l of these objects are red and m black where $l + m = n$.

This experiment has $\binom{N}{n}$ equally likely outcomes. There are $\binom{L}{l}$ ways of selecting l out of the L red objects and $\binom{M}{m}$ ways of selecting m out of the M black objects. Hence, as in (2-40),

$$p = \frac{\binom{L}{l}\binom{M}{m}}{\binom{N}{n}} = \frac{\dfrac{L!}{l!(L-l)!}\,\dfrac{M!}{m!(M-m)!}}{\dfrac{N!}{n!(N-n)!}} \tag{2-41}$$

■

Example 2.24

We deal a 5-card poker hand out of a 52-card well-shuffled deck. Find the probability p that the hand contains 3 spades.

This is a special case of (2-41) with $N = 52$, $L = 13$, and $l = 3$ if we identify the 13 spades with the red objects and the other 39 cards with the black objects. With $M = 39$, $n = 5$, and $m = 2$, (2-41) yields

$$p = \frac{\binom{13}{3}\binom{39}{2}}{\binom{52}{5}} = .082$$

■

In certain applications involving a large number of outcomes, it is not possible to determine empirically the probabilities p_i of the elementary events. In such cases, a theoretical model is formed by using the equally likely assumption leading to (2-33) as a *working hypothesis*. The hypothesis is accepted if its consequences agree with experimental observations. The following is an important applications from statistical mechanics.

Example 2.25

We place at random m particles in $n \geq m$ boxes. Find the probability p that the particles are located in m preselected boxes, one in each box.

The solution to this problem depends on what we consider as outcomes. We shall analyze the following celebrated models.

(a) *Maxwell-Boltzmann.* We assume that the m particles are *distinct,* and we consider as outcomes all possible ways of placing them into the n boxes. There are n choices for each particle; hence, the number N of outcomes equals n^m. The number $N_{\mathcal{A}}$ of favorable outcomes equals the $m!$ ways of placing the m particles into the m preselected boxes (permutations of m objects). Thus $N = n^m$, $N_{\mathcal{A}} = m!$, and (2-33) yields

$$p = \frac{m!}{n^m}$$

(b) *Bose-Einstein.* We now assume that the m particles are *identical.* In this case, N equals the number of ways of placing m identical objects into n boxes. As we know from (2-12) and from Fig. 2.8, this number equals $\binom{n+m-1}{m}$. There is, of course, only one way of placing the m particles in the m preselected boxes. Hence,

$$p = \frac{1}{\binom{n+m-1}{m}} = \frac{m!(n-1)!}{(n+m-1)!}$$

(c) *Fermi-Dirac.* We assume again that the particles are *identical* and that we place *only one* particle in each box. In this case, the number of possibilities equals the number $\binom{n}{m}$ of combinations of n objects taken m at a time. Of these, only one is favorable; hence,

$$p = \frac{1}{\binom{n}{m}} = \frac{m!(n-m)!}{n!}$$

One might argue, as indeed it was argued in the early years of statistical mechanics, that only the first of these solutions may be accepted. However, in the absence of direct or indirect experimental evidence, no single model is logically correct. The models are actually only *hypotheses;* the physicist accepts the particular model whose consequences agree with observations. ■

NONCOUNTABLE OUTCOMES We now consider experiments consisting of a noncountable number of outcomes. We assume, first, that $\mathcal{S}$ consists of all points on the real line

$$\mathcal{S} = \{-\infty < t < \infty\} \tag{2-42}$$

This case arises often: system failure, telephone calls, birth and death, arrival times, and many others. The events of this experiment are all intervals $\{t_1 \le t \le t_2\}$ and their unions and intersections (see "Fundamental Note" later in this discussion). The elementary events are of the form $\{t_i\}$ where t_i is any point on the t-axis, and their number is noncountable. Unlike the case of countable elements, $\mathcal{S}$ is not specified in terms of the probabilities of the elementary events. In fact, it is possible that $P\{t_i\} = 0$ for every outcome t_i

even though $\mathscr{S}$ is the union of all elementary events. This is not in conflict with (2-20) because in the equation, the events $\mathscr{A}_i$ are countable.

We give next a set of events whose probabilities specify $\mathscr{S}$ completely. To facilitate the clarification of the underlying concept of density, we shall use the mass interpretation of probability. If the experiment $\mathscr{S}$ has a countable number of outcomes ζ_i, the probabilities $p_i = P\{\zeta_i\}$ of the elementary events $\{\zeta_i\}$ can be viewed as *point masses* (Fig. 2.11). If $\mathscr{S}$ is noncountable as in (2-42) and $P\{\zeta_i\} = 0$ for every ζ_i, the probability masses are distributed along the axis and can be specified in terms of the density function $\alpha(t)$ defined as follows. The mass in an interval (t_1, t_2) equals the probability of the event $\{t_1 \le t \le t_2\}$. Thus

$$P\{t_1 \le t \le t_2\} = \int_{t_1}^{t_2} \alpha(t)\,dt \tag{2-43}$$

The function $\alpha(t)$ can be interpreted as a limit involving probabilities. As we see from (2-43), if Δt is sufficiently small, $P\{t_1 \le t \le t_1 + \Delta t\} \simeq \alpha(t_1)\Delta t$, and in the limit

$$\alpha(t_1) = \lim_{\Delta t \to 0} \frac{P\{t_1 \le t \le t_1 + \Delta t\}}{\Delta t} \tag{2-44}$$

We maintain that the function $\alpha(t)$ specifies the experiment $\mathscr{S}$ completely. Indeed, any event $\mathscr{A}$ of $\mathscr{S}$ is a set $\mathscr{D}$ of points that can be written as a countable union of nonoverlapping (disjoint) intervals (see "Fundamental Note"). From this and (2-20) it follows that $P(\mathscr{A})$ equals the area under the curve $\alpha(t)$ in the region $\mathscr{D}$. As we see from (2-30), the area of $\alpha(t)$ in any interval is positive; hence, $\alpha(t) \ge 0$ for any t. Furthermore, its total area equals $P(\mathscr{S}) = 1$. Hence, $\alpha(t)$ is such that

$$\alpha(t) \ge 0 \qquad \int_{-\infty}^{\infty} \alpha(t)\,dt = 1 \tag{2-45}$$

Note The function $\alpha(t)$ is related to but conceptually different from the density of a random variable, a concept to be developed in Chapter 4.

In (2-45) we assumed that $\mathscr{S}$ is the set of all points on the entire line. In many cases, however, the given set $\mathscr{S}$ is only a region $\mathscr{R}$ of the axis. In such cases, $\alpha(t)$ is specified only for $t \in \mathscr{R}$ and its area in the region $\mathscr{R}$ equals 1. The following is an important special case.

Figure 2.11

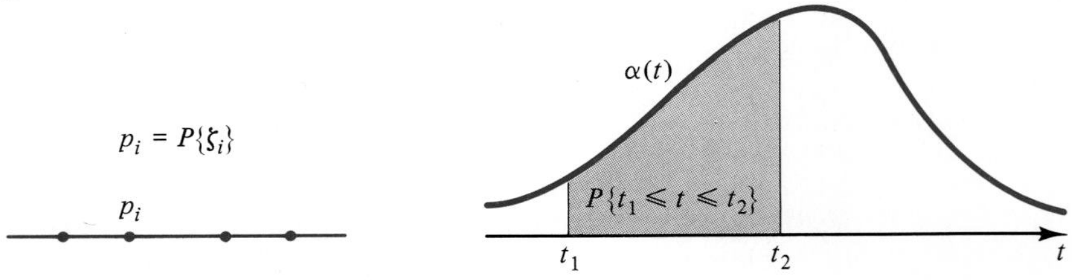

We shall say that $\mathscr{S}$ is a set of *random points* in the interval (a, b) if it consists of all points in this interval and $\alpha(t)$ is *constant*. In this case [see (2-45)], $\alpha(t) = 1/(b - a)$ and

$$P\{t_1 \le t \le t_2\} = \int_{t_1}^{t_2} \alpha(t)\, dt = \frac{t_2 - t_1}{b - a} \qquad a \le t_1, t_2 \le b \tag{2-46}$$

Fundamental Note. In the definition of a probability space, we have assumed tacitly that all subsets of $\mathscr{S}$ are events. We can do so if $\mathscr{S}$ is countable, but if it is noncountable, we cannot assign probabilities to all its subsets consistent with axiom III. For this reason, events of $\mathscr{S}$ are all subsets that can be expressed as countable unions and intersections of intervals. That this does not include all subsets of $\mathscr{S}$ is not easy to show. However, this is only of mathematical interest. For most applications, sets that are not countable unions or intersections of intervals are of no interest.

Empirical Interpretation of $\alpha(t)$. As we know, the probabilities of the events of a model can be evaluated empirically as relative frequencies. If, therefore, a model parameter can be expressed as the probability of an event, it can be so evaluated. The function $\alpha(t)$ is not a probability; however, its integral is [see (2-43)]. This fact leads to the following method for evaluating $\alpha(t)$.

To find $\alpha(t)$ for $t = t_i$, we form the event

$$\mathscr{A}_i = \{t_i \le t \le t_i + \Delta t\}$$

At a single trial, we observe an outcome t. If the observed t is in the interval $(t_i, t_i + \Delta t)$, the event $\mathscr{A}_i$ occurs. Denoting by Δn_i the number of such occurrences at n trials, we conclude from (2-43) and (1-1) that

$$P(\mathscr{A}_i) = \int_{t_i}^{t_i + \Delta t} \alpha(t)\, dt \simeq \frac{\Delta n_i}{n} \tag{2-47}$$

If Δt is sufficiently small, the integral equals $\alpha(t_i)\Delta t$; hence,

$$\alpha(t_i)\,\Delta t \simeq \frac{\Delta n_i}{n} \tag{2-48}$$

This expresses $\alpha(t_i)$ in terms of the number Δn_i of observed outcomes in the interval $(t_i, t_i + \Delta t)$.

Example 2.26

We denote by t the age of a person when he dies, ignoring the part of the population that live more than 100 years. The outcomes of the resulting experiment are all points in the interval (0, 100). The experimental model is thus specified in terms of a function $\alpha(t)$ defined for every t in this interval. This function can be determined from (2-48) if sufficient data are available. We shall asume that

$$\alpha(t) = 3 \times 10^{-9} t^2 (100 - t)^2 \qquad 0 \le t \le 100 \tag{2-49}$$

(see Fig. 2.12). The probability that a person will die between the ages of 60 and 70 equals

$$P\{60 \le t \le 70\} = 3 \times 10^{-9} \int_{60}^{70} t^2 (100 - t)^2\, dt = .154$$

The probability that a person is alive at 60 equals

$$P\{t > 60\} = 3 \times 10^{-9} \int_{60}^{100} t^2 (100 - t)^2\, dt = .317$$

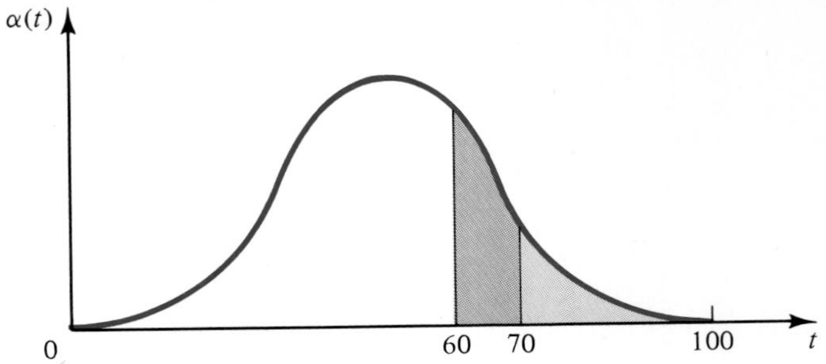

Figure 2.12

Thus according to this model, 15.4% of the population dies between the ages of 60 and 70 and 31.7% is alive at age 60. ■

Example 2.27

Consider a radioactive substance emitting particles at various times t_i. We observe the emissions starting at $t = 0$, and we denote by t_1 the time of emission of the first particle (Fig. 2.13). Find the probability p that t_1 is less than a given time t_0.

In this experiment, $\mathcal{S}$ is the axis $t > 0$. We shall assume that

$$\alpha(t) = \lambda e^{-\lambda t} \qquad t \geq 0 \tag{2-50}$$

From this and (2-43) it follows that

$$p = P\{t_1 < t_0\} = \lambda \int_0^{t_0} e^{-\lambda t}\, dt = 1 - e^{-\lambda t_0}$$ ■

Points on the Plane Experiments involving points on the plane or in space can be treated similarly. Suppose, for example, that the experimental outcomes are pairs of numbers (x, y) on the entire plane

$$\mathcal{S} = \{-\infty < x, y < \infty\} \tag{2-51}$$

or in certain subset of the plane. Events of this experiment are all rectangles and their countable unions and intersections. This includes all nonpathological plane regions $\mathcal{D}$. To complete the specification of $\mathcal{S}$, we must assign probabilities to these events. We can do as in (2-43): We select a positive function $\alpha(x, y)$, and we assign to the event $\{(x, y) \in \mathcal{D}\}$ the probability

$$P\{(x, y) \in \mathcal{D}\} = \iint_{\mathcal{D}} \alpha(x, y)\, dx\, dy \tag{2-52}$$

The function $\alpha(x, y)$ can be interpreted as surface mass density.

Figure 2.13

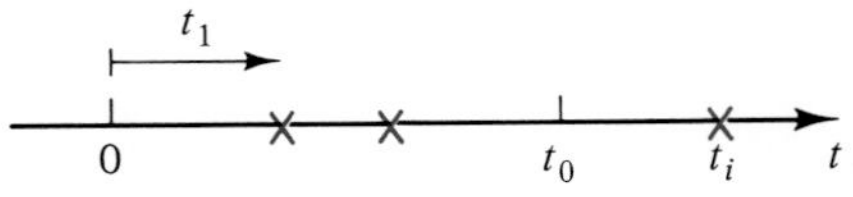

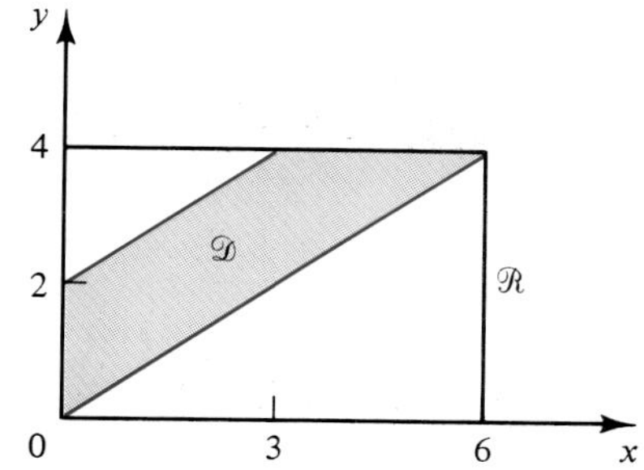

Figure 2.14

Example 2.28

A point is selected at random from the rectangle $\mathscr{R}$ of Fig. 2.14. Find the probability p that it is taken from the trapezoidal region $\mathscr{D}$.

The model of this experiment consists of all points in $\mathscr{R}$. The assumption that the points are selected at random is equivalent to the model assumption that the probability density $\alpha(x, y)$ is constant. The area of $\mathscr{R}$ equals 24; hence, $\alpha(x, y) = 1/24$. And since the area of $\mathscr{D}$ equals 9, we conclude from (2-52) that

$$p = \frac{1}{24}\iint_{\mathscr{D}} dx\,dy = \frac{9}{24}$$ ∎

2-3 Conditional Probability and Independence

Given an event $\mathscr{M}$ such that $P(\mathscr{M}) \neq 0$, we form the ratio $P(\mathscr{A} \cap \mathscr{M})/P(\mathscr{M})$ where $\mathscr{A}$ is any event of $\mathscr{S}$. This ratio is denoted by $P(\mathscr{A}|\mathscr{M})$ and is called the "*conditional probability* of $\mathscr{A}$ assuming $\mathscr{M}$." Thus

$$P(\mathscr{A}|\mathscr{M}) = \frac{P(\mathscr{A} \cap \mathscr{M})}{P(\mathscr{M})} \tag{2-53}$$

The significance of this important concept will be appreciated in the course of our development.

Empirical Interpretation. We repeat the experiment n times, and we denote by $n_{\mathscr{M}}$ and $n_{\mathscr{A}\cap\mathscr{M}}$ the number of occurrences of the events $\mathscr{M}$ and $\mathscr{A} \cap \mathscr{M}$, respectively. If n is large, then [see (1-1)]

$$P(\mathscr{M}) \simeq \frac{n_{\mathscr{M}}}{n} \qquad P(\mathscr{A} \cap \mathscr{M}) \simeq \frac{n_{\mathscr{A}\cap\mathscr{M}}}{n}$$

Hence,

$$\frac{P(\mathscr{A} \cap \mathscr{M})}{P(\mathscr{M})} \simeq \frac{n_{\mathscr{A}\cap\mathscr{M}}/n}{n_{\mathscr{M}}/n} = \frac{n_{\mathscr{A}\cap\mathscr{M}}}{n_{\mathscr{M}}}$$

and (2-53) yields

$$P(\mathcal{A}|\mathcal{M}) \simeq \frac{n_{\mathcal{A}\cap\mathcal{M}}}{n_{\mathcal{M}}} \qquad (2\text{-}54)$$

The event $\mathcal{A} \cap \mathcal{M}$ occurs iff $\mathcal{M}$ and $\mathcal{A}$ occur; hence, $n_{\mathcal{A}\cap\mathcal{M}}$ is the number of occurrences of $\mathcal{A}$ in the subsequence of trials in which $\mathcal{M}$ occurs. This leads to the relative frequence interpretation of $P(\mathcal{A}|\mathcal{M})$: The conditional probability of $\mathcal{A}$ assuming $\mathcal{M}$ is nearly equal to the relative frequency of the occurrence of $\mathcal{A}$ in the subsequence of trials in which $\mathcal{M}$ occurs. This is true if not only n but also $n_{\mathcal{M}}$ is large.

Example 2.29

Given a fair die, we shall determine the probability of 2 assuming *even*. This is the conditional probability of the elementary even $\mathcal{A} = \{f_2\}$ assuming $\mathcal{M} = \{\text{even}\}$. Clearly, $\mathcal{A} \subset \mathcal{M}$; hence, $\mathcal{A} \cap \mathcal{M} = \mathcal{A}$. Furthermore, $P(\mathcal{A}) = 1/6$ and $P(\mathcal{M}) = 3/6$; hence,

$$P(f_2/\text{even}) = \frac{P(\mathcal{A} \cap \mathcal{M})}{P(\mathcal{M})} = \frac{1/6}{1/2} = \frac{1}{3}$$

Thus the relative frequency of the occurrence of 2 in the subsequence of trials in which *even* shows equals 1/3. ■

Example 2.30

In the mortality experiment (Example 2.26), we wish to find the probability that a person will die between the ages of 60 and 70, assuming that the person is alive at 60.

Our problem is to find $P(\mathcal{A}|\mathcal{M})$ where

$$\mathcal{A} = \{60 < t \leq 70\} \qquad \mathcal{M} = \{t > 60\}$$

As we have seen, $P(\mathcal{A}) = .154$ and $P(\mathcal{M}) = .317$. Since $\mathcal{A} \cap \mathcal{M} = \mathcal{A}$, we conclude that

$$P(\mathcal{A}|\mathcal{M}) = \frac{P(\mathcal{A})}{P(\mathcal{M})} = \frac{.154}{.317} = .486$$

Thus 15% of all people die between the ages of 60 and 70. However, 48.6% of the people that are alive at 60 die between the ages of 60 and 70. ■

Example 2.31

A box contains 3 white balls and 2 red balls. We select 2 balls in succession. Find the probability p that the first ball is white and the second red.

The probability of the event $\mathcal{W}_1 = \{\text{white first}\}$ equals 3/5. After the removal of the white ball there remain 2 white and 2 red balls. Hence, the conditional probability of the event $\mathcal{R}_2 = \{\text{red second}\}$ assuming $\mathcal{W}_1$ equals 2/4. And since $\mathcal{W}_1 \cap \mathcal{R}_2$ is the event {white first, red second}, we conclude from (2-53) that

$$p = P(\mathcal{W}_1 \cap \mathcal{R}_2) = P(\mathcal{R}_2|\mathcal{W}_1)P(\mathcal{W}_1) = \frac{2}{4} \times \frac{3}{5} = \frac{6}{20}$$

Next let us find a direct solution. The experiment has 20 outcomes, namely, the $P_2^5 = 5 \times 4$ permutations

$$w_1w_2,\ w_1w_3,\ w_1r_1,\ \ldots,\ r_2w_2,\ r_2w_3,\ r_2r_1$$

of the 5 objects w_1, w_2, w_3, r_1, r_2 taken 2 at a time. The elementary events are equally likely, and their probability equals 1/20. The event {white first, red second} consists of the 6 outcomes

$$w_1r_1,\ w_2r_1,\ w_3r_1,\ w_1r_2,\ w_2r_2,\ w_3r_3$$

Hence, its probability equals 6/20. ■

In a number of applications, the available information (data), although sufficient to specify the model, does not lead directly to the determination of the probabilities of its events but is used to determine conditional probabilities. The next example is an illustration.

Example 2.32

We are given two boxes. Box 1 contains 2 red and 8 white cards; box 2 contains 9 red and 6 white cards. We select *at random* one of the boxes, and we pick *at random* one of its cards. Find the probability p that the selected card is red.

The outcomes of this experiment are the 25 cards contained in both boxes. We denote by $\mathcal{B}_1$ the event consisting of the 10 cards in box 1 and by $\mathcal{B}_2$ the event consisting of the 15 cards in box 2 in Fig. 2.15. From the assumption that a box is selected at random we conclude that

$$P(\mathcal{B}_1) = P(\mathcal{B}_2) = \frac{1}{2} \tag{2-55}$$

The event $\mathcal{R} = \{\text{red}\}$ consists of 11 outcomes. Our problem is to find its probability. We cannot do so directly because the 25 outcomes of $\mathcal{S}$ are not equally likely. They are, however, conditionally equally likely, subject to the condition that a box is selected. As a model concept this means that

$$P(\mathcal{R}|\mathcal{B}_1) = \frac{2}{10} \qquad P(\mathcal{R}|\mathcal{B}_2) = \frac{9}{15} \tag{2-56}$$

Equations (2-55) and (2-56) are not derived. They are *assumptions* about the model based on the hypothesis that the box and the card are selected at random. Using these assumptions, we shall derive $P(\mathcal{R})$ deductively.

Since

$$P(\mathcal{R}|\mathcal{B}_1) = \frac{P(\mathcal{R} \cap \mathcal{B}_1)}{P(\mathcal{B}_1)} \qquad P(\mathcal{R}|\mathcal{B}_2) = \frac{P(\mathcal{R} \cap \mathcal{B}_2)}{P(\mathcal{B}_2)}$$

it follows from (2-55) and (2-56) that

$$P(\mathcal{R} \cap \mathcal{B}_1) = \frac{2}{10} \times \frac{1}{2} = \frac{1}{10} \qquad P(\mathcal{R} \cap \mathcal{B}_2) = \frac{9}{15} \times \frac{1}{2} = \frac{3}{10}$$

The events $\mathcal{R} \cap \mathcal{B}_1$ and $\mathcal{R} \cap \mathcal{B}_2$ are mutually exclusive (see Fig. 2.15), and their union equals the event $\mathcal{R}$. Hence,

$$P(\mathcal{R}) = P(\mathcal{R} \cap \mathcal{B}_1) + P(\mathcal{R} \cap \mathcal{B}_2) = \frac{4}{10}$$

Thus if we pick at random a card from a randomly selected box, in 40% of the trials the selected card will be red. ■

Figure 2.15

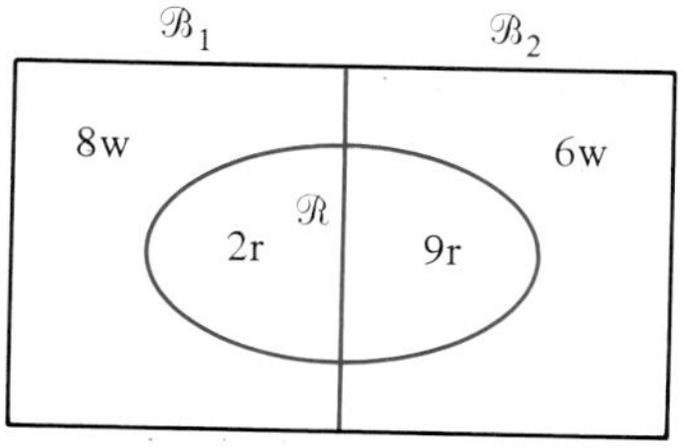

From (2-53) it follows that

$$P(\mathcal{A} \cap \mathcal{B}) = P(\mathcal{A}|\mathcal{B})P(\mathcal{B}) \tag{2-57}$$

Repeated application of this yields

$$P(\mathcal{A} \cap \mathcal{B} \cap \mathcal{C}) = P(\mathcal{A}|\mathcal{B} \cap \mathcal{C})\, P\,(\mathcal{B} \cap \mathcal{C}) = P(\mathcal{A}|\mathcal{B} \cap \mathcal{C})P(\mathcal{B}|\mathcal{C})P(\mathcal{C})$$

This is the *chain rule* for conditional probabilities and can be readily generalized (see Problem 2-14).

Fundamental Property We shall now examine the properties of the numbers $P(\mathcal{A}|\mathcal{M})$ for a *fixed* $\mathcal{M}$ as $\mathcal{A}$ ranges over all the events of $\mathcal{S}$. We maintain that these numbers are, indeed, probabilities; that is, they satisfy the axioms. To do so, we must prove the following:

I. $$P(\mathcal{A}|\mathcal{M}) \geq 0 \tag{2-58}$$

II. $$P(\mathcal{S}|\mathcal{M}) = 1 \tag{2-59}$$

III. If $\mathcal{A} \cap \mathcal{B} = \varnothing$, then

$$P(\mathcal{A} \cup \mathcal{B}|\mathcal{M}) = P(\mathcal{A}|\mathcal{M}) + P(\mathcal{B}|\mathcal{M}) \tag{2-60}$$

■ ***Proof.*** Equation (2-58) follows readily from (2-17) because $\mathcal{A} \cap \mathcal{M}$ is an event. Equation (2-59) is a consequence of that fact that $\mathcal{S} \cap \mathcal{M} = \mathcal{M}$. To prove (2-60), we observe that if the sets $\mathcal{A}$ and $\mathcal{B}$ are disjoint, their subsets $\mathcal{A} \cap \mathcal{M}$ and $\mathcal{B} \cap \mathcal{M}$ are also disjoint (Fig. 2.16). And since $(\mathcal{A} \cup \mathcal{B}) \cap \mathcal{M} = (\mathcal{A} \cap \mathcal{M}) \cup (\mathcal{B} \cap \mathcal{M})$, we conclude from (2-19) that

$$P(\mathcal{A} \cup \mathcal{B}|\mathcal{M}) = \frac{P[(\mathcal{A} \cup \mathcal{B}) \cap \mathcal{M}]}{P(\mathcal{M})} = \frac{P(\mathcal{A} \cap \mathcal{M})}{P(\mathcal{M})} + \frac{P(\mathcal{B} \cap \mathcal{M})}{P(\mathcal{M})}$$

and (2-60) results.

The foregoing shows that conditional probabilities can be used to create from a given experiment $\mathcal{S}$ a new experiment conditioned on an event $\mathcal{M}$ of $\mathcal{S}$. This experiment has the same outcomes ζ_i and events $\mathcal{A}_i$ as the original experiment $\mathcal{S}$, but its probabilities equal $P(\mathcal{A}_i|\mathcal{M})$. These probabilities specify a new experiment because, as we have just shown, they satisfy the axioms.

Figure 2.16

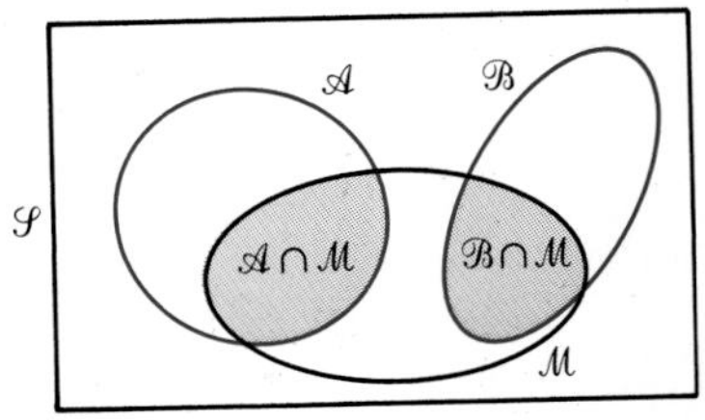

Note, finally, that $P(\mathcal{A}|\mathcal{M})$ can be given the following mass interpretation. In the Venn diagram (Fig. 2.16), the event $\mathcal{A} \cap \mathcal{M}$ is the part of $\mathcal{A}$ in $\mathcal{M}$, and $P(\mathcal{A}|\mathcal{M})$ is the mass in that region normalized by the factor $1/P(\mathcal{M})$.

Total Probability and Bayes' Theorem

In Example 2.32, we expressed the probability $P(\mathcal{R})$ of the event $\mathcal{R}$ in terms of the known conditional probabilities $P(\mathcal{R}|\mathcal{B}_1)$ and $P(\mathcal{R}|\mathcal{B}_2)$. The following is an important generalization.

Suppose that $[\mathcal{A}_1, \ldots, \mathcal{A}_m]$ is a partition of $\mathcal{S}$ consisting of m events as shown in Fig. 2.17. We maintain that the probability $P(\mathcal{B})$ of an arbitrary event $\mathcal{B}$ of $\mathcal{S}$ can be written as a sum:

$$P(\mathcal{B}) = P(\mathcal{B}|\mathcal{A}_1)P(\mathcal{A}_1) + \cdots + P(\mathcal{B}|\mathcal{A}_m)P(\mathcal{A}_m) \tag{2-61}$$

■ ***Proof.*** Clearly [see (2-4)],

$$\mathcal{B} = \mathcal{B} \cap \mathcal{S} = \mathcal{B} \cap (\mathcal{A}_1 \cup \cdots \cup \mathcal{A}_m) = (\mathcal{B} \cap \mathcal{A}_1) \cup \cdots \cup (\mathcal{B} \cap \mathcal{A}_m)$$

But the events $\mathcal{B} \cap \mathcal{A}_i$ are mutually exclusive because the events $\mathcal{A}_i$ are mutually exclusive. Hence [see (2-20)],

$$P(\mathcal{B}) = P(\mathcal{B} \cap \mathcal{A}_1) + \cdots + P(\mathcal{B} \cap \mathcal{A}_m) \tag{2-62}$$

And since $P(\mathcal{B} \cap \mathcal{A}_i) = P(\mathcal{B}|\mathcal{A}_i)P(\mathcal{A}_i)$, (2-61) follows.

This is called the theorem of *total probability*. It is used to evaluate the probability $P(\mathcal{B})$ of an event $\mathcal{B}$ if its conditional probabilities $P(\mathcal{B}|\mathcal{A}_i)$ are known.

Example 2.33

In a political poll, the following results are recorded: Among all voters, 70% are male and 30% are female. Among males, 40% are Republican and 60% Democratic. Among females, 45% are Republican and 55% are Democratic. Find the probability that a voter selected at random is Republican.

In this experiment, $\mathcal{S} = \{rm, rf, dm, df\}$. We form the events $\mathcal{M}$ = {male}, $\mathcal{F}$ = {female}, and $\mathcal{R}$ = {Republican}. The results of the poll yield the following data

$$P(\mathcal{M}) = .70 \qquad P(\mathcal{F}) = .30$$
$$P(\mathcal{R}|\mathcal{M}) = .40 \qquad P(\mathcal{R}|\mathcal{F}) = .45$$

Hence [see (2-61)],

$$P(\mathcal{R}) = P(\mathcal{R}|\mathcal{M})P(\mathcal{M}) + P(\mathcal{R}|\mathcal{F})P(\mathcal{F}) = .415$$

■

Figure 2.17

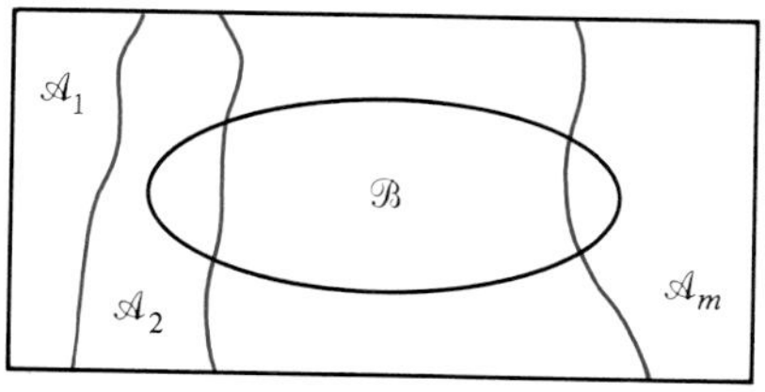

■ ***Bayes' Theorem.*** We show next that

$$P(\mathscr{A}_i|\mathscr{B}) = \frac{P(\mathscr{B}|\mathscr{A}_i)P(\mathscr{A}_i)}{P(\mathscr{B}|\mathscr{A}_1)P(\mathscr{A}_1) + \cdots + P(\mathscr{B}|\mathscr{A}_m)P(\mathscr{A}_m)} \tag{2-63}$$

This is an important result known as *Bayes' theorem*. It expresses the *posterior* probabilities $P(\mathscr{A}_i|\mathscr{B})$ of the events $\mathscr{A}_i$ in terms of their *prior* probabilities $P(\mathscr{A}_i)$. Its significance will become evident later.

■ ***Proof.*** From (2-53) it follows that

$$P(\mathscr{A}_i|\mathscr{B}) = \frac{P(\mathscr{A}_i \cap \mathscr{B})}{P(\mathscr{B})} \qquad P(\mathscr{B}|\mathscr{A}_i) = \frac{P(\mathscr{A}_i \cap \mathscr{B})}{P(\mathscr{A}_i)}$$

Hence,

$$P(\mathscr{A}_i|\mathscr{B}) = \frac{P(\mathscr{B}|\mathscr{A}_i)P(\mathscr{A}_i)}{P(\mathscr{B})} \tag{2-64}$$

Inserting (2-61) into (2-64), we obtain (2-63).

Example 2.34

We have two coins: coin A is fair with $P\{\text{heads}\} = 1/2$, and coin B is loaded with $P\{\text{heads}\} = 2/3$. We pick one of the coins at random, we toss it, and heads shows. Find the probability that we picked the fair coin.

This experiment has four outcomes:

$$\mathscr{S} = \{ah, at, bh, bt\}$$

For example, *ah* is the outcome "the fair coin is tossed and heads shows." We form the events

$$\mathscr{A} = \{\text{fair coin}\} = \{ah, ht\} \qquad \mathscr{B} = \{\text{loaded coin}\} = \{bh, bt\}$$

$$\mathscr{H} = \{\text{heads shows}\} = \{ah, bh\}$$

The assumption that the coin is picked at random yields $P(\mathscr{A}) = P(\mathscr{B}) = 1/2$. The probability of *heads* assuming that coin A is picked equals $P(\mathscr{H}|\mathscr{A}) = 1/2$. Similarly, $P(\mathscr{H}|\mathscr{B}) = 2/3$. This completes the specification of the model. Our problem is to find the probability $P(\mathscr{A}|\mathscr{H})$ that we picked the fair coin assuming that heads showed. To do so, we use (2-63):

$$P(\mathscr{A}|\mathscr{H}) = \frac{P(\mathscr{H}|\mathscr{A})P(\mathscr{A})}{P(\mathscr{H}|\mathscr{A})P(\mathscr{A}) + P(\mathscr{H}|\mathscr{B})P(\mathscr{B})} = \frac{\frac{1}{2} \times \frac{1}{2}}{\left(\frac{1}{2} \times \frac{1}{2}\right) + \left(\frac{2}{3} \times \frac{1}{2}\right)} = \frac{3}{7}$$ ■

In the next example, pay particular attention to the distinction between the determination of the model data based on the description of the experiment and the results that follow deductively from the axioms.

Example 2.35

We have four boxes. Box 1 contains 2,000 components, of which 1,900 are good and 100 are defective. Box 2 contains 500 components, of which 300 are good and 200 defective. Boxes 3 and 4 each contain 1,000 components, of which 900 are good and

100 defective. We select *at random* one of the boxes and pick *at random* a single component.

(a) Find the probability p_a that the selected component is defective.

(b) The selected component is defective; find the probability p_b that it came from box 2.

Model Specification The space $\mathcal{S}$ of this experiment has 4,000 good (g) elements forming the event $\mathcal{G}$ and 500 defective (d) elements forming the event $\mathcal{D}$. The elements in each box form the events

$$\mathcal{B}_1 = \{1{,}900g,\ 100d\} \qquad \mathcal{B}_2 = \{300g,\ 200d\}$$
$$\mathcal{B}_3 = \{900g,\ 100d\} \qquad \mathcal{B}_4 = \{900g,\ 100d\}$$

The information that the boxes are selected at random leads to the assumption that the four events have the same probability. Hence [see (2-39)],

$$P(\mathcal{B}_1) = P(\mathcal{B}_2) = P(\mathcal{B}_3) = P(\mathcal{B}_4) = \frac{1}{4} \tag{2-65}$$

The random selection of a component from the ith box leads to the assumption that all elements in that box are conditionally equally likely. From this it follows that the conditional probability $P(\mathcal{D}|\mathcal{B}_i)$ that a component taken from the ith box is defective equals the proportion of defective components in that box. This is the extension of (2-32) to conditional probabilities, and it yields

$$\begin{aligned} P(\mathcal{D}|\mathcal{B}_1) &= \frac{100}{2{,}000} = .05 \qquad & P(\mathcal{D}|\mathcal{B}_2) &= \frac{200}{500} = .4 \\ P(\mathcal{D}|\mathcal{B}_3) &= \frac{100}{1{,}000} = .1 \qquad & P(\mathcal{D}|\mathcal{B}_4) &= \frac{100}{1{,}000} = .1 \end{aligned} \tag{2-66}$$

Deduction

(a) From (2-62) and the foregoing it follows that the probability $P(\mathcal{D})$ that the selective component is defective equals

$$P(\mathcal{D}) = .05 \times \frac{1}{4} + .4 \times \frac{1}{4} + .1 \times \frac{1}{4} + .1 \times \frac{1}{4} = .1625$$

(b) The probability p_b that the defective component came from box 2 equals $P(\mathcal{B}_2|\mathcal{D})$. Hence [see (2-64)],

$$P(\mathcal{B}_2|\mathcal{D}) = \frac{P(\mathcal{D}|\mathcal{B}_2)P(\mathcal{B}_2)}{P(\mathcal{D})} = \frac{.4 \times .25}{.1625} = .615$$

Thus the *prior* probability $P(\mathcal{B}_2)$ of selecting box $\mathcal{B}_2$ equals .25; the *posterior* probability, assuming that the component is defective, equals .615. ■

Empirical Interpretation. We perform the experiment n times. In 25% of the trials, box 2 is selected. If we consider only the $n_{\mathcal{D}}$ trials in which the selected part is defective, then in 61.5% of such trials it came from box 2.

Example 2.36

In a large city, it is established that 0.5% of the population has contracted AIDS. The available tests give the correct diagnosis for 80% of healthy persons and for 98% of sick persons. A person is tested and found sick. Find the probability that the diagnosis is wrong, that is, that the person is actually healthy.

We introduce the events

$$\mathscr{A} = \text{healthy} \qquad \mathscr{B} = \text{tested healthy}$$
$$\mathscr{C} = \text{sick} = \overline{\mathscr{A}} \qquad \mathscr{D} = \text{tested sick} = \overline{\mathscr{B}}$$

The unknown probability is $P(\mathscr{A}|\mathscr{D})$. From the description of the problem it follows that

$$P(\mathscr{A}) = .995 \qquad P(\mathscr{C}) = .005$$
$$P(\mathscr{D}|\mathscr{A}) = .20 \qquad P(\mathscr{D}|\mathscr{C}) = .98$$

Hence [see (2-61)],

$$P(\mathscr{D}) = P(\mathscr{D}|\mathscr{A})P(\mathscr{A}) + P(\mathscr{D}|\mathscr{C})P(\mathscr{C}) = .2039$$

This is the probability that a person selected at random will test sick. Inserting into (2-64), we conclude that

$$P(\mathscr{A}|\mathscr{D}) = \frac{P(\mathscr{D}|\mathscr{A})P(\mathscr{A})}{P(\mathscr{D})} = \frac{.1990}{.2039} = .976$$ ■

Independent Events

Two events $\mathscr{A}$ and $\mathscr{B}$ are called *statistically independent* if the probability of their intersection equals the product of their probabilities:

$$P(\mathscr{A} \cap \mathscr{B}) = P(\mathscr{A})P(\mathscr{B}) \tag{2-67}$$

The word *statistical* will usually be omitted. As we see from (2-53), if the events $\mathscr{A}$ and $\mathscr{B}$ are independent, then

$$P(\mathscr{A}|\mathscr{B}) = P(\mathscr{A}) \qquad P(\mathscr{B}|\mathscr{A}) = P(\mathscr{B}) \tag{2-68}$$

The concept of independence is fundamental in the theory of probability. However, this is not apparent from the definition. Why should a relationship of the form of (2-67) merit special consideration? The importance of the concept will become apparent in the context of repeated trials and combined experiments. Let us examine briefly independence in the context of relative frequencies.

Empirical Interpretation. The probability $P(\mathscr{A})$ of the event $\mathscr{A}$ equals the relative frequency $n_{\mathscr{A}}/n$ of the occurrence of $\mathscr{A}$ in n trials. The conditional probability $P(\mathscr{A}|\mathscr{B})$ of $\mathscr{A}$ assuming $\mathscr{B}$ equals the relative frequence of the occurrence of $\mathscr{A}$ in the subsequence of $n_{\mathscr{B}}$ trials in which $\mathscr{B}$ occurs [see (2-54)]. If the events $\mathscr{A}$ and $\mathscr{B}$ are independent, then $P(\mathscr{A}|\mathscr{B}) = P(\mathscr{A})$; hence,

$$\frac{n_{\mathscr{A}}}{n} \simeq \frac{n_{\mathscr{A}\cap\mathscr{B}}}{n_{\mathscr{B}}} \tag{2-69}$$

Thus if the events $\mathscr{A}$ and $\mathscr{B}$ are independent, the relative frequency of the occurrence of $\mathscr{A}$ in a sequence of n trials equals its relative frequency in the subsequence of $n_{\mathscr{B}}$ trials in which $\mathscr{B}$ occurs. This agrees with our heuristic understanding of independence.

Example 2.37

In this example we use the notion of independence to investigate the possible connection between smoking and lung cancer. We conduct a survey among the following four groups: cancer patients who are smokers (cs), cancer patients who are nonsmokers ($c\bar{s}$), healthy smokers ($\bar{c}s$), healthy nonsmokers ($\bar{c}\bar{s}$). The results of the survey show that

$$P(cs) = p_1 \qquad P(c\bar{s}) = p_2 \qquad P(\bar{c}s) = p_3 \qquad P(\bar{c}\bar{s}) = p_4$$

We next form the events

$$\mathscr{C} = \{\text{cancer patients}\} = \{cs, c\bar{s}\} \qquad \mathscr{D} = \{\text{smokers}\} = \{\bar{c}s, cs\}$$
$$\mathscr{C} \cap \mathscr{D} = \{\text{cancer patients, smokers}\} = \{cs\}$$

Clearly,

$$P(\mathscr{C}) = p_1 + p_2 \qquad P(\mathscr{D}) = p_1 + p_3 \qquad P(\mathscr{C} \cap \mathscr{D}) = p_1$$

If $P(\mathscr{C} \cap \mathscr{D}) \neq P(\mathscr{C})P(\mathscr{D})$, that is, if $p_1 \neq (p_1 + p_2)(p_1 + p_3)$, the events {cancer patients} and {smokers} are statistically dependent.

Note that this reasoning does not lead to the conclusion that there is a causal relationship between lung cancer and smoking. Both factors might result from a common cause (work habits, for example) that has not been considered in the experimental model. ■

Example 2.38

Two trains, X and Y, arrive at a station at random between 0:00 and 0:20 A.M. The times of their arrival are independent. Train X stops for 5 minutes, and train Y stops for 4 minutes.

(a) Find the probability p_1 that train X arrives before train Y.

(b) Find the probability p_2 that the trains meet.

(c) Assuming that the trains meet, find the probability p_3 that train X arrived before train Y.

Model Specification An outcome of this experiment is a pair of numbers (x, y) where x and y are the arrival times of train X and train Y, respectively. The resulting space $\mathscr{S}$ is the set of points in the square of Fig. 2.18a. The event

$$\mathscr{A} = \{\text{X arrives in the interval } (t_1, t_2)\} = \{t_1 \leq x \leq t_2\}$$

is a vertical strip as shown. The assumption that x is a random number in the interval $(0, 20)$ yields

$$P(\mathscr{A}) = \frac{t_2 - t_1}{20}$$

The event

$$\mathscr{B} = \{\text{Y arrives in the interval } (t_3, t_4)\} = \{t_3 \leq y \leq t_4\}$$

is a horizontal strip, and its probability equals

$$P(\mathscr{B}) = \frac{t_4 - t_3}{20}$$

The event $\mathscr{A} \cap \mathscr{B}$ is a rectangle as shown, and its probability equals

$$P(\mathscr{A} \cap \mathscr{B}) = P(\mathscr{A})P(\mathscr{B}) = \frac{(t_4 - t_3)(t_2 - t_1)}{20 \times 20}$$

This is the model form of the assumed independence of the arrival times. Thus the probability that (x, y) is in a rectangular set equals the area of the rectangle divided by 400. And since any event $\mathscr{D}$ can be expressed as a countable union of disjoint rectangles, we conclude that

$$P(\mathscr{D}) = \frac{\text{area of } \mathscr{D}}{400}$$

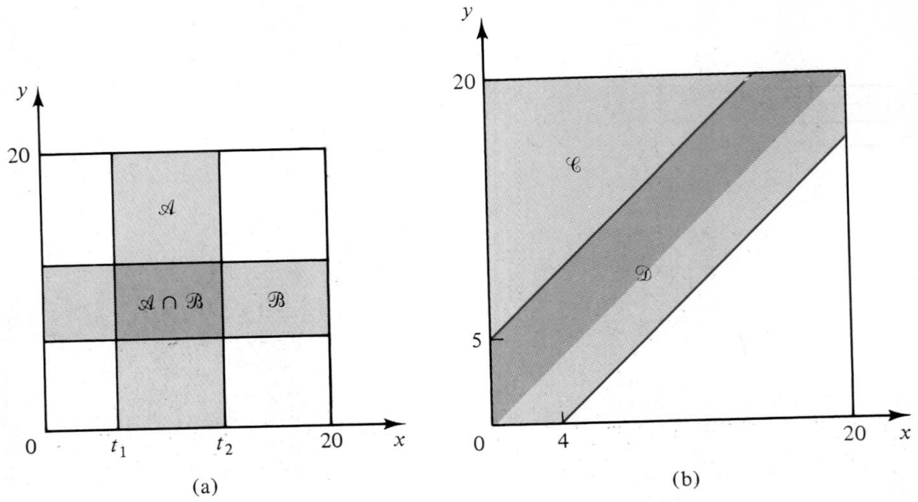

Figure 2.18

This concludes the specification of $\mathcal{S}$. We should stress again that the relationships are not derived; they are assumptions based on the description of the problem.

Deduction

(a) We wish to find the probability p_1 of the event $\mathcal{C} = \{\text{X arrives before Y}\}$. This event consists of all points in $\mathcal{S}$ such that $x \leq y$. Thus $\mathcal{C}$ is a triangle, and its area equals 200. Hence,

$$p_1 = P(\mathcal{C}) = \frac{200}{400} = .500$$

(b) The trains meet iff $x \leq y + 4$, because train Y stops for 4 minutes, and $y \leq x + 5$, because train X stops for 5 minutes. Thus the trains meet iff the event $\mathcal{D} = \{-5 \leq x - y \leq 4\}$ occurs. This event is the region of Fig. 2.18b consisting of two trapezoids, and its area equals 159.5. Hence,

$$p_2 = P(\mathcal{D}) = \frac{159.5}{400} = .399$$

(c) The probability that X arrives before Y (i.e., that event $\mathcal{C}$ occurred) assuming that the trains met (i.e., that event $\mathcal{D}$ occurred) equals $P(\mathcal{C}|\mathcal{D})$. Clearly, $\mathcal{C} \cap \mathcal{D}$ is a trapezoid, and its area equals 72. Hence,

$$p_3 = P(\mathcal{C}|\mathcal{D}) = \frac{P(\mathcal{C} \cap \mathcal{D})}{P(\mathcal{D})} = \frac{72}{159.5} = .451$$

This example demonstrates the value of the precise model specification in the solution of a probabilistic problem. ■

Example 2.39

Here we shall use the notion of independence to complete the specification of an experiment formed by combining two unrelated experiments.

We are given two experiments. The first is a fair die specified by the model

$$\mathcal{S}_1 = \{f_1, \ldots, f_6\} \qquad P\{f_i\} = \frac{1}{6}$$

and the second is a fair coin specified by the model

$$\mathcal{S}_2 = \{h, t\} \qquad P\{h\} = P\{t\} = \frac{1}{2}$$

We perform both experiments, and we wish to find the probability p that 5 shows on the die and *heads* on the coin. If we make the reasonable assumption that the two experiments are independent, it would appear from (2-67) that we should conclude that

$$p = \frac{1}{6} \times \frac{1}{2} = \frac{1}{12} \tag{2-70}$$

This conclusion is correct; it does not, however, follow from (2-67). In that equation, as in our entire development, we dealt only with subsets of a *single* space. To accept (2-70) not merely as a heuristic statement but as a conclusion that follows logically from the axioms in a single probabilistic model, we must construct a new experiment $\mathcal{S}$ in which 5 and *heads* are subsets. This is done as follows.

The space $\mathcal{S}$ of the new experiment consists of 12 outcomes, namely, all pairs of objects that we can form taking one from $\mathcal{S}_1$ and one from $\mathcal{S}_2$:

$$\mathcal{S} = \{f_1h, f_2h, \ldots, f_5t, f_6t\}$$

Thus $\mathcal{S} = \mathcal{S}_1 \times \mathcal{S}_2$ is the Cartesian product of the sets $\mathcal{S}_1$ and $\mathcal{S}_2$ [see (2-5)]. In this experiment, 5 is an event $\mathcal{A}$ consisting of the two outcomes f_5h and f_5t; *heads* is an event $\mathcal{B}$ consisting of the six outcomes $f_1h, \ldots, f_6h$. Thus

$$\mathcal{A} = \{5\} = \{f_5h, f_5t\} \qquad \mathcal{B} = \{\text{heads}\} = \{f_1h, \ldots, f_6h\}$$

To complete the specification of $\mathcal{S}$, we must assign probabilities to its subsets. Since {5} and {heads} are events in the original experiments, we must set

$$P(\mathcal{A}) = \frac{1}{6} \qquad P(\mathcal{B}) = \frac{1}{2}$$

In the experiment $\mathcal{S}$, the event {5 on the die and heads on the coin} is the intersection $\mathcal{A} \cap \mathcal{B}$ $\{f_5h\}$ of the events $\mathcal{A}$ and $\mathcal{B}$. To find the probability of this event, we use the independence of $\mathcal{A}$ and $\mathcal{B}$. This yields $P(\mathcal{A} \cap \mathcal{B}) = 1/6 \times 1/2 = 1/12$, in agreement with (2-70). ■

We show next that if the events $\mathcal{A}$ and $\mathcal{B}$ are independent, the events $\overline{\mathcal{A}}$ and $\mathcal{B}$ are also independent:

$$\text{If } P(\mathcal{A} \cap \mathcal{B}) = P(\mathcal{A})P(\mathcal{B}), \qquad \text{then} \qquad P(\overline{\mathcal{A}} \cap \mathcal{B}) = P(\overline{\mathcal{A}})P(\mathcal{B}) \tag{2-71}$$

■ ***Proof.*** As we know,

$$\mathcal{A} \cup \overline{\mathcal{A}} = \mathcal{S} \qquad \mathcal{B} = \{\mathcal{A} \cap \mathcal{B}) \cup (\overline{\mathcal{A}} \cap \mathcal{B})$$

Hence, $P(\mathcal{A}) + P(\overline{\mathcal{A}}) = 1$ and $P(\mathcal{B}) = P(\mathcal{A} \cap \mathcal{B}) + P(\overline{\mathcal{A}} \cap \mathcal{B})$. From this it follows that

$$\begin{aligned} P(\overline{\mathcal{A}} \cap \mathcal{B}) &= P(\mathcal{B}) - P(\mathcal{A} \cap \mathcal{B}) = P(\mathcal{B}) - P(\mathcal{A})P(\mathcal{B}) \\ &= [1 - P(\mathcal{A})]P(\mathcal{B}) = P(\overline{\mathcal{A}})P(\mathcal{B}) \end{aligned}$$

and (2-71) results.

We can similarly show that the events $\overline{\mathcal{A}}$ and $\overline{\mathcal{B}}$ are also independent.

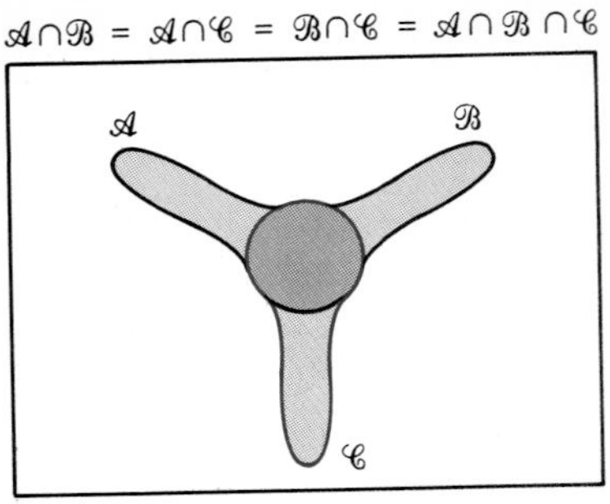

Figure 2.19

Generalization The events $\mathscr{A}_1, \ldots, \mathscr{A}_m$ are called mutually statistically independent, or, simply, independent if the probability of the intersection of any number of them in a group equals the product of probabilities of each event in that group. For example, if the events $\mathscr{A}_1$, $\mathscr{A}_2$, and $\mathscr{A}_3$ are independent, then

$$P(\mathscr{A}_1 \cap \mathscr{A}_2) = P(\mathscr{A}_1)P(\mathscr{A}_2)$$

$$P(\mathscr{A}_1 \cap \mathscr{A}_3) = P(\mathscr{A}_1)P(\mathscr{A}_3) \qquad P(\mathscr{A}_2 \cap \mathscr{A}_3) = P(\mathscr{A}_2)P(\mathscr{A}_3) \tag{2-72}$$

$$P(\mathscr{A}_1 \cap \mathscr{A}_2 \cap \mathscr{A}_3) = P(\mathscr{A}_1)P(\mathscr{A}_2)P(\mathscr{A}_3) \tag{2-73}$$

Thus the events $\mathscr{A}_1$, $\mathscr{A}_2$, $\mathscr{A}_3$ are independent, if they are independent in pairs *and* their intersection satisfies (2-73). As the next example shows, three events might be independent in pairs but not mutually independent.

Example 2.40 The events $\mathscr{A}$, $\mathscr{B}$, and $\mathscr{C}$ are such that (Fig. 2.19)

$$P(\mathscr{A}) = P(\mathscr{B}) = P(\mathscr{C}) = \frac{1}{5} \qquad P(\mathscr{A} \cap \mathscr{B} \cap \mathscr{C}) = \frac{1}{25}$$

$$P(\mathscr{A} \cap \mathscr{B}) = P(\mathscr{A} \cap \mathscr{C}) = P(\mathscr{B} \cap \mathscr{C}) = \frac{1}{25}$$

This shows that they are independent in pairs. However, they are not mutually independent because they do not satisfy (2-73). ■

Problems

2-1 Show that: **(a)** if $\mathscr{A} \cup \mathscr{B} = \mathscr{A} \cap \mathscr{B}$, then $\mathscr{A} = \mathscr{B}$; **(b)** $\mathscr{A} \cup (\mathscr{B} \cap \mathscr{C}) \subset (\mathscr{A} \cup \mathscr{B}) \cap \mathscr{C}$; **(c)** if $\mathscr{A}_1 \subset \mathscr{A}$, $\mathscr{B}_1 \subset \mathscr{B}$, and $\mathscr{A} \cap \mathscr{B} = \varnothing$, then $\mathscr{A}_1 \cap \mathscr{B}_1 = \varnothing$.

2-2 If $\mathscr{S} = \{-\infty < t < \infty\}$, $\mathscr{A} = \{4 \le t \le 8\}$, $\mathscr{B} = \{7 \le t \le 10\}$, find $\mathscr{A} \cup \mathscr{B}$, $\mathscr{A} \cap \mathscr{B}$, $\overline{\mathscr{A}}$, and $\mathscr{A} \cap \overline{\mathscr{B}}$.

2-3 The set $\mathscr{S}$ consists of the 10 integers 1 to 10. Find the number of its subsets that contain the integers 1, 3, and 7.

2-4 (De Morgan's law) Using Venn diagrams, show that $\overline{\mathscr{A} \cup \mathscr{B}} = \overline{\mathscr{A}} \cap \overline{\mathscr{B}}$ and $\overline{\mathscr{A} \cap \mathscr{B}} = \overline{\mathscr{A}} \cup \overline{\mathscr{B}}$.

2-5 Express the following statements concerning the three events $\mathscr{A}$, $\mathscr{B}$, and $\mathscr{C}$ in terms of the occurrence of a single event: **(a)** none occurs; **(b)** only one occurs; **(c)** at least one occurs; **(d)** at most two occur; **(e)** two and only two occur; **(f)** $\mathscr{A}$ and $\mathscr{B}$ occur but $\mathscr{C}$ does not occur.

2-6 If $P(\mathscr{A}) = .6$, $P(\mathscr{B}) = .3$, and $P(\mathscr{A} \cap \mathscr{B}) = .2$, find the probabilities of the events $\mathscr{A} \cup \mathscr{B}$, $\overline{\mathscr{A}} \cup \overline{\mathscr{B}}$, $\mathscr{A} \cap \overline{\mathscr{B}}$. and $\overline{\mathscr{A}} \cap \overline{B}$.

2-7 Show that: **(a)** $P(\mathscr{A} \cup \mathscr{B} \cup \mathscr{C}) = P(\mathscr{A}) + P(\mathscr{B}) + P(\mathscr{C}) - P(\mathscr{A} \cap \mathscr{B}) - P(\mathscr{A} \cap \mathscr{C}) - P(\mathscr{B} \cap \mathscr{C}) + P(\mathscr{A} \cap \mathscr{B} \cap \mathscr{C})$; **(b)** (Boole's inequality) $P(\mathscr{A}_1 \cup \cdots \cup \mathscr{A}_n) \le P(\mathscr{A}_1) + \cdots + P(\mathscr{A}_n)$.

2-8 Show that: **(a)** if $P(\mathscr{A}) = P(\mathscr{B}) = 1$, then $P(\mathscr{A} \cap \mathscr{B}) = 1$; **(b)** $P^2(\mathscr{A} \cap \mathscr{B}) \le P(\mathscr{A})P(\mathscr{B})$ and $P(\mathscr{A} \cap \mathscr{B}) \le [P(\mathscr{A}) + P(\mathscr{B})]/2$.

2-9 Find the probability that a bridge hand consists only of cards from 2 to 10.

2-10 In a raffle, 100 tickets numbered 1 to 100 are sold. Seven are picked at random, and each wins a prize. **(a)** Find the probability that ticket number 27 wins a prize. **(b)** Find the probability that each of the numbers 5, 15, and 40 wins a prize.

2-11 A box contains 100 fuses, 10 of which are defective. We pick 20 at random and test them. Find the probability that two will be defective.

2-12 If $P(\mathscr{A}) = .6$, $P(\mathscr{A} \cup \mathscr{B}) = .8$, and $P(\mathscr{A}|\mathscr{B}) = .5$, find $P(\mathscr{B})$.

2-13 Show that $P(\mathscr{A}) = P(\mathscr{A}|\mathscr{M})P(\mathscr{M}) + P(\mathscr{A}|\overline{\mathscr{M}})P(\overline{\mathscr{M}})$.

2-14 Show that $P(\mathscr{A} \cap \mathscr{B} \cap \mathscr{C} \cap \mathscr{D}) = P(\mathscr{A}|\mathscr{B} \cap \mathscr{C} \cap \mathscr{D})P(\mathscr{B}|\mathscr{C} \cap \mathscr{D})P(\mathscr{C}|\mathscr{D})P(\mathscr{D})$.

2-15 Show that: **(a)** if $\mathscr{A} \cap \mathscr{B} = \varnothing$, then

$$P(\mathscr{B}|\overline{\mathscr{A}}) = \frac{P(\mathscr{A})}{1 - P(\mathscr{A})} \qquad \text{and} \qquad P(\mathscr{B}|\mathscr{A} \cup \mathscr{B}) = \frac{P(\mathscr{B})}{P(\mathscr{A}) + P(\mathscr{B})}$$

(b) $P(\overline{\mathscr{A}}|\mathscr{M}) = 1 - P(\mathscr{A}|\mathscr{M})$.

2-16 We receive 100 bulbs of type A and 200 bulbs of type B. The probability that a bulb will last more than three months is .6 if it is type A and .8 if it is type B. The bulbs are mixed, and one is picked at random. Find the probability p_1 that it will last more than three months; if it does, find the probability p_2 that it is type A.

2-17 The duration t of a telephone call is an element of the space $\mathscr{S} = \{t \ge 0\}$ specified in terms of the function $\alpha(t) = \dfrac{1}{c} e^{-t/c}$, $c = 5$ minutes, as in (2-43). Find the probability of the events $\mathscr{A} = \{0 \le t \le 10\}$ and $\mathscr{B} = \{t \ge 5\}$; find the conditional probability $P(\mathscr{A}|\mathscr{B})$.

2-18 The events $\mathscr{A}$ and $\mathscr{B}$ are independent, and $\mathscr{B} \subset \mathscr{A}$. Find $P(\mathscr{A})$.

2-19 Can two events be independent and mutually exclusive?

2-20 Two houses A and B are far apart. The probability that in the next decade A will burn down equals 10^{-3} and that B will burn down equals 2×10^{-3}. Find the probability p_1 that at least one and the probability p_2 that both will burn down.

2-21 Show that if the events $\mathscr{A}$, $\mathscr{B}$, and $\mathscr{C}$ are independent, then (a) the events $\overline{\mathscr{A}}$ and $\mathscr{C}$ are independent; (b) the events, $\mathscr{A}$, $\mathscr{B}$, and $\overline{\mathscr{C}}$ are independent.

2-22 Show that 11 equations are needed to establish the independence of four events; generalize to n events.

2-23 Show that four events are independent iff they are independent in pairs and each is independent of the intersection of any of the others.

2-24 A string of Christmas lights consists of 50 independent bulbs connected in series; that is, the lights are on if all bulbs are good. The probability that a bulb is defective equals .01. Find the probability p that the lights are on.

3

Repeated Trials

Repeated trials have two interpretations. The first is empirical: We equate the probability $P(\mathscr{A})$ of an event $\mathscr{A}$ defined on an experiment $\mathscr{S}$ to the ratio $n_{\mathscr{A}}/n$ where $n_{\mathscr{A}}$ is the number of successes of $\mathscr{A}$ in n repetitions of the underlying physical experiment. The second is conceptual: We form a new experiment $\mathscr{S}_n = \mathscr{S} \times \cdots \times \mathscr{S}$ the elements of which are sequences $\xi_1 \ldots \xi_n$ where ξ_i is any one of the elements of $\mathscr{S}$. In the first interpretation n is large; in the second, n is arbitrary. In this chapter, we use the second interpretation of repeated trials and determine the probabilities of various events in the experiment $\mathscr{S}_n$.

3-1 Dual Meaning of Repeated Trials

In Chapter 1, we used the notion of repeated trials to establish the relationship between a theoretical model and a real experiment. This was based on the approximation $P(\mathscr{A}) \simeq n_{\mathscr{A}}/n$ relating the model parameter $P(\mathscr{A})$ to the observed ratio $n_{\mathscr{A}}/n$. In this chapter, we give an entirely different interpretation to the notion of repeated trials. To be concrete, we start with the coin experiment.

REPEATED TOSSES OF A COIN The experiment of the single toss of a coin is specified in terms of the space $\mathcal{S} = \{h, t\}$ and the probabilities of its elementary events

$$P\{h\} = p \qquad P\{t\} = q \tag{3-1}$$

Suppose that we wish to determine the probabilities of various events involving n tosses of the coin. For example, the probability that in 10 tosses, 7 heads will show. To do so, we must form a new model $\mathcal{S}_n$ the outcomes of which are sequences of the form

$$\xi_1 \ldots \xi_i \ldots \xi_n \tag{3-2}$$

where ξ_i is h or t. The space $\mathcal{S}_n$ so formed is written in the form

$$\mathcal{S}_n = \mathcal{S} \times \cdots \times \mathcal{S} \tag{3-3}$$

and it is called *Cartesian product*. This is a reminder of the fact that the elements of $\mathcal{S}_n$ are sequences as in (3-2) where ξ_i is one of the elements of $\mathcal{S}$. Clearly, there are 2^n such sequences; hence, $\mathcal{S}_n$ has 2^n elements.

The experiment $\mathcal{S}_n$ cannot be specified in terms of $\mathcal{S}$ alone. Additional information concerning the multiple tosses must be known. We shall presently show that if the tosses are *independent,* the model $\mathcal{S}_n$ is completely specified in terms of $\mathcal{S}$. Independence in the context of a real coin means that the outcome of a particular toss is not affected by the outcomes of the preceding tosses. This is in general a reasonable assumption. In the context of a theoretical model, independence will be interpreted in the sense of (2-67). As preparation, we discuss first the special case $n = 3$.

Example 3.1

A coin tossed $n = 3$ times generates the space $\mathcal{S}_3 = \mathcal{S} \times \mathcal{S} \times \mathcal{S}$ consisting of the $2^3 =$ 8 outcomes

$$hhh,\ hht,\ hth,\ htt,\ thh,\ tht,\ tth,\ ttt$$

We introduce the events

$$\mathcal{H}_i = \{\text{heads at the } i\text{th toss}\} \qquad \mathcal{T}_i = \{\text{tails at the } i\text{th toss}\} \tag{3-4}$$

and we assign to these events the probabilities

$$P(\mathcal{H}_i) = p \qquad P(\mathcal{T}_i) = q \tag{3-5}$$

This is consistent with (3-2). Using (3-5) and the independence of tosses, we shall determine the probabilities of the elementary events of $\mathcal{S}_3$. Each of the events $\mathcal{H}_i$ and $\mathcal{T}_i$ consists of four outcomes. For example,

$$\mathcal{H}_1 = \{hhh, hht, hth, htt\} \qquad \mathcal{T}_1 = \{thh, tht, tth, ttt\}$$

The elementary event $\{hhh\}$ can be written as the intersection of the events $\mathcal{H}_1$, $\mathcal{H}_2$, $\mathcal{H}_3$; hence,

$$P\{hhh\} = P\{\mathcal{H}_1 \cap \mathcal{H}_2 \cap \mathcal{H}_3\}$$

From the independence of the tosses and (2-73) it follows that

$$P(\mathcal{H}_1 \cap \mathcal{H}_2 \cap \mathcal{H}_3\} = P(\mathcal{H}_1)P(\mathcal{H}_2)P(\mathcal{H}_3) = p^3$$

hence, the probability of the event $\{hhh\}$ equals p^3. Proceeding similarly, we can determine the probability of all elementary events of $\mathcal{S}_3$. The result is

$$P\{hhh\} = p^3 \qquad P\{hht\} = p^2q \qquad P\{hth\} = p^2q \qquad P\{htt\} = pq^2$$
$$P\{thh\} = p^2q \qquad P\{tht\} = pq^2 \qquad P\{tth\} = pq^2 \qquad P\{ttt\} = p^3$$

Thus the probability of an elementary event equals $p^k q^{3-k}$ where k is the number of heads. This completes the specification of $\mathcal{S}_3$. To find the probability of any event in $\mathcal{S}_3$, we add the probabilities of its elementary events as in (2-30). For example, the event $\mathcal{A}$ = {two heads in any order} consists of the three outcomes hht, hth, thh; hence,

$$P(\mathcal{A}) = P\{hht\} + P\{hth\} + P\{thh\} = 3p^2q \qquad (3\text{-}6)$$ ■

A coin tossed n times generates the space $\mathcal{S}_n$ consisting of 2^n outcomes. Its elementary events are of the form $\mathcal{B}$ = {k heads in a specific order}. We shall show that

$$P\{k \text{ heads in a specific order}\} = p^k q^{n-k} \qquad (3\text{-}7)$$

We introduce the events $\mathcal{H}_i$ and $\mathcal{T}_i$ as in (3-4) and assign to them the probabilities p and q as in (3-5). To prove (3-7), it suffices to express the elementary event $\mathcal{B}$ as an intersection of the events $\mathcal{H}_i$ and $\mathcal{T}_i$. Suppose, to be concrete, that $\mathcal{B} = \{hth \ldots t\}$. In this case,

$$\{hth \ldots h\} = \mathcal{H}_1 \cap \mathcal{T}_2 \cap \mathcal{H}_3 \cap \cdots \cap \mathcal{H}_n$$

where the right side contains k events of the form $\mathcal{H}_i$ and $n - k$ events of the form $\mathcal{T}_i$. From the independence of the tosses and (2-73), it follows that the probability of the right side equals $p^k q^{n-k}$ as in (3-7).

Example 3.2

Find the probability p_a that the first 10 tosses of a fair coin will show heads and the probability p_b that the first 9 will be heads and the next one tails.

In the experiment $\mathcal{S}_{10}$, the events

$$\mathcal{A} = \{10 \text{ heads in a row}\} \qquad \mathcal{B} = \{9 \text{ heads then tails}\}$$

are elementary. With $p = q = 1/2$ and $n = 10$, (3-7) yields

$$P(\mathcal{A}) = P(\mathcal{B}) = \frac{1}{2^{10}}$$

Thus, contrary to a common impression, the events $\mathcal{A}$ and $\mathcal{B}$ are equally rare. ■

Note Equation (3-7) seems heuristically obvious: We have k heads and $n - k$ tails; the probability for heads equals p and for tails q; the tosses are independent; hence, (3-7) must be true. However, although heuristics lead in this case to a correct conclusion, it is essential that we phrase the problem in terms of a single model and interpret independence in terms of events satisfying (2-67).

We shall now show that the probability that we get k heads (and $n - k$ tails) in any order equals

$$P\{k \text{ heads in any order}\} = \binom{n}{k} p^k q^{n-k} \qquad (3\text{-}8)$$

■ ***Proof.*** The event {k heads in any order} consists of all outcomes formed with k heads and $n - k$ tails. The number of such outcomes equals the number of ways that we can place k heads and $n - k$ tails on a line. As we have shown in Section 2-1, this equals the combinations $C_k^n = \binom{n}{k}$ of n

objects taken k at a time [see (2-10)]. Multiplying by the probability $p^k q^{n-k}$ of each elementary event, we obtain (3-8).

For $n = 3$ and $k = 2$, (3-8) yields

$$P\{2 \text{ heads in any order}\} = \binom{3}{2} p^2 q = 3p^2 q$$

in agreement with (3-6).

Example 3.3

A fair coin is tossed 10 times. Find the probability that heads will show 5 times.

In this problem,

$$p = q = \frac{1}{2} \qquad n = 10 \qquad k = 5$$

and (3-8) yields

$$P\{5 \text{ heads in any order}\} = \binom{10}{5} \times \frac{1}{2^5} \times \frac{1}{2^5} = \frac{252}{1{,}024}$$

■

Example 3.4

We have two coins as in Example 2.34. Coin A is fair, and coin B is loaded with $P\{h\} = 2/3$. We pick one of the coins at random, toss it 10 times, and observe that heads shows 4 times. Find the probability that we picked the fair coin.

The space of this experiment is a Cartesian product $\mathcal{S}_0 \times \mathcal{S}_{10}$ where $\mathcal{S}_0 = \{a, b\}$ is the selection of a coin and $\mathcal{S}_{10}$ is the toss of a coin 10 times. Thus this experiment has 2×2^{10} outcomes. We introduce the events

$$\mathcal{A} = \{\text{coin A tossed 10 times}\} \qquad \mathcal{B} = \{\text{coin B tossed 10 times}\}$$
$$\mathcal{D} = \{\text{4 heads in any order}\}$$

Our problem is to find the conditional probability $P(\mathcal{A}|\mathcal{D})$.

From the randomness of the coin selection, it follows that $P(\mathcal{A}) = P(\mathcal{B}) = 1/2$. If $\mathcal{A}$ is selected, the probability $P(\mathcal{D}|\mathcal{A})$ that 4 heads will show is given by (3-8) with $p = q = 1/2$. Thus

$$P(\mathcal{D}|\mathcal{A}) = \binom{10}{4} \times \frac{1}{2^4} \times \frac{1}{2^6} \qquad P(\mathcal{D}|\mathcal{B}) = \binom{10}{4} \times \left(\frac{2}{3}\right)^4 \times \left(\frac{1}{3}\right)^6$$

Inserting into Bayes' theorem (2-63), we obtain

$$P(\mathcal{A}|\mathcal{D}) = \frac{P(\mathcal{D}|\mathcal{A})P(\mathcal{A})}{P(\mathcal{D}|\mathcal{A})P(\mathcal{A}) + P(\mathcal{D}|\mathcal{B})P(\mathcal{B})} = \frac{1}{1 + 2^{14}/3^{10}} = .783$$

■

Example 3.5

A coin with $P\{h\} = p$ is tossed n times. Find the probability p_a that at the first $n - 1$ tosses, heads shows $k - 1$ times, and at the nth toss, heads shows.

First Solution In n tosses, the probability of k heads equals $p^k q^{n-k}$. There are $\binom{n-1}{k-1}$ ways of obtaining $k - 1$ heads at the first $n - 1$ trials and heads at the nth trial; hence,

$$p_a = \binom{n-1}{k-1} p^k q^{n-k} \tag{3-9}$$

Second Solution The probability of $k - 1$ heads in $n - 1$ trials equals

$$\binom{n-1}{k-1} p^{k-1} q^{(n-1)-(k-1)} \tag{3-10}$$

The probability of heads at the nth trial equals p. Multiplying (3-10) by p (independent tosses), we obtain (3-9).

For $k = 1$, equation (3-9) yields

$$p_a = pq^{n-1} \tag{3-11}$$

This is the probability that heads will show at the nth toss but not before. ■

Note Example 3.5 can be given a different interpretation: If we toss the coin an infinite number of times, the probability that heads will show at the nth toss but not before equals pq^{n-1}. In this interpretation, the underlying experiment is the space $\mathcal{S}_\infty$ of the infinitely many tosses.

Probability Tree In Fig. 3.1 we give a graphical representation of repeated trials. The two horizontal segments under the letter $\mathcal{S}_1$ represent the two elements h and t of the experiment $\mathcal{S}_1 = \mathcal{S}$. The probabilities of $\{h\}$ and $\{t\}$ are shown at the end of each segment. The four horizontal segments under the letter $\mathcal{S}_2$ represent the four elements of the space $\mathcal{S}_2 = \mathcal{S} \times \mathcal{S}$ of the toss of a coin twice. The probabilities of the corresponding elementary events are shown at the end of each segment. Proceeding similarly, we can form a tree representing the experiments $\mathcal{S}_1, \ldots, \mathcal{S}_n$ for any n.

Figure 3.1

$\mathcal{S}_1 = \{h, t\}$

$\mathcal{S}_2 = \{hh, ht, th, tt\}$

$\mathcal{S}_3 = \{hhh, hht, hth, htt, thh, tht, tth, ttt\}$

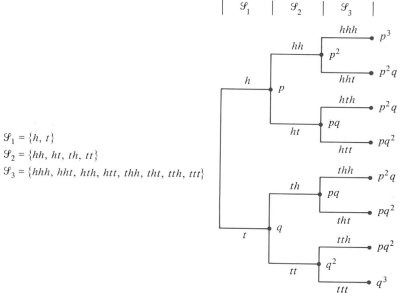

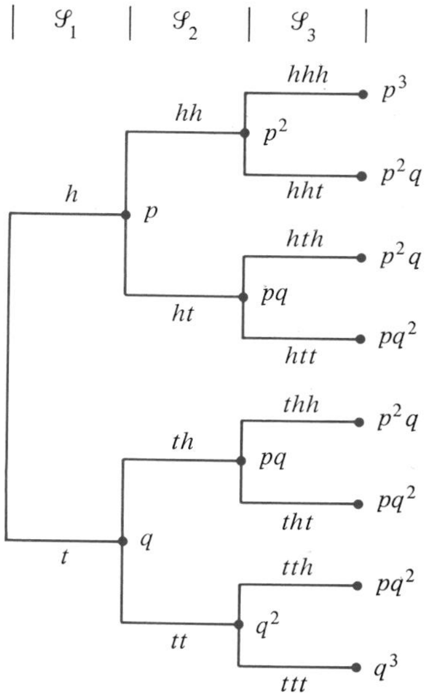

Dual Meaning of Repeated Trials The concept of repeated trials has two fundamentally different interpretations. The first is empirical, the second conceptual. We shall explain the difference in the context of the coin experiment.

In the first interpretation, the experimental model is the toss of a coin *once*. The space $\mathcal{S}$ consists of the two outcomes h and t. A trial is a single toss of the coin. The experiment is completely specified in terms of the probabilities $P\{h\} = p$ and $P\{t\} = q$ of its elementary events $\{h\}$ and $\{t\}$. Repeated trials are thus used to determine p and q empirically: We toss the real coin n times, and we set $p \simeq n_h/n$ where n_h is the observed number of heads. This is the *empirical* version of repeated trials. The approximation $p \simeq n_h/n$ is based on the assumption that n is *sufficiently large*.

In the second interpretation, the experiment model is the toss of a coin n times, where n is any number. The space $\mathcal{S}_n$ is now the Cartesian product $\mathcal{S}_n = \mathcal{S} \times \cdots \times \mathcal{S}$ consisting of 2^n outcomes of the form $hth \ldots h$. A single trial is the toss of the coin n times. This is the *conceptual* interpretation of repeated trials. In this interpretation, all statements are *exact*, and they hold for any n, large or small. If we wish to give a relative frequency interpretation to the probabilities in the space $\mathcal{S}_n$, we must repeat the experiment of the n tosses of the coin a large number of times and apply (1-1).

3-2

Bernoulli Trials

Using the coin experiment as an illustration, we have shown that if $\mathcal{S} = \{\zeta_1, \zeta_2\}$ is an experiment consisting of the two outcomes ζ_1, ζ_2 and it is repeated n times, the probability that ζ_1 will show k times in a specific order equals $p^k q^{n-k}$, and the probability that it will show k times in any order equals

$$p_n(k) = \binom{n}{k} p^k q^{n-k} \qquad p = P\{\zeta_1\}$$

The following is an important generalization.

Suppose that $\mathcal{S}$ is an experiment consisting of the elements ζ_i. Repeating $\mathcal{S}$ n times, we obtain a new experiment

$$\mathcal{S}_n = \mathcal{S} \times \mathcal{S} \times \cdots \times \mathcal{S}$$

(Cartesian product). The outcomes of this experiment are sequences of the form

$$\xi_1 \xi_2 \ldots \xi_n \tag{3-12}$$

where ξ_i is any one of the elements ζ_i of $\mathcal{S}$.

Consider an event $\mathcal{A}$ of $\mathcal{S}$ with $P(\mathcal{A}) = p$. Clearly, the complement $\overline{\mathcal{A}}$ is also an event with $P(\overline{\mathcal{A}}) = 1 - p = q$, and $[\mathcal{A}, \overline{\mathcal{A}}]$ is a partition of $\mathcal{S}$. The ith element ξ_i of the sequence (3-12) is an element of either $\mathcal{A}$ or $\overline{\mathcal{A}}$. Each sequence of the form (3-12) thus generates a sequence

$$\mathcal{B} = \mathcal{A}\overline{\mathcal{A}}\mathcal{A} \ldots \mathcal{A} \tag{3-13}$$

where we place $\mathcal{A}$ at the ith position if $\mathcal{A}$ occurs at the ith trial, that is, if $\xi_i \in \mathcal{A}$; otherwise, we place $\overline{\mathcal{A}}$. All sequences of the form of (3-13) are events of the experiment $\mathcal{S}_n$.

Clearly, $\mathcal{B}$ is an event in the space $\mathcal{S}_n$; we shall determine its probability under the assumption that the repeated trials of $\mathcal{S}$ are independent. From this assumption it follows as in (3-7) that

$$P(\mathcal{A}\overline{\mathcal{A}} \ldots \mathcal{A}) = P(\mathcal{A})P(\overline{\mathcal{A}}) \cdots P(\mathcal{A}) = pq \cdots p \qquad (3\text{-}14)$$

If in the sequence $\mathcal{B}$ the event $\mathcal{A}$ appears k times, then the event $\overline{\mathcal{A}}$ appears $n - k$ times and the right side of (3-14) equals $p^k q^{n-k}$. Thus

$$P\{\mathcal{A} \text{ occurs } k \text{ times in a specific order}\} = p^k q^{n-k} \qquad (3\text{-}15)$$

We shall next determine the probability of the event

$$\mathcal{D} = \{\mathcal{A} \text{ occurs } k \text{ times in any order}\}$$

■ ***Fundamental Theorem.*** In n independent trials, the probability

$$p_n(k) = P\{\mathcal{A} \text{ occurs } k \text{ times in any order}\}$$

that the event $\mathcal{A}$ will occur k times (and the event $\overline{\mathcal{A}}$ $n - k$ times) in any order equals

$$p_n(k) = \binom{n}{k} p^k q^{n-k} \qquad (3\text{-}16)$$

■ ***Proof.*** There are $C_k^n = \binom{n}{k}$ events of the form $\{\mathcal{A}$ occurs k times in a specific order$\}$, namely, the ways in which we can place $\mathcal{A}$ k times and $\overline{\mathcal{A}}$ $n - k$ times in a row [see (2-10)]. Furthermore, all these events are mutually exclusive, and their union equals the event $\{\mathcal{A}$ occurs k times in any order$\}$. Hence, (3-16) follows from (3-15).

Example 3.6

A fair die is rolled seven times. Find the probability $p_7(2)$ that 4 will show twice.

The original experiment $\mathcal{S}$ is the single roll of the die and $\mathcal{A} = \{f_4\}$. Thus

$$P(\mathcal{A}) = \frac{1}{6} \qquad P(\overline{\mathcal{A}}) = \frac{5}{6}$$

With $n = 7$ and $k = 2$, (3-16) yields

$$p_7(2) = \frac{7!}{2!5!}\left(\frac{1}{6}\right)^2\left(\frac{5}{6}\right)^5 = .234$$ ■

Example 3.7

A pair of fair dice is rolled four times. Find the probability $p_4(0)$ that 11 will not show.

In this case, $\mathcal{S}$ is the single roll of two dice and $\mathcal{A} = \{f_5 f_6, f_6 f_5\}$. Thus

$$P(\mathcal{A}) = \frac{2}{36} \qquad P(\overline{\mathcal{A}}) = \frac{34}{36}$$

With $n = 4$ and $k = 0$, (3-16) yields

$$p_4(0) = \left(\frac{34}{36}\right)^4 = .796 \qquad \blacksquare$$

We discuss next a number of problems that can be interpreted as repeated trials.

Example 3.8

Twenty persons arrive in a store between 9:00 and 10:00 A.M. The arrival times are random and independent. Find the probability p_a that four of these persons arrive between 9:00 and 9:10.

This can be phrased as a problem in repeated trials. The original experiment $\mathcal{S}$ is the random arrival of one person and the repeated trials are the arrivals of the 20 persons. Clearly,

$$\mathcal{A} = \{\text{a specific person arrives between 9:00 and 9:10}\}$$

is an event in $\mathcal{S}$, and its probability equals

$$P(\mathcal{A}) = \frac{10}{60}$$

[see (2-46)] because the arrival time is random. In the experiment $\mathcal{S}_{20}$ of the 20 arrivals, the event {four people arrive between 9:00 and 9:10} is the same as the event {$\mathcal{A}$ occurs four times}. Hence,

$$p_a = p_{20}(4) = \frac{20!}{4!16!}\left(\frac{1}{6}\right)^4\left(\frac{5}{6}\right)^{16} = .02 \qquad \blacksquare$$

Example 3.9

In a lottery, 2,000 persons take part, each selecting at random a number between 1 and 1,000. The winning number is 253 (how this number is selected is immaterial). (a) Find the probability p_a that no one will win. (b) Find the probability p_b that two persons will win.

(a) Interpreting this as a problem in repeated trials, we consider as the original experiment $\mathcal{S}$ the random selection of a number N between 1 and 1,000. The space $\mathcal{S}$ has 1,000 outcomes, and the event $\mathcal{A} = \{N = 253\}$ is an elementary event with probability

$$P(\mathcal{A}) = P\{N = 253\} = .001$$

The selection of 2,000 numbers is the repetition of $\mathcal{S}$ 2,000 times. It follows, therefore, from (3-16) with $k = 0$ and $p = .001$ that

$$p_a = (.999)^{2{,}000} \simeq e^{-2} = .135$$

(b) If two persons win, $\mathcal{A}$ occurs twice. Hence,

$$p_b = \binom{2{,}000}{2}(.001)^2(.999)^{1{,}998} = .27 \qquad \blacksquare$$

Example 3.10

A box contains K white and $N - K$ black cards.

(a) *With replacements.* We pick a card at random, examine it, and put it back. We repeat this process n times. Find the probability p_a that k of the picked cards are white and $n - k$ are black.

Since the picked card is put back, the conditions of the experiment remain the same at each selection. We can therefore apply the results of repeated trials. The original experiment is the selection of a single card. The probability that the selected card is white equals K/N. With

$$p = \frac{K}{N} \qquad q = \frac{N-K}{N}$$

the probability that k out of the n selections are white equals

$$p_a = \binom{n}{k}\left(\frac{K}{N}\right)^k \left(1 - \frac{K}{N}\right)^{n-k} \tag{3-17}$$

(b) *Without replacements.* We again pick a card from the box, but this time we do not replace it. We repeat this process n times. Find the probability p_b that k of the selected cards are white.

This time we cannot use (3-16) because the conditions of the experiment change after each selection. To solve the problem, we proceed directly. Since we are interested only in the total number of white cards, the order in which they are selected is immaterial. We can consider, therefore, the n selections as a single outcome of our experiment. In this experiment, the possible outcomes are the number $\binom{N}{n}$ of ways of selecting n out of N objects. Furthermore, there are $\binom{K}{k}$ ways of selecting k out of K white cards, and $\binom{N-K}{n-k}$ ways of selecting $n - k$ out of the $N - K$ black cards. Hence [see also (2-41)], p_b is given by the *hypergeometric series*

$$p_b = \frac{\binom{pN}{k}\binom{N-pN}{n-k}}{\binom{N}{n}} \qquad p = \frac{K}{N} \tag{3-18}$$

Note that if $k \ll K$ and $n \ll N$, then after each drawing, the number of white and black cards in the box remains essentially constant; hence, $p_a \simeq p_b$. ■

We determine next the probability $P\{k_1 \le k \le k_2\}$ that the number k of successes of an event $\mathscr{A}$ in n trials is between k_1 and k_2.

■ ***Theorem***

$$P\{k_1 \le k \le k_2\} = \sum_{k=k_1}^{k_2} \binom{n}{k} p^k q^{n-k} \tag{3-19}$$

where $p = P(\mathcal{A})$.

■ ***Proof.*** Clearly, $\{k_1 \le k \le k_2\}$ is the union of the events

$$\mathcal{B}_k = \{\mathcal{A} \text{ occurs } k \text{ times in any order}\}$$

These events are mutually exclusive, and $P(\mathcal{B}_k) = p_n(k)$ as in (3-16). Hence, (3-19) follows from axiom III on page 33.

Example 3.11

An order of 1,000 parts is received. The probability that a part is defective equals .1. Find the probability p_a that the total number of defective parts does not exceed 110.

In the context of repeated trials, $\mathcal{S}$ is the arrival of a single part, and $\mathcal{A}$ = {the part is defective} is an event of $\mathcal{S}$ with $P(\mathcal{A}) = .1$. The arrival of 1,000 parts generates the space $\mathcal{S}_n$ of repeated trials, and p_a is the probability that the number of times $\mathcal{A}$ occurs is between 0 and 110. With

$$p = .1 \qquad k_1 = 0 \qquad k_2 = 110 \qquad n = 1{,}000$$

(3-19) yields

$$P\{0 \le k \le 110\} = \sum_{k=0}^{110} \binom{1{,}000}{k} (.1)^k (.9)^{1{,}000-k}$$ ■

Example 3.12

We receive a lot of N mass-produced objects. Of these, K are defective. We select at random n of these objects and inspect them. Based on the inspection results, we decide whether to accept or reject the lot. We use the following acceptance test.

Simple Sampling Suppose that among the n sampled objects, k are defective. We choose a number k_0, depending on the particular application, and we accept the lot if $k \le k_0$. If $k > k_0$, the lot is rejected. Show that the probability that a lot so tested is accepted equals

$$\frac{\displaystyle\sum_{k=0}^{k_0} \binom{pN}{k}\binom{N-pN}{n-k}}{\displaystyle\binom{N}{n}} \qquad p = \frac{K}{N} \tag{3-20}$$

■ ***Proof.*** The lot is accepted if $k \le k_0$. It suffices therefore to find the probability p_k that among the n sampled components, k will be defective. As we have shown in Example 3.10, p_k is given by the hypergeometric series (3-18). Summing for k from 0 to k_0, we obtain (3-20). ■

We shall now examine the behavior of the numbers

$$p_n(k) = \binom{n}{k} p^k q^{n-k} \qquad \binom{n}{k} = \frac{n!}{k!(n-k)!}$$

for fixed n, as k increased from 0 to n (see also Problem 3-8).

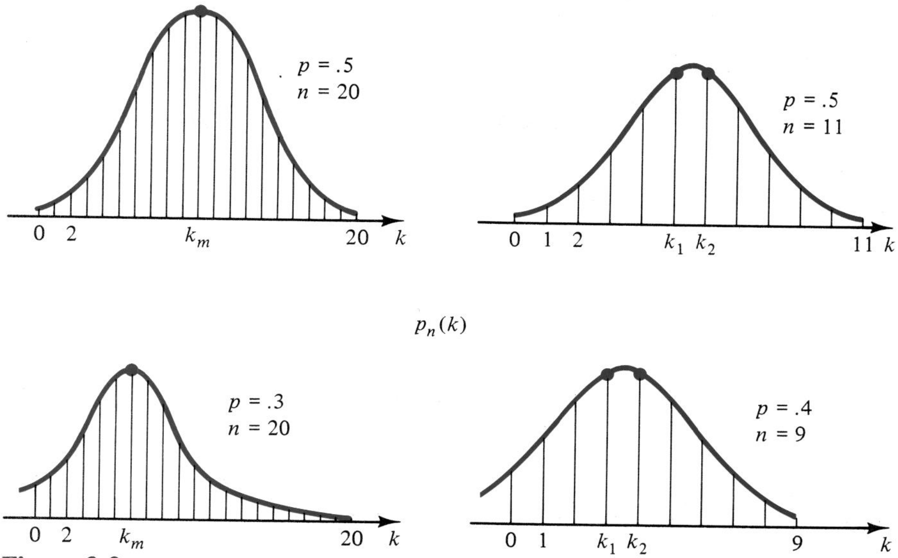

Figure 3.2

If $p = 1/2$, then $p_n(k)$ is proportional to the binomial coefficients $\binom{n}{k}$:

$$p_n(k) = \binom{n}{k} \frac{1}{2^n}$$

In this case, $p_n(k)$ is symmetrical about the midpoint $n/2$ of the interval $(0, n)$. If n is even, it has a *single* maximum for $k = k_m = n/2$; if n is odd, it has *two* maxima

$$k = k_1 = \frac{n-1}{2} \qquad \text{and} \qquad k = k_2 = \frac{n+1}{2}$$

If $p \neq 1/2$, then $p_n(k)$ is not symmetrical; its maximum is reached for $k \simeq np$. Precisely, if $(n + 1)p$ is not an integer, then $p_n(k)$ has a single maximum for $k = k_m = [(n + 1)p]$.* If $(n + 1)p$ is an integer, then $p_n(k)$ has two maxima:

$$k = k_1 = (n + 1)p - 1 \qquad \text{and} \qquad k = k_2 = (n + 1)p$$

These results are illustrated in Fig. 3.2 for the following cases:

1. $n = 20$ $\quad p = .5 \quad k_m = np = 10$
2. $n = 11$ $\quad p = .5 \quad k_1 = (n - 1)p = 5 \quad k_2 = (n + 1)p = 6$
3. $n = 20$ $\quad p = .3 \quad (n + 1)p = 6.3 \quad k_m = [6.3] = 6$
4. $n = 9$ $\quad p = .4 \quad (n + 1)p = 4 \quad k_1 = 3 \quad k_2 = 4$

The significance of the curves shown is explained in Section 3-3.

* $[x]$ means "the largest integer smaller than x."

3-3 Asymptotic Theorems

In repeated trials, we are faced with the problem of evaluating the probabilities

$$p_n(k) = \frac{n(n-1)\cdots(n-k+1)}{1\cdot 2\cdots k}\, p^k q^{n-k} \tag{3-21}$$

and their sum (3-19). For large n, this is a complicated task. In the following, we give simple approximations.

The Normal Curves We introduce the function

$$g(x) = \frac{1}{\sqrt{2\pi}}\, e^{-x^2/2} \tag{3-22}$$

and its integral

$$G(x) = \frac{1}{\sqrt{2\pi}} \int_{-\infty}^{x} e^{-y^2/2}\, dy \tag{3-23}$$

These functions are called (standard) *normal* or *Gaussian* curves. As we show later, they are used extensively in the theory of probability and statistics.

Clearly,

$$g(-x) = g(x)$$

Furthermore (see the appendix to this chapter),

$$\frac{1}{\sqrt{2\pi}} \int_{-\infty}^{\infty} e^{-x^2/2}\, dx = 1 \tag{3-24}$$

From this and the evenness of $g(x)$ it follows that

$$G(1) = 1 \qquad G(0) = \frac{1}{2} \qquad G(-x) = \frac{1}{2} - G(x) \tag{3-25}$$

In Fig. 3.3 we show the functions $g(x)$ and $G(x)$, and in Table* 1a we tabulate $G(x)$ for $0 \le x \le 3$. For $x > 3$ we can use the approximation

$$G(x) \simeq 1 - \frac{1}{x}\, g(x) \tag{3-26}$$

De Moivre–Laplace Theorem

It can be shown that for large n, $p_n(k)$ can be approximated by the samples of the normal curve $g(x)$ properly scaled and shifted:

$$\binom{n}{k} p^k q^{n-k} \simeq \frac{1}{\sigma\sqrt{2\pi}}\, e^{-(k-\eta)^2/2\sigma^2} \tag{3-27}$$

* All tables are at the back of the book.

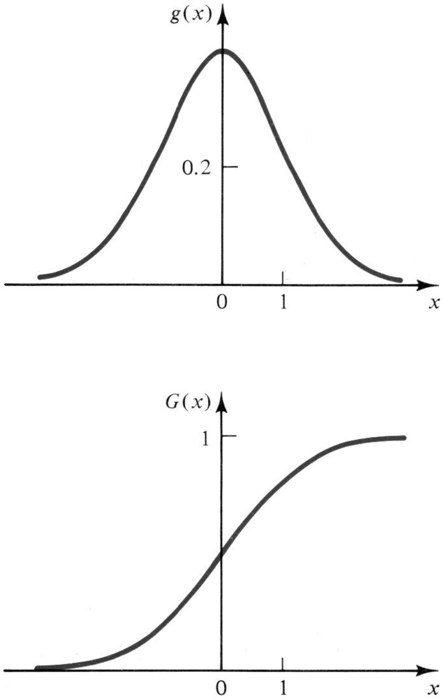

Figure 3.3

where

$$\eta = np \qquad \sigma = \sqrt{npq} \tag{3-28}$$

In the next chapter, we give another interpretation to the function $g(x)$ (density) and to the constants η (mean) and σ (standard deviation).

The result in (3-27) is known as the *De Moivre–Laplace theorem*. We shall not give its rather difficult proof here; we shall only comment on the range of values of n and k for which the approximation is satisfactory. The standard normal curve takes significant values only for $|x| < 3$; for $|x| > 3$, it is negligible. The scaled and shifted version of the interval $(-3, 3)$ is the interval $D = (\eta - 3\sigma, \eta + 3\sigma)$. The rule-of-thumb condition for the validity of (3-27) is as follows: If the interval D is in the interior of the interval $(0, n)$, (3-27) is a satisfactory approximation for every k in the interval D. In other words, (3-27) can be used if

$$0 < np - 3\sqrt{npq} < k < np + 3\sqrt{npq} < n \tag{3-29}$$

Note, finally, that the approximation is best if $p_n(k)$ is nearly symmetrical, that is, if p is close to .5, and it deteriorates if p is close to 0 or to 1.

We list the exact values of $p_n(k)$ and its approximate values obtained from (3-27) for $n = 8$ and for $p = .5$. As we see, even for such moderate values of n, the approximation error is small.

k	0	1	2	3	4	5	6	7	8
$p_8(k)$	.004	.031	.109	.219	.273	.219	.109	.031	.004
approx.	.005	.030	.104	.220	.282	.220	.104	.030	.005

Example 3.13

A fair coin is tossed 100 times. Find the probability that heads will show $k = 50$ and $k = 45$ times.

In this case,

$$n = 100 \qquad p = q = .5 \qquad np = 50 \qquad npq = 25$$

Condition (3-29) yields $0 < 50 - 15 < k < 50 + 15 < 100$; hence, the approximation

$$p_n(k) \simeq \frac{1}{5\sqrt{2\pi}} e^{-(k-50)^2/50}$$

is satisfactory provided that k is between 35 and 65. Thus

$$P\{k = 50\} \simeq \frac{1}{5\sqrt{2\pi}} = .08 \qquad P\{k = 45\} = \frac{1}{5\sqrt{2\pi}} e^{-1/2} = .048 \qquad \blacksquare$$

As we have seen, the probability that the number k of successes of an event $\mathcal{A}$ in n trials is between k_1 and k_2 equals

$$P\{k_1 \le k \le k_2\} = \sum_{k=k_1}^{k_2} \binom{n}{k} p^k q^{n-k}$$

Using the De Moivre–Laplace theorem (3-27), we shall give an approximate expression for this sum in terms of the normal curve $G(x)$ defined in (3-23).

■ ***Theorem***

$$\sum_{k=k_1}^{k_2} \binom{n}{k} p^k q^{n-k} \simeq G\left(\frac{k_2 - np}{\sqrt{npq}}\right) - G\left(\frac{k_1 - np}{\sqrt{npq}}\right) \qquad (3\text{-}30)$$

provided that k_1 *or* k_2 satisfies (3-29) and

$$\sigma = \sqrt{npq} \gg 1 \qquad (3\text{-}31)$$

■ ***Proof.*** From (3-27) it follows that

$$\sum_{k=k_1}^{k_2} \binom{n}{k} p^k q^{n-k} \simeq \frac{1}{\sqrt{2\pi}\sigma} \sum_{k=k_1}^{k_2} e^{-(k-\eta)^2/2\sigma^2}$$

The right side is the sum of the $k_2 - k_1 + 1$ samples of the function

$$f(x) = \frac{1}{\sigma\sqrt{2\pi}} e^{-(x-\eta)^2/2\sigma^2} = \frac{1}{\sigma} g\left(\frac{x-\eta}{\sigma}\right)$$

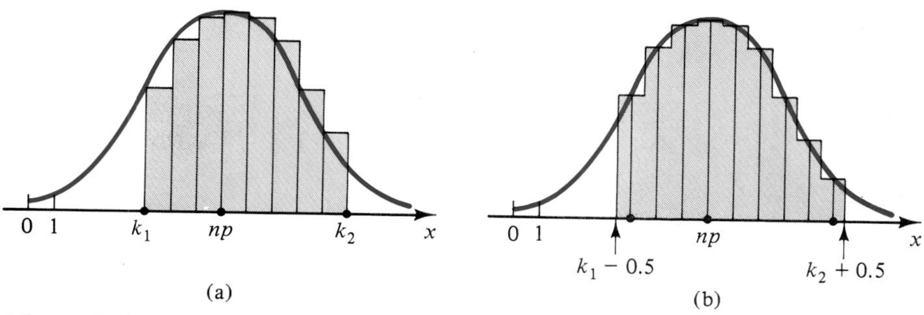

Figure 3.4

for $x = k_1, k_1 + 1, \ldots, k_2$. Since $\sigma \gg 1$ by assumption, the function $f(x)$ is nearly constant in the interval $(k, k + 1)$ of unit length (Fig. 3.4a); hence, its area in that interval is nearly $f(k)$. Thus

$$\frac{1}{\sigma\sqrt{2\pi}} \sum_{k=k_1}^{k_2} e^{-(k-\eta)^2/2\sigma^2} \simeq \frac{1}{\sigma\sqrt{2\pi}} \int_{k_1}^{k_2} e^{-(x-\eta)^2/2\sigma^2}\,dx \tag{3-32}$$

With the transformation

$$y = \frac{x - \eta}{\sigma} \qquad dx = \sigma\,dy$$

the right side of (3-32) equals [see (3-23)]

$$\frac{1}{\sqrt{2\pi}} \int_{(k_1-\eta)/\sigma}^{(k_2-\eta)/\sigma} e^{-y^2/2}\,dy = G\left(\frac{k_2 - \eta}{\sigma}\right) - G\left(\frac{k_1 - \eta}{\sigma}\right) \tag{3-33}$$

and (3-30) results.

The approximation (3-27) on which (3-30) is based holds only if k is in the interval $(\eta - 3\sigma, \eta + 3\sigma)$. However, outside the interval, the numbers $p_n(k)$ and the corresponding values of the exponential in (3-27) are negligible compared to the terms inside the interval. Hence, (3-7) holds so long as the sum in (3-27) contains terms in the interval $(\eta - 3\sigma, \eta + 3\sigma)$.

Note in particular that if $k_1 = 0$, then

$$G\left(\frac{k_1 - \eta}{\sigma}\right) = G\left(-\frac{np}{\sqrt{npq}}\right) < G(-3) \simeq .001 \simeq 0$$

because [see (3-29)] $np > 3\sqrt{npq}$; hence,

$$\sum_{k=0}^{k_2} \binom{n}{p} p^k q^{n-k} \simeq G\left(\frac{k_2 - np}{\sqrt{npq}}\right) \tag{3-34}$$

provided that k_2 is between $\eta - 3\sigma$ and $\eta + 3\sigma$.

Example 3.14

The probability that a voter in an election is a Republican equals .4. Find the probability that among 1,000 voters, the number k of Republicans is between 370 and 430.

This is a problem in repeated trials where $\mathcal{A} = \{r\}$ is an event in the experiment

$\mathcal{S}$ and $P(\mathcal{A}) = p = .4$. With

$$n = 1{,}000 \qquad k_1 = 370 \qquad k_2 = 430 \qquad np = 400 \qquad npq = 240$$

(3-34) yields

$$P\{370 \le k \le 430\} \simeq G\left(\frac{30}{\sqrt{240}}\right) - G\left(-\frac{30}{\sqrt{240}}\right) = .951 \qquad ■$$

Example 3.15

We receive an order of 10,000 parts. The probability that a part is defective equals .1. Find the probability that the number k of defective parts does not exceed, 1,100.

With

$$n = 10{,}000 \qquad k_2 = 1{,}100 \qquad p = .1 \qquad np = 1{,}000 \qquad npq = 900$$

(3-34) yields

$$P\{k \le 1{,}000\} = G\left(\frac{100}{\sqrt{900}}\right) = .999 \qquad ■$$

Correction In (3-32), we approximated the sum of the $k_2 - k_1 + 1$ terms on the left to the area of $f(x)$ in $k_2 - k_1$ intervals of length 1. If $k_2 - k_1 \gg 1$, the additional term on the left that is ignored is negligible, and the approximation (3-30) is satisfactory. For moderate values of $k_2 - k_1$, however, a better approximation results if the integration limits k_1 and k_2 on the right of (3-32) are replaced by $k_1 - 1/2$ and $k_2 + 1/2$. This yields the improved approximation

$$\sum_{k=k_1}^{k_2} \binom{n}{p} p^k q^{n-k} \simeq G\left(\frac{k_2 + 0.5 - np}{\sqrt{npq}}\right) - G\left(\frac{k_1 - 0.5 - np}{\sqrt{npq}}\right) \qquad (3\text{-}35)$$

of (3-30), obtained by replacing the normal curve $g(x)$ by the inscribed staircase function of Fig. 3.4b.

THE LAW OF LARGE NUMBERS According to the empirical interpretation of the probability p of an event $\mathcal{A}$, we should expect with near certainty that the number k of successes of $\mathcal{A}$ in n trials be close to np provided that n is large enough. Using (3-30), we shall give a probabilistic interpretation of this expectation in the context of the model $\mathcal{S}_n$ of repeated trials.

As we have shown, the most likely value of k is, as expected, np. However, not only is the probability [see (3-27)]

$$P\{k = np\} \simeq \frac{1}{\sqrt{npq}} \qquad (3\text{-}36)$$

that k equals np not close to 1, but it tends to zero as $n \to \infty$. What is almost certain is not that k equals np but that the ratio k/n is arbitrarily close to p for n large enough. This is the essence of the law of large numbers. A precise formulation follows.

■ ***Theorem.*** For any positive ε,

$$P\{p - \varepsilon \leq \frac{k}{n} \leq p + \varepsilon\} > .997 \tag{3-37}$$

provided that $n > 9pq/\varepsilon^2$.

■ ***Proof.*** With $k_1 = (p - \varepsilon)n$, $k_2 = (p + \varepsilon)n$, (3-30) yields

$$P\{(p - \varepsilon)n < k < (p + \varepsilon)n\} \simeq G\left[\frac{(p + \varepsilon)n - np}{\sqrt{npq}}\right] - G\left[\frac{(p - \varepsilon)n - np}{\sqrt{npq}}\right]$$

Hence,

$$P\left\{p - \varepsilon \leq \frac{k}{n} \leq p + \varepsilon\right\} = G\left(\varepsilon\sqrt{\frac{n}{pq}}\right) - G\left(-\varepsilon\sqrt{\frac{n}{pq}}\right) = 2G\left(\varepsilon\sqrt{\frac{n}{pq}}\right) - 1 \tag{3-38}$$

If $n > 9pq/\varepsilon^2$, then $\varepsilon\sqrt{n/pq} > 3$ and

$$2G\left(\varepsilon\sqrt{\frac{n}{pq}}\right) - 1 > 2G(3) - 1 \simeq 2 \times .9986 - 1 \simeq .997$$

and (3-37) results.

Note, finally, that

$$2G\left(\varepsilon\sqrt{\frac{n}{pq}}\right) - 1 \underset{n\to\infty}{\longrightarrow} 2G(\infty) - 1 = 1$$

Since this is true for any ε, we conclude with near certainty that k/n tends to p as $n \to \infty$. This is, in a sense, the theoretical justification of the empirical interpretation (1-1) of probability.

Example 3.16

(a) A fair coin is tossed 900 times. Find the probability that the ratio k/n is between .49 and .51.

In this problem, $n = 900$, $\varepsilon = .01$, $\varepsilon\sqrt{n/pq} = .6$, and (3-38) yields

$$P\left\{.49 < \frac{k}{900} < .51\right\} = 2G(.6) - 1 = .4515$$

(b) Find n such that the probability that k/n is between .49 and .51 is .95.

In this case, (3-38) yields

$$P\left\{.49 < \frac{k}{n} < .51\right\} = 2G(.02\sqrt{n}) - 1 = .95$$

Thus n is the solution of the equation $G(.02\sqrt{n}) = .975$. From Table 1 we see that

$$G(1.95) = .9744 < .975 < G(2.00) = .97725$$

Using linear interpolation, we conclude that [see (4-21)] $.02\sqrt{n} \simeq 1.96$; hence, $n \simeq 9{,}600$. ■

Generalized Bernoulli Trials

In Section 3-2 we determined the probability $p_n(k)$ that an event $\mathcal{A}$ of an experiment $\mathcal{S}$ will succeed k times in n independent repetitions of $\mathcal{S}$. Clearly, if $\mathcal{A}$ succeeds k times, its complement succeeds $n - k$ times. The fundamental theorem (3-16) can thus be phrased as follows:

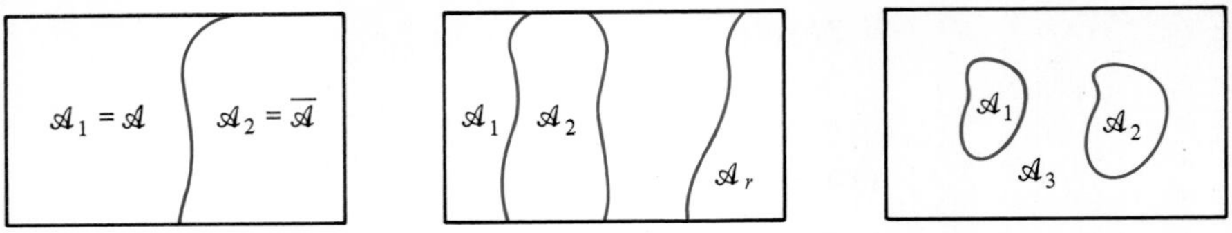

(a) (b) (c)

Figure 3.5

The events $\mathcal{A}_1 = \mathcal{A}$, $\mathcal{A}_2 = \overline{\mathcal{A}}$ form a partition of $\mathcal{S}$ (Fig. 3.5a), and their probabilities equal $p_1 = p$ and $p_2 = 1 - p$, respectively. In the space $\mathcal{S}_n$ of repeated trials, the probability of the event $\{\mathcal{A}_1$ occurs $k_1 = k$ times and $\mathcal{A}_2$ occurs $k_2 = n - k$ times$\}$ equals

$$p_n(k) = p_n(k_1, k_2) = \frac{n!}{k_1!k_2!} p_1^{k_1} p_2^{k_2} \tag{3-39}$$

Our purpose now is to extend this result to arbitrary partitions.

We are given a partition $A = [\mathcal{A}_1, \ldots, \mathcal{A}_r]$ of the space $\mathcal{S}$, consisting of the r events $\mathcal{A}_i$ (Fig. 3.5b). The probabilities of these events are r numbers $p_i = P(\mathcal{A}_i)$ such that

$$p_1 + \cdots + p_r = 1$$

Repeating the experiment $\mathcal{S}$ n times, we obtain the experiment $\mathcal{S}_n$ of repeated trials. A single outcome of $\mathcal{S}_n$ is a sequence

$$\xi_1 \xi_2 \ldots \xi_n$$

where ξ_j is an element of $\mathcal{S}$. Since A is a partition, the element ξ_j belongs to one and only one of the events $\mathcal{A}_i$. Thus at a particular trial, only one of the events of A occurs. Denoting by k_i the number of occurrences of $\mathcal{A}_i$ in n trials, we conclude that

$$k_1 + \cdots + k_r = n$$

At the jth trial, the probability that the event $\mathcal{A}_i$ occurs equals p_i. Furthermore, the trials are independent by assumption. Introducing the event

$$\mathcal{B} = \{\mathcal{A}_i \text{ occurs } k_i \text{ times in a specific order}\}$$

we obtain

$$P(\mathcal{B}) = p_1^{k_1} \cdots p_r^{k_r} \tag{3-40}$$

This is the extension of (3-15) to arbitrary partitions.

We shall next determine the probability $p_n(k_1, \ldots, k_r)$ of the event

$$\mathcal{D} = \{\mathcal{A}_i \text{ occurs } k_i \text{ times in any order}\}$$

■ ***Theorem***

$$p_n(k_1, \ldots, k_r) = \frac{n!}{k_1! \cdots k_r!} p_1^{k_1} \cdots p_r^{k_r} \tag{3-41}$$

■ ***Proof.*** For a specific set of numbers $k_1, \ldots, k_r$, all events of $\mathcal{S}_n$ of the form $\mathcal{B}$ have the same probability. Furthermore, they are mutually exclusive, and their union is the event $\mathscr{D}$. To find $P(\mathscr{D})$ it suffices to find the total number N of such events. Clearly, N equals the number $C^n_{k_1,\ldots,k_r}$ of combinations of n objects grouped in r classes, with k_i objects in the ith class. As we have shown in (2-14), the number of such combinations equals

$$C^n_{k_1,\ldots,k_r} = \frac{n!}{k_1! \cdots k_r!}$$

Multiplying by $P(\mathscr{B})$, we obtain (3-41).

■ ***Corollary.*** We are given two mutually exclusive events $\mathscr{A}_1$ and $\mathscr{A}_2$ (Fig. 3.5c) with $p_1 = P(\mathscr{A}_1)$ and $p_2 = P(\mathscr{A}_2)$. We wish to find the probability p_a that in n trials, the event $\mathscr{A}_1$ occurs k_1 times and the event $\mathscr{A}_2$ occurs k_2 times.

To solve this problem, we introduce the event

$$\mathscr{A}_3 = \overline{\mathscr{A}_1 \cup \mathscr{A}_2}$$

This event occurs $k_3 = n - (k_1 + k_2)$ times and its probability equals $p_3 = 1 - p_1 - p_2$. Furthermore, the events $\mathscr{A}_1$, $\mathscr{A}_2$, $\mathscr{A}_3$ form a partition of $\mathscr{S}$. Hence,

$$p_a = p_n(k_1, k_2, k_3) = \frac{n!}{k_1!k_2!k_3!} p_1^{k_1} p_2^{k_2} p_3^{k_3} \tag{3-42}$$

Example 3.17

A fair die is rolled 10 times. Find the probability p_a that f_1 shows 3 times and even shows 5 times.

In this case, $\mathscr{A}_1 = \{f_1\}$, $\mathscr{A}_2 = \{\text{even}\}$

$$n = 10 \qquad k_1 = 3 \qquad k_2 = 5 \qquad p_1 = \frac{1}{6} \qquad p_2 = \frac{3}{6}$$

Clearly, the events $\mathscr{A}_1$ and $\mathscr{A}_2$ are mutually exclusive. We can therefore apply (3-42) with

$$\mathscr{A}_3 = \overline{\mathscr{A}}_1 \cap \overline{\mathscr{A}}_2 = \{f_3, f_5\} \qquad p_3 = \frac{2}{6} \qquad k_3 = 2$$

This yields

$$p_{10}(3, 5, 2) = \frac{10!}{3!5!2!}\left(\frac{1}{6}\right)^3\left(\frac{3}{6}\right)^5\left(\frac{2}{6}\right)^2 = .4$$

■

3-4 Rare Events and Poisson Points

We shall now examine the asymptotic behavior of the probabilities

$$p_n(k) = \frac{n(n-1)\cdots(n-k+1)}{k!} p^k q^{n-k}$$

under the assumption that the event $\mathscr{A}$ is *rare,* that is, that $p \ll 1$. If n is so large that $np \gg 1$, we can use the De Moivre–Laplace approximation (3-27). If, however, $n \gg 1$ but np is of the order of 1—for example, if $n = 1{,}000$ and $p = .002$—then the approximation no longer holds. In such cases, the following important result can be used.

■ ***Poisson Theorem.*** If $p \ll 1$ and $n \gg 1$, then, for k of the order of np,

$$\binom{n}{k} p^k q^{n-k} \simeq e^{-np} \frac{(np)^k}{k!} \tag{3-43}$$

■ ***Proof.*** The probabilities $p_n(k)$ take significant values for k near np. Since $p \ll 1$ we can use the approximations $np \ll n$, $k \ll n$:

$$n(n-1) \cdots (n-k+1) \simeq n^k \qquad q = 1 - p \simeq e^{-p} \qquad q^{n-k} \simeq q^n \simeq e^{-np}$$

This yields

$$\frac{n(n-1) \cdots (n-k+1)}{k!} p^k q^{n-k} \simeq \frac{n^k}{k!} p^k e^{-np} \tag{3-44}$$

and (3-43) results.

We note that (3-43) is only an approximation. The formal theorem can be phrased as a limit:

$$\binom{n}{k} p^k q^{n-k} \to e^{-a} \frac{a^k}{k!} \tag{3-45}$$

as $n \to \infty$, $p \to 0$, and $np \to a$. The proof is based on a refinement of the approximation (3-44).

Example 3.18

An order of 2,000 parts is received. The probability that a part is defective equals 10^{-3}. Find the probability p_a that no component is defective and the probability p_b that there are at most 3 defective components.

With $n = 2{,}000$, $p = 10^{-3}$, $np = 2$, (3-43) yields

$$p_a = P\{k = 0\} = q^n = (1 - 10^{-3})^{2{,}000} \simeq e^{-2} = .406$$

$$p_b = P\{k \le 3\} = \sum_{k=0}^{3} \binom{n}{k} p^k q^{n-k} \simeq \sum_{k=0}^{3} e^{-np} \frac{(np)^k}{k!} = e^{-2}\left(1 + \frac{2}{1} + \frac{2^2}{2!} + \frac{2^3}{3!}\right) = .857 \quad ■$$

Generalization Consider the r events $\mathscr{A}_i$ of the partition A of Fig. 3.5. We shall examine the asymptotic behavior of the probabilities $p_n(k_1, \ldots, k_r)$ in (3-41) under the assumption that the first $r - 1$ of these events are rare, that is, that

$$p_1 \ll 1, \ldots, p_{r-1} \ll 1$$

Proceeding as in (3-44), we obtain the approximation

$$p_n(k_1, \ldots, k_r) = \frac{n!}{k_1! \cdots k_r} p_1^{k_1} \cdots p_r^{k_r} \simeq \prod_{i=1}^{r-1} e^{-a_i} \frac{a_i^{k_i}}{k_i!} \tag{3-46}$$

where $a_1 = np_1, \ldots, a_{r-1} = np_{r-1}$.

This approximation holds for k_i of the order np_i, and it is an equality in the limit as $n \to \infty$.

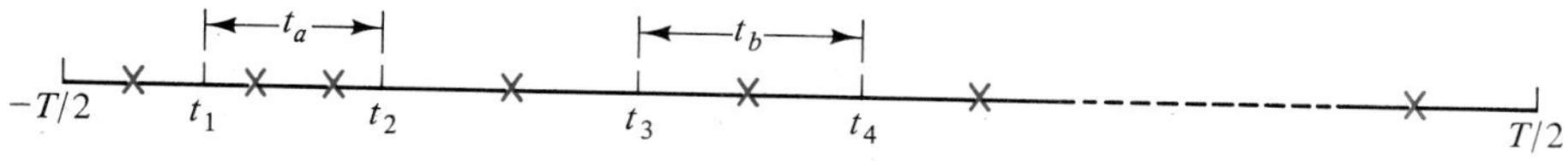

Figure 3.6

Poisson Points

The Poisson approximation is of particular importance in problems involving random points in time or space. This includes radioactive emissions, telephone calls, and traffic accidents. In these and other fields, the points are generated by a model $\mathcal{S}_n = \mathcal{S} \times \cdots \times \mathcal{S}$ of repeated trials where $\mathcal{S}$ is the experiment of the random selection of a single point. In the following, we discuss the resulting model as $n \to \infty$, starting with the single-point experiment $\mathcal{S}$.

We are given an interval $(-T/2, T/2)$ (Fig. 3.6), and we place a point in this interval at random. The probability that the point is in the interval (t_1, t_2) of length $t_a = t_2 - t_1$ equals t_a/T [see (2-46)]. We thus have an experiment $\mathcal{S}$ with outcomes all points in the interval $(-T/2, T/2)$.

The outcomes of the experiment $\mathcal{S}_n$ of the repeated trials of $\mathcal{S}$ are n points in the interval $(-T/2, T/2)$. We shall find the probability $p_n(k)$ that k of these points are in the interval (t_1, t_2). Clearly, $p_n(k)$ is the probability of the event $\{k \text{ points in } t_a\}$. This event occurs if the event $\{t_1 \le t \le t_2\}$ of the experiment $\mathcal{S}$ occurs k times. Hence,

$$P\{k \text{ points in } t_a\} = \binom{n}{k} p^k q^{n-k} \qquad p = \frac{t_a}{T} \tag{3-47}$$

We shall now assume that $n \gg 1$ and $T \gg t_a$. In this case, $\{t_1 \le t \le t_2\}$ is a rare event, and (3-44) yields

$$P\{k \text{ points in } t_a\} \simeq e^{-nt_a/T} \frac{(nt_a/T)^k}{k!} \tag{3-48}$$

This, we repeat, holds only if $t_a \ll T$ and $k \ll n$. Note that this probability depends not on n and T separately but only on their ratio

$$\lambda = \frac{n}{T} \tag{3-49}$$

This ratio will be called the *density* of the points. As we see from (3-49), λ equals the average number of points per unit of time.

Next we increase n and T, keeping the ratio n/T constant. The limit of $\mathcal{S}_n$ as $n \to \infty$ will be denoted by $\mathcal{S}_\infty$ and will be called the experiment of *Poisson points* with density λ. Clearly, a single outcome of $\mathcal{S}_n$ is a set of n points in the interval $(-T/2, T/2)$; hence, a single outcome of $\mathcal{S}_\infty$ is a set of infinitely many points on the entire t axis. In the experiment $\mathcal{S}_\infty$, the probability that k points will be in an interval (t_1, t_2) of length t_a equals

$$P\{k \text{ points in } t_a\} = e^{-\lambda t_a}\frac{(\lambda t_a)^k}{k!} \qquad k = 0, 1, \ldots \tag{3-50}$$

This follows from (3-48) because the right side depends only on the ratio n/T.

Nonoverlapping Intervals Consider the nonoverlapping intervals (t_1, t_2) and (t_3, t_4) of Fig. 3.6 with lengths $t_a = t_2 - t_1$ and $t_b = t_4 - t_3$. In the experiment $\mathscr{S}$ of placing a single point in the interval $(-T/2, T/2)$, the events $\mathscr{A}_1 = \{\text{the point is in } t_a\}$, $\mathscr{A}_2 = \{\text{the point is in } t_b\}$, and $\mathscr{A}_3 = \{\overline{\mathscr{A}}_1 \cap \overline{\mathscr{A}}_2\} = \{$the point is outside the intervals t_a and $t_b\}$ form a partition, and their probabilities equal

$$P(\mathscr{A}_1) = \frac{t_a}{T} \qquad P(\mathscr{A}_2) = \frac{t_b}{T} \qquad P(\mathscr{A}_3) = 1 - \frac{t_a}{T} - \frac{t_b}{T}$$

respectively. In the space $\mathscr{S}_n$ of placing n points in the interval $(-T/2, T/2)$, the event $\{k_a \text{ in } t_a, k_b \text{ in } t_b\}$ occurs iff $\mathscr{A}_1$ occurs $k_1 = k_a$ times, $\mathscr{A}_2$ occurs $k_2 = k_b$ times, and $\mathscr{A}_3$ occurs $k_3 = n - k_a - k_b$ times. Hence [see (3-46)],

$$P\{k_a \text{ in } t_a, k_b \text{ in } t_b\} = \frac{n!}{k_1!k_2!k_3!}\left(\frac{t_a}{T}\right)^{k_1}\left(\frac{t_b}{T}\right)^{k_2}\left(1 - \frac{t_a}{T} - \frac{t_b}{T}\right)^{k_3} \tag{3-51}$$

From this and (3-46) it follows that if

$$n \to \infty \qquad T \to \infty \qquad \frac{n}{T} = \lambda \tag{3-52}$$

then

$$P\{k_a \text{ in } t_a, k_b \text{ in } t_b\} = e^{-\lambda t_a}\frac{(\lambda t_a)^{k_a}}{k_a!}\, e^{-\lambda t_b}\frac{(\lambda t_b)^{k_b}}{k_b!} \tag{3-53}$$

Note from (3-51) that if T is finite, the events $\{k_a \text{ in } t_a\}$ and $\{k_b \text{ in } t_b\}$ are not independent, however, for $T \to \infty$, these events are independent because then [see (3-53) and (3-50)]

$$P\{k_a \text{ in } t_a, k_b \text{ in } t_b\} = P\{k_a \text{ in } t_a\}P\{k_b \text{ in } t_b\} \tag{3-54}$$

Summary Starting from an experiment involving n points randomly placed in a finite interval, we constructed a model $\mathscr{S}_\infty$ consisting of infinitely many points on the entire axis, with the following properties:

1. The number of points in an interval of length t_a is an event $\{k_a \text{ in } t_a\}$ the probability of which is given by (3-50).
2. If two intervals t_a and t_b are nonoverlapping, the events $\{k_a \text{ in } t_a\}$ and $\{k_b \text{ in } t_b\}$ are independent.

These two properties and the parameter λ specify completely the model $\mathscr{S}_\infty$ of Poisson points.

Appendix

Area under the Normal Curve

We shall show that

$$\int_{-\infty}^{\infty} e^{-\alpha x^2}\,dx = \sqrt{\frac{\pi}{\alpha}} \qquad \alpha > 0 \tag{3A-1}$$

■ ***Proof.*** We set

$$I = \int_{-\infty}^{\infty} e^{-\alpha x^2}\,dx = \int_{-\infty}^{\infty} e^{-\alpha y^2}\,dy$$

This yields

$$I^2 = \int_{-\infty}^{\infty}\int_{-\infty}^{\infty} e^{-\alpha(x^2+y^2)}\,dx\,dy$$

In the differential ring ΔR of Fig. 3.7, the integrand equals $e^{-\alpha r^2}$ because $x^2 + y^2 = r^2$. From this it follows that the integral in ΔR equals $e^{-\alpha r^2}$ times the area $2\pi r\,dr$ of the ring. Integrating the resulting product for r from 0 to ∞ and setting $r^2 = z$, we obtain

$$I^2 = \int_0^{\infty} 2\pi r e^{-\alpha r^2}\,dr = \pi \int_0^{\infty} e^{-\alpha z}\,dz = \frac{\pi}{\alpha}$$

and (3A-1) results.

Figure 3.7

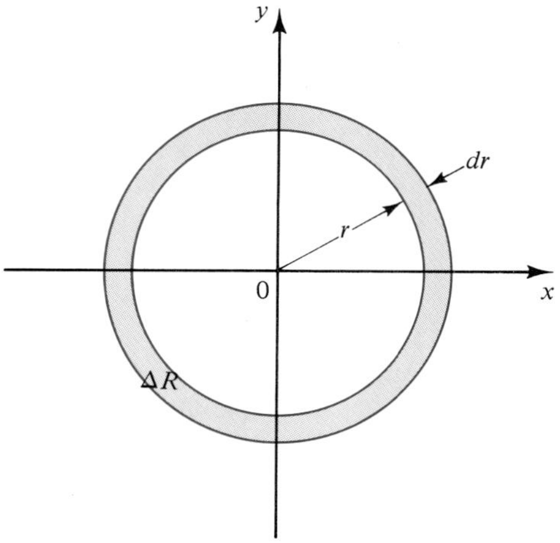

Problems

3-1 A fair coin is tossed 8 times. **(a)** Find the probability p_1 that heads shows once. **(b)** Find the probability p_2 that heads shows at the sixth toss but not earlier.

3-2 A fair coin is tossed 10 times. Find the probability p that at the eighth toss but not earlier heads shows for the second time.

3-3 A fair coin is tossed four times. Find the number of outcomes of the space $\mathcal{S}_4$ and of the event $\mathcal{A}$ = {heads shows three times in any order}; find $P(\mathcal{A})$.

3-4 Two fair dice are rolled 10 times. Find the probability p_1 that 7 will show twice; find the probability p_2 that 11 will show once.

3-5 Two fair dice are rolled three times. Find the number of outcomes of the space $\mathcal{S}_3$ and of the events $\mathcal{A}$ = {7 shows at the second roll}, $\mathcal{B}$ = {11 does not show}; find $P(\mathcal{A})$ and $P(\mathcal{B})$.

3-6 A string of Christmas lights consists of 20 bulbs connected in series. The probability that a bulb is defective equals .01. **(a)** Find the probability p_1 that the string of lights works. **(b)** If it does not work, we replace one bulb at a time with a good bulb until the string works; find the probability p_2 that the string works when the fifth bulb is replaced. **(c)** Find the probability p_3 that we need at most five replacements until the string works.

3-7 **(a)** A shipment consists of 100 units, and the probability that a unit is defective is .2. We select 12 units; find the probability p_1 that 3 of them are defective. **(b)** A shipment consists of 80 good and 20 defective components. We select at random 12 units; find the probability p_2 that 3 of them are defective.

3-8 Compute the probabilities $p_n(k) = \binom{n}{k} p^k q^{n-k}$ for: **(a)** $n = 20$, $p = .3$, $k = 0, 1, \ldots, 20$; **(b)** $n = 9$, $p = .4$, $k = 0, 1, \ldots, 9$. **(c)** Show that the ratio $r(k)$ of two consecutive values of $p_n(k)$ equals

$$r(k) = \frac{p_n(k-1)}{p_n(k)} = \frac{kq}{(n-k+1)p}$$

and that $r(k)$ increases as k increases from 0 to n. Using this, find the values of k for which $p_n(k)$ is maximum.

3-9 Using the approximation (3-27), find $p_n(k)$ for $n = 900$, $p = .4$, and $k = 340$, 350, 360, and 370.

3-10 The probability that a salesman completes a sale is .2. In one day, he sees 100 customers. Find the probability p that he will complete 16 sales.

3-11 Compute the two sides of (3-18): **(a)** for $n = 20$, $p = .4$, and k from 5 to 11; **(b)** for $n = 20$, $p = .2$, and k from 1 to 7.

3-12 A fair coin is tossed n times, and heads shows k times. Find the smallest number n such that $P\{.49 \le k/n \le .51\} > .95$.

3-13 A fair die is rolled 720 times, and 5 shows k times. Find the probability that k is a number in the interval (96, 144).

3-14 The probability that a passenger smokes is .3. A plane has 189 passengers, of whom k are smokers. Find the probability p that $45 \le k \le 67$: **(a)** using the approximation (3-30); **(b)** using the approximation (3-35).

3-15 Of all highway accidents, 52% are minor, 30% are serious, and 18% are fatal. In one day, 10 accidents are reported. Using (3-42), find the probability p that two of the reported accidents are serious and one is fatal.

3-16 The probability that a child is handicapped is 10^{-3}. Find the probability p_1 that in a school of 1,800 children, one is handicapped, and the probability p_2 that more than two children are handicapped: **(a)** exactly; **(b)** using the approximation (3-43).

3-17 We receive a 12-digit number by teletype. The probability that a digit is printed wrong equals 10^{-3}. Find the probability p that one of the digits is wrong: **(a)** exactly; **(b)** using the Poisson approximation.

3-18 The probability that a driver has an accident in a month is .01. Find the probability p_1 that in one year he will have one accident and the probability p_2 that he will have at least one accident: **(a)** exactly; **(b)** using (3-43).

3-19 Particles emitted from a radioactive substance form a Poisson set of points with $\lambda = 1.7$ per second. Find the probability p that in 2 seconds, fewer than five particles will be emitted.

4 The Random Variable

A random variable is a function $\mathbf{x}(\zeta)$ with domain the set $\mathcal{S}$ of experimental outcomes ζ and with range a set of numbers. Thus $\mathbf{x}(\zeta)$ is a model concept, and all its properties follow from the properties of the experiment $\mathcal{S}$. A function $\mathbf{y} = g(\mathbf{x})$ of the random variable $\mathbf{x}$ is a composite function $\mathbf{y}(\zeta) = g(\mathbf{x}(\zeta))$ with domain the set $\mathcal{S}$. Some authors define random variables as functions with domain the real line. The resulting theory is consistent and operationally equivalent to ours. We feel strongly, however, that it is conceptually preferable to interpret $\mathbf{x}$ as a function defined within an abstract space $\mathcal{S}$, even if $\mathcal{S}$ is not explicitly used. This approach avoids the use of infinitely dimensional spaces, and it leads to a unified theory.

4-1 Introduction

We have dealt so far with experimental outcomes, events, and probabilities. The outcomes are various objects that can be identified somehow, for example, "heads," "red," "the queen of spades." We have also considered experiments the outcomes of which are numbers, for example, "time of a call," "IQ of a child"; however, in the study of events and probabilities, the numerical character of such outcomes is only a way of identifying them. In this chapter we introduce a new concept. We assign to each outcome ζ of an experiment $\mathcal{S}$ a number $\mathbf{x}(\zeta)$. This number could be the gain or loss in a game of chance, the size of a product, the voltage of a random source, or any other quantity of interest. We thus establish a relationship between the elements ζ_i of the set $\mathcal{S}$ and various numbers $\mathbf{x}(\zeta_i)$. In other words, we form a function with domain the set $\mathcal{S}$ of abstract objects ζ_i and with range a set of numbers. Such a function will be called a *random variable*.

Example 4.1

The die experiment has six outcomes. To the outcome f_i we assign the number $10i$. We have thus formed a function $\mathbf{x}$ such that $\mathbf{x}(f_i) = 10i$. In the same experiment, we form another function $\mathbf{y}$ such that $\mathbf{y}(f_i) = 0$ if i is odd and $\mathbf{y}(f_i) = 1$ if i is even. In the following table, we list the functions so constructed.

	f_1	f_2	f_3	f_4	f_5	f_6
$\mathbf{x}(f_i)$	10	20	30	40	50	60
$\mathbf{y}(f_i)$	0	1	0	1	0	1

The domain of both functions is the set $\mathcal{S}$. The range of $\mathbf{x}$ consists of the six numbers 10, . . . , 60. The range of $\mathbf{y}$ consists of the two numbers 0 and 1. ■

To clarify the concept of a random variable, we review briefly the notion of a function.

Meaning of a Function As we know, a function $x = x(t)$ is a rule of correspondence between the values of t and x. The independent variable t takes numerical values forming a set $\mathcal{S}_t$ on the t-axis called the *domain* of the function. To every t in $\mathcal{S}_t$ we assign, according to some rule, a number $x(t)$ to the dependent variable x. The values of x form a set $\mathcal{S}_x$ on the x-axis called the *range* of the function. Thus a function is a *mapping* of the set $\mathcal{S}_t$ on the set $\mathcal{S}_x$. The rule of correspondence between t and x could be a table, a curve, or a formula, for example, $x(t) = t^2$.

The notation $x(t)$ used to represent a function has two meanings. It means the particular number $x(t)$ corresponding to a specific t; it also means the function $x(t)$, namely, the entire mapping of the set $\mathcal{S}_t$ on the set $\mathcal{S}_x$. To avoid this ambiguity, we shall denote the mapping by x, leaving its dependence on t understood.

Generalization The definition of a function can be phrased as follows: We are given two sets of *numbers* $\mathcal{S}_t$ and $\mathcal{S}_x$. To every $t \in \mathcal{S}_t$ we assign a number $x(t)$ belonging to the set $\mathcal{S}_x$. This leads to the following generalization.

We are given two sets of *objects* $\mathcal{S}_\alpha$ and $\mathcal{S}_\beta$ consisting of the arbitrary elements α and β, respectively:

$$\alpha \in \mathcal{S}_\alpha \qquad \beta \in \mathcal{S}_\beta$$

We say that β is a function of α if to every element α of the set $\mathcal{S}_\alpha$ we make correspond one element β of the set $\mathcal{S}_\beta$. The set $\mathcal{S}_\alpha$ is called the domain of the function and the set $\mathcal{S}_\beta$ its range. Suppose that $\mathcal{S}_\alpha$ is the set of all children in a community and $\mathcal{S}_\beta$ the set of their mothers. The pairing of a child with its mother is a function. We note that to a given α there corresponds a single β. However, more than one element of the set $\mathcal{S}_\alpha$ might be paired with the same β (a child has only one mother, but a mother might have more than one child). Thus the number N_β of elements of the set $\mathcal{S}_\beta$ is equal to or smaller than the number N_α of the elements of the set $\mathcal{S}_\alpha$. If the correspondence is one-to-one, then $N_\alpha = N_\beta$.

The Random Variable

A random variable (RV) represents a process of assigning to every outcome ζ of an experiment $\mathcal{S}$ a number $\mathbf{x}(\zeta)$. Thus an RV $\mathbf{x}$ is a function with domain the set $\mathcal{S}$ of experimental outcomes and range a set of numbers (Fig. 4.1). *All RVs will be written in boldface letters.* The notation $\mathbf{x}(\zeta)$ will indicate the number assigned to the specific outcome ζ, and the notation $\mathbf{x}$ will indicate the entire function, that is, the rule of correspondence between the elements ζ of $\mathcal{S}$ and the numbers $\mathbf{x}(\zeta)$ assigned to these elements. In Example 4.1, $\mathbf{x}$ indicates the table pairing the six faces of the die with the six numbers 10, . . . , 60. The domain of this function is the set $\mathcal{S} = \{f_1, \ldots, f_6\}$, and its range is the set $\{10, \ldots, 60\}$. The expression $\mathbf{x}(f_2)$ is the number 20. In the same example, $\mathbf{y}$ indicates the correspondence between the six faces f_i and the two numbers 0 and 1. The range of $\mathbf{y}$ is therefore the set $\{0, 1\}$. The expression $\mathbf{y}(f_2)$ is the number 1 (Fig. 4.2).

Events Generated by RVS In the study of RVs, questions of the following form arise: What is the probability that the RV $\mathbf{x}$ is less than a given number

Figure 4.1

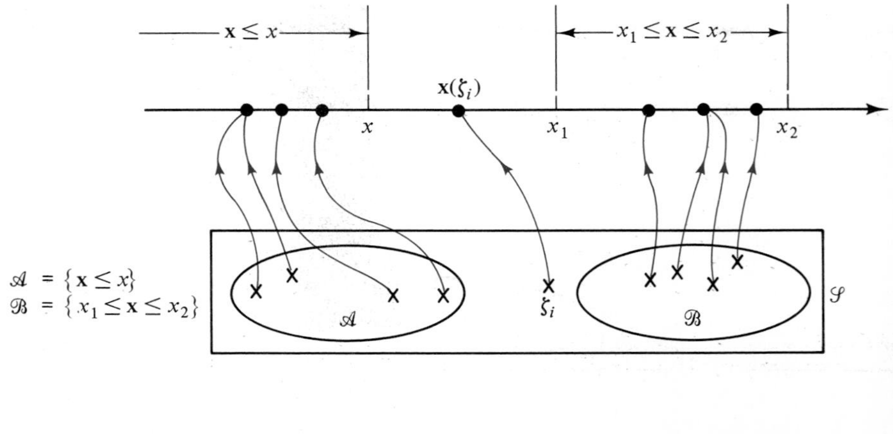

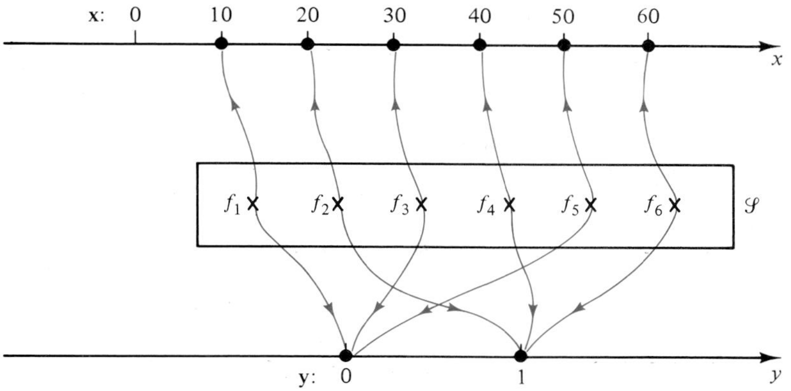

Figure 4.2

x? What is the probability that $\mathbf{x}$ is between the numbers x_1 and x_2? We might, for example, wish to find the probability that the height $\mathbf{x}$ of a person selected at random will not exceed certain bounds. As we know, probabilities are assigned only to events; to answer such questions, we must therefore express the various conditions imposed on $\mathbf{x}$ as events.

We start with the determination of the probability that the RV $\mathbf{x}$ does not exceed a specific number x. To do so, we introduce the notation

$$\mathcal{A} = \{\mathbf{x} \le x\}$$

This notation specifies an event $\mathcal{A}$ consisting of all outcomes ζ such that $\mathbf{x}(\zeta) \le x$. We emphasize that $\{\mathbf{x} \le x\}$ is not a set of numbers; it is a set $\mathcal{A}$ of experimental outcomes (Fig. 4.1). The probability $P(\mathcal{A})$ of this set is the probability that the RV $\mathbf{x}$ does not exceed the number x.

The notation

$$\mathcal{B} = \{x_1 < \mathbf{x} < x_2\}$$

specifies an event $\mathcal{B}$ consisting of all outcomes ζ such that the corresponding values $\mathbf{x}(\zeta)$ of the RV $\mathbf{x}$ are between the numbers x_1 and x_2.

Finally,

$$\mathcal{C} = \{\mathbf{x} = x_0\}$$

is an event consisting of all outcomes ζ such that the value $\mathbf{x}(\zeta)$ of $\mathbf{x}$ equals the number x_0.

Example 4.2

We shall illustrate with the RVs $\mathbf{x}$ and $\mathbf{y}$ of Example 4.1. The set $\{\mathbf{x} \le 35\}$ consists of the elements f_1, f_2, and f_3 because $\mathbf{x}(f_i) \le 35$ only if $i = 1, 2$, or 3. The set of $\{\mathbf{x} \le 5\}$ is empty because there is no outcome such that $\mathbf{x}(f_i) \le 5$. The set $\{20 \le \mathbf{x} \le 35\}$ consists of the outcomes f_2 and f_3 because $20 \le \mathbf{x}(f_i) \le 35$ only if $i = 2$ or 3. The set $\{\mathbf{x} = 40\}$ consists of the element f_4 because $\mathbf{x}(f_i) = 40$ only if $i = 4$. Finally, $\{\mathbf{x} = 35\}$ is the empty set because there is no outcome such that $\mathbf{x}(f_i) = 35$.

Similarly, $\{\mathbf{y} < 0\}$ is the empty set because there is no outcome such that $\mathbf{y}(f_i) < 0$. The set $\{\mathbf{y} < 1\}$ consists of the outcomes f_1, f_3, and f_5 because $\mathbf{y}(f_i) < 1$ for $i = 1, 3$, or 5. Finally, $\{\mathbf{y} \le 1\}$ is the certain event because $\mathbf{y}(f_i) \le 1$ for every f_i. ■

In the definition of an RV, the numbers $\mathbf{x}(\zeta)$ assigned to $\mathbf{x}$ can be finite or infinite. We shall assume, however, that the set of outcomes ζ such that $\mathbf{x}(\zeta) = \pm\infty$ has zero probability:

$$P\{\mathbf{x} = \infty\} = 0 \qquad P\{\mathbf{x} = -\infty\} = 0 \tag{4-1}$$

With this mild restriction, the definition of an RV is complete.

4-2 The Distribution Function

It appears from the definition of an RV that to determine the probability that $\mathbf{x}$ takes values in a set I of the x-axis, we must first determine the event $\{\mathbf{x} \in I\}$ consisting of all outcomes ζ such that $\mathbf{x}(\zeta)$ is in the set I. To do so, we need to know the underlying experiment $\mathcal{S}$. However, as we show next, this is not necessary. To find $P\{\mathbf{x} \in I\}$ it suffices to know the distribution of the RV $\mathbf{x}$. This is a function $F_x(x)$ of x defined as follows.

Given a number x, we form the event $\mathcal{A}_x = \{\mathbf{x} \le x\}$. This event depends on the number x; hence, its probability is a function of x. This function is denoted by $F_x(x)$ and is called the *cumulative distribution function* (c.d.f.) of the RV $\mathbf{x}$. For simplicity, we shall call it the distribution function or just the distribution of $\mathbf{x}$.

■ ***Definition.*** The *distribution* of the RV $\mathbf{x}$ is the function

$$F_x(x) = P\{\mathbf{x} \le x\} \tag{4-2}$$

defined for every x from $-\infty$ to ∞.

In the notation $F_x(x)$, the subscript x identifies the RV $\mathbf{x}$ and the independent variable x specifies the event $\{\mathbf{x} \le x\}$. The variable x could be replaced by any other variable. Thus $F_x(w)$ equals the probability of the event $\{\mathbf{x} \le w\}$. The distributions of the RVs $\mathbf{x}$, $\mathbf{y}$, and $\mathbf{z}$ are denoted by $F_x(x)$, $F_y(y)$, and $F_z(z)$, respectively. If, however, there is no fear of ambiguity, the subscripts will be omitted and all distributions will be identified by their independent variables. In this notation, the distributions of the RVs $\mathbf{x}$, $\mathbf{y}$, and $\mathbf{z}$ will be $F(x)$, $F(y)$, and $F(z)$, respectively. Several illustrations follow.

Example 4.3

(a) In the fair-die experiment, we define the RV $\mathbf{x}$ such that $\mathbf{x}(f_i) = 10i$ as in Example 4.1. We shall determine its distribution $F_x(x)$ for every x from $-\infty$ to ∞. We start with specific values of x

$$F_x(200) = P\{\mathbf{x} \le 200\} = P(\mathcal{S}) = 1$$

$$F_x(45) = P\{\mathbf{x} \le 45\} = P\{f_1, f_2, f_3, f_4\} = \frac{4}{6}$$

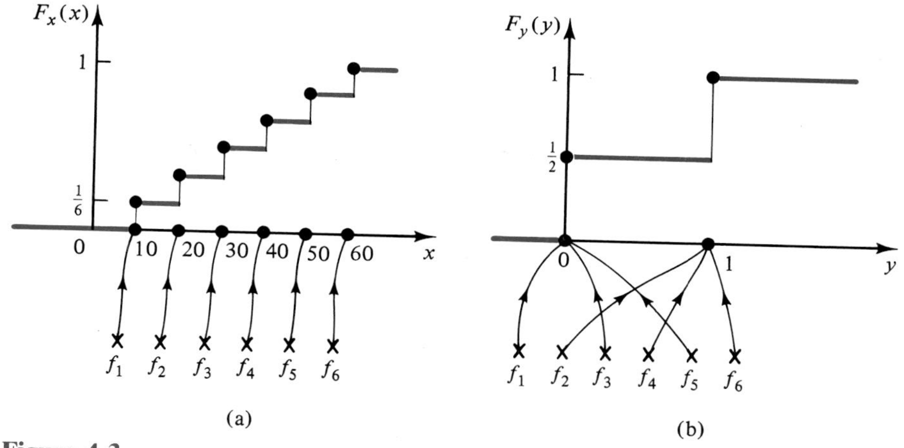

Figure 4.3

$$F_x(30) = P\{\mathbf{x} \le 30\} = P\{f_1, f_2, f_3\} = \frac{3}{6}$$

$$F_x(29.99) = P\{\mathbf{x} \le 29.99\} = P\{f_1, f_2\} = \frac{2}{6}$$

$$F_x(10.1) = P\{\mathbf{x} \le 10.1\} = P\{f_1\} = \frac{1}{6}$$

$$F_x(5) = P\{\mathbf{x} \le 5\} = P(\varnothing) = 0$$

Proceeding similarly for any x, we obtain the staircase function $F_x(x)$ of Fig. 4.3a.

(b) In the same experiment, the RV $\mathbf{y}$ is such that

$$\mathbf{y}(f_i) = \begin{cases} 0 & i = 1, 3, 5 \\ 1 & i = 2, 4, 6 \end{cases}$$

In this case,

$$F_y(15) = P\{\mathbf{y} \le 15\} = P(\mathcal{S}) = 1$$
$$F_y(1) = P\{\mathbf{y} \le 1\} = P(\mathcal{S}) = 1$$
$$F_y(0) = P\{\mathbf{y} \le 0\} = P\{f_1, f_3, f_5\} = \frac{3}{6}$$
$$F_y(-20) = P\{\mathbf{y} \le -20\} = P(\varnothing) = 0$$

The function $F_y(y)$ is shown in Fig. 4.3b ■

Example 4.4

In the coin experiment,

$$\mathcal{S} = \{h, t\} \qquad P\{h\} = p \qquad P\{t\} = q$$

We form the RV $\mathbf{x}$ such that

$$\mathbf{x}(h) = 1 \qquad \mathbf{x}(t) = 0$$

The distribution of $\mathbf{x}$ is the staircase function $F(x)$ of Fig. 4.4. Note, in particular, that

$$F(4) = P\{\mathbf{x} \le 4\} = P(\mathcal{S}) = 1 \qquad F(0.9) = P(\mathbf{x} \le 0.9\} = P\{t\} = q$$
$$F(0) = P(\mathbf{x} \le 0\} = P\{t\} = q \qquad F(-5) = P\{\mathbf{x} \le -5\} = P(\varnothing) = 0$$

■

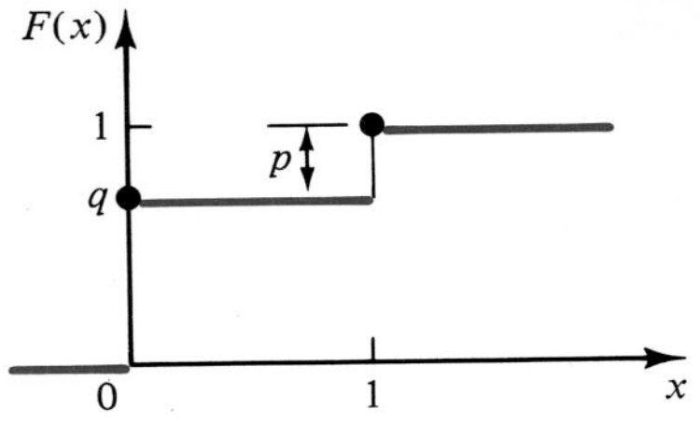

Figure 4.4

Example 4.5

Telephone calls occurring at random and uniformly in the interval $(0, T)$ specify an experiment $\mathscr{S}$ the outcomes of which are all points in this interval. The probability that t is in the interval (t_1, t_2) equals [see (2-46)]

$$P\{t_1 \le t \le t_2\} = \frac{t_2 - t_1}{T}$$

We introduce the RV $\mathbf{x}$ such that

$$\mathbf{x}(t) = t \qquad 0 \le t \le T$$

In this example, the variable t has a double meaning. It is the outcome of the experiment $\mathscr{S}$ and the corresponding value $\mathbf{x}(t) = t$ of the RV $\mathbf{x}$. We shall show that the distribution of $\mathbf{x}$ is a ramp, as in Fig. 4.5. To do so, we must find the probability of the event $\{\mathbf{x} \le x\}$ for every x.

Suppose, first, that $x > T$. In this case, $\mathbf{x}(t) \le x$ for every t in the interval $(0, T)$ because $\mathbf{x}(t) = t$; hence,

$$F(x) = P\{\mathbf{x}(t) \le x\} = P\{0 \le t \le T\} = 1 \qquad x > T$$

If $0 \le x \le T$, then $\mathbf{x}(t) \le x$ for every t in the interval $(0, x)$; hence,

$$F(x) = P\{\mathbf{x} \le x\} = P\{0 \le t \le x\} = \frac{x}{T} \qquad 0 \le x \le T$$

Finally, if $x < 0$, then $\{\mathbf{x}(t) \le x\} = \varnothing$ because $\mathbf{x}(t) = t \ge 0$ for every t in $\mathscr{S}$; hence,

$$F(x) = P\{\mathbf{x} < x\} = P(\varnothing) = 0 \qquad x < 0$$

■

Example 4.6

The experiment $\mathscr{S}$ consists of all points in the interval $(0, \infty)$. The events of $\mathscr{S}$ are all intervals (t_1, t_2) and their unions. To specify $\mathscr{S}$, it suffices, therefore, to know the probability of the event $\{t_1 \le t \le t_2\}$ for every t_1 and t_2 in $\mathscr{S}$. This probability can be

Figure 4.5

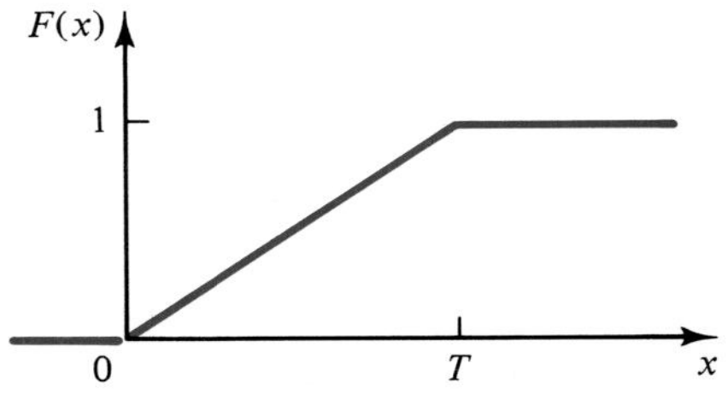

specified in terms of a function $\alpha(t)$ as in (2-43):

$$P\{t_1 \le t \le t_2\} = \int_{t_1}^{t_2} \alpha(t)\,dt$$

We shall assume that

$$\alpha(t) = 2e^{-2t}$$

This yields

$$P\{0 \le t \le t_0\} = 2\int_0^{t_0} e^{-2t}\,dt = 1 - e^{-2t_0} \tag{4-3}$$

(a) We form an RV $\mathbf{x}$ such that

$$\mathbf{x}(t) = t \qquad t \ge 0$$

Thus, as in Example 4.5, t is an outcome of the experiment $\mathcal{S}$ and the corresponding value of the RV $\mathbf{x}$. We shall show (Fig. 4.6a) that

$$F(x) = \begin{cases} 1 - e^{-2x} & x \ge 0 \\ 0 & x < 0 \end{cases} \tag{4-4}$$

If $x \ge 0$, then $\mathbf{x}(t) \le x$ for every t in the interval $(0, x)$; hence,

$$F_x(x) = P\{\mathbf{x} \le x\} = P\{0 \le t \le x\} = 1 - e^{-2x}$$

If $x < 0$, then $\{\mathbf{x} \le x\} = \varnothing$ because $\mathbf{x}(t) \ge 0$ for every t in $\mathcal{S}$; hence,

$$F_x(x) = P\{\mathbf{x} \le x\} = P(\varnothing) = 0$$

Note that whereas (4-3) has a meaning only for $t_0 \ge 0$, (4-4) is defined for all x.

(b) In the same experiment, we define the RV $\mathbf{y}$ such that

$$\mathbf{y}(t) = \begin{cases} 1 & 0 \le t \le 0.5 \\ 0 & t > 0.5 \end{cases}$$

Thus $\mathbf{y}$ takes the values 0 and 1 and

$$P\{\mathbf{y} = 0\} = P\{0 \le t \le 0.5\} = 1 - e^{-1}$$

$$P\{\mathbf{y} = 1\} = P\{t > 0.5\} = 2\int_{0.5}^{\infty} e^{-2t}\,dt = e^{-1}$$

From this it follows (Fig. 4.6b) that

$$F(y) = \begin{cases} 1 & y \ge 1 \\ 1 - e^{-1} & 0 \le y \le 1 \\ 0 & y < 0 \end{cases}$$

■

Figure 4.6

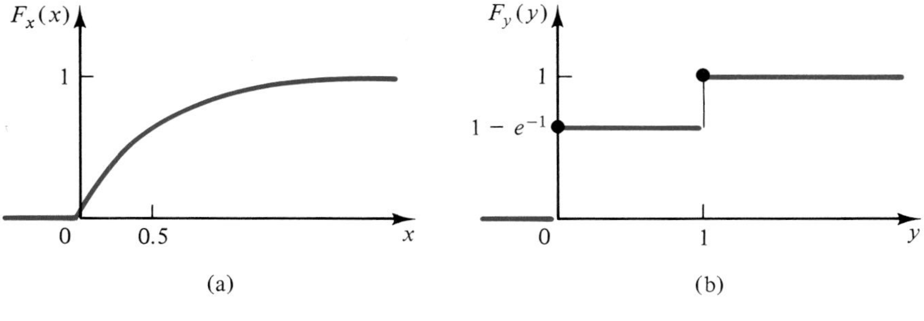

It is clear from the foregoing examples that if the experiment $\mathcal{S}$ consists of finitely many outcomes, $F(x)$ is a staircase function. This is also true if $\mathcal{S}$ consists of infinitely many outcomes but $\mathbf{x}$ takes finitely many values.

PROPERTIES OF DISTRIBUTIONS The following properties are simple consequences of (4-1) and (4-2).

1. $$F(-\infty) = P\{\mathbf{x} = -\infty\} = 0 \qquad F(\infty) = P\{\mathbf{x} \le \infty\} = 1 \tag{4-5}$$

2. The function $F(x)$ is monotonically increasing; that is,
$$\text{if} \quad x_1 < x_2 \quad \text{then} \quad F(x_1) \le F(x_2) \tag{4-6}$$

■ ***Proof.*** If $x_1 < x_2$ and ζ is such that $\mathbf{x}(\zeta) \le x_1$, then $\mathbf{x}(\zeta) \le x_2$; hence, the event $\{\mathbf{x} < x_1\}$ is a subset of the event $\{\mathbf{x} \le x_2\}$. This yields
$$F(x_1) = P\{\mathbf{x} < x_1\} \le P\{\mathbf{x} < x_2\} = F(x_2) \tag{4-7}$$
From (4-5) and (4-7) it follows that
$$0 \le F(x) \le 1 \tag{4-8}$$
Furthermore,
$$\text{if} \quad F(x_0) = 0 \quad \text{then} \quad F(x) = 0 \quad \text{for every} \quad x \le x_0 \tag{4-9}$$

3. $$P\{\mathbf{x} > x\} = 1 - F(x) \tag{4-10}$$

■ ***Proof.*** For a specific ζ, the events $\{\mathbf{x} \le x\}$ and $\{\mathbf{x} > x\}$ are mutually exclusive, and their union equals $\mathcal{S}$. Hence,
$$P\{\mathbf{x} \le x\} + P\{\mathbf{x} > x\} = P(\mathcal{S}) = 1$$

4. $$P\{x_1 < \mathbf{x} \le x_2\} = F(x_2) - F(x_1) \tag{4-11}$$

■ ***Proof.*** The events $\{\mathbf{x} \le x_1\}$ and $\{x_1 < \mathbf{x} \le x_2\}$ are mutually exclusive, and their union is the event $\{\mathbf{x} \le x_2\}$:
$$\{\mathbf{x} \le x_1\} \cup \{x_1 < \mathbf{x} \le x_2\} = \{\mathbf{x} \le x_2\}$$
This yields
$$P\{\mathbf{x} \le x_1\} + P\{x_1 < \mathbf{x} \le x_2\} = P\{\mathbf{x} \le x_2\}$$
and (4-11) results.

5. The function $F(x)$ might be continuous or discontinuous. We shall examine its behavior at or near a discontinuity point. Consider first the die experiment of Example 4.3a. Clearly, $F(x)$ is discontinuous at $x = 30$, and
$$F(29.99) = \frac{2}{6} \qquad F(30) = \frac{3}{6} \qquad F(30.01) = \frac{3}{6}$$
Thus the value of $F(x)$ for $x = 30$ equals its value for x near 30 to the right but is different from its value for x near 30 to the left. Furthermore, the discontinuity jump of $F(x)$ from 2/6 to 3/6 equals the probability $P\{\mathbf{x} = 30\} = 1/6$. We maintain that this is true in general.

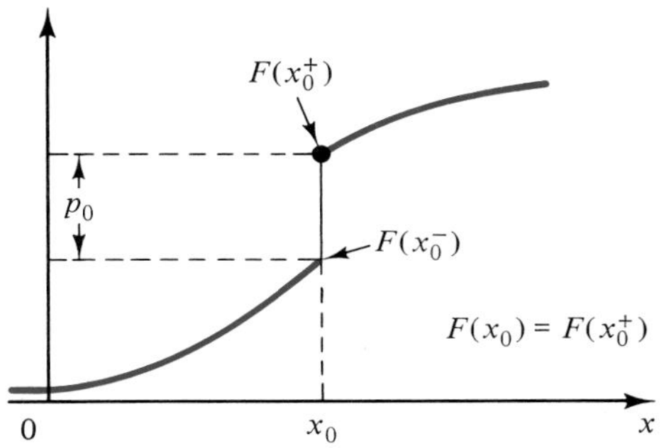

Figure 4.7

Suppose that $F(x)$ is discontinuous at the point $x = x_0$. We denote by $F(x_0^+)$ and $F(x_0^-)$ the limit of $F(x)$ as x approaches x_0 from the right and left, respectively (Fig. 4.7). The difference

$$p_0 = F(x_0^+) - F(x_0^-) \qquad (4\text{-}12)$$

is the "discontinuity jump" of $F(x)$ at the point x_0. As we see from the figure,

$$P\{\mathbf{x} < x_0\} = F(x_0^-) \qquad P\{\mathbf{x} \le x_0\} = F(x_0^+) \qquad (4\text{-}13)$$

This shows that

$$F(x_0) = F(x_0^+) \qquad (4\text{-}14)$$

Note, finally, that the events $\{\mathbf{x} < x_0\}$ and $\{\mathbf{x} = 0\}$ are mutually exclusive, and their union equals $\{\mathbf{x} \le x\}$. This yields

$$P\{\mathbf{x} < x_0\} + P\{\mathbf{x} = x_0\} = P\{\mathbf{x} \le x_0\}$$

From this and (4-13) it follows that

$$P\{\mathbf{x} = x_0\} = F(x_0) - F(x_0^-) = p_0 \qquad P\{\mathbf{x} < x_0\} = F(x_0^-) \qquad (4\text{-}15)$$

Continuous, Discrete, and Mixed Type RVS We shall say that an RV $\mathbf{x}$ is of *continuous type* if its distribution is continuous for every x (Fig. 4.8). In this case, $F(x^-) = F(x) = F(x^+)$; hence,

$$P\{\mathbf{x} = x\} = 0 \qquad (4\text{-}16)$$

Figure 4.8

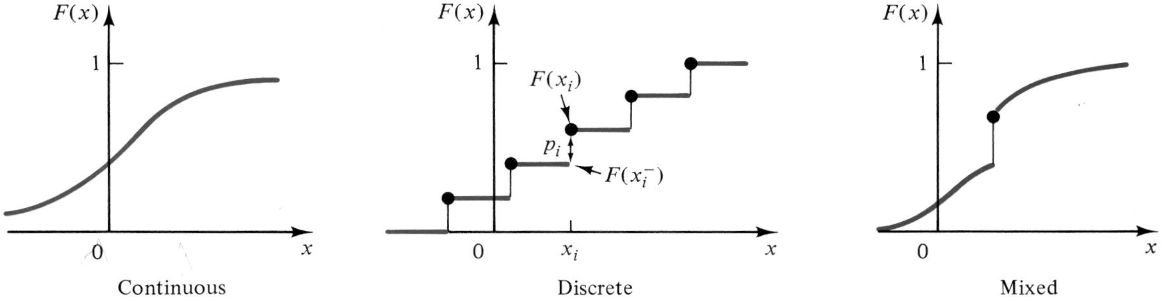

for every x. Thus if $\mathbf{x}$ is of continuous type, the probability that it equals a specific number x is zero for every x. We note that in this case,

$$P\{\mathbf{x} \le x\} = P\{\mathbf{x} < x\} = F(x)$$
$$P\{x_1 \le \mathbf{x} \le x_2\} = P\{x_1 < \mathbf{x} \le x_2\} = F(x_2) - F(x_1) \tag{4-17}$$

We shall say that an RV $\mathbf{x}$ is of *discrete type* if its distribution is a staircase function. Denoting by x_i the discontinuity points of $F(x)$ and by p_i the jumps at x_i, we conclude as in (4-12) and (4-14) that

$$P\{\mathbf{x} = x_i\} = F(x_i) - F(x_i^-) = p_i \tag{4-18}$$

Since $F(-\infty) = 0$ and $F(\infty) = 1$, it follows that if $F(x)$ has N steps, then

$$p_1 + \cdots + p_N = 1 \tag{4-19}$$

Thus if $\mathbf{x}$ is of discrete type, it takes the values x_i with probabilities p_i. It might take also other values; however, the set of the corresponding outcomes has zero probability.

We shall say that an RV $\mathbf{x}$ is of *mixed type* if its distribution is discontinuous but not a staircase.

If an experiment $\mathcal{S}$ has finitely many outcomes, any RV $\mathbf{x}$ defined on $\mathcal{S}$ is of discrete type. However, an RV $\mathbf{x}$ might be of discrete type even if $\mathcal{S}$ has infinitely many outcomes. The next example is an illustration.

Example 4.7 Suppose that $\mathcal{A}$ is an event of an arbitrary experiment $\mathcal{S}$. We shall say that $\mathbf{x}_{\mathcal{A}}$ is the *zero-one* RV associated with the event $\mathcal{A}$ if

$$\mathbf{x}_{\mathcal{A}}(\zeta) = \begin{cases} 1 & \zeta \subset \mathcal{A} \\ 0 & \zeta \notin \mathcal{A} \end{cases}$$

Thus $\mathbf{x}_{\mathcal{A}}$ takes the values 0 and 1 (Fig. 4.9), and

$$P\{\mathbf{x}_{\mathcal{A}} = 1\} = P(\mathcal{A}) = p \qquad P\{\mathbf{x}_{\mathcal{A}} = 0\} = P(\bar{\mathcal{A}}) = 1 - p$$ ■

The Percentile Curve The distribution function equals the probability $u = F(x)$ that the RV $\mathbf{x}$ does not exceed a given number x. In many cases, we are faced with the inverse problem. We are given u and wish to find the value x_u of x such that $P\{\mathbf{x} \le x_u\} = u$. Clearly, x_u is a number that depends on u, and it

Figure 4.9

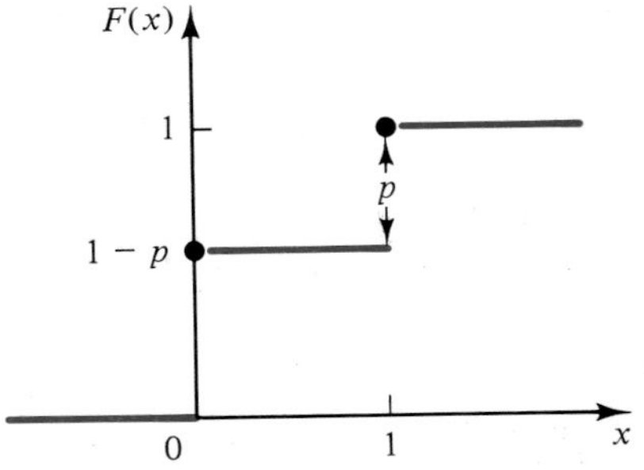

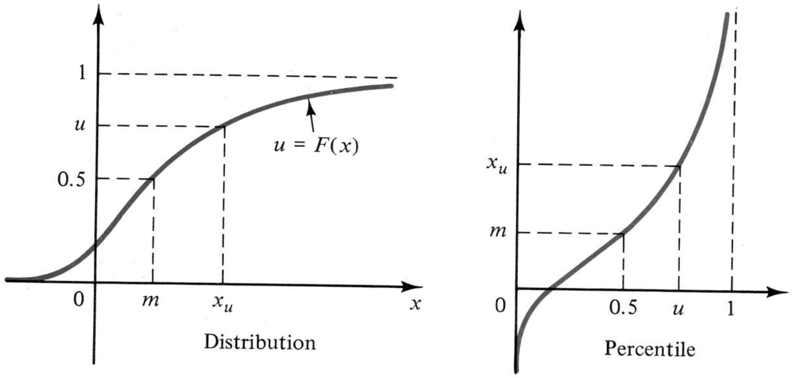

Figure 4.10

is found by solving the equation

$$F(x_u) = u \tag{4-20}$$

Thus x_u is a function of u called the *u-percentile* (or *quantile* or *fractile*) of the RV $\mathbf{x}$. Empirically, this means that $100u\%$ of the observed values of $\mathbf{x}$ do not exceed the number x_u. The function x_u is the inverse of the function $u = F(x)$. To find its graph, we interchange the axes of the $F(x)$ curve (Fig. 4.10). The domain of x_u is the interval $0 \le u \le 1$, and its range is the x-axis $-\infty \le x \le \infty$.

Note that if the function $F(x)$ is tabulated, we use interpolation to find the values of x_u for specific values of u. Suppose that u is between the tabulated numbers u_a and u_b:

$$F(x_a) = u_a < u < u_b = F(x_b)$$

The corresponding x_u is obtained by the straight line approximation

$$x_u \simeq x_a + \frac{x_b - x_a}{u_b - u_a}(u - u_a) \tag{4-21}$$

of $F(x)$ in the interval (x_a, x_b). In Fig. 4.11, we demonstrate the determina-

Figure 4.11

x	$F(x)$	u	x_u
1.60	.94520	.95	1.64
1.65	.95053		
1.95	.97441	.975	1.96
2.00	.97725		
2.30	.98928	.99	2.33
2.35	.99061		

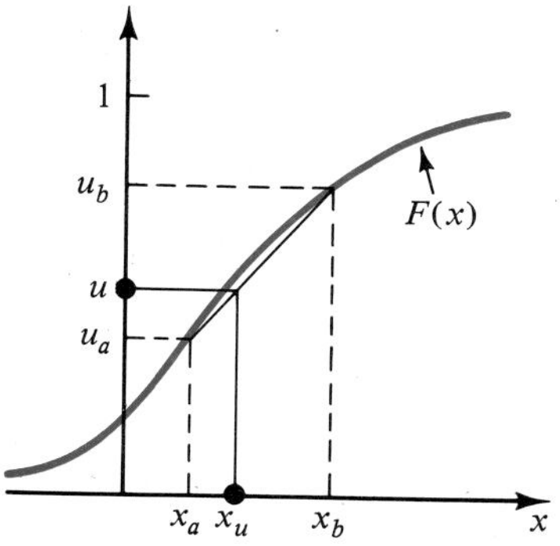

tion of x_u for u = .95, .975, and .99 where we use for $F(x)$ the standard normal curve $G(x)$ [see (3-23)].

Median The .5-percentile $x_{.5}$ is of particular interest. It is denoted by m and is called the *median* of **x**. Thus

$$F(m) = .5 \qquad m = x_{.5}$$

The Empirical Distribution. We shall now give the relative frequency interpretation of the function $F(x)$. To do so, we perform the experiment n times and denote by ξ_i the observed outcome at the ith trial. We thus obtain a sequence

$$\xi_1, \ldots, \xi_i, \ldots, \xi_n \tag{4-22}$$

of n outcomes where ξ_i is one of the elements ζ of $\mathscr{S}$. The RV **x** provides a rule for assigning to each element ζ of $\mathscr{S}$ a number $\mathbf{x}(\zeta)$. The sequence of outcomes (4-22) therefore generates the sequence of numbers

$$x_1, \ldots, x_i, \ldots, x_n \tag{4-23}$$

where $x_i = \mathbf{x}(\xi_i)$ is the value of the RV **x** at the ith trial. We place the numbers x_i on the x-axis and form a staircase function $F_n(x)$ consisting of n steps, as in Fig. 4.12. The steps are located at the points x_i (identified by dots), and their height equals $1/n$. The first step is at the smallest value x_{min} of x_i and the last at the largest value x_{max}. Thus

$$F_n(x) = 0 \qquad \text{for} \qquad x < x_{min}$$
$$F_n(x) = 1 \qquad \text{for} \qquad x \geq x_{max}$$

The function $F_n(x)$ so constructed is called the *empirical distribution* of the RV **x**.

As n increases, the number of steps increases, and their height $1/n$ tends to zero. We shall show that for sufficiently large n,

$$F_n(x) \simeq F(x) \tag{4-24}$$

in the sense of (1-1). We denote by n_x the number of trials such that $x_i \leq x$. Thus n_x is the number of steps of $F_n(x)$ to the left of x; hence,

$$F_n(x) = \frac{n_x}{n} \tag{4-25}$$

Figure 4.12

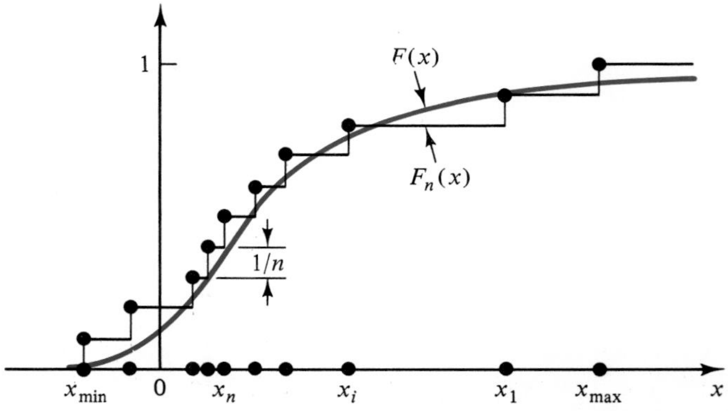

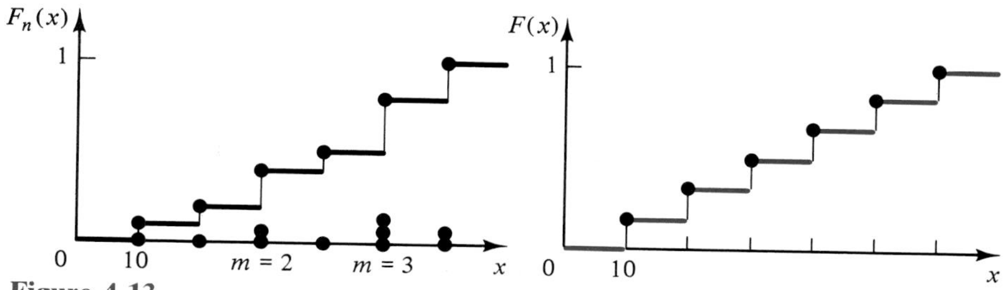

Figure 4.13

As we know, $\{\mathbf{x} \le x\}$ is an event with probability $F(x)$. This event occurs at the ith trial iff $x_i \le x$. From this it follows that n_x is the number of successes of the event $\{\mathbf{x} \le x\}$ in n trials. Applying (1-1) to this event, we obtain

$$F(x) = P\{\mathbf{x} \le x\} \simeq \frac{n_x}{n} = F_n(x)$$

Thus the empirical function $F_n(x)$ can be used to estimate the conceptual function $F(x)$ (see also Section 9-4).

In the construction of $F_n(x)$, we assumed that the numbers x_i are all different. This is most likely the case if $F(x)$ is continuous. If, however, the RV $\mathbf{x}$ is of discrete type taking the N values c_k, then $x_i = c_k$ for some k. In this case, the steps of $F_n(x)$ are at the points c_k, and the height of each step equals m/n where m is the multiplicity of the numbers x_i that equal c_k (Fig. 4.13).

Example 4.8

We roll a fair die 10 times and observe the outcomes

$$f_1 f_5 f_6 f_4 f_5 f_2 f_6 f_3 f_5 f_3$$

The corresponding values of the RV $\mathbf{x}$ defined as in Example 4.1 are

10 50 60 40 50 20 60 30 50 30

In Fig. 4.13 we show the distribution $F(x)$ and the empirical distribution $F_n(x)$. ■

The Empirical Percentile (Quetelet Curve). Using the n numbers x_i in (4-23), we form n segments of length x_i. We place them in line parallel to the y-axis in order of increasing length, distance $1/n$ apart (Fig. 4.14). If, for example, $\mathbf{x}$ is the length of pine needles, the segments are n needles selected at random. We then form a polygon the corners of which are the endpoints of the segments. For sufficiently large n, this polygon approaches the u-percentile curve x_u of the RV $\mathbf{x}$.

The Density Function

We can use the distribution function to determine the probability $P\{\mathbf{x} \in R\}$ that the RV $\mathbf{x}$ takes values in an arbitrary region R of the real axis. To do so, we express R as a union of nonoverlapping intervals and apply (4-11). We show next that the result can be expressed in terms of the derivative $f(x)$ of $F(x)$. We shall assume, first, that $F(x)$ is continuous and that its derivative exists nearly everywhere.

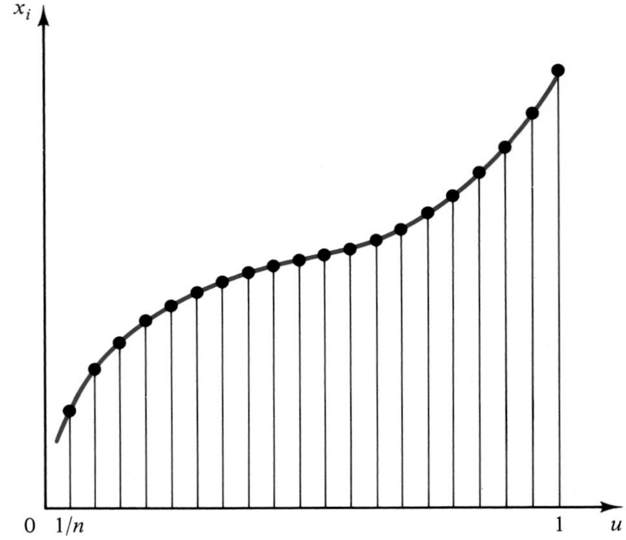

Figure 4.14

■ ***Definition.*** The derivative

$$f(x) = \frac{dF(x)}{dx} \tag{4-26}$$

of $F(x)$ is called the *probability density function* (p.d.f.) or the *frequency function* of the RV $\mathbf{x}$. We discuss next various properties of $f(x)$.

Since $F(x)$ increases as x increases, we conclude that

$$f(x) \geq 0 \tag{4-27}$$

Integrating (4-26) from x_1 to x_2, we obtain

$$F(x_2) - F(x_1) = \int_{x_1}^{x_2} f(x)\,dx \tag{4-28}$$

With $x_2 = -\infty$, this yields

$$F(x) = \int_{-\infty}^{x} f(\xi)\,d\xi \tag{4-29}$$

because $F(-\infty) = 0$. Setting $x = \infty$, we obtain

$$\int_{-\infty}^{\infty} f(x)\,dx = 1 \tag{4-30}$$

Note, finally, that

$$\int_{x_1}^{x_2} f(x)\,dx = P\{x_1 \leq \mathbf{x} \leq x_2\} \tag{4-31}$$

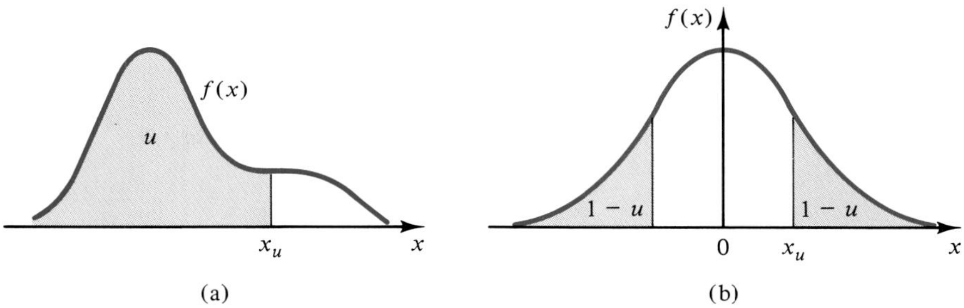

Figure 4.15

This follows from (4-28) and (4-17). Thus the area of $f(x)$ in an interval (x_1, x_2) equals the probability that $\mathbf{x}$ is in this interval.

If $x_1 = x$, $x_2 = x + \Delta x$, and Δx is sufficiently small, the integral in (4-31) is approximately equal to $f(x)\Delta x$; hence,

$$P\{x \le \mathbf{x} \le x + \Delta x\} \simeq f(x)\Delta x \tag{4-32}$$

From this it follows that the density $f(x)$ can be defined directly as a limit involving probabilities:

$$f(x) = \lim_{\Delta x \to 0} \frac{P\{x \le \mathbf{x} \le x + \Delta x\}}{\Delta x} \tag{4-33}$$

With x_u the u-percentile of $\mathbf{x}$, (4-29) yields (Fig. 4.15a)

$$u = F(x_u) = \int_{-\infty}^{x_u} f(x)\,dx$$

Note, finally, that if $f(x)$ is an *even* function—that is, if $f(-x) = f(x)$ (Fig. 4.15b)—then

$$1 - F(x) = F(-x) \qquad x_{1-u} = -x_u \tag{4-34}$$

From this and the table in Fig. 4.11 it follows that if $F(x)$ is a normal distribution, then

$$x_{.05} = -x_{.95} = -1.6 \qquad x_{.01} = -x_{.99} = -2.3$$

DISCRETE TYPE RVS Suppose now that $F(x)$ is a staircase function with discontinuities at the points x_k. In this case, the RV $\mathbf{x}$ takes the values x_k with probability

$$P\{\mathbf{x} = x_k\} = p_k = F(x_k) - F(x_k^-) \tag{4-35}$$

The numbers p_k will be represented graphically either in terms of $F(x)$ or by vertical segments at the points x_k with height equal to p_k (Fig. 4.16). Occasionally, we shall also use the notation

$$p_k = f(x_k)$$

to specify the probabilities p_k. The function $f(x)$ so defined will be called *point density*. It should be understood that its values $f(x_k)$ are not the derivatives of $F(x)$; they equal the discontinuity jumps of $F(x)$.

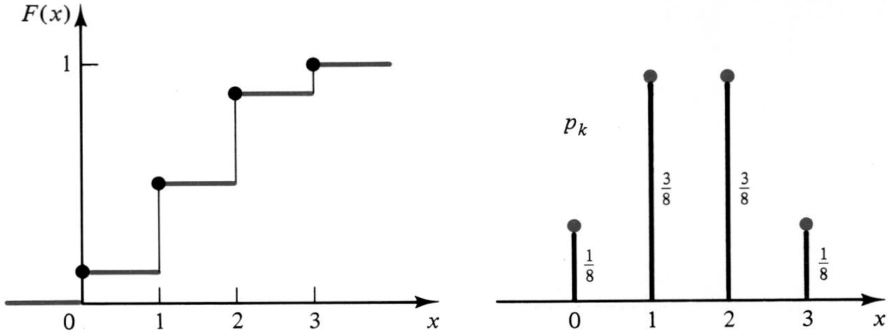

Figure 4.16

Example 4.9

The experiment $\mathcal{S}$ is the toss of a fair coin three times. In this case, $\mathcal{S}$ has eight outcomes, as in Example 3.1. We define the RV $\mathbf{x}$ such that its value at a specific outcome equals the number of heads in that outcome. Thus $\mathbf{x}$ takes the values 0, 1, 2, and 3, and

$$P\{\mathbf{x} = 0\} = \frac{1}{8} \qquad P\{\mathbf{x} = 1\} = \frac{3}{8} \qquad P\{\mathbf{x} = 2\} = \frac{3}{8} \qquad P\{\mathbf{x} = 3\} = \frac{1}{8}$$

In Fig. 4.16, we show its distribution $F(x)$ and the probabilities p_k. ■

The Empirical Density (Histogram). We have performed an experiment n times, and we obtained the n values x_i of the RV $\mathbf{x}$. In Fig. 4.12, we placed the numbers x_i on the x-axis and formed the empirical curve $F_n(x)$. In many cases, this is too detailed; what is needed is not the exact values of x_i but their number in various intervals of the x-axis. For example, if $\mathbf{x}$ represents yearly income, we might wish to know only the number of persons in various income brackets. To display such information graphically, we proceed as follows.

We divide the x-axis into intervals of length Δ, and we denote by n_k the number of points x_i that are in the kth interval. We then form a staircase function $f_n(x)$, as in Fig. 4.17. The kth step is in the kth interval $(c_k, c_k + \Delta)$, and

Figure 4.17

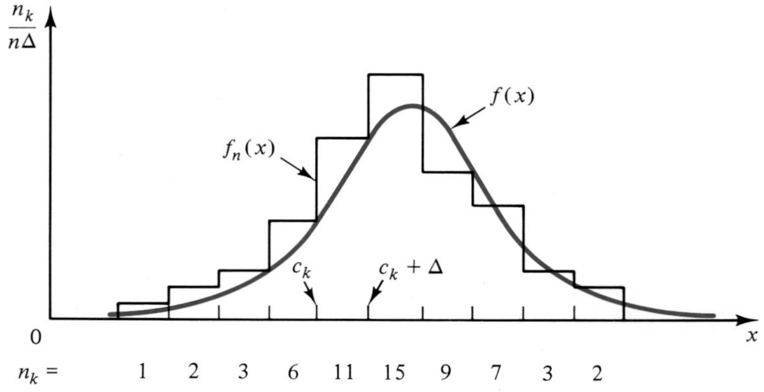

its height equals $n_k/n\Delta$. Thus

$$f_n(x) = \frac{n_k}{n\Delta} \qquad c_k \le x \le c_k + \Delta \tag{4-36}$$

The function $f_n(x)$ is called the *histogram* of the RV $\mathbf{x}$. The histogram is used to describe economically the data x_i. We show next that if n is large and Δ small, $f_n(x)$ approaches the density $f(x)$:

$$f_n(x) \simeq f(x) \tag{4-37}$$

Indeed, the event $\{c_k \le \mathbf{x} < c_k + \Delta\}$ occurs n_k times in n trials, and its probability equals $f(c_k)\Delta$ [see (4-32)]. Hence,

$$f(c_k)\Delta \simeq P\{c_k \le \mathbf{x} < c_k + \Delta\} \simeq \frac{n_k}{n} = f_n(x)\Delta$$

Probability Mass Density In Section 2-2 we interpreted the probability $P(\mathscr{A})$ of an event $\mathscr{A}$ as mass associated with $\mathscr{A}$. We shall now give a similar interpretation of the distribution and the density of an RV $\mathbf{x}$. The function $F(x)$ equals the probability of the event $\{\mathbf{x} \le x\}$; hence, $F(x)$ can be interpreted as mass along the x-axis from $-\infty$ to x. The difference $F(x_2) - F(x_1)$ is the mass in the interval (x_1, x_2), and the difference $F(x + \Delta x) - F(x) \simeq f(x)\Delta x$ is the mass in the interval $(x, x + \Delta x)$. From this it follows that $f(x)$ can be interpreted as mass density. If $\mathbf{x}$ is of discrete type, taking the values x_k with probability p_k, then the probabilities are point masses p_k located at x_k. Finally, if $\mathbf{x}$ is of mixed type, it has distributed masses with density $f(x)$, where $F'(x)$ exists, and point masses at the discontinuities of $F(x)$.

4-3 Illustrations

Now we shall introduce various RVs with specified distributions. It might appear that to do so, we need to start with the specification of the underlying experiment. We shall show, however, that this is not necessary. Given a distribution $\Phi(x)$, we shall construct an experiment $\mathscr{S}$ and an RV $\mathbf{x}$ such that its distribution equals $\Phi(x)$.

FROM THE DISTRIBUTION TO THE MODEL We are given a function $\Phi(x)$ having all the properties of a distribution: It increases monotonically from 0 to 1 as x increases from $-\infty$ to ∞, and it is continuous from the right. Using this function, we construct an experimental model $\mathscr{S}$ as follows.

The outcomes of $\mathscr{S}$ are all points on the t-axis. The events of $\mathscr{S}$ are all intervals and their unions and intersections. The probability of the event $\{t_1 \le t \le t_2\}$ equals

$$P\{t_1 \le t \le t_2\} = \Phi(t_2) - \Phi(t_1) \tag{4-38}$$

This completes the specification of $\mathscr{S}$.

We next form an RV $\mathbf{x}$ with domain the space $\mathscr{S}$ and distribution the given function $\Phi(x)$. To do so, we set

$$\mathbf{x}(t) = t \tag{4-39}$$

Thus t has a dual meaning: It is an element of $\mathcal{S}$ identified by the letter t, and it is the value of the RV $\mathbf{x}$ corresponding to this element. For a given x, the event $\{\mathbf{x} \leq x\}$ consists of all elements t such that $\mathbf{x}(t) \leq x$. Since $\mathbf{x}(t) = t$, we conclude that $\{\mathbf{x} \leq x\} = \{t \leq x\}$; hence [see (4-38)],

$$F(x) = P\{\mathbf{x} \leq x\} = P\{t \leq x\} = \Phi(x) \tag{4-40}$$

Note that $\mathcal{S}$ is the entire t-axis even if $\Phi(x)$ is a staircase function. In this case, however, all probability masses are at the discontinuity points of $\Phi(x)$. All other points of the t-axis form a set with zero probability.

We have thus constructed an experiment specified in terms of an arbitrary function $\Phi(x)$. This experiment is, of course, only a theoretical model. Whether it can be used as the model of a real experiment is another matter. In the following illustrations, we shall often identify also various physical problems generating specific idealized distributions.

Fundamental Note. From the foregoing construction it follows that in the study of a *single* RV $\mathbf{x}$, we can avoid the notion of an abstract space. We can assume in all cases that the underlying experiment $\mathcal{S}$ is the real line and its outcomes are the value x of $\mathbf{x}$. This approach is taken by many authors. We believe, however, that it is preferable to differentiate between experimental outcomes and values of the RV $\mathbf{x}$ and to interpret all RVs as functions with domain an abstract set of objects, limiting the real line to special cases. One reason we do so is to make clear the conceptual difference between outcomes and RVs. The other reason involves the study of several, possibly noncountably many RVs (stochastic processes). If we use the real line approach, we must consider spaces with many, possibly infinitely many, coordinates. It is conceptually much simpler, it seems to us, to define all RVs as functions with domain an abstract set $\mathcal{S}$.

We shall use the following notational simplifications. First, we introduce the *step function* (Fig. 4.18):

$$U(x) = \begin{cases} 1 & x \geq 0 \\ 0 & x < 0 \end{cases} \tag{4-41}$$

This function will be used to identify distributions that equal zero for $x < 0$. For example, $f(x) = 2e^{-2x}U(x)$ will mean that $f(x) = 2e^{-2x}$ for $x \geq 0$ and $f(x) = 0$ for $x < 0$.

Figure 4.18

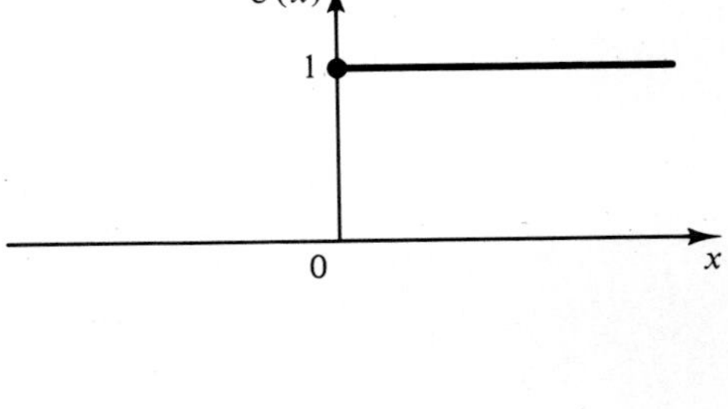

The notation

$$f(x) \sim \phi(x) \tag{4-42}$$

will mean that $f(x) = \gamma\phi(x)$ where γ is a factor that does not depend on x. If $f(x)$ is a density, then γ can be found from (4-30).

Normal We shall say that an RV $\mathbf{x}$ is *standard normal* or *Gaussian* if its density is the function (Fig. 4.19)

$$g(x) = \frac{1}{\sqrt{2\pi}} e^{-x^2/2}$$

introduced in (3-22). The corresponding distribution is the function

$$G(x) = \frac{1}{\sqrt{2\pi}} \int_{-\infty}^{x} e^{-\xi^2/2}\, d\xi$$

From the evenness of $g(x)$ it follows that

$$G(-x) = 1 - G(x) \tag{4-43}$$

Shifting and scaling $g(x)$, we obtain the general normal curves

$$f(x) = \frac{1}{\sigma\sqrt{2\pi}} e^{-(x-\eta)^2/2\sigma^2} = \frac{1}{\sigma} g\left(\frac{x-\eta}{\sigma}\right) \tag{4-44}$$

$$\boxed{F(x) = \frac{1}{\sigma\sqrt{2\pi}} \int_{-\infty}^{x} e^{-(\xi-\eta)^2/2\sigma^2}\, d\xi = G\left(\frac{x-\eta}{\sigma}\right)} \tag{4-45}$$

We shall use the notation $N(\eta, \sigma)$ to indicate that the RV $\mathbf{x}$ is normal, as in (4-44). Thus $N(0, 1)$ indicates that $\mathbf{x}$ is standard normal.

From (4-45) it follows that if $\mathbf{x}$ is $N(\eta, \sigma)$, then

$$P\{x_1 \le \mathbf{x} \le x_2\} = F(x_2) - F(x_1) = G\left(\frac{x_2-\eta}{\sigma}\right) - G\left(\frac{x_1-\eta}{\sigma}\right) \tag{4-46}$$

Figure 4.19

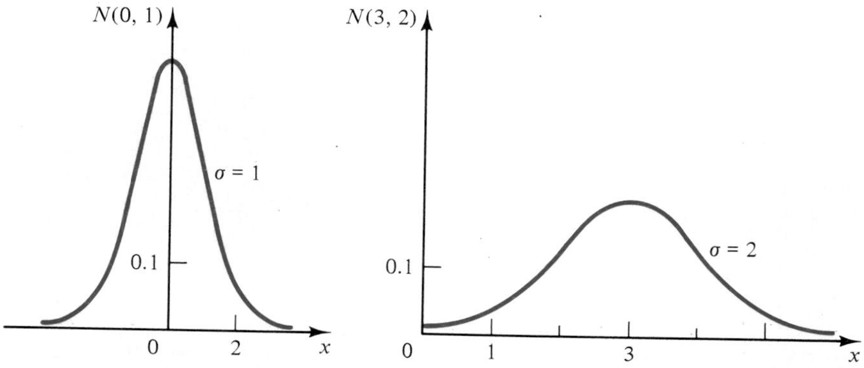

With $x_1 = \eta - k\sigma$, and $x_2 = \eta + k\sigma$, (4-46) yields

$$P\{\eta - k\sigma \le \mathbf{x} \le \eta + k\sigma\} = G(k) - G(-k) = 2G(k) - 1 \qquad (4\text{-}47)$$

This is the area of the normal curve (4-44) in the interval $(\eta - k\sigma, \eta + k\sigma)$.

The following special cases are of particular interest. As we see from Table 1a,

$$G(1) = .8413 \qquad G(2) = .9772 \qquad G(3) = .9987$$

Inserting into (4-47), we obtain

$$\begin{aligned} P\{\eta - \sigma < \mathbf{x} \le \eta + \sigma\} &\simeq .683 \\ P\{\eta - 2\sigma < \mathbf{x} \le \eta + 2\sigma\} &\simeq .954 \\ P\{\eta - 3\sigma < \mathbf{x} \le \eta + 3\sigma\} &\simeq .997 \end{aligned} \qquad (4\text{-}48)$$

We note further (see Fig. 4.11) that

$$\begin{aligned} P\{\eta - 1.96\sigma < \mathbf{x} \le \eta + 1.96\sigma\} &= .95 \\ P\{\eta - 2.58\sigma < \mathbf{x} \le \eta + 2.58\sigma\} &= .99 \\ P\{\eta - 3.29\sigma < \mathbf{x} \le \eta + 3.29\sigma\} &= .999 \end{aligned} \qquad (4\text{-}49)$$

In Fig. 4.20, we show the areas under the $N(\eta, \sigma)$ curve for the intervals in (4-48) and (4-49).

The normal distribution is of central importance in the theory and the applications of probability. It is a reasonable approximation of empirical distributions in many problems, and it is used even in cases involving RVs with domain a finite interval (a, b). In such cases, the approximation is possible if the normal curve is suitably truncated and scaled, or, if its area is negligible outside the interval (a, b).

Figure 4.20

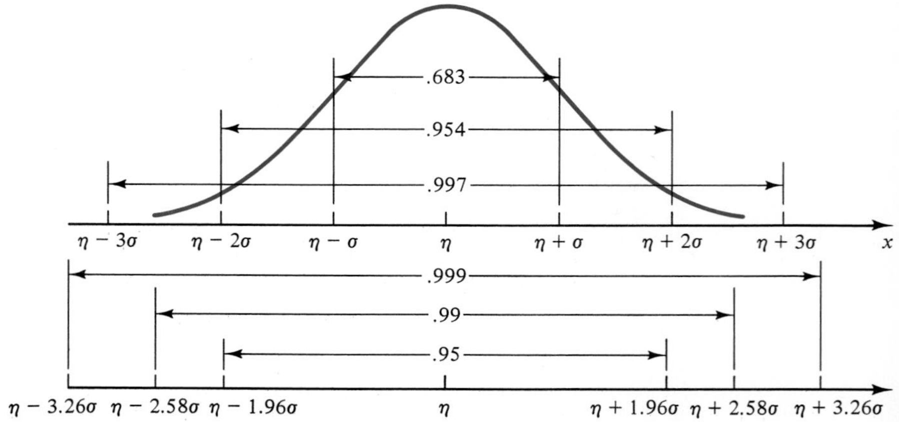

Example 4.10

The diameter of cylinders coming out of a production line are the values of a normal RV with $\eta = 10$ cm, $\sigma = 0.05$ cm.

(a) We set as tolerance limits the points 9.9, 10.1, and we reject all units outside the interval

$$(9.9,\ 10.1) = (\eta - 2\sigma,\ \eta + 2\sigma)$$

Find the percentage of the rejected units.

As we see from (4-48), $P\{9.9 < \mathbf{x} \le 10.1\} = .954$; hence, 4.6% of the units are rejected.

(b) We wish to find a tolerance interval $(10 - c,\ 10 + c)$ such that only 1% of the units will be rejected.

From (4-49) it follows that $P\{10 - c < \mathbf{x} \le 10 + c\} = .99$ for $c = 2.58\sigma = .129$ cm. Thus if we increase the size of the tolerance interval from 0.2 cm to 0.258 cm, we decrease the number of rejected components from 4.6% to 1%. ■

Uniform We shall say that an RV $\mathbf{x}$ is *uniform* (or uniformly distributed) in the interval $(a - c/2,\ a + c/2)$ if

$$f(x) = \begin{cases} \dfrac{1}{c} & a - \dfrac{c}{2} \le x \le a + \dfrac{c}{2} \\ 0 & \text{elsewhere} \end{cases}$$

The corresponding distribution is a ramp, as shown in Fig. 4.21.

Gamma The RV $\mathbf{x}$ has a *gamma distribution* if

$$f(x) = \gamma x^{b-1} e^{-cx} U(x) \qquad b > 0 \qquad c > 0 \tag{4-50}$$

The constant γ can be expressed in terms of the following integral:

$$\Gamma(\alpha) = \int_0^\infty y^{\alpha-1} e^{-y}\, dy \tag{4-51}$$

This integral converges for $\alpha > 0$, and it is called the *gamma function*. Clearly [see (3A-1)],

$$\begin{aligned} \Gamma(1) &= \int_0^\infty e^{-y}\, dy = 1 \\ \Gamma\left(\frac{1}{2}\right) &= \int_0^\infty \frac{1}{\sqrt{y}} e^{-y}\, dy = 2\int_0^\infty e^{-z^2}\, dz = \sqrt{\pi} \end{aligned} \tag{4-52}$$

Figure 4.21

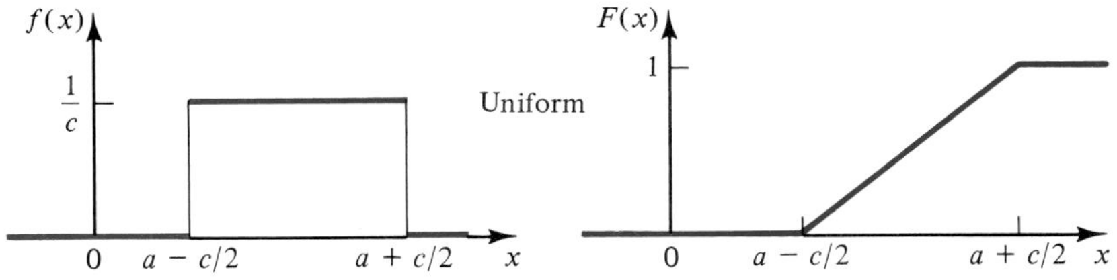

Replacing α by $\alpha + 1$ in (4-51) and integrating by parts, we obtain

$$\Gamma(\alpha + 1) = \alpha \int_0^\infty y^{\alpha-1} e^{-y}\, dy = \alpha\Gamma(\alpha) \tag{4-53}$$

This shows that if we know $\Gamma(\alpha)$ for $1 < \alpha < 2$ (Fig. 4.22), we can determine it recursively for any $\alpha > 0$. Note, in particular, that if $\alpha = n$ is an integer,

$$\Gamma(n + 1) = n\Gamma(n) = n(n - 1) \cdots \Gamma(1) = n!$$

For this reason, the gamma function is called *generalized factorial.*

With $y = cx$, (4-51) yields

$$\gamma \int_0^\infty x^{b-1} e^{-cx}\, dx = \frac{\gamma}{c^b} \int_0^\infty y^{b-1} e^{-y}\, dy = \frac{\gamma}{c^b}\Gamma(b) \tag{4-54}$$

And since the area of $f(x)$ equals 1, we conclude that

$$\gamma = \frac{c^b}{\Gamma(b)}$$

The gamma density has extensive applications. The following special cases are of particular interest.

Chi-square

$$f(x) = \frac{1}{2^{n/2}\Gamma\left(\frac{n}{2}\right)} x^{(n/2)-1} e^{-x/2} U(x) \qquad n\text{: integer}$$

Of central interest in statistics.

Erlang

$$f(x) = \frac{c^n}{(n-1)!} x^{n-1} e^{-cx} U(x) \qquad n\text{: integer}$$

Used in queueing theory, traffic, radioactive emission.

Figure 4.22

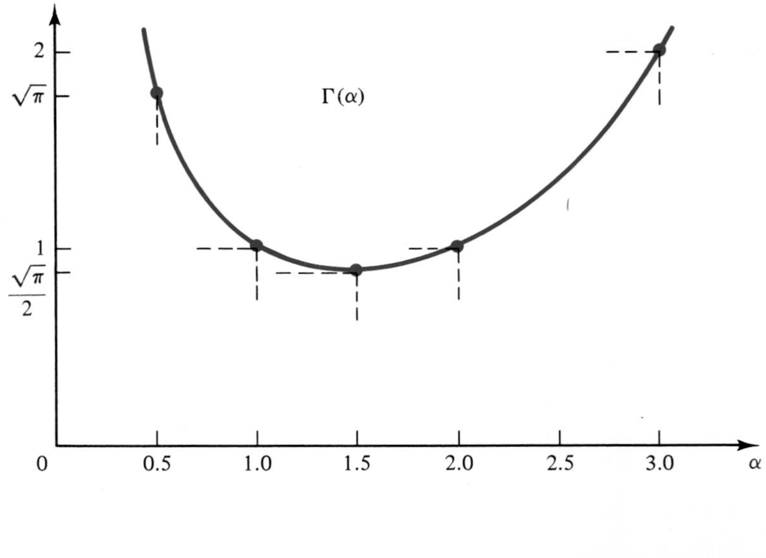

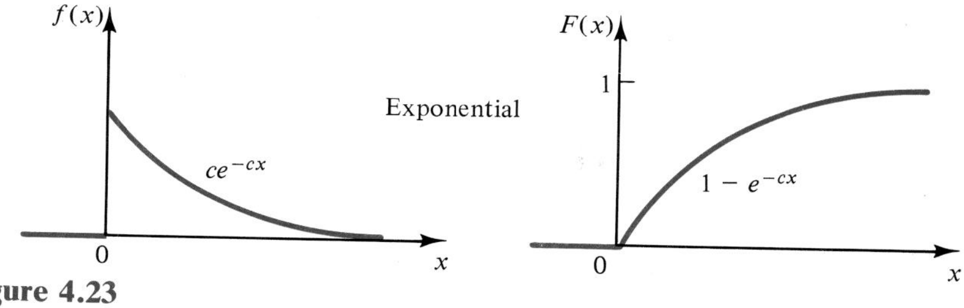

Figure 4.23

Exponential (Fig. 4.23)

$$f(x) = ce^{-cx}U(x) \qquad F(x) = (1 - e^{-cx})U(x)$$

Important in the study of Poisson points.

Cauchy We shall introduce this density in terms of the following experiment. A particle leaves the origin in a free motion. Its path is a straight line forming an angle θ with the horizontal axis (Fig. 4.24). The angle θ is selected at random in the interval $(-\pi/2, \pi/2)$. This specifies an experiment $\mathcal{S}$ the outcomes of which are all points in that interval. The probability of the event $\{\theta_1 \le \theta \le \theta_2\}$ equals

$$P\{\theta_1 \le \theta \le \theta_2\} = \frac{\theta_2 - \theta_1}{\pi}$$

as in (2-46). In this experiment, we define an RV $\mathbf{x}$ such that

$$\mathbf{x}(\theta) = a \tan \theta$$

Thus $\mathbf{x}(\theta)$ equals the ordinate of the point of intersection of the particle with the vertical line of Fig. 4.24. Clearly, the event $\{\mathbf{x} < x\}$ consists of all outcomes θ in the interval $(-\pi/2, \phi)$ where $x = a \tan \phi$; hence,

$$F(x) = P\{\mathbf{x} \le x\} = P\left\{-\frac{\pi}{2} \le \theta \le \phi\right\} = \frac{\phi + \pi/2}{\pi} = \frac{1}{2} + \frac{1}{\pi}\arctan\frac{x}{a}$$

Differentiating, we obtain the *Cauchy density:*

$$f(x) = \frac{a/\pi}{x^2 + a^2} \tag{4-55}$$

Figure 4.24

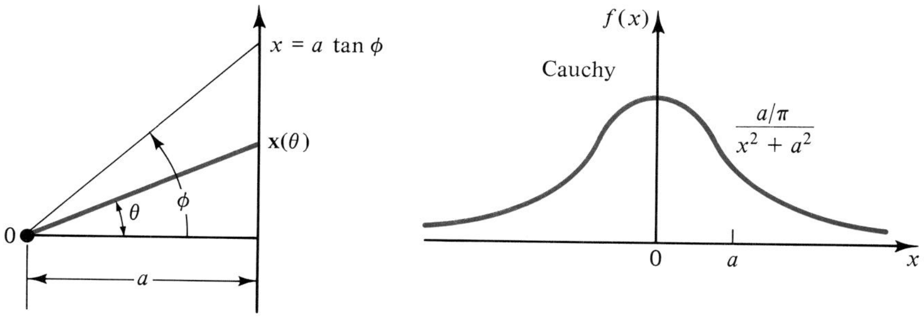

Binomial We shall say that an RV $\mathbf{x}$ has a *binomial distribution* of order n if it takes the values 0, 1, . . . , n with probabilities

$$P\{\mathbf{x} = k\} = \binom{n}{k} p^k q^{n-k} \qquad k = 0, 1, \ldots, n \tag{4-56}$$

Thus $\mathbf{x}$ is a discrete type RV, and its distribution $F(x)$ is a staircase function

$$F(x) = \sum_{k \le x} \binom{n}{k} p^k q^{n-k} \qquad q = 1 - p \tag{4-57}$$

with discontinuities at the points $x = k$ (Fig. 4.25). The density $f(x)$ of $\mathbf{x}$ is different from zero only for $x = k$, and

$$f(k) = \binom{n}{k} p^k q^{n-k} \qquad k = 0, 1, \ldots, n$$

The binomial distribution originates in the experiment $\mathcal{S}_n$ of repeated trials if we define an RV $\mathbf{x}$ equal to the number of successes of an event $\mathcal{A}$ in n trials. Suppose, first, that $\mathcal{S}_n$ is the experiment of the n tosses of a coin. An outcome of this experiment is a sequence

$$\xi = \xi_1 \cdots \xi_n$$

where ξ_i is h or t. We define $\mathbf{x}$ such that $\mathbf{x}(\xi) = k$ where k is the number of heads in ξ. Thus $\{\mathbf{x} = k\}$ is the event $\{k \text{ heads}\}$; hence, (4-56) follows from (3-8). Suppose, next, that $\mathcal{S}_n$ is the space of the n repetitions of an arbitrary experiment $\mathcal{S}$ and that $\mathcal{A}$ is an event of $\mathcal{S}$ with $P(\mathcal{A}) = p$. In this case, we set $\mathbf{x}(\xi) = k$ if ξ is an element of the event $\{\mathcal{A} \text{ occurs } k \text{ times}\}$, and (4-56) follows from (3-16).

Large n As we have seen in Chapter 3 (De Moivre–Laplace theorem), the binomial probabilities $f(k)$ approach the samples of a normal curve with

$$\eta = np \qquad \sigma = \sqrt{npq} \tag{4-58}$$

Thus [see (3-27)]

$$f(k) \simeq \frac{1}{\sqrt{2\pi npq}} e^{-(k-np)^2/2\sqrt{npq}} \qquad F(x) \simeq G\left(\frac{x - np}{\sqrt{npq}}\right) \tag{4-59}$$

Figure 4.25

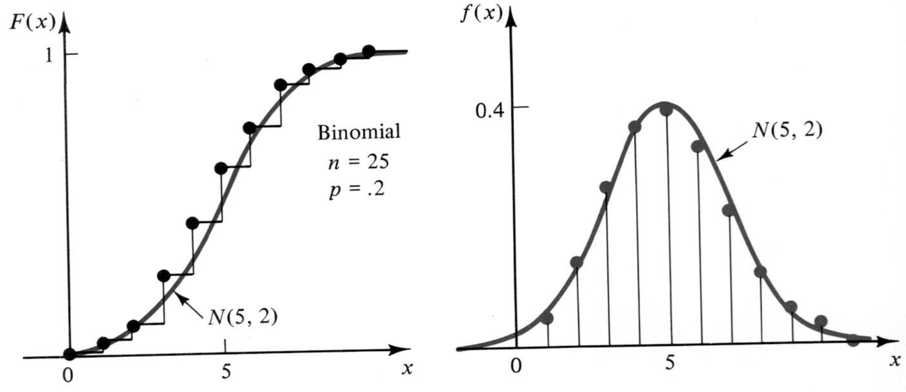

Note, however, the difference between a binomial and a normal RV. A binomial RV is of discrete type, and the function $f(x)$ is defined at $x = k$ only. Furthermore, $f(x)$ is not a density; its values $f(k)$ are probabilities.

The approximation (4-59) is satisfactory even for moderate values of n. In Fig. 4.25, we show the functions $F(x)$ and $f(x)$ and their normal approximations for $n = 25$ and $p = .2$. In this case, $\eta = np = 5$ and $\sigma = \sqrt{npq} = 2$. In the following table, we show the exact values of $f(k)$ and the corresponding values of the normal density $N(5, 2)$.

k	0	1	2	3	4	5	6	7	8	9	10	11
$f(k)$	.004	.024	.071	.136	.187	.196	.163	.111	.062	.029	.012	.004
$N(5, 2)$	.009	.027	.065	.121	.176	.199	.176	.121	.065	.027	.009	.002

Poisson An RV $\mathbf{x}$ has a *Poisson distribution* with parameter a if it takes the values 0, 1, 2, . . . with probability

$$P\{\mathbf{x} = k\} = e^{-a}\frac{a^k}{k!} \qquad k = 0, 1, 2, \ldots \tag{4-60}$$

The distribution of $\mathbf{x}$ is a staircase function

$$F(x) = e^{-a}\sum_{k\le x}\frac{a^k}{k!} \tag{4-61}$$

The discontinuity jumps of $F(x)$ form the sequence (Fig. 4.26)

$$f(k) = e^{-a}\frac{a^k}{k!} \qquad k = 0, 1, 2, \ldots \tag{4-62}$$

depending on the parameter a. We maintain that if $a < 1$, then $f(k)$ is maximum for $k = 0$ and decreases monotonically as $k \to \infty$. If $a > 1$ and a is

Figure 4.26

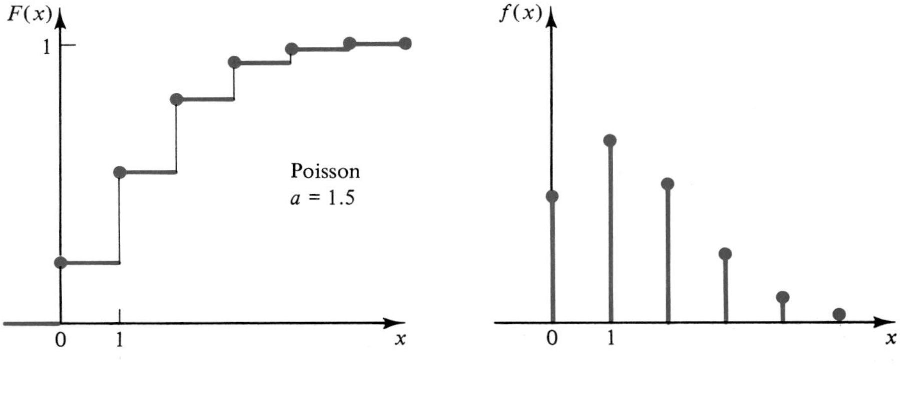

not an integer, $f(k)$ has a single maximum for $k = [a]$. If $a > 1$ and a is an integer, $f(k)$ has two maxima: for $k = a - 1$ and for $k = a$.

All this follows readily if we form the ratio of two consecutive terms in (4-62):

$$\frac{f(k-1)}{f(k)} = \frac{a^{k-1}(k-1)!}{a^k/k!} = \frac{k}{a}$$

Large *a* If $a \gg 1$, the Poisson distribution approaches a normal distribution with $\eta = a$, $\sigma = \sqrt{a}$:

$$e^{-a}\frac{a^k}{k!} \simeq \frac{1}{\sqrt{2\pi a}}e^{-(k-a)^2/2a} \qquad a \gg 1 \tag{4-63}$$

This is a consequence of the following: The binomial distribution (4-55) approaches the Poisson distribution with $a = np$ if $n \gg 1$ and $p \ll 1$ [see Poisson theorem (3-43)]. If $n \gg 1$, $p \ll 1$, and $np \gg 1$, both distributions tend to a normal curve, as in (4-63).

Poisson Points In Section 3-4 we introduced the space $\mathcal{S}_\infty$ of Poisson points specified in terms of the following properties:

1. The probability that there are k points in an interval (t_1, t_2) of length $t_a = t_2 - t_1$ equals

$$e^{-\lambda t_a}\frac{(\lambda t_a)^k}{k!} \qquad k = 0, 1, \ldots \tag{4-64}$$

   where λ is the "density" of the points.
2. If (t_1, t_2) and (t_3, t_4) are two nonoverlapping intervals, the events $\{k_a$ points in $(t_1, t_2)\}$ and $\{k_b$ points in $(t_3, t_4)]$ are independent.

Given an interval (t_1, t_2) as here, we define the RV $\mathbf{x}$ as follows: An outcome ζ of $\mathcal{S}_\infty$ is an infinite set of points on the real axis. If k of these points are in the interval (t_1, t_2), then $\mathbf{x}(\zeta) = k$. From (4-64) it follows that this RV is Poisson-distributed with parameter $a = \lambda t_a$ where λ is the density of the points and $t_a = t_2 - t_1$.

In the next example we show the relationship between Poisson points and exponential distributions.

Example 4.11 Given a set of Poisson points, identified by dots in Fig. 4.27, we select an arbitrary point 0 and denote by $\mathbf{w}$ the distance from 0 to the first Poisson point to the right of 0. We have thus created an RV $\mathbf{w}$ depending on the set of the Poisson points. We

Figure 4.27

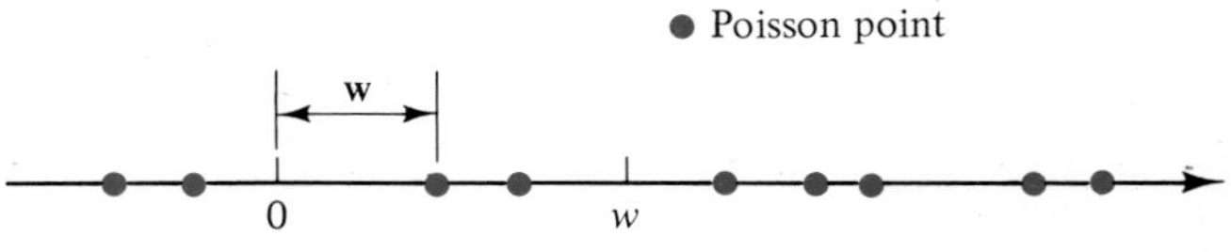

maintain that the RV **w** has an exponential density:

$$f_w(w) = \lambda e^{-\lambda w} U(w) \qquad F_w(w) = (1 - e^{-\lambda w})U(w) \tag{4-65}$$

■ ***Proof.*** It suffices to find the probability $P\{\mathbf{w} \le w\}$ of the event $\{\mathbf{w} \le w\}$ where w is a specified positive number. Clearly, $\mathbf{w} \le w$ iff there is at least one Poisson point in the interval $(0, w)$. We denote by **x** the number of Poisson points in the interval $(0, w)$. As we know, the RV **x** is Poisson-distributed with parameter λw. Hence,

$$P\{\mathbf{x} = 0\} = e^{-\lambda w} \qquad w > 0$$

And since $\{\mathbf{w} \le w\} = \{\mathbf{x} \ge 1\}$, we conclude that

$$F_w(w) = P\{\mathbf{w} \le w\} = P\{\mathbf{x} \ge 1\} = 1 - P\{\mathbf{x} = 0\} = 1 - e^{-\lambda w}$$

Differentiating, we obtain (4-65). ■

Geometric An RV **x** has a *geometric distribution* if it takes the values 1, 2, 3, . . . with probability

$$P\{\mathbf{x} = k\} = pq^{k-1} \qquad k = 1, 2, 3, \ldots \tag{4-66}$$

where $q = 1 - p$. This is a geometric sequence, and

$$\sum_{k=1}^{\infty} pq^{k-1} = \frac{p}{1-q} = 1$$

The geometric distribution has its origin in the following application of Bernoulli trials (see Example 3.5): Consider an event $\mathscr{A}$ of an experiment $\mathscr{S}$ with $P(\mathscr{A}) = p$. We repeat $\mathscr{S}$ an infinite number of times, and we denote by **x** the number of trials until the event $\mathscr{A}$ occurs for the *first* time. Clearly, **x** is an RV defined in the space $\mathscr{S}_\infty$ of Bernoulli trials, and, as we have shown in (3-11), it has a geometric distribution.

Hypergeometric The RV **x** has a *hypergeometric* distribution if it takes the values 0, 1, . . . , n with probabilities

$$P\{\mathbf{x} = k\} = \frac{\binom{K}{k}\binom{N-K}{n-k}}{\binom{N}{n}} \qquad k = 0, 1, \ldots, n \tag{4-67}$$

where N, K, and n are given numbers such that

$$n \le K \le N$$

Example 4.12

A set contains K red objects and $N - K$ black objects. We select $n \le K$ of these objects and denote by **x** the number of red objects among the n selections. As we see from (2-41), the RV **x** so formed has a hypergeometric distribution. ■

Example 4.13

We receive a shipment of 1,000 units, 200 of which are defective. We select at random from this shipment 25 units, test them, and accept the shipment if the defective units are at most 4. Find the probability p that the shipment is accepted.

The number of defective components is a hypergeometric RV **x** with

$$N = 1{,}000 \qquad K = 200 \qquad n = 25$$

The shipment is accepted if $\mathbf{x} \le 4$; hence,

$$p = \sum_{k=0}^{4} P\{\mathbf{x} = k\} = \frac{\sum_{k=0}^{4} \binom{200}{k}\binom{800}{25-k}}{\binom{1{,}000}{25}} = .419$$

This result is used in quality control. ■

4-4 Functions of One Random Variable

Recall from calculus that a *composite function* $y(t) = g(x(t))$ is a function $g(x)$ of another function $x(t)$. The domain of $y(t)$ is the t-axis. A function of an RV is an extension of this concept to functions with domain a probability space $\mathcal{S}$.

Given a function $g(x)$ of the real variable x and an RV **x** with domain the space $\mathcal{S}$, we form the composite function

$$\mathbf{y}(\zeta) = g(\mathbf{x}(\zeta)) \tag{4-68}$$

This function defines an RV **y** with domain the set $\mathcal{S}$. For a specific $\zeta_i \in \mathcal{S}$, the value $\mathbf{y}(\zeta_i)$ of the RV **y** so formed is given by $y_i = g(x_i)$, where $x_i = \mathbf{x}(\zeta_i)$ is the corresponding value of the RV **x** (Fig. 4.28). We have thus constructed a function $\mathbf{y} = g(\mathbf{x})$ of the RV **x**.

Distribution of $g(\mathbf{x})$

We shall express the distribution $F_y(y)$ of the RV **y** so formed in terms of the distribution $F_x(x)$ of the RV **x** and the function $g(x)$. We start with an example.

Figure 4.28

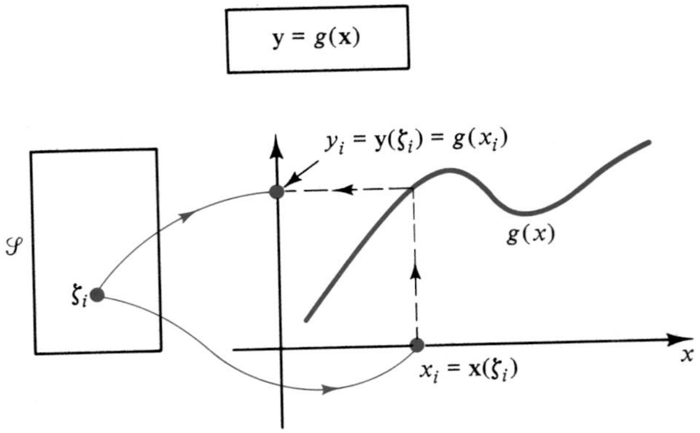

We shall find the distribution of the RV

$$\mathbf{y} = \mathbf{x}^2$$

starting with the determination of $F_y(4)$. Clearly, $\mathbf{y} \le 4$ iff $-2 \le \mathbf{x} \le 2$; hence, the events $\{\mathbf{y} \le 4\}$ and $\{-2 \le \mathbf{x} \le 2\}$ are equal. This yields

$$F_y(4) = P\{\mathbf{y} \le 4\} = P\{-2 \le \mathbf{x} \le 2\} = F_x(2) - F_x(-2)$$

We shall next find $F_y(-3)$. The event $\{\mathbf{y} \le -3\}$ consists of all outcomes such that $\mathbf{y}(\zeta) \le -3$. This event has no elements because $\mathbf{y}(\zeta) = \mathbf{x}^2(\zeta) \ge 0$ for every ζ; hence,

$$F_y(-3) = P\{\mathbf{y} \le -3\} = P(\varnothing) = 0$$

Suppose, finally, that $y \ge 0$ but is otherwise arbitrary. The event $\{\mathbf{y} \le y\}$ consists of all outcomes ζ such that the values $\mathbf{y}(\zeta)$ of the RV $\mathbf{y}$ are on the portion of the parabola $g(x) = x^2$ below the horizontal line L_y of Fig. 4.29. This event consists of all outcomes ζ such that $\{\mathbf{x}^2(\zeta) \le y\}$ where, we repeat, y is a specific positive number. Hence, if $y \ge 0$, then

$$F_y(y) = P\{\mathbf{y} \le y\} = P\{-\sqrt{y} \le \mathbf{x} \le \sqrt{y}\} = F_x(\sqrt{y}) - F_x(-\sqrt{y}) \quad (4\text{-}69)$$

If $y < 0$, then $\{\mathbf{y} \le y\}$ is the impossible event, and

$$F_y(y) = P\{\mathbf{y} \le y\} = P(\varnothing) = 0 \quad (4\text{-}70)$$

With $F_y(y)$ so determined, the density $f_y(y)$ of y is obtained by differentiating $F_y(y)$. Since

$$\frac{dF_x(\sqrt{y})}{dy} = \frac{1}{2\sqrt{y}} f_x(\sqrt{y})$$

Figure 4.29

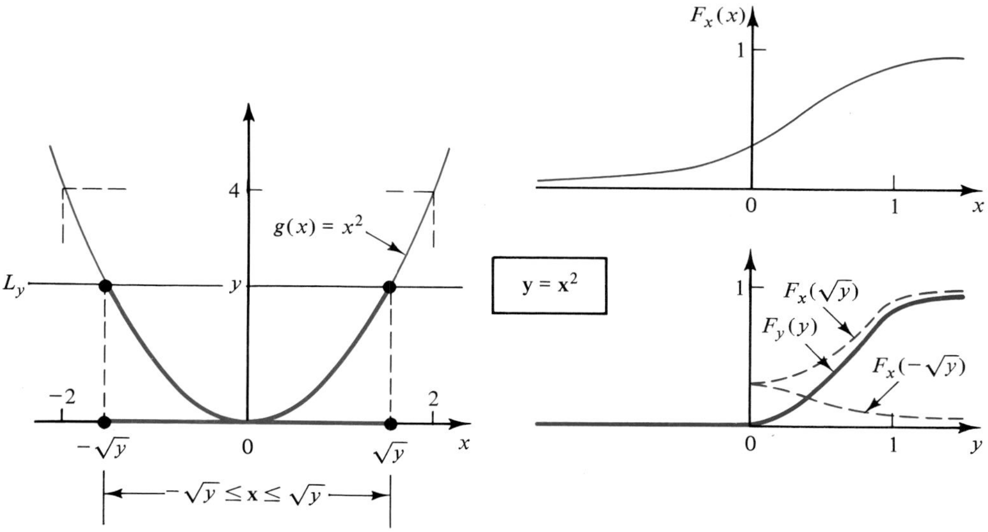

(4-69) and (4-70) yield

$$f_y(y) = \begin{cases} \dfrac{1}{2\sqrt{y}} f_x(\sqrt{y}) + \dfrac{1}{2\sqrt{y}} f_x(-\sqrt{y}) & y > 0 \\ 0 & y < 0 \end{cases} \tag{4-71}$$

Example 4.14

Suppose that

$$f_x(x) = \frac{1}{\sigma\sqrt{2\pi}} e^{-x^2/2\sigma^2} \qquad \text{and} \qquad \mathbf{y} = \mathbf{x}^2$$

Since $f_x(-x) = f_x(x)$, (4-71) yields

$$f_y(y) = \frac{1}{\sqrt{y}} f_x(\sqrt{y})U(y) = \frac{1}{\sigma\sqrt{2\pi y}} e^{-y/2\sigma^2} U(y)$$

■

We proceed similarly for an arbitrary $g(x)$. To find $F_y(y)$ for a specific y, we find the set I_y of points on the x-axis such that $g(x) \le y$. If $x \in I_y$, then $g(x)$ is below the horizontal line L_y (heavy in Fig. 4.30) and $\mathbf{y} \le y$. Hence,

$$F_y(y) = P\{\mathbf{y} \le y\} = P\{\mathbf{x} \in I_y\} \tag{4-72}$$

Thus to find $F_y(y)$ for a specific y, it suffices to find the set I_y such that $g(x) \le y$ and the probability that $\mathbf{x}$ is in this set. For y as in Fig. 4.30, I_y is the union of the half line $x \le x_1$ and the interval $x_2 \le x \le x_3$; hence, for that value of y,

$$\begin{aligned} F_y(y) &= P\{\mathbf{x} \in I_y\} = P\{\mathbf{x} \le x_1\} + P\{x_2 \le \mathbf{x} \le x_3\} \\ &= F_x(x_1) + F_x(x_3) - F_x(x_2) \end{aligned}$$

The function $g(x)$ of Fig. 4.30 is between the horizontal lines $y = y_a$ and $y = y_b$:

$$y_b < g(x) < y_a \qquad \text{for every } x \tag{4-73}$$

Since $\mathbf{y}(\zeta)$ is a point on that curve, we conclude that if $y > y_a$, then $\mathbf{y}(\zeta) \le y$ for every $\zeta \in \mathcal{S}$ and if $y < y_b$, then there is no ζ such that $\mathbf{y}(\zeta) \le y$. Hence,

Figure 4.30

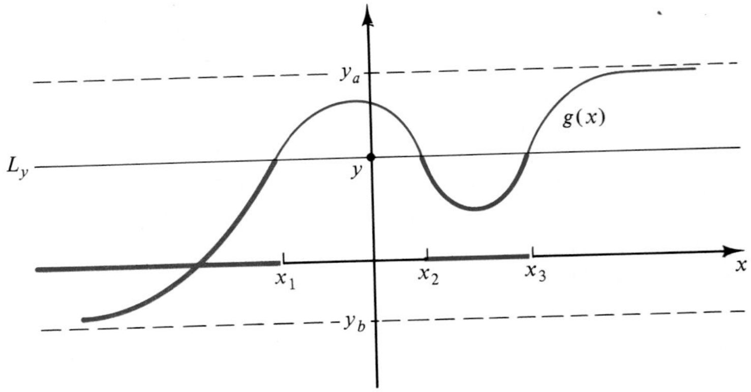

$$F_y(y) = \begin{cases} P(\mathcal{S}) = 1 & y > y_a \\ P(\varnothing) = 0 & y < y_b \end{cases} \tag{4-74}$$

This holds for any $g(x)$ satisfying (4-73).

Let us look at several examples. In all cases, the RV $\mathbf{x}$ is of continuous type. The examples are introduced not only because they illustrate the determination of the distribution $F_y(y)$ of the RV $\mathbf{y} = g(\mathbf{x})$ but also because the underlying reasoning contributes to the understanding of the meaning of an RV. In the determination of $F_y(y)$, it is essential to differentiate between the RV $\mathbf{y}$ and the number y.

Illustrations

1. *Linear transformation.* (See Fig. 4.31).

(a)
$$g(x) = \frac{x}{2} + 3$$

Clearly, $\mathbf{y} \le y$ iff $\mathbf{x} \le x_a = 2y - 6$ (Fig. 4.31a); hence,

$$F_y(y) = P\{\mathbf{x} \le x_a\} = F_x(2y - 6)$$

(b)
$$g(x) = -\frac{x}{2} + 8$$

In this case, $\mathbf{y} \le y$ iff $\mathbf{x} \ge x_b = 16 - 2y$ (Fig. 4.31b); hence,

$$F_y(y) = P\{\mathbf{x} \ge x_b\} = 1 - F_x(16 - 2y)$$

2. *Limiter.* (See Fig. 4.32.)

$$g(x) = \begin{cases} a & x > a \\ x & -a \le x \le a \\ -a & x < -a \end{cases}$$

If $y > a$, then $\{\mathbf{y} \le y\} = \mathcal{S}$; hence, $F_y(y) = 1$.

If $-a \le y \le a$, then $\{\mathbf{y} \le y\} = \{\mathbf{x} \le y\}$; hence, $F_y(y) = F_x(y)$.

If $y < -a$, then $\{\mathbf{y} \le y\} = \varnothing$; hence, $F_y(y) = 0$.

In this example, $F_y(y)$ is discontinuous at $y = \pm a$; hence, the RV $\mathbf{y}$ is of the mixed type, and

$$P\{\mathbf{y} = a\} = P\{\mathbf{x} \ge a\} = 1 - F_x(a)$$
$$P\{\mathbf{y} = -a\} = P\{\mathbf{x} \le -a\} = F_x(-a)$$

Figure 4.31

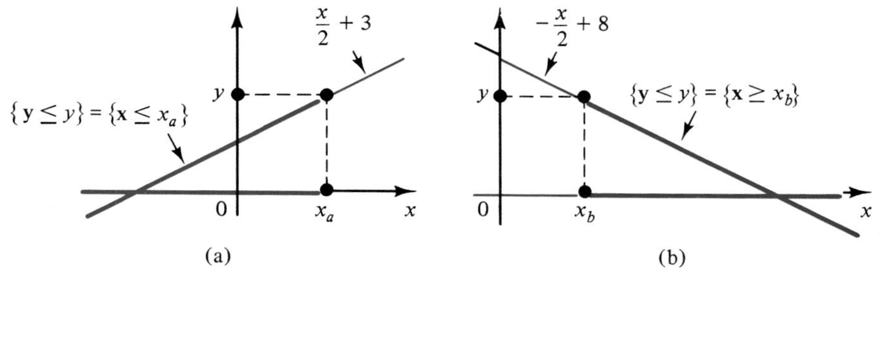

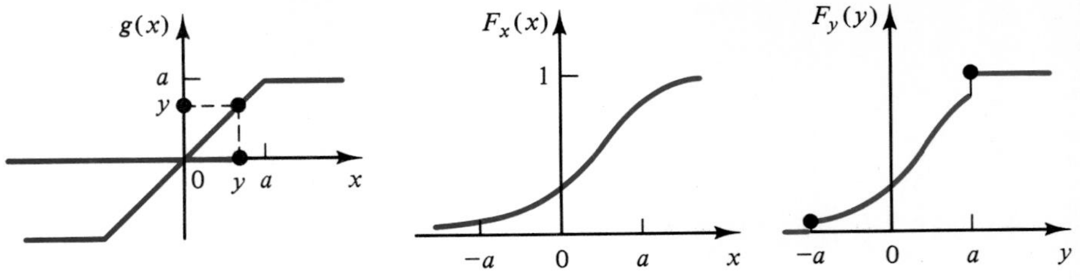

Figure 4.32

3. *Dead zone*. (See Fig. 4.33.)

$$g(x) = \begin{cases} x - c & x > c \\ 0 & -c \le x \le c \\ x + c & x < -c \end{cases}$$

If $y \ge 0$, then $\{\mathbf{y} \le y\} = \{\mathbf{x} \le y + c\}$; hence, $F_y(y) = F_x(y + c)$.
If $y < 0$, then $\{\mathbf{y} \le y\} = \{\mathbf{x} \le y - c\}$; hence, $F_y(y) = F_x(y - c)$.
Again, $F_y(y)$ is discontinuous at $y = 0$; hence, the RV $\mathbf{y}$ is of mixed type, and

$$P\{\mathbf{y} = 0\} = P\{-c < \mathbf{x} < c\} = F_x(c) - F_x(-c)$$

4. *Discontinuous transformation*. (See Fig. 4.34.)

$$g(x) = \begin{cases} x + c & x \ge 0 \\ x - c & x < 0 \end{cases}$$

If $y \ge c$, then $\{\mathbf{y} \le y\} = \{\mathbf{x} \le y - c\}$; hence, $F_y(y) = F_x(y - c)$.
If $-c < y < c$, then $\{\mathbf{y} \le y\} = \{\mathbf{x} \le 0\}$; hence, $F_y(y) = F_x(0)$.
If $y < -c$, then $\{\mathbf{y} \le y\} = \{\mathbf{x} \le y + c\}$; hence, $F_y(y) = F_x(y + c)$.

5. In this illustration, we select for $g(x)$ the distribution $F_x(x)$ of the RV $\mathbf{x}$ (Fig. 4.35). We shall show that the resulting RV

$$\mathbf{y} = F_x(\mathbf{x})$$

is uniform in the interval (0, 1) for any $F_x(x)$.
The function $g(x) = F_x(x)$ is between the lines $y = 0$ and $y = 1$.

Figure 4.33

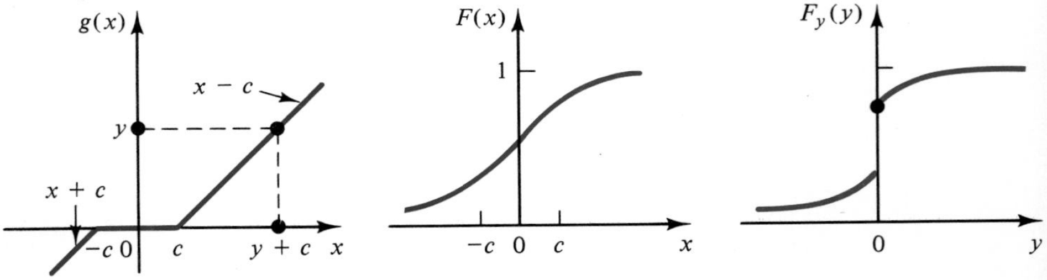

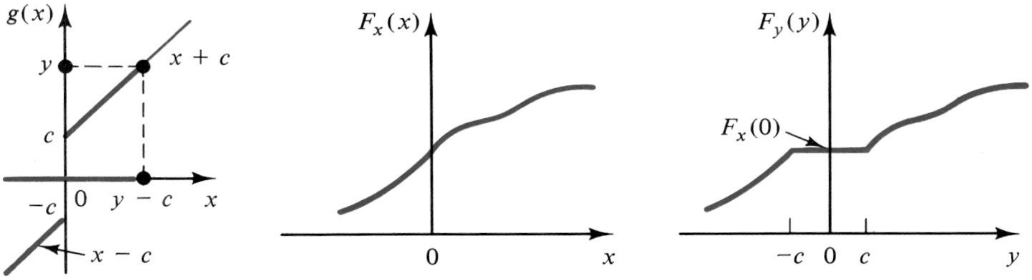

Figure 4.34

Hence [see (4-74)],

$$F_y(y) = \begin{cases} 1 & y > 1 \\ 0 & y < 0 \end{cases} \tag{4-75}$$

If $0 \le y \le 1$, then $\{\mathbf{y} \le y\} = \{\mathbf{x} \le x_y\}$ where x_y is such that $F_x(x_y) = y$. Thus x_y is the y-percentile of $\mathbf{x}$, and

$$F_y(y) = P\{\mathbf{x} \le x_y\} = y \qquad 0 \le y \le 1 \tag{4-76}$$

Illustration 5 is used to construct an RV $\mathbf{y}$ with uniform distribution starting from an RV $\mathbf{x}$ with an arbitrary distribution. Using a related approach, we can construct an RV $\mathbf{y} = g(\mathbf{x})$ with a specified distribution starting with an RV $\mathbf{x}$ with an arbitrary distribution (see Section 8-3).

Density of $g(\mathbf{x})$

If we know $F_y(y)$, we can find $f_y(y)$ by differentiation. In many cases, however, it is simpler to express the density $f_y(y)$ of the RV $\mathbf{y} = g(\mathbf{x})$ directly in terms the density $f_x(x)$ of $\mathbf{x}$ and the function $g(x)$. To do so, we form the equation

$$g(x) = y \tag{4-77}$$

where y is a specific number, and we solve for x. The solutions of this equation are the abscissas x_i of the intersection points of the horizontal line L_y (Fig. 4.36) with the curve $g(x)$.

Figure 4.35

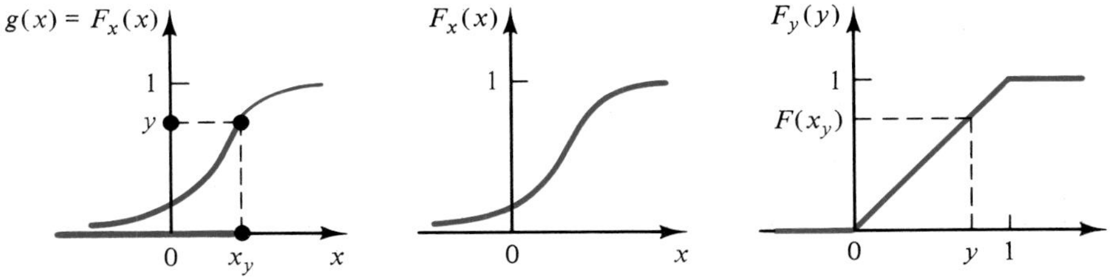

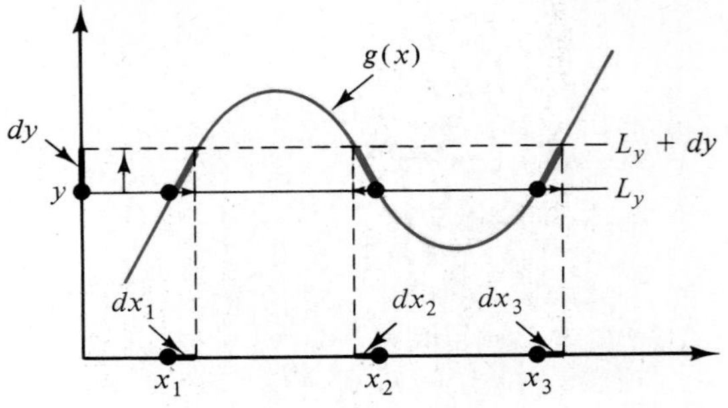

Figure 4.36

■ ***Fundamental Theorem.*** For a specific y, the density $f_y(y)$ is given by

$$f_y(y) = \frac{f_x(x_1)}{|g'(x_1)|} + \cdots + \frac{f_x(x_i)}{|g'(x_i)|} + \cdots = \sum_i \frac{f_x(x_i)}{|g'(x_i)|} \tag{4-78}$$

where x_i are the roots of (4-77):

$$y = g(x_1) = \cdots = g(x_i) = \cdots \tag{4-79}$$

and $g'(x_i)$ are the derivatives of $g(x)$ at $x = x_i$.

■ ***Proof.*** To avoid generalities, we assume that the equation $y = g(x)$ has three roots, as in Fig. 4.36. As we know [see (4-32)],

$$f_y(y)\,dy = P\{y < \mathbf{y} < y + dy\} \tag{4-80}$$

Clearly, $\mathbf{y}$ is between y and $y + dy$ iff $\mathbf{x}$ is in any one of the intervals

$$(x_1, x_1 + dx_1) \qquad (x_2 - |dx_2|, x_2) \qquad (x_3, x_3 + dx_3)$$

where $dx_1 > 0$, $dx_2 < 0$, $dx_3 > 0$. Hence,

$$P\{y < \mathbf{y} < y + dy\} = P\{x_1 < \mathbf{x} < x_1 + dx_1\} + P\{x_2 - |dx_2| < \mathbf{x} < x_2\} + P\{x_3 < \mathbf{x} < x_3 + dx_3\} \tag{4-81}$$

This is the probability that $\mathbf{y}$ is in the three segments of the curve $g(x)$ between the lines L_y and $L_y + dy$. The terms on the right of (4-81) equal

$$P\{x_1 < \mathbf{x} < x_1 + dx_1\} = f_x(x_1)\,dx_1 \qquad dx_1 = \frac{1}{g'(x_1)}dy$$

$$P\{x_2 - |dx_2| < \mathbf{x} < x_2\} = f_x(x_2)|dx_2| \qquad dx_2 = \frac{1}{g'(x_2)}dy$$

$$P\{x_3 < \mathbf{x} < x_3 + dx_3\} = f_x(x_3)\,dx_3 \qquad dx_3 = \frac{1}{g'(x_3)}dy$$

Inserting into (4-81), we obtain

$$f_y(y)\,dy = \frac{f_x(x_1)}{g'(x_1)}dy + \frac{f_x(x_2)}{|g'(x_2)|}dy + \frac{f_x(x_3)}{g'(x_3)}dy$$

and (4-78) results.

Thus to find $f_y(y)$ for a specific y, we find the roots x_i of the equation $y = g(x)$ and insert into (4-78). The numbers x_i depend, of course, on y.

We assumed that the equation $y = g(x)$ has at least one solution. If for some values of y the line L_y does not intersect $g(x)$, then $f_y(y) = 0$ for these values of y.

Note, finally, that if $g(x) = y_0$ for every x in an interval (a, b) as in Fig. 4.33, then $F_y(y)$ is discontinuous at $y = y_0$.

Illustrations

1.
$$g(x) = ax + b$$

The equation $y = ax + b$ has a single solution $x_1 = (y - b)/a$ for every y. Furthermore, $g'(x) = a$; hence,

$$f_y(y) = \frac{1}{|a|} f_x(x_1) = \frac{1}{|a|} f_x\left(\frac{y - b}{a}\right) \tag{4-82}$$

Thus $f_y(y)$ is obtained by shifting and scaling the density $f_x(x)$ of x.

2.
$$g(x) = \frac{1}{x}$$

The equation $y = 1/x$ has a single solution $x_1 = 1/y$ for every y. Furthermore,

$$g'(x) = \frac{-1}{x^2} \qquad g'(x_1) = -y^2$$

Hence,

$$f_y(y) = \frac{1}{y^2} f_x\left(\frac{1}{y}\right) \tag{4-83}$$

3.
$$g(x) = ax^2 \qquad a > 0$$

If $y > 0$, the equation $y = ax^2$ has two solutions: $x_1 = \sqrt{y/a}$ and $x_2 = -\sqrt{y/a}$ (Fig. 4.29). Furthermore,

$$g'(x) = 2ax \qquad g'(x_1) = 2\sqrt{ay} \qquad g'(x_2) = -2\sqrt{ay}$$

Hence,

$$f_y(y) = \frac{1}{2\sqrt{ay}} f_x\left(\sqrt{\frac{y}{a}}\right) + \frac{1}{2\sqrt{ay}} f_x\left(-\sqrt{\frac{y}{a}}\right) \tag{4-84}$$

If $y < 0$, the equation $y = ax^2$ has no real roots; hence, $f_y(y) = 0$ [see also (4-71)].

Example 4.15

The RV **x** is uniform in the interval (5, 10), as in Fig. 4.37, and $\mathbf{y} = 4\mathbf{x}^2$. In this case, $f_x(-\sqrt{y/4}) = 0$ for every $y > 0$ and

$$f_x\left(\sqrt{\frac{y}{4}}\right) = \begin{cases} .2 & 5 < \sqrt{\frac{y}{4}} < 10 \\ 0 & \text{elsewhere} \end{cases}$$

Hence,

$$f_y(y) = \begin{cases} \dfrac{.05}{\sqrt{y}} & 100 < y < 400 \\ 0 & \text{elsewhere} \end{cases}$$

■

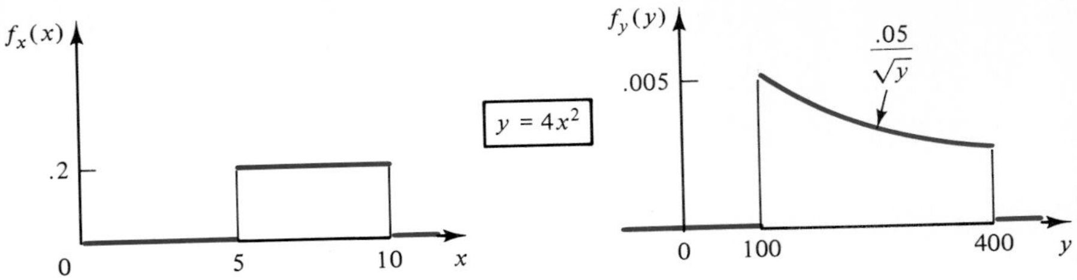

Figure 4.37

4. $$g(x) = \sin x$$

If $|y| > 1$, the equation $y = \sin x$ has no real solutions; hence, $f_y(y) = 0$. If $|y| < 1$, it has infinitely many solutions x_i

$$y = \sin x_i \qquad x_i = \arcsin y$$

Furthermore,

$$g'(x_i) = \cos x_i = \sqrt{1 - \sin^2 x_i} = \sqrt{1 - y^2}$$

and (4-78) yields

$$f_y(y) = \frac{1}{\sqrt{1 - y^2}} \sum_i f_x(x_i) \tag{4-85}$$

where $f_x(x_i)$ are the values of the density $f_x(x)$ of $\mathbf{x}$ (Fig. 4.38).

Suppose now that the RV $\mathbf{x}$ is uniform in the interval $(-\pi, \pi)$. In this case,

$$f_x(x_i) = \begin{cases} \dfrac{1}{2\pi} & -\pi < x_i < \pi \\ 0 & \text{otherwise} \end{cases}$$

Clearly, there are two and only two roots x_i in the interval $(-\pi, \pi)$

Figure 4.38

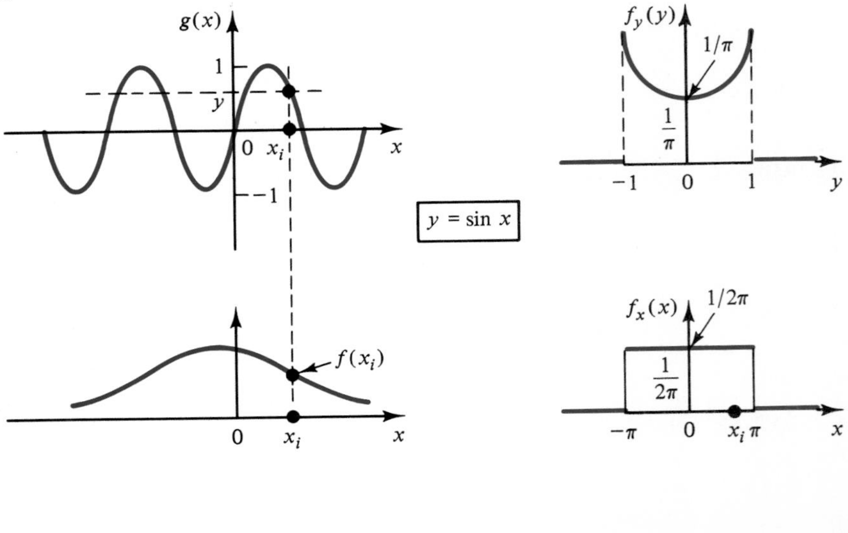

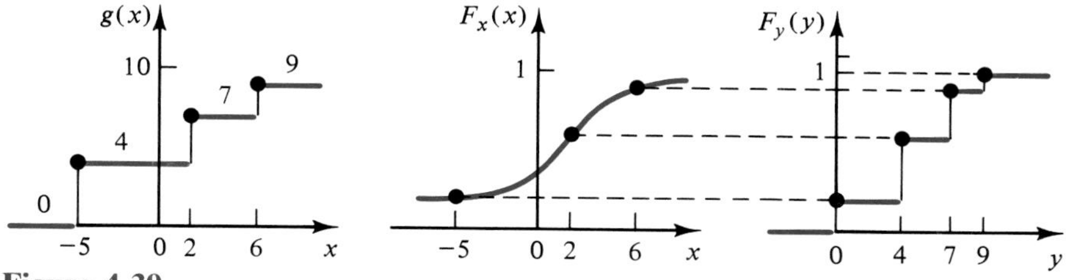

Figure 4.39

for any y; hence, the sum in (4-85) equals $2/2\pi$. This yields

$$f_y(y) = \begin{cases} \dfrac{1/\pi}{\sqrt{1 - y^2}} & |y| < 1 \\ 0 & |y| > 1 \end{cases} \tag{4-86}$$

5.
$$g(x) = F_x(x)$$

as in Fig. 4.35. Clearly, $g(x)$ is between the lines $y = 0$ and $y = 1$; hence, $f_y(y) = 0$ for $y < 0$ and $y > 1$. For $0 \le y \le 1$, the equation $y = F_x(x)$ has a single solution $x_1 = x_y$ where x_y is the y-percentile of the RV $\mathbf{x}$; that is, $y = F(x_y)$. Furthermore, $g'(x) = F'_x(x) = f_x(x)$; hence,

$$f_y(y) = \frac{f_x(x_1)}{g'(x_1)} = \frac{f_x(x_y)}{f_x(x_y)} = 1 \qquad 0 \le y \le 1$$

Thus the RV $\mathbf{y}$ is uniform in the interval (0, 1) for any $F_x(x)$ [see also (4-76)].

Point Masses Suppose that $\mathbf{x}$ is a discrete type RV taking the values x_k with probability p_k. In this case, the RV $\mathbf{y} = g(\mathbf{x})$ is also of discrete type, taking the values $y_k = g(x_k)$. If $y_k = g(x_k)$ only for one x_k, then

$$P\{\mathbf{y} = y_k\} = P\{\mathbf{x} = x_k\} = p_k = f_x(x_k)$$

If, however, $\mathbf{y} = y_k$ for $\mathbf{x} = x_a$ and $\mathbf{x} = x_b$—that is, if $y_k = g(x_a) = g(x_b)$—then the event $\{\mathbf{y} = y_k\}$ is the union of the events $\{\mathbf{x} = x_a\}$ and $\{\mathbf{x} = x_b\}$; hence,

$$P\{\mathbf{y} = y_k\} = P\{\mathbf{x} = x_a\} + P\{\mathbf{x} = x_b\} = p_a + p_b$$

Note, finally, that if $\mathbf{x}$ is of continuous type but $g(x)$ is a staircase function with discontinuities at the points x_k, then $\mathbf{y}$ is of discrete type, taking the values $y_k = g(x_k)$. In Fig. 4.39, for example,

$$P\{\mathbf{y} = 7\} = P\{2 < \mathbf{x} \le 6\} = F_x(6) - F_x(2) = .3$$

This is the discontinuity jump of $F_y(y)$ at $y = 7$.

4-5 Mean and Variance

The properties of an RV $\mathbf{x}$ are completely specified in terms of its distribution. In many cases, however, it is sufficient to specify $\mathbf{x}$ only partially in terms of certain parameters. For example, knowledge of the percentiles x_u of $\mathbf{x}$, given only in increments of .1, is often adequate. The most important numerical parameters of $\mathbf{x}$ are its mean and its variance.

Mean

The *mean* or *expected value* or *statistical average* of an RV $\mathbf{x}$ is, by definition, the center of gravity of the probability masses of $\mathbf{x}$ (Fig. 4.40). This number will be denoted by $E\{\mathbf{x}\}$ or η_x or η.

If $\mathbf{x}$ is of continuous type with density $f(x)$ (distributed masses), then

$$E\{\mathbf{x}\} = \int_{-\infty}^{\infty} xf(x)\,dx \tag{4-87}$$

If $\mathbf{x}$ is of discrete type taking the values x_k with probabilities $p_k = f(x_k)$ (point masses), then

$$E\{\mathbf{x}\} = \sum_k x_k f(x_k) \tag{4-88}$$

A *constant* c can be interpreted as an RV $\mathbf{x}$ taking the value c for every ζ. Applying (4-88), we conclude that

$$E\{c\} = cP\{\mathbf{x} = c\} = c$$

Empirical Interpretation. We repeat the experiment n times and denote by x_i the resulting values of the RV $\mathbf{x}$ [see also (4-23)]. We next form the *arithmetic average*

$$\bar{x} = \frac{x_1 + \cdots + x_n}{n} = \sum_i x_i \frac{1}{n} \tag{4-89}$$

of these numbers. We maintain that $\bar{x}$ tends to the *statistical average* $E\{\mathbf{x}\}$ of $\mathbf{x}$ as n increases:

$$\bar{x} \simeq E\{\mathbf{x}\} \qquad n \text{ large} \tag{4-90}$$

Figure 4.40

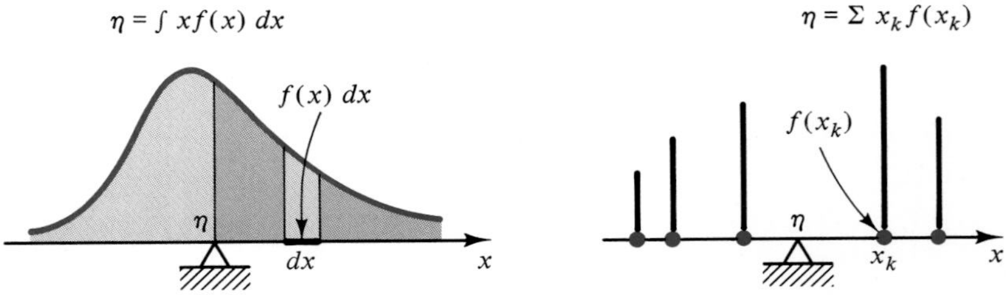

We divide the x-axis into intervals of length Δ as in Fig. 4.17 and denote by n_k the number of x_i between c_k and $c_k + \Delta$. If Δ is small, then $x_i \simeq c_k$ for $c_k \le x_i \le c_k + \Delta$. Introducing this approximation into (4-89), we obtain

$$\bar{x} \simeq \sum_k c_k \frac{n_k}{n} = \sum_k c_k f_n(c_k)\,\Delta \tag{4-91}$$

where $f_n(c_k) = n_k/n\Delta$ is the histogram of $\mathbf{x}$ [see (4-36)]. As we have shown in (4-37), $f_n(x) \simeq f(x)$ for small Δ; hence, the last sum in (4-91) tends to the integral in (4-87) as $n \to \infty$; this yields (4-90).

We note that x_i/n is the area under the empirical percentile curve in an interval of length $1/n$ (Fig. 4.41); hence, the sum in (4-89) equals the total area of the empirical percentile.

From the foregoing it follows that the mean $E\{\mathbf{x}\}$ of our RV $\mathbf{x}$ equals the algebraic area under its percentile curve x_u, namely,

$$\int_{-\infty}^{\infty} xf(x)\,dx = \int_0^1 x_u\,du \qquad F(x_u) = u$$

Interchanging the x and $F(x)$ axes, we conclude that $E\{\mathbf{x}\}$ equals the difference of the areas of the regions ACD and OAB of Fig. 4.41. This yields

$$E\{\mathbf{x}\} = \int_0^{\infty} R(x)\,dx - \int_{-\infty}^{0} F(x)\,dx \tag{4-92}$$

where $R(x) = 1 - F(x)$.

Equation (4-92) can be established directly from (4-87) if we use integration by parts.

Symmetrical Densities If $f(x)$ is an even function, that is, if $f(-x) = f(x)$, then $E\{\mathbf{x}\} = 0$. More generally, if $f(x)$ is symmetrical about the number a, that is, if

$$f(a + x) = f(a - x) \qquad \text{then} \qquad E\{\mathbf{x}\} = a$$

This follows readily from (4-87) or, directly, from the mass interpretation of the mean.

A density that has no center of symmetry is called *skewed*.

Figure 4.41

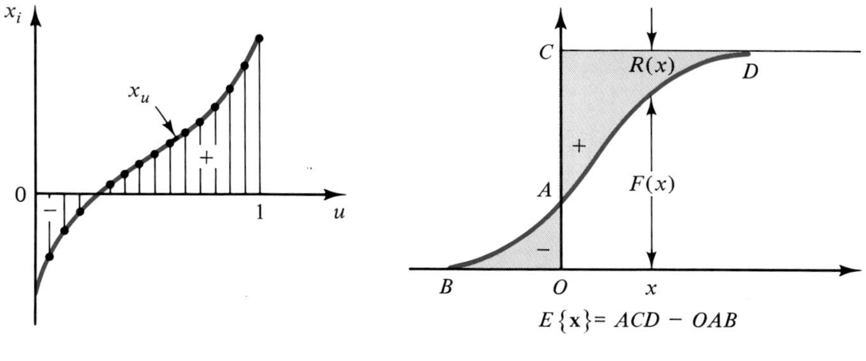

MEAN OF $g(\mathbf{x})$ Given an RV $\mathbf{x}$ and a function $g(x)$, we form the RV

$$\mathbf{y} = g(\mathbf{x})$$

As we see from (4-87) and (4-88), the mean of $\mathbf{y}$ equals

$$E\{\mathbf{y}\} = \int_{-\infty}^{\infty} y f_y(y)\,dy \qquad \text{or} \qquad \sum_k y_k f_y(y_k) \tag{4-93}$$

for the continuous and the discrete case, respectively. To determine $E\{\mathbf{y}\}$ from (4-93), we need to find $f_y(y)$. In the following, we show that $E\{\mathbf{y}\}$ can be expressed directly in terms of $f_x(x)$ and $g(x)$.

■ ***Fundamental Theorem***

$$E\{g(\mathbf{x})\} = \int_{-\infty}^{\infty} g(x) f_x(x)\,dx \qquad \text{or} \qquad \sum_k g(x_k) f_x(x_k) \tag{4-94}$$

■ ***Proof.*** The discrete case follows readily: If $y_k = g(x_k)$ for only a single x_k, then

$$\sum y_k P\{\mathbf{y} = y_k\} = \sum g(x_k) P\{\mathbf{x} = x_k\}$$

If $y_k = g(x_k)$ for several x_k, we add the corresponding terms on the right.

To prove the continuous case, we assume that $g(x)$ is the curve of Fig. 4.36. Clearly,

$$\{y < \mathbf{y} < y + dy\}$$
$$= \{x_1 < \mathbf{x} < x_1 + dx\} \cup \{x_2 < \mathbf{x} < x_2 + dx_2\} \cup \{x_3 < \mathbf{x} < x_3 + dx_3\}$$

where, in contrast to (4-81), all differentials are positive. Multiplying by $y = g(x_1) = g(x_2) = g(x_3)$ the probabilities of both sides and using (4-81), we obtain

$$y f_y(y)\,dy = g(x_1) f_x(x_1)\,dx_1 + g(x_2) f_x(x_2)\,dx_2 + g(x_3) f_x(x_3)\,dx_3$$

Thus to each differential of the integral in (4-93) corresponds one or more nonoverlapping differentials of the integral in (4-94). As dy covers the y-axis, each corresponding dx covers the x-axis; hence, the two integrals are equal.

We shall verify this theorem with an example.

Example 4.16

The RV $\mathbf{x}$ is uniform in the interval (1, 3) (Fig. 4.42), and $\mathbf{y} = 2\mathbf{x} + 1$. As we see from (4-82),

$$f_y(y) = \frac{1}{2} f_x\left(\frac{y-1}{2}\right)$$

From this it follows that $\mathbf{y}$ is uniform in the interval (3, 7) and

$$E\{\mathbf{y}\} = \int_{-\infty}^{\infty} y f_y(y)\,dy = .25 \int_3^7 y\,dy = 5$$

This agrees with (4-94) because $g(x) = 2x + 1$ and

$$\int_{-\infty}^{\infty} g(x) f_x(x)\,dx = .5 \int_1^3 (2x + 1)\,dx = 5$$

■

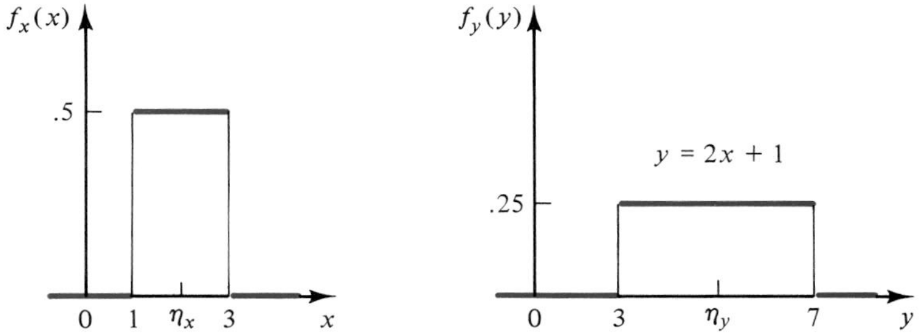

Figure 4.42

Linearity From (4-94) it follows that

$$E\{a\mathbf{x}\} = a\int_{-\infty}^{\infty} xf_x(x)\,dx = aE\{\mathbf{x}\} \tag{4-95}$$

$$\begin{aligned} E\{g_1(\mathbf{x}) + \cdots + g_n(\mathbf{x})\} &= \int_{-\infty}^{\infty} [g_1(x) + \cdots + g_n(x)]\,f_x(x)\,dx \\ &= E\{g_1(\mathbf{x})\} + \cdots + E\{g_n(\mathbf{x})\} \end{aligned} \tag{4-96}$$

Thus the mean of the sum of n functions $g_i(\mathbf{x})$ of the RV $\mathbf{x}$ equals the sum of their means.

Note, in particular, that $E\{b\} = b$ and

$$E\{a\mathbf{x} + b\} = aE\{\mathbf{x}\} + b \tag{4-97}$$

In Example 4.16, $\eta_x = 2$ and $\eta_y = 5 = 2\eta_x + 1$, in agreement with (4-97).

Variance

The *variance* or *dispersion* of an RV $\mathbf{x}$ is by definition the central moment of inertia of its probability masses. This number is denoted by σ^2. Thus for distributed masses,

$$\sigma^2 = \int_{-\infty}^{\infty} (x - \eta)^2 f(x)\,dx \tag{4-98}$$

and for point masses,

$$\sigma^2 = \sum_k (x_k - \eta)^2 f(x_k) \tag{4-99}$$

From (4-98) and (4-94) it follows that

$$\sigma^2 = E\{(\mathbf{x} - \eta)^2\} = E\{\mathbf{x}^2\} - 2\eta E\{\mathbf{x}\} + \eta^2$$

where $\eta = E\{\mathbf{x}\}$. This yields

$$E\{\mathbf{x}^2\} = \sigma^2 + \eta^2 \tag{4-100}$$

This has a familiar mass interpretation: Clearly, $E\{\mathbf{x}^2\}$ is the moment of inertia of the probability masses with respect to the origin and η^2 is the moment of inertia of a unit point mass located at $x = \eta$. Thus (4-100) states that the moment of inertia with respect to the origin equals the sum of the central moment of inertia, plus the moment with respect to the origin, of a unit mass at $x = \eta$. In probability theory, $E\{\mathbf{x}\}$ is called the *first moment,* $E\{\mathbf{x}^2\}$ the *second moment,* and σ^2 the *second central moment* (see also Section 5-2). The square root σ of σ^2 is called the *standard deviation* of the RV $\mathbf{x}$.

We show next that the variance σ_y^2 of the RV $\mathbf{y} = a\mathbf{x} + b$ equals

$$\sigma_y^2 = a^2\sigma_x^2 \tag{4-101}$$

We know that $\eta_y = a\eta_x + b$; hence,

$$\sigma_y^2 = E\{(\mathbf{y} - \eta_y)^2\} = E\{[(a\mathbf{x} + b) - (a\eta_x + b)]^2\} = E\{a^2(\mathbf{x} - \eta_x)^2\}$$

and (4-101) follows from (4-95).

Note, in particular, that the variance of $\mathbf{x} + b$ equals the variance $\mathbf{x}$. This shows that a shift of the origin has no effect on the variance.

As we show in (4-113), the variance is a measure of the concentration of the probability masses near their mean. In fact, if $\sigma = 0$, all masses are at a single point because then $f(x) = 0$ for $x \neq \eta$ [see (4-98)]. Thus

$$\text{if} \quad \sigma = 0 \quad \text{then} \quad \mathbf{x} = \eta = \text{constant} \tag{4-102}$$

in the sense that the probability that $\mathbf{x} \neq \eta$ is zero.

Note, in particular, that

$$\text{if} \quad E\{\mathbf{x}^2\} = 0 \quad \text{then} \quad \mathbf{x} = 0 \tag{4-103}$$

Illustrations

We determine next the mean and the variance of various distributions. Note that it is often simpler to determine σ^2 from (4-100).

Normal The density

$$f(z) = \frac{1}{\sqrt{2\pi}} e^{-z^2/2}$$

of a standard normal RV $\mathbf{z}$ is even; hence, $\eta_z = 0$. We shall show that the variance of $\mathbf{z}$ equals 1:

$$\sigma_z^2 = \frac{1}{\sqrt{2\pi}} \int_{-\infty}^{\infty} z^2 e^{-z^2/2}\, dz = 1 \tag{4-104}$$

To do so, we differentiate the identity [see (3A-1)]

$$\int_{-\infty}^{\infty} e^{-\alpha z^2}\, dz = \sqrt{\pi}(\alpha^{-1/2})$$

with respect to α. This yields

$$\int_{-\infty}^{\infty} (-z^2)e^{-\alpha z^2}\,dz = -\frac{1}{2}\sqrt{\pi}(\alpha^{-3/2})$$

Setting $\alpha = 1/2$, we obtain (4-104).

We next form the RV $\mathbf{x} = a\mathbf{z} + b$. As we see from (4-82), if $a > 0$, then

$$f_x(x) = \frac{1}{a} f_z\left(\frac{x-b}{a}\right) = \frac{1}{a\sqrt{2\pi}}\, e^{-(x-b)^2/2a^2}$$

Thus $\mathbf{x}$ is $N(b, a)$, and [see (4-100)]

$$\eta_x = b \qquad \sigma_x = a \tag{4-105}$$

This justifies the use of the letters η and σ in the definition (4-44) of a general normal density.

Uniform Suppose that $\mathbf{x}$ is uniform in the interval $(a - c/2, a + c/2)$, as in Fig. 4.21. In this case, a is the center of symmetry of $f(x)$, and since the location of the origin is irrelevant in the determination of σ^2, we conclude that

$$\eta = a \qquad \sigma^2 = \frac{1}{c}\int_{-c/2}^{c/2} x^2\,dx = \frac{c^2}{12} \tag{4-106}$$

Exponential If

$$f(x) = ce^{-cx}U(x)$$

then

$$E\{\mathbf{x}\} = c\int_0^{\infty} xe^{-cx}\,dx = \frac{1}{c} \qquad E\{\mathbf{x}^2\} = c\int_0^{\infty} x^2e^{-cx}\,dx = \frac{2}{c^2}$$

Hence,

$$\eta = \frac{1}{c} \qquad \sigma^2 = E\{\mathbf{x}^2\} - \eta^2 = \frac{1}{c^2} \tag{4-107}$$

We continue with discrete type RVs.

Zero-One The RV $\mathbf{x}$ takes the values 1 and 0 with

$$P\{\mathbf{x} = 1\} = p \qquad P\{\mathbf{x} = 0\} = q = 1 - p$$

In this case,

$$E\{\mathbf{x}\} = 0 \times q + 1 \times p = p \qquad E\{\mathbf{x}^2\} = 0^2 \times q + 1^2 \times p = p$$

Hence,

$$\eta = p \qquad \sigma^2 = p - p^2 = pq \tag{4-108}$$

Geometric The RV $\mathbf{x}$ takes the values 1, 2, . . . with

$$P\{\mathbf{x} = k\} = pq^{k-1} \qquad k = 1, 2, \ldots$$

We shall show that

$$\eta = \frac{1}{p} \qquad \sigma^2 = \frac{q}{p^2} \tag{4-109}$$

■ ***Proof.*** Differentiating the geometric series

$$\sum_{k=1}^{\infty} q^k = \frac{q}{1-q}$$

twice, we obtain

$$\sum_{k=1}^{\infty} kq^{k-1} = \frac{1}{(1-q)^2} \qquad \sum_{k=1}^{\infty} k(k-1)q^{k-2} = \frac{2}{(1-q)^3}$$

From this it follows that

$$E\{\mathbf{x}\} = p \sum_{k=1}^{\infty} kq^{k-1} = \frac{p}{(1-q)^2}$$

$$E\{\mathbf{x}^2\} = p \sum_{k=1}^{\infty} k^2q^{k-1} = \frac{2pq}{(1-q)^3} + \frac{p}{(1-q)^2} = \frac{1+q}{p^2}$$

and (4-109) results [see (4-100)].

From (4-109) it follows that the expected number of tosses until heads shows for the first time (see Example 3.5) equals $1/p$.

Poisson The RV $\mathbf{x}$ takes the values 0, 1, . . . with

$$P\{\mathbf{x} = k\} = e^{-a} \frac{a^k}{k!} \qquad k = 0, 1, \ldots$$

We shall show that

$$\eta = a \qquad \sigma^2 = a \tag{4-110}$$

■ ***Proof.*** Differentiating the identity

$$e^a = \sum_{k=0}^{\infty} \frac{a^k}{k!}$$

twice, we obtain

$$e^a = \sum_{k=0}^{\infty} k \frac{a^{k-1}}{k!} \qquad e^a = \sum_{k=0}^{\infty} k(k-1) \frac{a^{k-2}}{k!}$$

Hence,

$$E\{\mathbf{x}\} = e^{-a} \sum_{k=0}^{\infty} k \frac{a^k}{k!} = a$$

$$E\{\mathbf{x}^2\} = e^{-a} \sum_{k=0}^{\infty} k^2 \frac{a^k}{k!} = a^2 + a$$

and (4-110) results.

From (4-110) and (4-64) it follows that the expected number of Poisson points in an interval of length t_a equals λt_a. This shows that λ equals the expected number of points in a unit interval.

APPROXIMATE EVALUATION OF $E\{g(\mathbf{x})\}$ We wish to determine the mean of the RV $\mathbf{y} = g(\mathbf{x})$. To do so, we must know the density of $\mathbf{x}$. We shall show

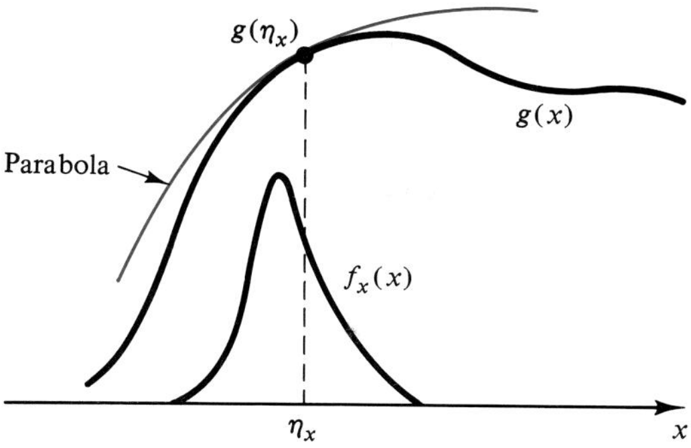

Figure 4.43

that if $g(x)$ is sufficiently smooth in the region $(\eta_x - c, \eta_x + c)$ where $f_x(x)$ takes significant values (Fig. 4.43), we can express $E\{g(\mathbf{x})\}$ in terms of the mean η_x and the variance σ_x^2 of $\mathbf{x}$.

Suppose, first, that $g(x) = ax + b$. In this case,

$$E\{g(\mathbf{x})\} = a\eta_x + b = g(\eta_x)$$

This suggests that if $g(x)$ is approximated by its tangent

$$g(x) \simeq g(\eta_x) + g'(\eta_x)(x - \eta_x)$$

in the interval $\eta_x \pm c$, then $E\{g(\mathbf{x})\} \simeq g(\eta_x)$. Indeed, since $E\{\mathbf{x} - \eta_x\} = 0$, the estimate

$$E\{g(\mathbf{x})\} \simeq g(\eta_x) + g'(\eta_x)E\{\mathbf{x} - \eta_x\} = g(\eta_x) \tag{4-111}$$

results. This estimate can be improved if we approximate $g(x)$ by a parabola

$$g(x) \simeq g(\eta_x) + g'(\eta_x)(x - \eta_x) + \frac{g''(\eta_x)}{2}(x - \eta_x)^2 \tag{4-112}$$

Since $E\{(\mathbf{x} - \eta_x)^2\} = \sigma_x^2$, we conclude that

$$E\{g(\mathbf{x})\} \simeq g(\eta_x) + g''(\eta_x)\frac{\sigma_x^2}{2} \tag{4-113}$$

This is an approximation based on the truncation (4-112) of the Taylor series expansion of $g(x)$ about the point $x = \eta_x$. The approximation can be further improved if we include more terms in (4-112). The result, however, involves knowledge of higher-order moments of $\mathbf{x}$.

Example 4.17

$$g(x) = \frac{1}{x}$$

In this case,

$$g''(x) = -\frac{2}{x^3} \qquad g(\eta_x) = \frac{1}{\eta_x} \qquad g''(\eta_x) = -\frac{2}{\eta_x^3}$$

and (4-113) yields

$$E\left\{\frac{1}{\mathbf{x}}\right\} \simeq \frac{1}{\eta_x} - \frac{1}{\eta_x^3}\sigma_x^2 = \frac{1}{\eta_x}\left[1 - \left(\frac{\sigma_x}{\eta_x}\right)^2\right] \qquad \blacksquare$$

TCHEBYCHEFF'S INEQUALITY The mean η of an RV $\mathbf{x}$ is the center of gravity of its masses. The variance σ^2 is their moment of inertia with respect to η. We shall show that a major proportion of the masses is concentrated in an interval of the order of σ centered at η.

Consider the points $a = \eta - k\sigma$, $b = \eta + k\sigma$ of Fig. 4.44 where k is an arbitrary constant. If we remove all masses between a and b (Fig. 4.44b) and replace the masses $p_1 = P\{\mathbf{x} \le \eta - k\sigma\}$ to the left of a by a point mass p_1 at a and the masses $p_2 = P\{\mathbf{x} \ge \eta + k\sigma\}$ to the right of b by a point mass p_2 at b (Fig. 4.44c), the resulting moment of inertia with respect to η will be smaller than σ^2:

$$p_1(k\sigma)^2 + p_2(k\sigma^2) < \sigma^2$$

Hence,

$$p_1 + p_2 \le \frac{1}{k^2}$$

From this it follows that

$$P\{|\mathbf{x} - \eta| \ge k\sigma\} \le \frac{1}{k^2} \tag{4-114}$$

or, equivalently,

$$P\{\eta - k\sigma < \mathbf{x} < \eta + k\sigma\} \ge 1 - \frac{1}{k^2} \tag{4-115}$$

where equality holds iff the original masses consist of two equal point masses.

We have thus shown that the probability masses $p_1 + p_2$ outside the interval $(\eta - k\sigma, \eta + k\sigma)$ are smaller than $1/k^2$ for any k and for any $F(x)$. Note, in particular, that if $1/k^2 = .05$—that is, if $k = 4.47$—then at most 5% of all masses are outside the interval $\eta \pm 4.47\sigma$ regardless of the form of

Figure 4.44

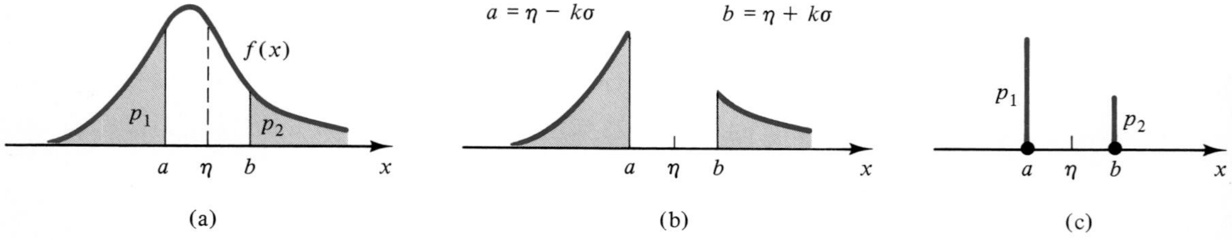

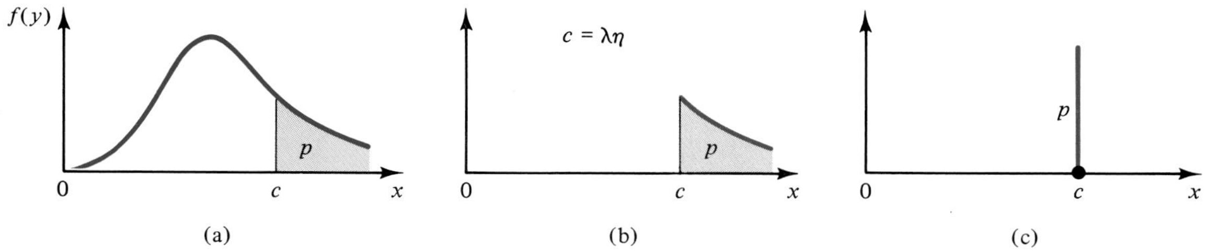

Figure 4.45

$F(x)$. If $F(x)$ is known, tighter bounds can be obtained; if, for example, $\mathbf{x}$ is normal, then [see (4-50)] 5% of all masses are outside the interval $\eta \pm 1.96\sigma$.

Markoff's Inequality Suppose now that $\mathbf{x} \geq 0$. In this case, $F(x) = 0$ for $x < 0$; hence, all masses are on the positive axis. We shall show that a major proportion of the masses is concentrated in an interval $(0, c)$ of the order of η.

Consider the point $c = \lambda\eta$ of Fig. 4.45 where λ is an arbitrary constant. If we remove all masses to the left of c (Fig. 4.45b) and replace the masses $p = P\{\mathbf{x} \geq \lambda\eta\}$ to the right of c by a point mass p at c, the moment with respect to the origin will decrease from η to pc; hence, $p\lambda\eta < \eta$. This yields

$$P\{\mathbf{x} \geq \lambda\eta\} < \frac{1}{\lambda} \tag{4-116}$$

for any λ and for any $F(x)$.

Note that (4-114) is a special case of (4-116). Indeed, the RV $(\mathbf{x} - \eta)^2$ is positive and its mean equals σ^2. Applying (4-116) to the RV $(\mathbf{x} - \eta)^2$, we conclude with $\lambda = k^2$ that

$$P\{|\mathbf{x} - \eta|^2 \geq k^2\sigma^2\} \leq \frac{1}{k^2}$$

and (4-114) results because the left side equals $P\{|\mathbf{x} - \eta| \geq k\sigma\}$.

Problems

4-1 The roll of two fair dice specifies an experiment with 36 outcomes $f_i f_k$. The RVs $\mathbf{x}$ and $\mathbf{y}$ are such that

$$\mathbf{x}(f_i f_k) = i + k \qquad \mathbf{y}(f_i f_k) = \begin{cases} 5 & \text{if } i + k = 7 \\ 20 & \text{if } i + k = 11 \end{cases}$$

and $y(f_i f_k) = 0$ otherwise. **(a)** Find the probabilities of the events $\{6.5 < \mathbf{x} < 10\}$, $\{\mathbf{x} > 2\}$, $\{5 < \mathbf{y} < 25\}$, $\{\mathbf{y} < 4\}$. **(b)** Sketch the distributions $F_x(x)$, $F_y(y)$ and the point densities $f_x(x)$, $f_y(y)$.

4-2 The RV $\mathbf{x}$ is normal with $\eta = 10$, $\sigma = 2$. Find the probabilities of the following events: $\{\mathbf{x} > 14\}$, $\{8 < \mathbf{x} < 14\}$, $\{|\mathbf{x} - 10| < 2\}$, $\{\mathbf{x} - 5 < 7 < \mathbf{x} - 1\}$.

4-3 The SAT scores of students in a school district are modeled by a normal RV with $\eta = 510$ and $\sigma = 50$. Find the percentage of students scoring between 500 and 600.

4-4 The distribution of **x** equals $F(x) = (1 - e^{-2x})U(x)$. Find $P\{\mathbf{x} > .5\}$, $P\{.3 < \mathbf{x} < .7\}$. Find the u-percentiles of **x** for $u = .1, .2, \ldots, .9$.

4-5 Given $F(5) = .940$ and $F(6) = .956$, find the $x_{0.95}$ percentile of **x** using linear interpolation.

4-6 If $F(x) = x/5$ for $0 \le x \le 5$, find $f(x)$ for all x. Find the corresponding u-percentiles x_u for $u = .1, .2, \ldots, .9$.

4-7 The following is a list of the observed values x_i of an RV **x**:

−17 −9 −7 −6 −2 3 6 11 14 18 23 26 31 34 35 37 39 42 44 48 51 53 61

Sketch the empirical distribution $F_n(x)$ and the percentile curve x_u of **x**. Sketch the histogram $f_n(x)$ for $\Delta = 5$ and $\Delta = 10$.

4-8 The life length of a pump is modeled by an RV **x** with distribution $F(x) = (1 - e^{-cx})U(x)$. The median of **x** equals $x_{0.5} = 50$ months. **(a)** Under a warranty, a pump is replaced if it fails within five months. Find the percentage of units that are replaced under the warranty. **(b)** How long should the warranty last if under the warranty only 3% of the units are replaced?

4-9 **(a)** Show that if $f(x) = c^2xe^{-cx}U(x)$, then $F(x) = (1 - e^{-cx} - cxe^{-cx})U(x)$. **(b)** Find the distribution of the Erlang density (page 106).

4-10 The maximum electric power needed in a community is modeled by an RV **x** with density $c^2xe^{-cx}U(x)$ where $c = 5 \times 10^{-6}$ per kilowatt. **(a)** The power available is 10^6 kw; find the probability for a blackout. **(b)** Find the power needed so that the probability for a blackout is less than .005.

4-11 We wish to control the quality of resistors with nominal resistance $R = 1{,}000\Omega$. To do so, we measure all resistors and accept only the units with R between 960 and 1,040. Find the percentage of the units that are rejected. **(a)** If R is modeled by an $N(1000, 20)$ RV. **(b)** If R is modeled by an RV uniformly distributed in the interval (950, 1,050).

4-12 The probability that a manufactured product is defective equals p. We test each unit and denote by **x** the number of units tested until a defective one is found. **(a)** Find the distribution of **x**. **(b)** Find $P\{\mathbf{x} > 30\}$ if $p = .05$.

4-13 We receive a shipment of 200 units, of which 18 are defective. We test 10 units at random, and we model the number k of the units that test defective by an RV **x**. **(a)** Find the distribution of **x**. **(b)** Find $P\{\mathbf{x} = 2\}$.

4-14 (Pascal distribution) The RV **x** equals the number of tosses of a coin until heads shows k times. Show that

$$P\{\mathbf{x} = k\} = \binom{n-1}{k-1} p^k q^{n-k}$$

4-15 The number of daily accidents in a region is a Poisson RV **x** with parameter $a = 3$. **(a)** Find the probability that there are no accidents in a day. **(b)** Find the probability that there are more than four accidents. **(c)** Sketch $F(x)$.

4-16 Given an $N(1, 2)$ RV **x** and the functions (Fig. P4.16)

$$g_1(x) = \begin{cases} 0 & |x| \le 2 \\ x & |x| > 2 \end{cases} \qquad g_2(x) = \begin{cases} 2 & x > 2 \\ x & |x| \le 2 \\ -2 & x < -2 \end{cases} \qquad g_3(x) = \begin{cases} -1 & x < 0 \\ 1 & x > 0 \end{cases}$$

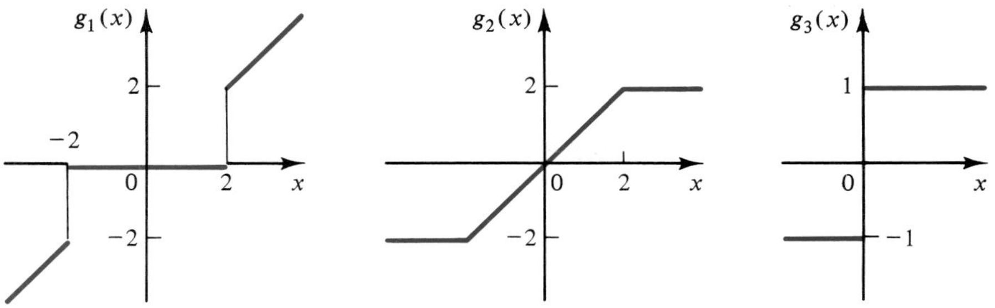

Figure P4.16

we form the RVS $\mathbf{y} = g_1(\mathbf{x})$, $\mathbf{z} = g_2(\mathbf{x})$, $\mathbf{w} = g_3(\mathbf{x})$. **(a)** Find their distributions. **(b)** Find the probabilities of the events $\{\mathbf{y} = 0\}$, $\{\mathbf{z} = 0\}$, and $\{\mathbf{w} = 0\}$.

4-17 The RV $\mathbf{x}$ is uniform in the interval (0, 6). Find and sketch the density of the RV $\mathbf{y} = -2\mathbf{x} + 3$.

4-18 The input to a system is the sum $\mathbf{x} = 10 + \boldsymbol{\nu}$ where $\boldsymbol{\nu}$ is an $N(0, 2)$ RV, and the output is $\mathbf{y} = \mathbf{x}^2$. Find $f_y(y)$ and $F_y(y)$.

4-19 Express the density $f_y(y)$ of the RV $\mathbf{y} = g(\mathbf{x})$ in terms of the density $f_x(x)$ of $\mathbf{x}$ for the following cases: **(a)** $g(x) = x^3$; **(b)** $g(x) = x^4$; **(c)** $g(x) = |x|$ (full-wave rectifier); **(d)** $g(x) = xU(x)$ (half-wave rectifier).

4-20 The base of a right triangle equals 5, and the adjacent angle is an RV $\boldsymbol{\theta}$ uniformly distributed in the interval $(0, \pi/4)$. Find the distribution of the length $\mathbf{b} = 5 \tan\boldsymbol{\theta}$ of the opposite side.

4-21 The RV $\mathbf{x}$ is uniform in the interval (0, 1). Show that the density of the RV $\mathbf{y} = -\ ln\mathbf{x}$ equals $e^{-y}U(y)$.

4-22 *Lognormal distribution.* The RV $\mathbf{x}$ is $N(\eta, \sigma)$ and $\mathbf{y} = e^{\mathbf{x}}$. Show that

$$f_y(y) = \frac{1}{\sigma y \sqrt{2\pi}} \exp\left\{-\frac{1}{2\sigma^2}(\ln y - \eta)\right\} U(y)$$

This density is called *lognormal*.

4-23 Given an RV $\mathbf{x}$ with distribution $F_x(x)$, we form the RV $\mathbf{y} = 2F_x(\mathbf{x}) + 3$. Find $f_y(y)$ and $F_y(y)$.

4-24 Find the constants a and b such that if $\mathbf{y} = a\mathbf{x} + b$ and $\eta_x = 5$, $\sigma_x = 2$, then $\eta_y = 0$, $\sigma_y = 1$.

4-25 A fair die is rolled once, and $\mathbf{x}$ equals the number of faces up. Find η_x and σ_x.

4-26 We roll two fair dice and denote by $\mathbf{x}$ the number of rolls until 7 shows. Find $E\{\mathbf{x}\}$.

4-27 A game uses two fair dice. To participate, you pay \$20 per roll. You win \$10 if even shows, \$42 if 7 shows, and \$102 if 11 shows. The game is fair if your expected gain is \$20. Is the game fair?

4-28 Show that for any c, $E\{(\mathbf{x} - c)^2\} = (\eta_x - c)^2 + \sigma_x^2$.

4-29 The resistance of a resistor is an RV $\mathbf{R}$ with mean 1,000Ω and standard deviation 20Ω. It is connected to a voltage source $V = 110$V. Find approximately the mean of the power $\mathbf{P} = V^2/\mathbf{R}$ dissipated in the resistor, using (4-113).

4-30 The RV **x** is uniform in the interval (9, 11) and $\mathbf{y} = x^3$. **(a)** Find $f_y(y)$; **(b)** find η_x and σ_x; **(c)** find the mean of **y** using three methods: (i) directly from (4-93); (ii) indirectly from (4-94); (iii) approximately from (4-113).

4-31 The RV **x** is $N(\eta, \sigma)$, and the function $g(x)$ is nearly linear in the interval $\eta - 3\sigma < x < y + 3\sigma$ with derivative $g'(x)$. Show that the RV $\mathbf{y} = g(\mathbf{x})$ is nearly normal with mean $\eta_y = g(\eta)$ and standard deviation $\sigma_y = |g'(\eta)|\sigma$.

4-32 Show that if $f_x(x) = e^{-x}U(x)$ and $\mathbf{y} = \sqrt{2\mathbf{x}}$ then $f_y(y) = ye^{-y^2/2}U(y)$.

5 Two Random Variables

Extending the concepts developed in Chapter 4 to two RVS, we associate a distribution, a density, an expected value, and a moment-generating function to the pair $(\mathbf{x}, \mathbf{y})$ where $\mathbf{x}$ and $\mathbf{y}$ are two RVS. In addition, we introduce the notion of independence of two RVS in terms of the independence of two events as defined in Chapter 2. Finally, we form the composite functions $\mathbf{z} = g(\mathbf{x}, \mathbf{y})$, $\mathbf{w} = h(\mathbf{x}, \mathbf{y})$ and express their joint distribution in terms of the joint distribution of the RVS $\mathbf{x}$ and $\mathbf{y}$.

5-1 The Joint Distribution Function

We are given two RVS $\mathbf{x}$ and $\mathbf{y}$ defined on an experiment $\mathscr{S}$. The properties of each of these RVS are completely specified in terms of their respective distributions $F_x(x)$ and $F_y(y)$. However, their joint properties—that is, the probability $P\{(\mathbf{x}, \mathbf{y}) \in D\}$ that the point $(\mathbf{x}, \mathbf{y})$ is in an arbitrary region D of the plane—cannot generally be determined if we know only $F_x(x)$ and $F_y(y)$. To find this probability, we introduce the following concept.

■ ***Definition.*** The *joint cumulative distribution function* or, simply, the *joint distribution* $F_{xy}(x, y)$ of the RVS $\mathbf{x}$ and $\mathbf{y}$ is the probability of the event

$$\{\mathbf{x} \le x, \mathbf{y} \le y\} = \{\mathbf{x} \le x\} \cap \{\mathbf{y} \le y\}$$

consisting of all outcomes ζ such that $\mathbf{x}(\zeta) \le x$ and $\mathbf{y}(\zeta) \le y$. Thus

$$F_{xy}(x, y) = P\{\mathbf{x} < x, \mathbf{y} \le y\} \tag{5-1}$$

The function $F_{xy}(x, y)$ is defined for every x and y; its subscripts will often be omitted.

The probability $P\{(\mathbf{x}, \mathbf{y}) \in D\}$ will be interpreted as mass in the region D. Clearly, $F(x, y)$ is the probability that the point $(\mathbf{x}, \mathbf{y})$ is in the quadrant D_0 of Fig. 5.1; hence, it equals the mass in D_0. Guided by this, we conclude that the masses in the regions D_1, D_2, and D_3 are given by

$$P\{\mathbf{x} \le x, y_1 < \mathbf{y} \le y_2\} = F(x, y_2) - F(x, y_1) \tag{5-2}$$

$$P\{x_1 < \mathbf{x} \le x_2, \mathbf{y} \le y\} = F(x_2, y) - F(x_1, y) \tag{5-3}$$

$$P\{x_1 < \mathbf{x} \le x_2, y_1 < \mathbf{y} \le y_2\} = F(x_2, y_2) - F(x_1, y_2) - F(x_2, y_1) + F(x_1, y_1) \tag{5-4}$$

respectively.

JOINT DENSITY If the RVS $\mathbf{x}$ and $\mathbf{y}$ are of continuous type, the probability masses in the xy plane can be described in terms of the function $f(x, y)$ specified by

$$P\{x < \mathbf{x} \le x + dx, y < \mathbf{y} \le y + dy\} = f(x, y)\,dx\,dy \tag{5-5}$$

This is the probability that the RVS $\mathbf{x}$ and $\mathbf{y}$ are in a differential rectangle of area $dx\,dy$. The function $f(x, y)$ will be called the *joint density* function of the RVS $\mathbf{x}$ and $\mathbf{y}$.

From (5-5) it follows that the probability that $(\mathbf{x}, \mathbf{y})$ is a point in an arbitrary region D of the plane equals

$$P\{(\mathbf{x}, \mathbf{y}) \subset D\} = \iint_D f(x, y)\,dx\,dy \tag{5-6}$$

Applying (5-6) to the region D_0 of Fig. 5.1, we obtain

$$F(x, y) = \int_{-\infty}^{y} \int_{-\infty}^{x} f(\alpha, \beta)\,d\alpha\,d\beta \tag{5-7}$$

This yields

$$f(x, y) = \frac{\partial^2 F(x, y)}{\partial x\,\partial y} \tag{5-8}$$

Note, finally, that

$$f(x, y) \ge 0 \qquad \int_{-\infty}^{\infty} \int_{-\infty}^{\infty} f(x, y)\,dx\,dy = F(\infty, \infty) = 1 \tag{5-9}$$

This follows from (5-5) and the fact that $\{\mathbf{x} \le \infty, \mathbf{y} \le \infty\}$ is the certain event.

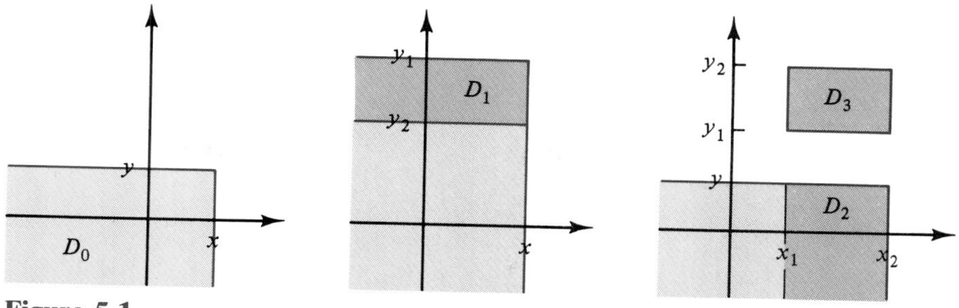

Figure 5.1

MARGINAL DISTRIBUTIONS In the study of the joint properties of several RVS, the distributions of each are called marginal. Thus $F_x(x)$ is the *marginal distribution* and $f_x(x)$ the *marginal density* of the RV **x**. As we show next, these functions can be expressed in terms of the joint distribution of **x** and **y**.

We maintain that

$$F_x(x) = F_{xy}(x, \infty) \qquad f_x(x) = \int_{-\infty}^{\infty} f_{xy}(x, y)\, dy \tag{5-10}$$

The first equation follows from the fact that $\{\mathbf{x} \le x\} = \{\mathbf{x} \le x, \mathbf{y} \le \infty\}$ is the event consisting of all outcomes such that the point $(\mathbf{x}, \mathbf{y})$ is on the left of the line L_x of Fig. 5.2a. To prove the second equation, we observe that $f_x(x)\,dx = P\{x \le \mathbf{x} \le x + dx\}$ is the probability mass in the shaded region ΔD of Fig. 5.2b. This yields

$$f_x(x)\, dx = \iint_{\Delta D} f_{xy}(x, y)\, dx\, dy = dx \int_{-\infty}^{\infty} f_{xy}(x, y)\, dy$$

We can show similarly that

$$F_y(y) = F_{xy}(\infty, y) \qquad f_y(y) = \int_{-\infty}^{\infty} f_{xy}(x, y)\, dx \tag{5-11}$$

Figure 5.2

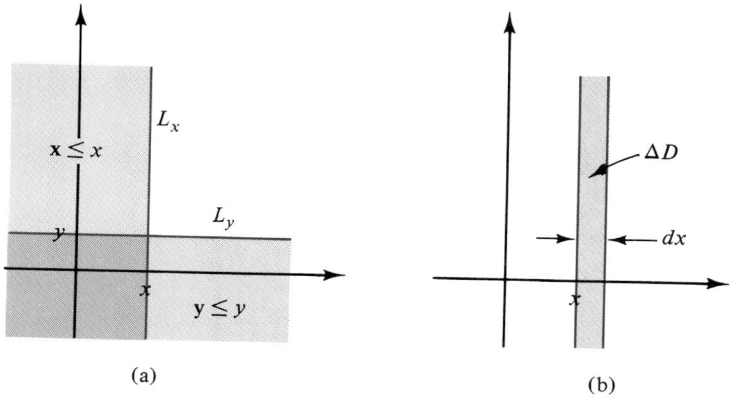

POINT AND LINE MASSES Suppose that the RVs **x** and **y** are of discrete type taking the values x_i and y_k, respectively, with *joint probabilities*

$$P\{\mathbf{x} = x_i, \mathbf{y} = y_k\} = f(x_i, y_k) = p_{ik} \tag{5-12}$$

In this case, the probability masses are at the points (x_i, y_k) and $f(x_i\, y_k)$ is a point density.

The *marginal probabilities*

$$P\{\mathbf{x} = x_i\} = f_x(x_i) = p_i \qquad P\{\mathbf{y} = y_k\} = f_y(y_k) = p_k \tag{5-13}$$

can be expressed in terms of the joint probabilities p_{ik}. Clearly, $P\{\mathbf{x} = x_i\}$ equals the sum of the masses of all points on the vertical line $x = x_i$, and $P\{\mathbf{y} = y_k\}$ equals the sum of the masses of all points on the horizontal line $y = y_k$. Hence,

$$p_i = \sum_k p_{ik} \qquad q_k = \sum_i p_{ik} \tag{5-14}$$

Note, finally, that

$$\sum_i p_i = \sum_k q_k = \sum_{i,k} p_{ik} = 1 \tag{5-15}$$

Example 5.1

A fair die rolled twice defines an experiment with 36 outcomes $f_i f_j$.

(a) The RVs **x** and **y** equal the number of faces up at the first and second roll, respectively:

$$\mathbf{x}(f_i f_j) = i \qquad \mathbf{y}(f_i, f_j) = j \qquad i, j = 1, \ldots, 6 \tag{5-16}$$

This yields the 36 points of Fig. 5.3a; the mass of each point equals 1/36.

(b) We now define **x** and **y** such that

$$\mathbf{x}(f_i f_j) = |i - j| \qquad \mathbf{y}(f_i f_j) = i + j$$

Thus **x** takes the six values 0, 1, . . . , 5 and **y** takes the 11 values 2, 3, . . . , 12. The corresponding marginal probabilities equal

$$p_i = \frac{6}{36}\ \frac{10}{36}\ \frac{8}{36}\ \frac{6}{36}\ \frac{4}{36}\ \frac{2}{36}$$

$$q_j = \frac{1}{36}\ \frac{2}{36}\ \frac{3}{36}\ \frac{4}{36}\ \frac{5}{36}\ \frac{6}{36}\ \frac{5}{36}\ \frac{4}{36}\ \frac{3}{36}\ \frac{2}{36}\ \frac{1}{36}$$

Figure 5.3

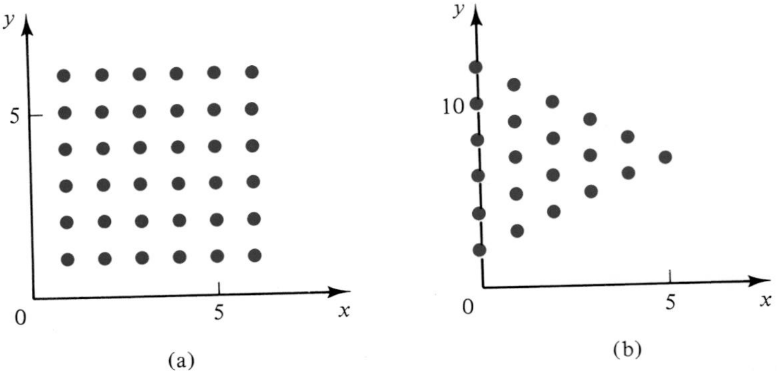

In this case, we have 21 points on the plane (Fig. 5.3b). There are six points on the line $x = 0$ with masses of 1/36; for example, the mass of the point (0, 4) equals 1/36 because $\{\mathbf{x} = 0, \mathbf{y} = 4\} = \{f_2, f_2\}$. The masses of all other points equal 2/36; for example, the mass of the point (3, 5) equals 2/36 because $\{\mathbf{x} = 3, \mathbf{y} = 5\} = \{f_1 f_4, f_4 f_1\}$. ■

In addition to continuous and discrete type RVs, joint distributions can be of mixed type involving distributed masses, point masses, and line masses of various kinds. We comment next on two cases involving line masses only. Suppose, first, that **x** is of discrete type, taking the values x_i, and **y** is of continuous type. In this case, all probability masses are on the vertical lines $x = x_i$ (Fig. 5.4a). The mass between the points y_1 and y_2 of the line $x = x_i$ equals the probability of the event $\{\mathbf{x} = x_i, y_1 < \mathbf{y} < y_2\}$. Suppose, finally, that **x** is of continuous type and $\mathbf{y} = g(\mathbf{x})$. If D is a region of the plane not containing any part of the curve $y = g(x)$, the probability that $(\mathbf{x}, \mathbf{y})$ is in this region equals zero. From this it follows that all masses are on the curve $y = g(x)$. In this case, the joint distribution $F(x, y)$ can be expressed in terms of the marginal distribution $F_x(x)$ and the function $g(x)$. For example, with x and y as in Fig. 5.4b, $F(x, y)$ equals the masses on the curve $y = g(x)$ inside the shaded area. This includes the masses on the left of the point A and between the points B and C. Hence,

$$F(x, y) = F_x(x_1) + F_x(x) - F_x(x_2)$$

Independent RVS

As we recall [see (2-67)], two events $\mathcal{A}$ and $\mathcal{B}$ are independent if

$$P(\mathcal{A} \cap \mathcal{B}) = P(\mathcal{A})P(\mathcal{B})$$

The notion of independence of two RVs is based on this.

■ ***Definition.*** We shall say that the RVs **x** and **y** are *statistically independent* if the events $\{\mathbf{x} \le x\}$ and $\{\mathbf{y} \le y\}$ are independent, that is, if

$$P\{\mathbf{x} \le x, \mathbf{y} \le y\} = P\{\mathbf{x} \le x\}P\{\mathbf{y} \le y\} \qquad (5\text{-}17)$$

Figure 5.4

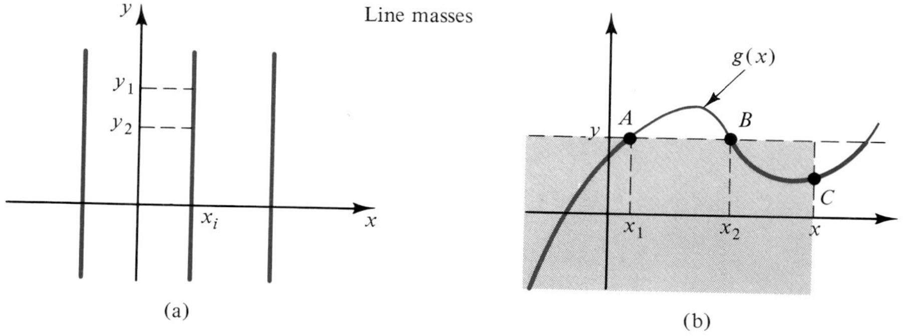

for any x and y. This yields

$$F_{xy}(x, y) = F_x(x)F_y(y) \tag{5-18}$$

Differentiating, we obtain

$$f_{xy}(x, y) = f_x(x)f_y(y) \tag{5-19}$$

Thus two RVs are independent if they satisfy (5-17) or (5-18) or (5-19). Otherwise, they are "statistically dependent."

From the definition it follows that the events $\{x_1 \le \mathbf{x} < x_2\}$ and $\{y_1 \le \mathbf{y} < y_2\}$ are independent; hence, the masses in the rectangle D_3 of Fig. 5.1 equals the product of the masses in the vertical strip $(x_1 \le x < x_2)$ times the masses in the horizontal strip $(y_1 \le y < y_2)$. More generally, if A and B are two point sets on the x-axis and the y-axis, respectively, the events $\{\mathbf{x} \in A\}$ and $\{\mathbf{y} \in B\}$ are independent.

Applying this reasoning to the events $\{\mathbf{x} = x_i\}$ and $\{\mathbf{y} = y_k\}$, we conclude that if $\mathbf{x}$ and $\mathbf{y}$ are two discrete type RVs as in (5-12) and independent, then

$$p_{ik} = p_i q_k \tag{5-20}$$

Note, finally, that if two RVs $\mathbf{x}$ and $\mathbf{y}$ are "functionally dependent," that is, if $\mathbf{y} = g(\mathbf{x})$, they cannot be (statistically) independent.

Independent Experiments Independent RVs are generated primarily by combined experiments. Suppose that $\mathcal{S}_1$ is an experiment with outcomes ζ_1 and $\mathcal{S}_2$ another experiment with outcomes ζ_2. Proceeding as in Section 3-1, we form the combined experiment (product space) $\mathcal{S} = \mathcal{S}_1 \times \mathcal{S}_2$. The outcomes of this experiment are $\xi_1\xi_2$ where ξ_1 is any one of the elements ζ_1 of $\mathcal{S}_1$ and ξ_2 is any one of the elements ζ_2 of $\mathcal{S}_2$. In the experiment $\mathcal{S}$, we define the RVs $\mathbf{x}$ and $\mathbf{y}$ such that $\mathbf{x}$ depends only on the outcomes of $\mathcal{S}_1$ and $\mathbf{y}$ depends only on the outcomes of $\mathcal{S}_2$. In other words, $\mathbf{x}(\xi_1\xi_2)$ depends only on ξ_1 and $\mathbf{y}(\xi_1\xi_2)$ depends only on ξ_2. We can show that if the experiments $\mathcal{S}_1$ and $\mathcal{S}_2$ are independent, the RVs $\mathbf{x}$ and $\mathbf{y}$ so formed are independent as well.

Consider, for example, the RVs $\mathbf{x}$ and $\mathbf{y}$ in (5-16). In this case, $\mathcal{S}_1$ is the first roll of the die and $\mathcal{S}_2$ the second. Furthermore, $P\{\mathbf{x} = i\} = 1/6$, $P\{\mathbf{y} = k\} = 1/6$. This yields

$$P = \{\mathbf{x} = i, \mathbf{y} = k\} = \frac{1}{36} = P\{\mathbf{x} = i\}P\{\mathbf{y} = k\}$$

Hence, in the space $\mathcal{S}$ of the two independent rolls of a die, the RVs $\mathbf{x}$ and $\mathbf{y}$ are independent.

ILLUSTRATIONS The RVs **x** and **y** are independent, **x** is uniform in the interval $(0, a)$, and **y** is uniform in the interval $(0, b)$.

$$f_x(x) = \begin{cases} \dfrac{1}{a} & 0 \le x \le a \\ 0 & \text{otherwise} \end{cases}$$

$$f_y(y) = \begin{cases} \dfrac{1}{b} & 0 \le y \le b \\ 0 & \text{otherwise} \end{cases}$$

Thus the probability that **x** is in the interval $(x, x + \Delta x)$ equals $\Delta x/a$, and the probability that **y** is in the interval $(y, y + \Delta y)$ equals $\Delta y/b$. From this and the independence of **x** and **y** it follows that the probability that the point $(\mathbf{x}, \mathbf{y})$ is in a rectangle with sides Δx and Δy included in the region $R = \{0 \le x \le a, 0 \le y \le b\}$ equals $\Delta x \Delta y/ab$. This leads to the conclusion that

$$f(x, y) = \begin{cases} \dfrac{1}{ab} & 0 \le x \le a,\ 0 \le y \le b \\ 0 & \text{otherwise} \end{cases} \qquad (5\text{-}21)$$

The probability that $(\mathbf{x}, \mathbf{y})$ is in a region D included in R equals the area of D divided by ab.

Example 5.2

A fine needle of length c is dropped "at random" on a board covered with parallel lines distance d apart (Fig. 5.5a). We wish to find the probability p that the needle intersects one of the lines. This experiment is called *Buffon's needle*.

We shall first explain the meaning of randomness: We denote by **x** the angle between the needle and the parallel lines and by **y** the distance from the center of the needle to the nearest line. We assume that the RVs **x** and **y** are independent and uniform in the intervals $(0, \pi/2)$ and $(0, d/2)$, respectively. Clearly, the needle intersects the lines iff

$$\mathbf{y} < \frac{c}{2} \sin \mathbf{x}$$

Figure 5.5

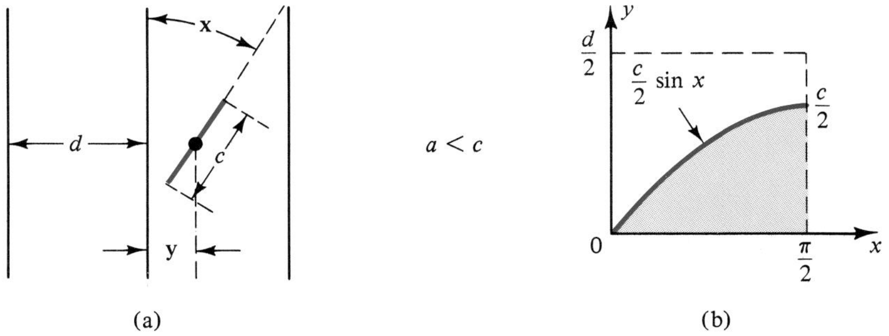

Hence, to find p, it suffices to find the area of the shaded region of Fig. 5.5b and use (5-21) with $a = \pi/2$ and $b = d/2$. Thus

$$p = \frac{4}{\pi d} \int_0^{c/2} \frac{c}{2} \cos x \, dx = \frac{2c}{\pi d}$$

The Monte Carlo Method From the empirical interpretation of probability it follows that if the needle is dropped n times, the number n_i of times it intersects one of the lines equals $n_i \simeq np = 2nc/\pi d$. This relationship is used to determine π empirically: If n is sufficiently large, $\pi \simeq 2nc/dn_i$. Thus we can evaluate approximately the deterministic number π in terms of averages of numbers obtained from a random experiment. This technique, known as the Monte Carlo method, is used in performing empirically various mathematical operations, especially estimating complicated integrals in several variables (see Section 8-3). ■

The RVS $\mathbf{x}$ and $\mathbf{y}$ are independent and normal, with densities

$$f_x(x) = \frac{1}{\sigma\sqrt{2\pi}} e^{-x^2/2\sigma^2} \qquad f_y(y) = \frac{1}{\sigma\sqrt{2\pi}} e^{-y^2/2\sigma^2}$$

In this case, (5-19) yields

$$f(x, y) = \frac{1}{2\pi\sigma^2} e^{-(x^2+y^2)/2\sigma^2} \tag{5-22}$$

Example 5.3

With $f(x, y)$ as in (5-22), find the probability p that the point $(\mathbf{x}, \mathbf{y})$ is in the circle $\sqrt{x^2 + y^2} \le a$.

As we see from (5-22) and (5-6),

$$P\{\sqrt{\mathbf{x}^2 + \mathbf{y}^2} < a\} = \frac{1}{2\pi\sigma^2} \iint_{\sqrt{x^2+y^2}\le a} e^{-(x^2+y^2)/2\sigma^2} \, dx \, dy$$

$$= \frac{1}{2\pi\sigma^2} \int_0^a 2\pi r \, e^{-r^2/2\sigma^2} \, dr = 1 - e^{-a^2/2\sigma^2}$$ ■

Circular Symmetry We shall say that a density function $f(x, y)$ has *circular symmetry* if it depends only on the distance of the point (x, y) from the origin, that is, if

$$f(x, y) = \phi(r) \qquad r = \sqrt{x^2 + y^2} \tag{5-23}$$

As we see from (5-22), if the RVS $\mathbf{x}$ and $\mathbf{y}$ are normal independent with zero mean and equal variance, their joint density has circular symmetry. We show next that the converse is also true.

■ ***Theorem.*** If the RVS $\mathbf{x}$ and $\mathbf{y}$ are independent and their joint density is circularly symmetrical as in (5-23), they are normal with zero mean and equal variance.

This remarkable theorem is another example of the importance of the normal distribution. It shows that normality is a consequence of independence and circular symmetry, conditions that are met in many applications.

■ ***Proof.*** From (5-23) and (5-19) it follows that

$$\phi(r) = f_x(x)f_y(y) \qquad r = \sqrt{x^2 + y^2} \tag{5-24}$$

We shall show that this identity leads to the conclusion that the functions $f_x(x)$ and $f_y(y)$ are normal. Differentiating both sides with respect to x and using the identity

$$\frac{\partial}{\partial x}\phi(r) = \frac{d\phi(r)}{dr}\frac{\partial r}{\partial x} = \frac{x}{r}\phi'(r)$$

we obtain

$$\frac{x}{r}\phi'(r) = f_x'(x)f_y(y) \tag{5-25}$$

As we see from (5-24), $x\phi(r) = xf_x(x)f_y(y)$. From this and (5-25) it follows that

$$\frac{1}{r}\frac{\phi'(r)}{\phi(r)} = \frac{1}{x}\frac{f_x'(x)}{f_x(x)}$$

The left side is a function only of $r = \sqrt{x^2 + y^2}$, and the right side is independent of y. Setting $x = 0$, we conclude that both sides are constant. Thus

$$\frac{1}{x}\frac{f_x'(x)}{f_x(x)} = \alpha = \text{constant} \qquad \frac{d}{dx}\ln f_x(x) = \alpha x$$

$$\ln f_x(x) = \frac{\alpha x^2}{2} + C \qquad f_x(x) = C\,e^{\alpha x^2/2}$$

Since $f_x(x)$ is a density, $\alpha < 0$; hence, $\mathbf{x}$ is normal. Reasoning similarly, we conclude that $\mathbf{y}$ is also normal with the same α.

Functions of Independent RVs Suppose now that $\mathbf{z}$ is a function of the RV $\mathbf{x}$ and that $\mathbf{w}$ is a function of the RV $\mathbf{y}$:

$$\mathbf{z} = g(\mathbf{x}) \qquad \mathbf{w} = h(\mathbf{x}) \tag{5-26}$$

We maintain that if the RVs $\mathbf{x}$ and $\mathbf{y}$ are statistically independent, the RVs $\mathbf{z}$ and $\mathbf{w}$ are also statistically independent.

■ ***Proof.*** We denote by A_z the set of values of x such that $g(x) \le z$ and by B_w the set of values y such that $h(y) \le w$. From this it follows that

$$\{\mathbf{z} \le z\} = \{\mathbf{x} \in A_z\} \qquad \{\mathbf{w} \le w\} = \{\mathbf{y} \in B_w\}$$

for any z and w. Since $\mathbf{x}$ and $\mathbf{y}$ are independent, the events $\{\mathbf{x} \in A_z\}$ and $\{\mathbf{y} \in B_w\}$ are independent; hence,

$$P\{\mathbf{z} \le z, \mathbf{w} \le w\} = P\{\mathbf{z} \le z\}P\{\mathbf{w} \le w\} \tag{5-27}$$

Note, for example, that the RVs $\mathbf{x}^2$ and $\mathbf{y}^3$ are independent.

5-2

Mean, Correlation, Moments

The properties of two RVs **x** and **y** are completely specified in terms of their joint distribution. In this section, we give a partial specification involving only a small number of parameters. As a preparation, we define the function $g(\mathbf{x}, \mathbf{y})$ of the RVs **x** and **y** and determine its mean.

Given a function $g(x, y)$, we form the composite function

$$\mathbf{z} = g(\mathbf{x}, \mathbf{y})$$

This function is an RV as in Section 4-1, with domain the set $\mathcal{S}$ of experimental outcomes. For a specific outcome $\zeta_i \in \mathcal{S}$, the value $\mathbf{z}(\zeta_i)$ of **z** equals $g(x_i, y_i)$ where x_i and y_i are the corresponding values of **x** and **y.** We shall express the mean of **z** in terms of the joint density $f(x, y)$ of the RVs **x** and **y.**

As we have shown in (4-94), the mean of the RV $g(\mathbf{x})$ is given by an integral involving the density $f_x(x)$. Expressing $f_x(x)$ in terms of $f(x, y)$ [see (5-10)], we obtain

$$E\{g(\mathbf{x})\} = \int_{\infty}^{\infty} g(x)f_x(x)\,dx = \int_{\infty}^{\infty}\int_{\infty}^{\infty} g(x)f(x, y)\,dx\,dy \tag{5-28}$$

We shall obtain a similar expression for $E\{g(\mathbf{x}, \mathbf{y})\}$.

The mean of the RV $\mathbf{z} = g(\mathbf{x}, \mathbf{y})$ equals

$$\eta_z = \int_{-\infty}^{\infty} zf_z(z)\,dz \tag{5-29}$$

To find η_z, we must find first the density of **z**. We show next that, as in (5-28), this can be avoided.

■ ***Theorem***

$$E\{g(\mathbf{x}, \mathbf{y})\} = \int_{-\infty}^{\infty}\int_{-\infty}^{\infty} g(x, y)f(x, y)\,dx\,dy \tag{5-30}$$

and if the RVs **x** and **y** are of discrete type as in (5-12),

$$E\{g(\mathbf{x}, \mathbf{y})\} = \sum_{i,k} g(x_i, y_k)f(x_i, y_k) \tag{5-31}$$

■ ***Proof.*** The theorem is an extension of (4-94) to two variables. To prove it, we denote by ΔD_z the region of the xy plane such that $z < g(x, y) < z + dz$. To each differential dz in (5-29) there corresponds a region ΔD_z in the xy plane where $g(x, y) \simeq z$ and

$$P\{z \le \mathbf{z} \le z + dz\} = P\{(\mathbf{x}, \mathbf{y}) \in \Delta D_z\}$$

As dz covers the z-axis, the corresponding regions ΔD_z are nonoverlapping and they cover the entire xy plane. Hence, the integrals in (5-29) and (5-30) are equal.

It follows from (5-30) and (5-31) that

$$E\{g_1(\mathbf{x}, \mathbf{y}) + \cdots + g_n(\mathbf{x}, \mathbf{y})\} = E\{g_1(\mathbf{x}, \mathbf{y})\} + \cdots + E\{g_n(\mathbf{x}, \mathbf{y})\} \quad (5\text{-}32)$$

as in (4-96) (linearity of expected values).

Note [see (5-28)] that

$$\eta_x = \int_{-\infty}^{\infty}\int_{-\infty}^{\infty} xf(x, y)\,dx\,dy \qquad \sigma_x^2 = \int_{-\infty}^{\infty}\int_{-\infty}^{\infty} (x - \eta_x)^2 f(x, y)\,dx\,dy$$

Thus (η_x, η_y) is the center of gravity of the probability masses on the plane. The variances σ_x^2 and σ_y^2 are measures of the concentration of these masses near the lines $x = \eta_x$ and $y = \eta_y$, respectively. We introduce next a fifth parameter that gives a measure of the linear dependence between **x** and **y**.

Covariance and Correlation

The *covariance* μ_{xy} of two RVS **x** and **y** is by definition the "mixed central moment":

$$\boxed{\operatorname{Cov}(\mathbf{x}, \mathbf{y}) = \mu_{xy} = E\{(\mathbf{x} - \eta_x)(\mathbf{y} - \eta_y)\}} \quad (5\text{-}33)$$

Since

$$E\{(\mathbf{x} - \eta_x)(\mathbf{y} - \eta_y)\} = E\{\mathbf{xy}\} - \eta_x E\{\mathbf{y}\} - \eta_y E\{\mathbf{x}\} + \eta_x\eta_y$$

(5-33) yields

$$\mu_{xy} = E\{\mathbf{xy}\} - E\{\mathbf{x}\}E\{\mathbf{y}\} \quad (5\text{-}34)$$

The ratio

$$\boxed{r_{xy} = \frac{\mu_{xy}}{\sigma_x\sigma_y}} \quad (5\text{-}35)$$

is called the *correlation coefficient* of the RVS **x** and **y**.

Note that r is the covariance of the centered and normalized RVS

$$\mathbf{x}_0 = \frac{\mathbf{x} - \eta_x}{\sigma_x} \qquad \mathbf{y}_0 = \frac{\mathbf{y} - \eta_y}{\sigma_y}$$

Indeed,

$$E\{\mathbf{x}_0\} = E\{\mathbf{y}_0\} = 0 \qquad \sigma_{x_0}^2 = E\{\mathbf{x}_0^2\} = 1 \qquad \sigma_{y_0}^2 = E\{\mathbf{y}_0^2\} = 1$$

$$E\{\mathbf{x}_0\mathbf{y}_0\} = E\left\{\frac{\mathbf{x} - \eta_x}{\sigma_x}\,\frac{\mathbf{y} - \eta_y}{\sigma_y}\right\} = \frac{\mu_{xy}}{\sigma_x\sigma_y} = r$$

Uncorrelatedness We shall say that the RVS $\mathbf{x}$ and $\mathbf{y}$ are *uncorrelated* if $r = 0$ or, equivalently, if

$$E\{\mathbf{xy}\} = E\{\mathbf{x}\}E\{\mathbf{y}\} \tag{5-36}$$

We shall next express the mean and the variance of the sum of two RVS in terms of the five parameters $\eta_x, \eta_y, \sigma_x, \sigma_y, \mu_{xy}$. Suppose that

$$\mathbf{z} = a\mathbf{x} + b\mathbf{y}$$

In this case, $\eta_z = a\eta_x + b\eta_x$ and

$$E\{(\mathbf{z} - \eta_z)^2\} = E\{[a(\mathbf{x} - \eta_x) + b(\mathbf{y} - \eta_y)]^2\}$$

Expanding the square and using the linearity of expected values, we obtain

$$\sigma_z^2 = a^2\sigma_x^2 + b^2\sigma_y^2 + 2ab\mu_{xy} \tag{5-37}$$

Note, in particular, that if $\mu_{xy} = 0$, then

$$\sigma_{x+y}^2 = \sigma_x^2 + \sigma_y^2 \tag{5-38}$$

Thus if two RVS are uncorrelated, the variance of their sum equals the sum of their variances.

Orthogonality We shall say that two RVS $\mathbf{x}$ and $\mathbf{y}$ are *orthogonal* if

$$E\{\mathbf{xy}\} = 0 \tag{5-39}$$

In this case, $E\{(\mathbf{x} + \mathbf{y})^2\} = E\{\mathbf{x}^2\} + E(\mathbf{y}^2\}$.

Orthogonality is closely related to uncorrelatedness: If the RVS $\mathbf{x}$ and $\mathbf{y}$ are uncorrelated, the *centered* RVS $\mathbf{x} - \eta_x$ and $\mathbf{y} - \eta_y$ are orthogonal.

Independence Recall that two RVS $\mathbf{x}$ and $\mathbf{y}$ are independent iff

$$f(x, y) = f_x(x)f_y(y) \tag{5-40}$$

■ ***Theorem.*** If two RVS are independent, they are uncorrelated.

■ ***Proof.*** If (5-40) is true,

$$E\{\mathbf{xy}\} = \int_{-\infty}^{\infty}\int_{-\infty}^{\infty} xyf(x, y)\,dx\,dy = \int_{-\infty}^{\infty}\int_{-\infty}^{\infty} xyf_x(x)f_y(y)\,dx\,dy$$

$$= \int_{-\infty}^{\infty} xf_x(x)\,dx \int_{-\infty}^{\infty} yf_y(y)\,dy = E\{\mathbf{x}\}E\{\mathbf{y}\}$$

The converse is true for normal RVS (see page 163) but generally it is not true: If two RVS are uncorrelated, they are not necessarily independent. Independence is a much stronger property than uncorrelatedness. The first is a point property; the second involves only averages. The following is an illustration of the difference.

Given two independent RVS $\mathbf{x}$ and $\mathbf{y}$, we form the RVS $g(\mathbf{x})$ and $h(\mathbf{y})$. As we know from (5-27), these RVS are also independent. From this and the theorem just given it follows that

$$E\{g(\mathbf{x})h(\mathbf{y})\} = E\{g(\mathbf{x})\}E\{h(\mathbf{y})\} \tag{5-41}$$

for any $g(x)$ and $g(y)$. This is not necessarily true if the RVs $\mathbf{x}$ and $\mathbf{y}$ are uncorrelated; functions of uncorrelated RVs are generally not uncorrelated.

SCHWARZ'S INEQUALITY We show next that the covariance of two RVs $\mathbf{x}$ and $\mathbf{y}$ cannot exceed the product $\sigma_x\sigma_y$. This is based on the following result, known as Schwarz's inequality:

$$E^2\{\mathbf{xy}\} \le E\{\mathbf{x}^2\}E\{\mathbf{y}^2\} \qquad (5\text{-}42)$$

Equality holds iff

$$\mathbf{y} = c_0\mathbf{x} \qquad (5\text{-}43)$$

■ ***Proof.*** With c an arbitrary constant, we form the mean of the RV $(\mathbf{y} - c\mathbf{x})^2$:

$$I(c) = E\{(\mathbf{y} - c\mathbf{x})^2\} = \int_{-\infty}^{\infty}\int_{-\infty}^{\infty} (y - cx)^2 f(x, y)\,dx\,dy \qquad (5\text{-}44)$$

Expanding the square and using the linearity of expected values, we obtain

$$I(c) = E\{\mathbf{y}^2\} - 2cE\{\mathbf{xy}\} + c^2E\{\mathbf{x}^2\} \qquad (5\text{-}45)$$

Thus $I(c)$ is a parabola (Fig. 5.6), and $I(c) \ge 0$ for every c; hence, the parabola cannot intersect the x-axis. Its discriminant D must therefore be negative:

$$D = 4E^2\{\mathbf{xy}\} - 4E\{\mathbf{x}^2\}E\{\mathbf{y}^2\} \le 0$$

and (5-42) results.

To prove (5-43), we observe that if $D = 0$, then $I(c)$ has a real root c_0; that is,

$$I(c_0) = E\{(\mathbf{y} - c_0\mathbf{x})^2\} = 0$$

This is possible only if the RV $\mathbf{y} - c_0\mathbf{x}$ equals zero [see (4-103)], as in (5-43).

■ ***Corollary.*** Applying (5-42) to the RVs $\mathbf{x} - \eta_x$ and $\mathbf{y} - \eta_y$, we conclude that

$$E^2\{(\mathbf{x} - \eta_x)(\mathbf{y} - \eta_y)\} \le E\{(\mathbf{x} - \eta_x)^2\}E\{(\mathbf{y} - \eta_y)^2\}$$

Figure 5.6

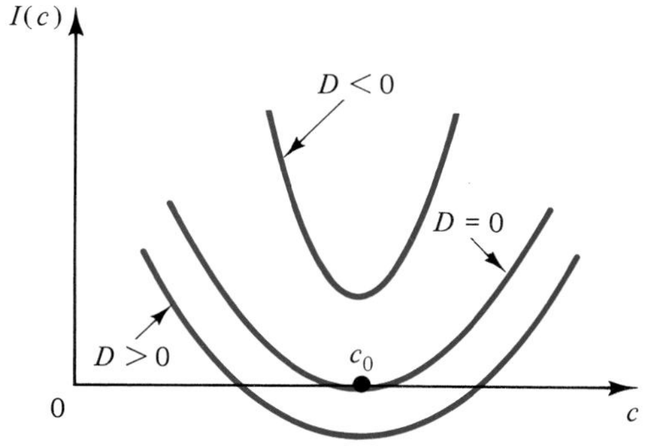

Hence,

$$\mu_{11}^2 \le \sigma_x \sigma_y \qquad |r| \le 1$$

Furthermore [see (5-43)], $|r| = 1$ iff

$$\mathbf{y} - \eta_y = c_0(\mathbf{x} - \eta_x) \tag{5-46}$$

To find c_0, we multiply both sides of (5-46) by $(\mathbf{x} - \eta_x)$ and take expected values. This yields

$$E\{(\mathbf{x} - \eta_x)(\mathbf{y} - \eta_y)\} = c_0 E\{(\mathbf{x} - \eta_x)^2\}$$

Hence,

$$c_0 = \frac{r\sigma_x\sigma_y}{\sigma_x^2} = \begin{cases} \dfrac{\sigma_y}{\sigma_x} & \text{if } r = 1 \\ -\dfrac{\sigma_y}{\sigma_x} & \text{if } r = -1 \end{cases}$$

We have thus reached the conclusion that if $|r| = 1$, then $\mathbf{y}$ is a linear function of $\mathbf{x}$ and all probability masses are on the line

$$y - \eta_y = \pm \frac{\sigma_y}{\sigma_x}(x - \eta_x) \qquad r = \pm 1$$

As $|r|$ decreases from 1 to 0, the masses move away from this line (Fig. 5.7).

Empirical Interpretation. We repeat the experiment n times and denote by x_i and y_i the observed values of the RVs $\mathbf{x}$ and $\mathbf{y}$ at the ith trial. At each of the n points (x_i, y_i), we place a point mass equal to $1/n$. The resulting pattern is the empirical mass distribution of $\mathbf{x}$ and $\mathbf{y}$. The joint empirical distribution $F_n(x, y)$ of these RVs is the sum of all masses in the quadrant D_0 of Fig. 5.1. The joint empirical density (two-dimensional histogram) is a function $f_n(x, y)$ such that the product $\Delta_1\Delta_2 f_n(x, y)$ equals the number of masses in a rectangle with sides Δ_1 and Δ_2 containing the point (x, y). For large n,

$$F_n(x, y) \simeq F(x, y) \qquad f_n(x, y) = f(x, y)$$

as in (4-24) and (4-37).

Figure 5.7

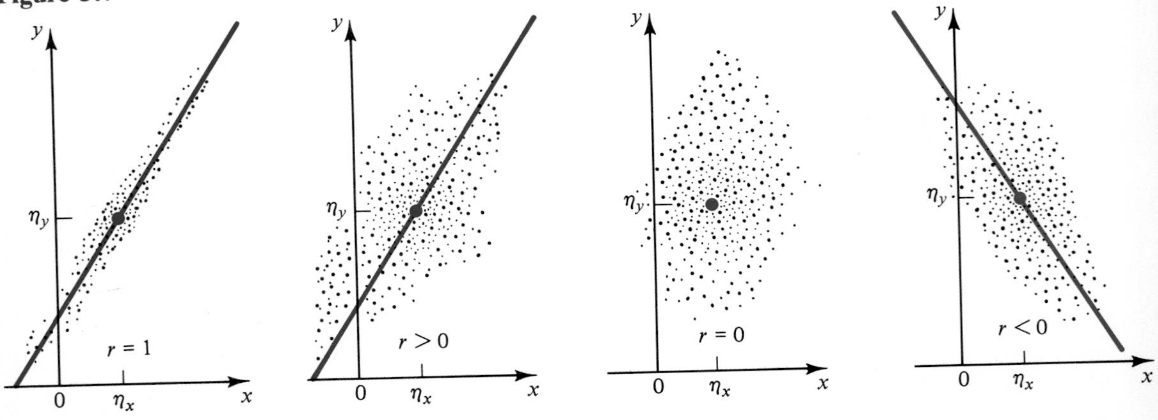

We next form the arithmetic averages $\bar{x}$ and $\bar{y}$ as in (4-89) and the empirical estimates

$$\bar{\sigma}_x^2 = \frac{1}{n}\sum_i (x_i - \bar{x})^2 \qquad \bar{\sigma}_y^2 = \frac{1}{n}\sum_i (y_i - \bar{y})^2$$

$$\bar{\mu}_{xy} = \frac{1}{n}\sum_i (x_i - \bar{x})(y_i - \bar{y})$$

of the parameters σ_x^2, σ_y^2, and μ_{xy}. The point $(\bar{x}, \bar{y})$ is the center of gravity of the empirical masses, $\bar{\sigma}_x^2$ is the second moment with respect to the line $x = \bar{x}$, and $\bar{\sigma}_y^2$ is the second moment with respect to the line $y = \bar{y}$. The ratio $\bar{\mu}_{xy}/\bar{\sigma}_x\bar{\sigma}_y$ is a measure of the concentration of the points near the line

$$y - \bar{y} = \pm \frac{\bar{\sigma}_y}{\bar{\sigma}_x}(x - \bar{x})$$

This line approaches the line (5-46) as $n \to \infty$. In Fig. 5.7 we show the empirical masses for various values of r.

LINEAR REGRESSION Regression is the determination of a function $y = \phi(x)$ "fitting" the probability masses in the xy plane. This can be phrased as a problem in estimation: Find a function $\phi(x)$ such that, if the RV $\mathbf{y}$ is estimated by $\phi(\mathbf{x})$, the estimation error $\mathbf{y} - \phi(\mathbf{x})$ is minimum in some sense. The general estimation problem is developed in Section 6-3 for two RVs and in Section 11-4 for an arbitrary number of RVs. In this section, we discuss the linear case; that is, we assume that $\phi(x) = a + bx$.

We are given two RVs $\mathbf{x}$ and $\mathbf{y}$, and we wish to find two constants a and b such that the line $y = a + bx$ is the best fit of "$\mathbf{y}$ on $\mathbf{x}$" in the sense of minimizing the mean-square value

$$e = E\{[\mathbf{y} - (a + b\mathbf{x})^2\} = \int_{-\infty}^{\infty}\int_{-\infty}^{\infty} [y - (a + bx)]^2 f(x, y)\,dx\,dy \qquad (5\text{-}48)$$

of the deviation $\nu = \mathbf{y} - (a + b\mathbf{x})$ of $\mathbf{y}$ from the straight line $a + b\mathbf{x}$ (Fig. 5.8).

Figure 5.8

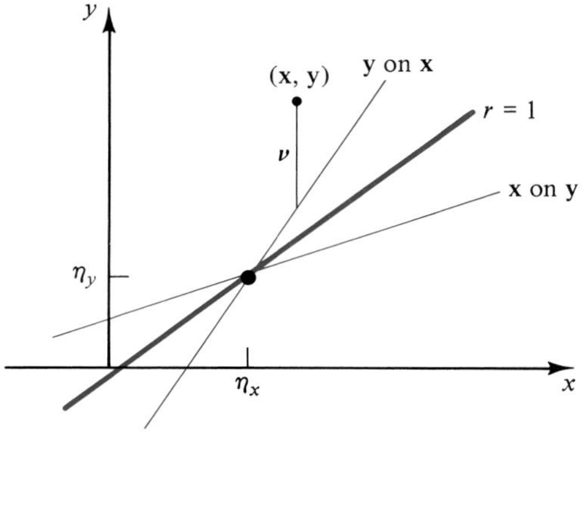

Clearly, e is a function of a and b, and it is minimum if

$$\begin{aligned}\frac{\partial e}{\partial a} &= -2\int_{-\infty}^{\infty}\int_{-\infty}^{\infty}[y - a + bx)]f(x, y)\,dx\,dy \\ &= -2E\{\mathbf{y} - (a + b\mathbf{x})\} = 0 \\ \frac{\partial e}{\partial b} &= -2\int_{-\infty}^{\infty}\int_{-\infty}^{\infty}[y - (a + bx)]xf(x, y)\,dx\,dy \\ &= -2E\{[\mathbf{y} - (a + b\mathbf{x})]\mathbf{x}\} = 0\end{aligned} \tag{5-49}$$

The first equation yields

$$\eta_y = a + b\eta_x$$

From this it follows that

$$\mathbf{y} - (a + b\mathbf{x}) = (\mathbf{y} - \eta_y) - b(\mathbf{x} - \eta_x) \qquad E\{(\mathbf{y} - \eta_y) - b(\mathbf{x} - \eta_x)\} = 0$$

Inserting into the second equation in (5-49), we obtain

$$E\{[(\mathbf{y} - \eta_y) - b(\mathbf{x} - \eta_x)](\mathbf{x} - \eta_x)\} = r\sigma_x\sigma_y - b\sigma_x^2 = 0$$

Hence, $b = r\sigma_y/\sigma_x$. We thus conclude that the linear LMS (least mean square) fit of $\mathbf{y}$ on $\mathbf{x}$ is the line

$$y - \eta_y = r\frac{\sigma_y}{\sigma_x}(x - \eta_x) \tag{5-50}$$

known as *regression line* (Fig. 5.8). This line passes through the point (η_x, η_y), and its slope equals $r\sigma_y/\sigma_x$. The LMS fit of "$\mathbf{x}$ on $\mathbf{y}$" is the line

$$x - \eta_x = r\frac{\sigma_x}{\sigma_y}(y - \eta_y)$$

passing through the point (η_x, η_y) with slope $\sigma_y/r\sigma_x$. If $|r| = 1$, the two regression lines coincide with the line (5-46).

APPROXIMATE EVALUATION OF $E\{g(\mathbf{x}, \mathbf{y})\}$ For the determination of the mean of $g(\mathbf{x}, \mathbf{y})$, knowledge of the joint density $f(x, y)$ is required. We shall show that if $g(x, y)$ is sufficiently smooth in the region D where $f(x, y)$ takes significant values, $E\{g(\mathbf{x}, \mathbf{y})\}$ can be expressed in terms of the parameters η_x, η_y, σ_x, σ_y, and μ_{xy}.

Suppose, first, that we approximate $g(x, y)$ by a plane:

$$g(x, y) \simeq g(\eta_x, \eta_y) + (x - \eta_x)\frac{\partial g}{\partial x} + (y - \eta_y)\frac{\partial g}{\partial y}$$

(all derivatives are evaluated at $x = \eta_x$ and $y = \eta_y$). Since $E\{\mathbf{x} - \eta_x\} = 0$ and $E\{\mathbf{y} - \eta_y\} = 0$, it follows that

$$E\{g(\mathbf{x}, \mathbf{y})\} \simeq g(\eta_x, \eta_y) \tag{5-51}$$

This estimate can be improved if we approximate $g(x, y)$ by a quadratic surface:

$$\begin{aligned}g(x, y) \simeq g(\eta_x, \eta_y) &+ (x - \eta_x)\frac{\partial g}{\partial x} + (y - \eta_y)\frac{\partial g}{\partial y} \\ &+ \frac{1}{2}(x - \eta_x)^2\frac{\partial^2 g}{\partial x^2} + (x - \eta_x)(y - \eta_y)\frac{\partial^2 g}{\partial x\,\partial y} + \frac{1}{2}(y - \eta_y)^2\frac{\partial^2 g}{\partial y^2}\end{aligned}$$

This yields

$$E\{g(\mathbf{x}, \mathbf{y})\} \simeq g(\eta_x, \eta_y) + \frac{1}{2}\left(\sigma_x^2 \frac{\partial^2 g}{\partial x^2} + 2\mu_{xy} \frac{\partial^2 g}{\partial x\, \partial y} + \sigma_y^2 \frac{\partial^2 g}{\partial y^2}\right) \qquad (5\text{-}52)$$

Example 5.4

$$g(x, y) = x^2 y$$

Clearly,

$$\frac{\partial^2 g}{\partial x^2} = 2y \qquad \frac{\partial^2 g}{\partial y^2} = 0 \qquad \frac{\partial^2 g}{\partial x\, \partial y} = 2x$$

Inserting into (5-52), we obtain

$$E\{\mathbf{x}^2\mathbf{y}\} \simeq \eta_x^2\eta_y + \eta_y\sigma_x^2 + 2\eta_x r\sigma_x\sigma_y$$

■

Moments and Moment Functions

The first moment η and the variance σ^2 give a partial characterization of the distribution $F(x)$ of an RV $\mathbf{x}$. We shall introduce higher-order moments and use them to improve the specification of $F(x)$.

■ ***Definitions.*** The mean

$$m_n = E\{\mathbf{x}^n\} = \int_{-\infty}^{\infty} x^n f(x)\, dx \qquad (5\text{-}53)$$

of $\mathbf{x}^n$ is the *moment* of order n of the RV $\mathbf{x}$. The mean

$$\mu_n = E\{(\mathbf{x} - \eta)^n\} = \int_{-\infty}^{\infty} (x - \eta)^n f(x)\, dx \qquad (5\text{-}54)$$

is the *central moment* of order n.

Clearly,

$$m_0 = \mu_0 = 1 \qquad m_1 = \eta \qquad \mu_1 = 0 \qquad \mu_2 = \sigma^2 \qquad m_2 = \mu_2 + \eta^2$$

Similarly, we define the absolute moments $E\{|x|^n\}$ and the moments $E\{(\mathbf{x} - a)^n\}$ with respect to an arbitrary point a.

Note that if $f(x)$ is even, then $\eta = 0$, $m_n = \mu_n$, and $\mu_{2n+1} = 0$; if $f(x)$ is symmetrical about the point a, then $\eta = a$ and $\mu_{2n+1} = 0$.

Example 5.5

We shall show that the central moments of a normal RV equal

$$\mu_n = \begin{cases} 1 \times 3 \cdots (n - 1)\sigma^n & n \text{ even} \\ 0 & n \text{ odd} \end{cases} \qquad (5\text{-}55)$$

■ ***Proof.*** Since $f(x)$ is symmetrical about its mean, it suffices to find the moments of the centered density

$$f(x) = \frac{1}{\sigma\sqrt{2\pi}} e^{-x^2/2\sigma^2}$$

Differentiating the normal integral (3-55) k times with respect to α we obtain

$$\int_{-\infty}^{\infty} z^{2k}e^{-\alpha z^2}\,dz = \frac{1 \times 3 \cdots (2k-1)}{2^k}\sqrt{\frac{\pi}{\alpha^{2k+1}}}$$

With $\alpha = 1/2\sigma^2$ and $n = 2k$, this equation yields

$$\frac{1}{\sigma\sqrt{2\pi}}\int_{-\infty}^{\infty} z^n e^{-z^2/2\sigma^2}\,dz = 1 \times 3 \cdots (n-1)\sigma^n$$

And since $\mu_n = 0$ for n odd, (5-55) results. ■

MOMENT-GENERATING FUNCTIONS The moment-generating function, or, simply, the *moment function*, of an RV $\mathbf{x}$ is by definition the mean of the function $e^{s\mathbf{x}}$. This function is denoted by $\Phi(s)$, and it is defined for every s for which $E\{e^{s\mathbf{x}}\}$ exists. Thus

$$\Phi(s) = E\{e^{s\mathbf{x}}\} = \int_{-\infty}^{\infty} e^{sx}f(x)\,dx \qquad (5\text{-}56)$$

and for discrete type RVs

$$\Phi(s) = \sum e^{sx_k}f(x_k) \qquad (5\text{-}57)$$

From the definition it follows that

$$E\{e^{(a\mathbf{x}+b)s}\} = e^{bs}E\{e^{as\mathbf{x}}\} = e^{bs}\Phi_x(as)$$

Hence, the moment function $\Phi_y(s)$ of the RV $\mathbf{y} = a\mathbf{x} + b$ equals

$$\Phi_y(s) = E\{e^{s(a\mathbf{x}+b)}\} = e^{bs}\Phi_x(as) \qquad (5\text{-}58)$$

Example 5.6

We shall show that the moment function of a $N(\eta, \sigma)$ RV equals

$$\Phi(s) = e^{\eta s}e^{\sigma^2 s^2/2} \qquad (5\text{-}59)$$

■ ***Proof.*** We shall find first the moment function $\Phi_0(s)$ of the RV $\mathbf{x}_0 = (\mathbf{x} - \eta)/\sigma$. Clearly, $\mathbf{x}_0$ is $N(0, 1)$; hence,

$$\Phi_0(s) = \frac{1}{\sqrt{2\pi}}\int_{-\infty}^{\infty} e^{sx}e^{-x^2/2}\,dx$$

Inserting the identity

$$sx - \frac{x^2}{2} = -\frac{1}{2}(x-s)^2 + \frac{s^2}{2}$$

into the integral, we obtain

$$\Phi_0(s) = e^{s^2/2}\int_{-\infty}^{\infty}\frac{1}{\sqrt{2\pi}}e^{-(x-s)^2/2}\,dx = e^{s^2/2}$$

because the last integral equals 1. And since $\mathbf{x} = \eta + \sigma\mathbf{x}_0$, (5-59) follows from (5-58). ■

Example 5.7

The RV $\mathbf{x}$ is Poisson-distributed with parameter a

$$P\{\mathbf{x} = k\} = e^{-a}\frac{a^k}{k!} \qquad k = 0, 1, \ldots$$

Inserting into (5-57), we obtain

$$\Phi(s) = e^{-a} \sum_{k=0}^{\infty} \frac{a^k}{k!} e^{sk} = e^{-a} e^{ae^s} \qquad (5\text{-}60) \ \blacksquare$$

We shall now relate the moments m_n of the RV $\mathbf{x}$ to the derivatives at the origin of its moment function. Other applications of the moment function include the evaluation of the density of certain functions of $\mathbf{x}$ and the determination of the density of the sum of independent RVs.

▪ ***Moment Theorem.*** We maintain that

$$E\{\mathbf{x}^n\} = m_n = \Phi^{(n)}(0) \qquad (5\text{-}61)$$

▪ ***Proof.*** Differentiating (5-56) and (5-57) n times with respect to s, we obtain

$$\Phi^{(n)}(s) = \int_{-\infty}^{\infty} x^n e^{sx} f(x)\,dx \qquad \text{and} \qquad \sum_k x_k^n e^{sx_k} f(x_k)$$

for continuous and discrete type RVs, respectively. With $s = 0$, (5-61) follows from (5-53).

▪ ***Corollary.*** With $n = 0$, 1, and 2, the theorem yields $\Phi(0) = 1$,

$$\Phi'(0) = m_1 = \eta \qquad \Phi''(0) = m_2 = \eta^2 + \sigma^2 \qquad (5\text{-}62)$$

We shall use (5-62) to determine the mean and the variance of the gamma and the binomial distribution.

Example 5.8

The moment-generating function of the gamma distribution

$$f(x) = \gamma x^{b-1} e^{-cx} U(x)$$

equals

$$\Phi(s) = \gamma \int_0^{\infty} x^{b-1} e^{-(c-s)x}\,dx = \frac{\gamma\Gamma(b)}{(c-s)^b} = \frac{c^b}{(c-s)^b} \qquad (5\text{-}63)$$

[see (4-54)]. Thus

$$\Phi^{(n)}(s) = \frac{b(b+1)\cdots(b+n-1)c^b}{(c-s)^{b+n}}$$

Hence,

$$\Phi^{(n)}(0) = E\{\mathbf{x}^n\} = \frac{b(b+1)\cdots(b+n-1)}{c^n} \qquad (5\text{-}64)$$

With $n = 1$ and $n = 2$, this yields

$$E\{\mathbf{x}\} = \frac{b}{c} \qquad E\{\mathbf{x}^2\} = \frac{b(b+1)}{c^2} \qquad \sigma^2 = \frac{b}{c^2} \qquad (5\text{-}65) \ \blacksquare$$

Example 5.9

The RV $\mathbf{x}$ takes the values 0, 1, . . . , n with

$$P\{\mathbf{x} = k\} = \binom{n}{k} p^k q^{n-k}$$

In this case,

$$\Phi(s) = \sum_{k=0}^{n} \binom{n}{k} p^k q^{n-k} e^{sk} = (pe^s + q)^n \tag{5-66}$$

Clearly,

$$\Phi'(s) = n(pe^s + q)^{n-1} pe^s$$
$$\Phi''(s) = n(n-1)(pe^s + q)^{n-2} p^2 e^{2s} + n(pe^s + q)^{n-1} pe^s$$
$$\Phi'(0) = np \qquad \Phi''(0) = n^2p^2 - np^2 + np$$

Hence,

$$\eta = np \qquad m_2 = n^2p^2 + npq \qquad \sigma^2 = npq \tag{5-67}$$ ■

The integral in (5-56) is also called the *Laplace transform* of the function $f(x)$. For an arbitrary $f(x)$, this integral exists in a vertical strip R of the complex s plane. If $f(x)$ is a density, R contains the $j\omega$-axis; hence, the function

$$\Phi(j\omega) = \int_{-\infty}^{\infty} e^{j\omega x} f(x)\,dx \tag{5-68}$$

exists for every real ω. In probability theory, $\Phi(j\omega)$ is called the *characteristic function* of the RV **x**. In general, it is called the *Fourier transform* of the function $f(x)$.

It can be shown that

$$f(x) = \frac{1}{2\pi} \int_{-\infty}^{\infty} e^{-j\omega x} \Phi(j\omega)\,d\omega \tag{5-69}$$

This important result, known as the *inversion formula,* shows that $f(x)$ is uniquely determined in terms of its moment function.

Note, finally, that $f(x)$ can be determined in terms of its moments. Indeed, the moments m_n equal the derivatives of $\Phi(s)$ at the origin, and $\Phi(s)$ is determined in terms of these derivatives if its Taylor expansion converges for every s. Inserting the resulting $\Phi(j\omega)$ into (5-69), we obtain $f(x)$.

JOINT MOMENTS Proceeding as in (5-53) and (5-54), we introduce the *joint moments*

$$m_{kr} = E\{\mathbf{x}^k \mathbf{y}^r\} = \int_{-\infty}^{\infty} \int_{-\infty}^{\infty} x^k y^r f(x, y)\,dx\,dy \tag{5-70}$$

and the *joint central moments*

$$\mu_{kr} = E\{(\mathbf{x} - \eta_x)^k (\mathbf{y} - \eta_y)^r\} = \int_{-\infty}^{\infty} \int_{-\infty}^{\infty} (x - \eta_x)^k (y - \eta_y)^r f(x, y)\,dx\,dy \tag{5-71}$$

Clearly,

$$m_{10} = \eta_x \qquad m_{01} = \eta_y \qquad m_{20} = \eta_x^2 + \sigma_x^2 \qquad m_{02} = \eta_y^2 + \sigma_y^2$$
$$\mu_{10} = 0 \qquad \mu_{01} = 0 \qquad \mu_{11} = \mu_{xy} \qquad \mu_{20} = \sigma_x^2 \qquad \mu_{02} = \sigma_y^2$$

The *joint moment function* $\Phi(s_1, s_2)$ of the RVs **x** and **y** is by definition

$$\Phi(s_1, s_2) = E\{e^{s_1\mathbf{x} + s_2\mathbf{y}}\} = \int_{-\infty}^{\infty} \int_{-\infty}^{\infty} e^{(s_1 x + s_2 y)} f(x, y)\,dx\,dy \tag{5-72}$$

Repeated differentiation yields

$$\frac{\partial^k \partial^r \Phi(0, 0)}{\partial x^k \partial y^r} = m_{kr} \tag{5-73}$$

This is the two-dimensional form of the moment theorem (5-61).

Denoting by $\Phi_x(s)$ and $\Phi_y(s)$ the moment functions of $\mathbf{x}$ and $\mathbf{y}$, we obtain the following relationship between marginal and joint moment functions:

$$\Phi_x(s_1) = E\{e^{s_1\mathbf{x}}\} = \Phi(s_1, 0) \qquad \Phi_y(s_2) = E\{e^{s_2\mathbf{y}}\} = \Phi(0, s_2) \tag{5-74}$$

Note, finally, that if the RVs $\mathbf{x}$ and $\mathbf{y}$ are *independent*, the RVs $e^{s_1\mathbf{x}}$ and $e^{s_2\mathbf{y}}$ are independent [see (5-41)]. From this it follows that

$$\Phi(s_1, s_2) = E\{e^{s_1\mathbf{x}}\}E\{e^{s_2\mathbf{y}}\} = \Phi_x(s_1)\Phi_y(s_2) \tag{5-75}$$

Example 5.10

From (5-75) and (5-59) it follows that if the RV $\mathbf{x}$ and $\mathbf{y}$ are normal and independent, with zero mean, then

$$\Phi(s_1, s_2) = \exp\left\{\frac{1}{2}(\sigma_x^2 s_1^2 + \sigma_y^2 s_2^2)\right\} \tag{5-76}$$ ■

5-3 Functions of Two Random Variables

Given two RVs $\mathbf{x}$ and $\mathbf{y}$ and two functions $g(x, y)$ and $h(x, y)$, we form the functions

$$\mathbf{z} = g(\mathbf{x}, \mathbf{y}) \qquad \mathbf{w} = h(\mathbf{x}, \mathbf{y})$$

These functions are composite with domain the set $\mathcal{S}$; that is, they are RVs. We shall express their joint distribution in terms of the joint distribution of the RVs $\mathbf{x}$ and $\mathbf{y}$.

We start with the determination of the marginal distribution $F_z(z)$ of the RV $\mathbf{z}$. As we know, $F_z(z)$ is the probability of the event $\{\mathbf{z} \le z\}$. To find $F_z(z)$, it suffices, therefore, to express the event $\{\mathbf{z} \le z\}$ in terms of the RVs $\mathbf{x}$ and $\mathbf{y}$. To do so, we introduce the region D_z of the xy plane such that $g(x, y) \le z$ (Fig. 5.9). Clearly, $\mathbf{z} \le z$ iff $g(\mathbf{x}, \mathbf{y}) \le z$; hence,

$$P\{\mathbf{z} \le z\} = P\{g(\mathbf{x}, \mathbf{y}) \le z\} = P\{(\mathbf{x}, \mathbf{y}) \in D_z\}$$

The region D_z is the projection on the xy plane of the part of the $g(x, y)$ surface below the plane $z =$ constant; the function $F_z(z)$ equals the probability masses in that region.

Example 5.11

The RVs $\mathbf{x}$ and $\mathbf{y}$ are normal and independent with joint density

$$f_{xy}(x, y) = \frac{1}{2\pi\sigma^2} e^{-(x^2+y^2)/2\sigma^2}$$

We shall determine the distribution of the RV

$$\mathbf{z} = +\sqrt{\mathbf{x}^2 + \mathbf{y}^2}$$

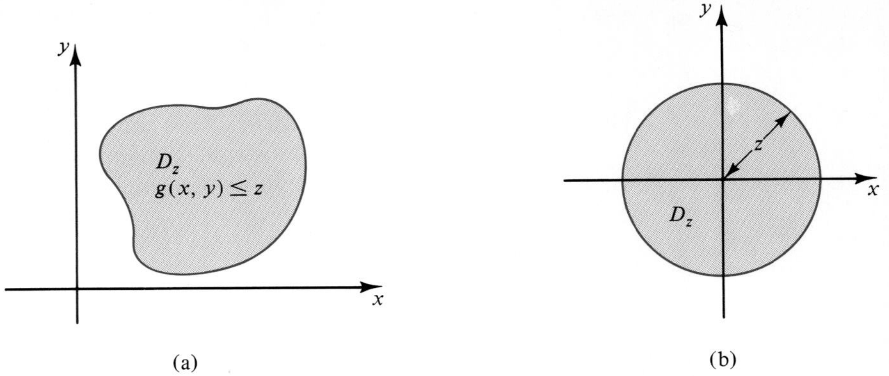

Figure 5.9

If $z \geq 0$, then D_z is the circle $g(x, y) = \sqrt{x^2 + y^2} \leq z$ shown in Fig. 5.9. Hence (see Example 5.2),

$$F_z(z) = \frac{1}{2\pi\sigma^2} \iint_{D_z} e^{-(x^2+y^2)/2\sigma^2}\, dx\, dy = 1 - e^{-z^2/2\sigma^2}$$

If $z < 0$, then $\{\mathbf{z} \leq z\} = \varnothing$; hence, $F_z(z) = 0$.

Rayleigh Density. Differentiating $F_z(z)$, we obtain the function

$$f_z(z) = \frac{z}{\sigma^2} e^{-z^2/2\sigma^2} U(z) \tag{5-77}$$

known as the Rayleigh density. ■

Example 5.12

(a) $\mathbf{z} = \max(\mathbf{x}, \mathbf{y})$

The region D_z of the xy plane such that $\max(x, y) \leq z$ is the set of points such that $x \leq z$ and $y \leq z$ (shaded in Fig. 5.10). The probability masses in that region equal $F_{xy}(z, z)$. Hence,

$$F_z(z) = F_{xy}(z, z) \tag{5-78}$$

(b) $\mathbf{z} = \min(\mathbf{x}, \mathbf{y})$

The region D_z of the xy plane such that $\min(x, y) \leq z$ is the set of points such that $x \leq z$ or $y \leq z$ (shaded in Fig. 5.10). The probability masses in that region equal the masses $F_x(z)$ to the left of the vertical line $x = z$ plus the masses $F_y(z)$ below the horizontal line $y = z$, minus the masses $F_{xy}(z, z)$ in the quadrant $(x \leq z, y \leq z)$. Hence,

$$F_z(z) = F_x(z) + F_y(z) - F_{xy}(z, z) \tag{5-79}$$ ■

Joint Density of g(x, y) and h(x, y)

We shall express the joint density $f_{zw}(z, w)$ of the RVs $\mathbf{z} = g(\mathbf{x}, \mathbf{y})$ and $\mathbf{w} = h(\mathbf{x}, \mathbf{y})$ in terms of the joint density $f_{xy}(x, y)$ of the RVs $\mathbf{x}$ and $\mathbf{y}$. The following theorem is an extension of (4-78) to two RVs, and it involves the *Jacobian* $J(x, y)$ of the transformation

$$z = g(x, y) \qquad w = h(x, y) \tag{5-80}$$

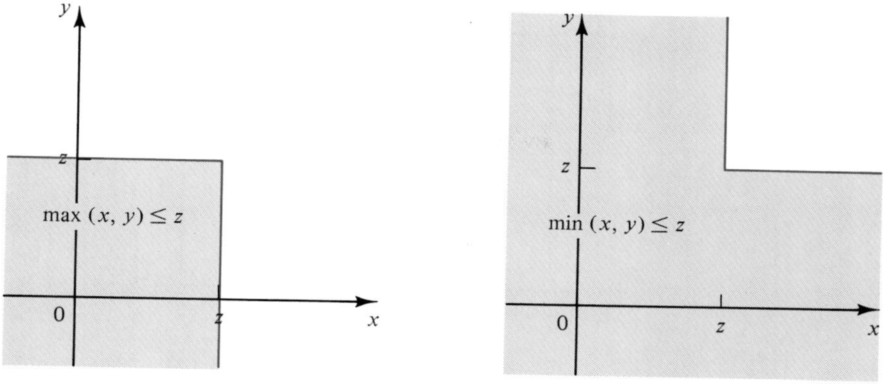

Figure 5.10

The function $J(x, y)$ is by definition the determinant

$$J(x, y) = \begin{vmatrix} \dfrac{\partial g(x, y)}{\partial x} & \dfrac{\partial g(x, y)}{\partial y} \\ \dfrac{\partial h(x, y)}{\partial x} & \dfrac{\partial h(x, y)}{\partial y} \end{vmatrix} \tag{5-81}$$

and it is used in the determination of two-dimensional integrals involving change of variables.

■ ***Fundamental Theorem.*** To find $f_{zw}(z, w)$, we solve the systems (5-80) for x and y. If this system has no real solutions in some region of the zw plane, $f_{zw}(z, w) = 0$ for every (z, w) in that region. Suppose, then, that (5-80) has one or more solutions (x_i, y_i), that is,

$$g(x_i, y_i) = z \qquad h(x_i, y_i) = w \tag{5-82}$$

In this case,

$$f_{zw}(z, w) = \frac{f_{xy}(x_1, y_1)}{|J(x_1, y_1)|} + \cdots + \frac{f_{xy}(x_i, y_i)}{|J(x_i, y_i)|} + \cdots \tag{5-83}$$

where (x_i, y_i) are all pairs (x_i, y_i) that satisfy (5-82). The number of such pairs and their values depend, of course, on the particular values of z and w.

■ ***Proof.*** The system (5-82) transforms the differential rectangle A of Fig. 5.11a into one or more differential parallelograms B_i of Fig. 5.11b. As we know from calculus, the area $|B_i|$ or the ith parallelogram equals the area $|A| = dz\,dw$ of the rectangle, divided by $J(x_i, y_i)$. Thus

$$\begin{aligned} P\{z \le \mathbf{z} \le z + dz,\ w \le \mathbf{w} \le w + dw\} &= f_{zw}(z, w)\,dz\,dw \\ P\{(\mathbf{x}, \mathbf{y}) \in B_i\} &= f_{xy}(x_i, y_i)|J^{-1}(x_i, y_i)|\,dz\,dw \end{aligned} \tag{5-84}$$

And since the event $\{z \le \mathbf{z} \le z + dz,\ w \le \mathbf{w} \le w + dw\}$ is the union of the disjoint events $\{(\mathbf{x}_i, \mathbf{y}_i) \in B_i\}$, (5-83) follows from (5-84).

With $f_{zw}(z, w)$ so determined, the marginal densities $f_z(z)$ and $f_w(z)$ can be obtained as in (5-10).

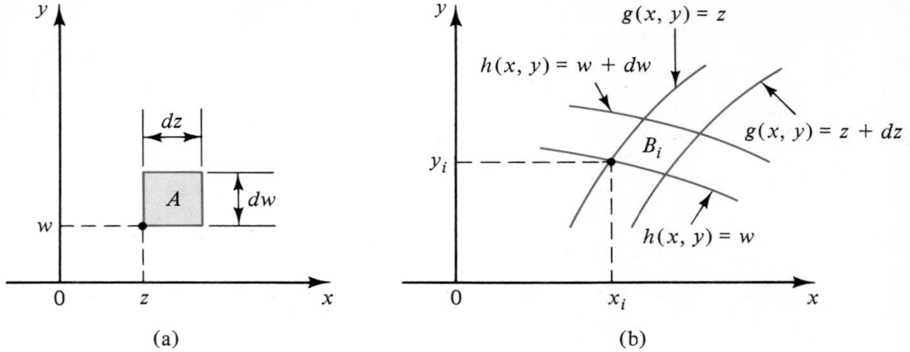

Figure 5.11

Auxiliary Variable We can use the preceding theorem to find the density $f_z(z)$ of *one* function $\mathbf{z} = g(\mathbf{x}, \mathbf{y})$ of two RVs $\mathbf{x}$ and $\mathbf{y}$. To do so, we introduce an auxiliary RV $\mathbf{w} = h(\mathbf{x}, \mathbf{y})$, we determine the joint density $f_{zw}(z, w)$ from (5-83) and the density $f_z(z)$ from (5-10):

$$f_z(z) = \int_{-\infty}^{\infty} f_{zw}(z, w)\,dw \tag{5-85}$$

The variable $\mathbf{w}$ is selected at our convenience. For example, we can set $\mathbf{w} = \mathbf{x}$ or $\mathbf{y}$. We continue with several illustrations.

Linear Transformations Suppose first that

$$z = ax + b \qquad w = cy + d$$

This system has a single solution

$$x_1 = \frac{z-b}{a} \qquad y_1 = \frac{w-d}{c}$$

for any z and w, and its Jacobian equals ac. Inserting into (5-83), we obtain

$$f_{zw}(z, w) = \frac{1}{|ac|} f_{xy}\left(\frac{z-b}{a}, \frac{w-d}{c}\right) \tag{5-86}$$

Suppose, next, that

In this case, the Jacobian equals

$$J(x, y) = \begin{vmatrix} a_1 & b_1 \\ a_2 & b_2 \end{vmatrix} = a_1b_2 - a_2b_1 = D$$

If $D = 0$, then $\mathbf{w} = a_2\mathbf{z}/b_1$; hence, the joint distribution of $\mathbf{z}$ and $\mathbf{w}$ consists of line masses and can be expressed in terms of $F_z(z)$. It suffices, therefore, to assume that $D \neq 0$. With this assumption, we have a single solution

$$x_1 = \frac{b_2z - b_1w}{D} \qquad y_1 = \frac{-a_2z + a_1w}{D}$$

and (5-83) yields

$$f_{zw}(z, w) = f_{xy}\left(\frac{b_2z - b_1w}{D}, \frac{-a_2z + a_1w}{D}\right)\frac{1}{|D|} \tag{5-88}$$

DISTRIBUTION OF x + y An important special case is the determination of the distribution of the sum $\mathbf{z} = \mathbf{x} + \mathbf{y}$. Introducing the auxiliary variable $\mathbf{w} = \mathbf{y}$, we obtain from (5-88) with $a_1 = b_1 = 1$, $a_2 = 0$, $b_2 = 1$,

$$f_{zw}(z, w) = f_{xy}(z - w, w) \tag{5-89}$$

Hence,

$$f_z(z) = \int_{-\infty}^{\infty} f_{xy}(z - w, w)\, dw \tag{5-90}$$

This result has the following mass interpretation. Clearly, $z \le \mathbf{z} \le z + dz$ iff $z \le \mathbf{x} + \mathbf{y} \le z + dz$, that is, iff the point $(\mathbf{x}, \mathbf{y})$ is in the shaded region ΔD_z of Fig. 5.12. Hence,

$$f_z(z)\, dz = P\{z \le \mathbf{x} + \mathbf{y} \le z + dz\} = P\{(\mathbf{x}, \mathbf{y}) \in \Delta D_z\} \tag{5-91}$$

Thus to find $f_z(z)$, it suffices to find the masses in the strip ΔD_z.

For discrete type RVs, we proceed similarly. Suppose that $\mathbf{x}$ and $\mathbf{y}$ take the values x_i and y_k, respectively. In this case, their sum $\mathbf{z} = \mathbf{x} + \mathbf{y}$ takes the values $\mathbf{z} = z_r = x_i + y_k$, and $P\{\mathbf{z} = z_r\}$ equals the sum of all point masses on the line $x + y = z_r$.

Example 5.13

The RVs $\mathbf{x}$ and $\mathbf{y}$ are independent, taking the values 1, 2, . . . , 6 with probability 1/6 (two fair dice rolled once). In this case, there are 36 point masses on the xy plane, as in Fig. 5.13, and each mass equals 1/36. The sum $\mathbf{z} = \mathbf{x} + \mathbf{y}$ takes the values 2, 3, . . . , 12. On the line $x + y = 5$, there are four point masses; hence, $P\{\mathbf{z} = 5\} = 4/36$. On the line $x + y = 12$, there is a single point; hence, $P\{\mathbf{z} = 12\} = 1/36$. ■

The determination of the density of the sum of two independent RVs is often simplified if we use moment functions. As we know, if the RVs $\mathbf{x}$ and $\mathbf{y}$ are independent, the RVs $e^{s\mathbf{x}}$ and $e^{s\mathbf{y}}$ are also independent; hence [see (5-75)],

$$E\{e^{s(\mathbf{x}+\mathbf{y})}\} = E\{e^{s\mathbf{x}}\}E\{e^{s\mathbf{y}}\}$$

Figure 5.12

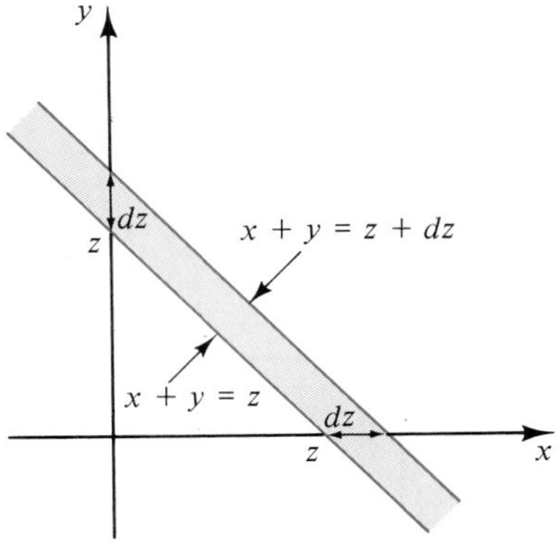

Figure 5.13

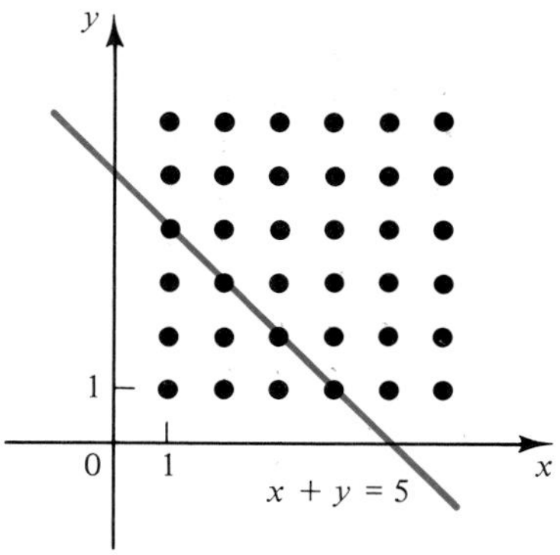

From this it follows that

$$\Phi_{x+y}(s) = \Phi_x(s)\Phi_y(s) \tag{5-92}$$

Thus the moment function of the sum of two independent RVs equals the product of their moment functions.

Example 5.14

If the RVs **x** and **y** are normal, then [see (5-59)]

$$\Phi_x(s) = \exp\left\{\eta_x s + \frac{1}{2}\sigma_x^2 s^2\right\} \qquad \Phi_y(s) = \exp\left\{\eta_y s + \frac{1}{2}\sigma_y^2 s^2\right\}$$

Inserting into (5-92), we obtain

$$\Phi_{x+y}(s) = \exp\left\{(\eta_x + \eta_y)s + \frac{1}{2}(\sigma_x^2 + \sigma_y^2)s^2\right\} \tag{5-93}$$

Clearly, (5-93) is the moment function of a normal RV; hence, the sum of two normal independent RVs is normal with $\eta_z = \eta_x + \eta_y$, $\sigma_z^2 = \sigma_x^2 + \sigma_y^2$. ■

Example 5.15

If the RVs **x** and **y** are Poisson, then [see (5-60)]

$$\Phi_x(s) = \exp\{a_x(e^s - 1)\} \qquad \Phi_y(s) = \exp\{a_y(e^s - 1)]$$

Inserting into (5-92), we obtain

$$\Phi_{x+y}(s) = \exp\{a_x + a_y)(e^s - 1)\} \tag{5-94}$$

Thus the sum of two independent Poisson RVs with parameters a_x and a_y, respectively, is a Poisson RV with parameter $a_x + a_y$. ■

Example 5.16

If the RVs **x** and **y** have a binomial distribution with the same p, then [see (5-66)]

$$\Phi_x(s) = (pe^s + q)^m \qquad \Phi_y(s) = (pe^s + q)^n$$

Hence,

$$\Phi_{x+y}(s) = (pe^s + q)^{m+n} \tag{5-95}$$

Thus the sum of two independent binomial RVs of order m and n, respectively, and with the same p is a binomial RV of order $m + n$. ■

Independence and Convolution We shall determine the density of the sum $\mathbf{z} = \mathbf{x} + \mathbf{y}$ of two independent continuous type RVs. In this case,

$$f_{xy}(x, y) = f_x(x)f_y(y)$$

Inserting into (5-90), we obtain

$$f_z(z) = \int_{-\infty}^{\infty} f_x(z - w)f_y(w)\,dw \tag{5-96}$$

This integral is called the *convolution* of the functions $f_x(x)$ and $f_y(y)$. We have thus reached the important conclusion that the density of the sum of two independent RVs equals the convolution of their respective densities.

Combining (5-96) and (5-92), we obtain the following mathematical result, known as the *convolution theorem*: The moment function $\Phi_z(s)$ of the convolution $f_z(z)$ of two densities $f_x(x)$ and $f_y(y)$ equals the product of their moment functions $\Phi_x(s)$ and $\Phi_y(s)$.

Clearly, if the range of $\mathbf{x}$ is the interval (a, b) and the range of $\mathbf{y}$ the interval (c, d), the range of their sum $\mathbf{z} = \mathbf{x} + \mathbf{y}$ is the interval $(a + c, b + d)$. From this it follows that if the functions $f_x(x)$ and $f_y(y)$ equal zero outside the intervals (a, b) and (c, d), respectively, their convolution $f_z(z)$ equals zero outside the interval $(a + c, b + d)$. Note in particular that if

$$f_x(x) = 0 \quad \text{for } x < 0 \qquad \text{and} \qquad f_y(y) = 0 \quad \text{for } y < 0$$

then

$$f_z(z) = \begin{cases} \int_0^z f_x(z - w) f_y(w)\,dw & z > 0 \\ 0 & z < 0 \end{cases} \tag{5-97}$$

because $f_x(z - w) = 0$ for $w > z$.

Example 5.17

From (5-97) it follows that if the RVs $\mathbf{x}$ and $\mathbf{y}$ are independent with densities

$$f_x(x) = \alpha e^{-\alpha x} U(x) \qquad f_y(y) = \alpha e^{-\alpha y} U(y)$$

then the density of their sum equals

$$f_z(z) = \alpha^2 U(z) \int_0^z e^{-\alpha(z - w)} e^{-\alpha w}\,dw = \alpha^2 z e^{-\alpha z} U(z)$$ ■

In the following example, we use the mass interpretation of $f_z(z)$ (see Fig. 5.12) to facilitate the evaluation of the convolution integral.

Example 5.18

The RVs $\mathbf{x}$ and $\mathbf{y}$ are independent, and each is uniformly distributed in the interval (0, c). We shall show that the density of their sum is a triangle as in Fig. 5.14.

Clearly, $f(x, y) = 1/c^2$ for every (x, y) in the square S (shaded area) and $f(x, y) = 0$ elsewhere. From this it follows that $f_z(z)\,dz$ equals $|\Delta D_z|/c^2$ where $|\Delta D_z|$ is the area of the region $z \le x + y \le z + dz$ inside the square S [see (5-91)]. As we see from Fig. 5.14,

$$|\Delta D_z| = \begin{cases} z\,dz & 0 \le z \le c \\ (2c - z)\,dz & c < z < 2c \end{cases}$$

Hence,

$$f_z(z) = \begin{cases} \dfrac{z}{c^2} & 0 \le z \le c \\[2ex] \dfrac{2c - z}{c^2} & c < z < 2c \end{cases}$$ ■

JOINTLY NORMAL DISTRIBUTIONS We define joint normality in terms of the normality of a single RV.

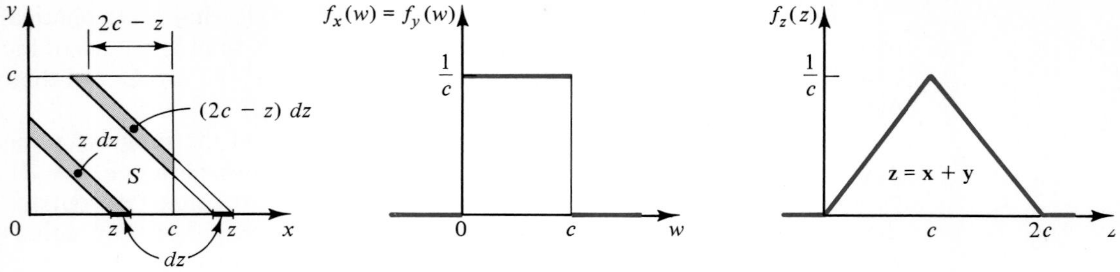

Figure 5.14

■ ***Definition.*** Two RVs $\mathbf{x}$ and $\mathbf{y}$ are jointly normal iff the sum

$$\mathbf{z} = a\mathbf{x} + b\mathbf{y}$$

is normal for every a and b.

We show next that this definition specifies completely the moment generating function

$$\Phi_{xy}(s_1, s_2) = E\{e^{s_1\mathbf{x}+s_2\mathbf{y}}\}$$

of $\mathbf{x}$ and $\mathbf{y}$. We assume for simplicity that $E\{\mathbf{x}\} = E\{\mathbf{y}\} = 0$.

■ ***Theorem.*** Two RVs $\mathbf{x}$ and $\mathbf{y}$ with zero mean are jointly normal iff

$$\Phi_{xy}(s_1, s_2) = \exp\left\{\frac{1}{2}(\sigma_1^2 s_1^2 + 2r\sigma_1\sigma_2 s_1 s_2 + \sigma_2^2 s_2^2)\right\} \qquad \text{(5-98)}$$

where σ_1 and σ_2 are the standard deviations of $\mathbf{x}$ and $\mathbf{y}$, respectively, and r is their correlation coefficient.

■ ***Proof.*** The moment function of the RV $\mathbf{z} = a\mathbf{x} + b\mathbf{y}$ equals [see (5-59)]

$$\Phi_z(s) = E\{e^{s\mathbf{z}}\} = \exp\left\{\frac{1}{2}\sigma_z^2 s^2\right\} \quad \text{where} \quad \sigma_z^2 = a^2\sigma_1^2 + 2abr\sigma_1\sigma_2 + b^2\sigma_2^2$$

Furthermore,

$$\Phi_z(s) = E\{e^{(a\mathbf{x}+b\mathbf{y})s}\} \qquad \Phi_z(1) = E\{e^{a\mathbf{x}+b\mathbf{y}}\} = \Phi_{xy}(a, b)$$

$$\Phi_z(1) = \exp\left\{\frac{1}{2}\sigma_z^2\right\} = \exp\left\{\frac{1}{2}(a^2\sigma_1^2 + 2abr\sigma_1\sigma_2 + b^2\sigma_2^2)\right\}$$

Setting $a = s_1$ and $b = s_2$ we obtain (5-98).

Conversely, if $\Phi_{xy}(s_1, s_2)$ is given by (5-98) and $\mathbf{z} = a\mathbf{x} + b\mathbf{y}$, then

$$\Phi_z(s) = \Phi_{xy}(as, bs) = \exp\left\{\frac{s^2}{2}(a^2\sigma_1^2 + 2abr\sigma_1\sigma_2 + b^2\sigma_2^2)\right\}$$

This shows that **z** is a normal RV, hence (definition) the RVs **x** and **y** are jointly normal.

Joint Density From (5-98) and (5-72) it follows that

$$f_{xy}(x, y) = \frac{1}{2\pi\sigma_1\sigma_2\sqrt{1 - r^2}} \exp\left\{-\frac{1}{2(1 - r^2)}\left(\frac{x^2}{\sigma_1^2} - 2r\frac{xy}{\sigma_1\sigma_2} + \frac{y^2}{\sigma_2^2}\right)\right\} \quad (5\text{-}99)$$

The proof is involved.

Joint normality can be defined directly: Two RVs **x** and **y** are jointly normal if their joint density equals $e^{-Q(x, y)}$ where $Q(x, y)$ is a positive quadratic function of x and y, that is

$$Q(x, y) = c_1x^2 + c_2xy + c_3y^2 + c_4x + c_5y + c_6 \geq 0$$

Expressing the five parameters η_1, η_2, σ_1, σ_2, and r in terms of c_i and using the fact that the integral of $f(x, y)$ equals 1, we obtain

$$f_{xy}(x, y) =$$

$$\frac{1}{2\pi D} \exp\left\{-\frac{1}{2D^2}[\sigma_2^2(x - \eta_1)^2 - 2r\sigma_1\sigma_2(x - \eta_1)(y - \eta_2) + \sigma_1^2(y - \eta_2)^2]\right\} \quad (5\text{-}100)$$

where $D = \sigma_1\sigma_2\sqrt{1 - r^2}$. Note that (5-100) reduces to (5-99) if $\eta_1 = \eta_2 = 0$.

Uncorrelatedness and Independence It follows from (5-100) that if $r = 0$ then

$$f_{xy}(x, y) = f_x(x)f_y(y)$$

This shows that if two RVs are jointly normal and uncorrelated, they are independent.

Marginal Normality It follows from (5-98) or directly from the definition with $a = 1$ and $b = 0$, that, if two RVs are jointly normal, they are marginally normal.

Linear Transformations If

$$\mathbf{z} = a\mathbf{x} + b\mathbf{y} \qquad \mathbf{w} = c\mathbf{x} + d\mathbf{y}$$

then [see (5-88)]

$$f_{zw}(z, w) = e^{-Q_1(z, w)} \qquad Q_1(z, w) = Q\left(\frac{dz - bw}{ad - bc}, \frac{-cz + aw}{ad - bc}\right)$$

If $Q(x, y)$ is a quadratic function of x and y, $Q_1(z, w)$ is a quadratic function of z and w. This leads to the following conclusion: If two RVs **z** and **w** are linear functions of two jointly normal RVs **x** and **y**, they are jointly normal. To find their joint density, it suffices therefore to know the five parameters η_z, η_w, σ_z, σ_w, and r_{zw}.

Example 5.19

The RVS **x** and **y** are normal with

$$\eta_x = 2 \qquad \eta_y = 0 \qquad \sigma_x = 1 \qquad \sigma_y = 2 \qquad r = \frac{3}{8}$$

Find the density of the RV $\mathbf{z} = 2\mathbf{x} + 3\mathbf{y}$.

As we know, **z** is $N(\eta_z, \sigma_z)$ with

$$\eta_z = 2\eta_x + 3\eta_y = 4 \qquad \sigma_z^2 = 4\sigma_x^2 + 9\sigma_y^2 + 12r\sigma_x\sigma_y = 49$$ ■

GENERAL TRANSFORMATIONS We conclude with an illustration of the fundamental theorem (5-83). We shall determine the distribution of the RV

$$\mathbf{z} = \frac{\mathbf{x}}{\mathbf{y}}$$

using as auxiliary variable the RV $\mathbf{w} = \mathbf{y}$. The system

$$z = \frac{x}{y} \qquad w = y$$

has a single solution $x_1 = zw$, $y_1 = w$, and its Jacobian equals

$$J(x, y) = \begin{vmatrix} \dfrac{1}{y} & -\dfrac{x}{y^2} \\ 0 & 1 \end{vmatrix} = \frac{1}{y}$$

Inserting into (5-83), we obtain

$$f_{zw}(z, w) = |w| f_{xy}(zw, w)$$

and (5-85) yields

$$f_z(z) = \int_{\infty}^{\infty} |w| f_{xy}(zw, w)\, dw \tag{5-101}$$

Example 5.20

The RVS **x** and **y** are jointly normal with zero mean as in (5-99). We shall show that their ratio $\mathbf{z} = \mathbf{x}/\mathbf{y}$ has a *Cauchy* density centered at $z = r\sigma_1/\sigma_2$:

$$f_z(z) = \frac{\sigma_1\sigma_2\sqrt{1 - r^2}}{\pi} \frac{1}{\sigma_2^2(z - r\sigma_1/\sigma_2)^2 + \sigma_1^2(1 - r^2)} \tag{5-102}$$

■ ***Proof.*** Since $f_{xy}(-x, -y) = f_{xy}(x, y)$, it follows from (5-101) that

$$f_z(z) = \frac{2}{2\pi\sigma_1\sigma_2\sqrt{1 - r^2}} \int_0^{\infty} w \exp\left\{-\frac{w^2}{2(1 - r^2)}\left(\frac{z^2}{\sigma_1^2} - 2r\frac{z}{\sigma_1\sigma_2} + \frac{1}{\sigma_2^2}\right)\right\} dw$$

With

$$a = \frac{1}{1 - r^2}\left(\frac{z^2}{\sigma_1^2} - 2r\frac{z}{\sigma_1\sigma_2} + \frac{1}{\sigma_2^2}\right)$$

the integral equals

$$\int_0^{\infty} we^{-aw^2/2}\, dw = \int_0^{\infty} e^{-at}\, dt = \frac{1}{a}$$

and (5-102) follows.

Integrating (5-102), we obtain the distribution

$$F_z(z) = \int_{-\infty}^{z} f_z(\alpha)\, d\alpha = \frac{1}{2} + \frac{\sigma_2 z - r\sigma_1}{\sigma_1\sqrt{1 - r^2}} \tag{5-103}$$ ■

Problems

5-1 If $f(x, y) = \gamma e^{-2x^2 - 8y^2}$, find: **(a)** the marginal densities $f_x(x)$ and $f_y(y)$; **(b)** the constant γ; **(c)** the probability $p = P\{\mathbf{x} \le .5, \mathbf{y} \le .5\}$.

5-2 **(a)** Express $F_{xy}(x, y)$ in terms of $F_x(x)$ if: $\mathbf{y} = 2\mathbf{x}$; $\mathbf{y} = -2\mathbf{x}$; $\mathbf{y} = \mathbf{x}^2$. **(b)** Express the probability $P\{\mathbf{x} \le x, \mathbf{y} > y\}$ in terms of $F_{xy}(x, y)$.

5-3 The RVs $\mathbf{x}$ and $\mathbf{y}$ are of discrete type and independent, taking the values $\mathbf{x} = n$, $n = 0, \ldots, 3$ and $\mathbf{y} = m$, $m = 0, \ldots, 5$ with $P\{\mathbf{x} = n\} = 1/4$, $P\{\mathbf{y} = m\} = 1/6$. Find and sketch the point density of their sum $\mathbf{z} = \mathbf{x} + \mathbf{y}$.

5-4 Show that

$$F^2(x, y) \le F_x(x)F_y(y) \qquad F(x, y) \le \frac{F_x(x) + F_y(y)}{2}$$

5-5 **(a)** Show that if the RVs $\mathbf{x}$ and $\mathbf{y}$ are independent and $F_x(w) = F_y(w)$ for every w, then $P\{\mathbf{x} \le \mathbf{y}\} = P\{\mathbf{x} \ge \mathbf{y}\}$. **(b)** Show that if the RVs $\mathbf{x}$ and $\mathbf{z} = \mathbf{x}/\mathbf{y}$ are independent, then $E\{\mathbf{x}^3/\mathbf{y}^3\} = E\{\mathbf{x}^3\}/E\{\mathbf{y}^3\}$.

5-6 Show that if $\eta_x = \eta_y$, $\sigma_x = \sigma_y$, and $r_{xy} = 1$, then $\mathbf{x} = \mathbf{y}$ in probability.

5-7 Show that if $m_n = E\{\mathbf{x}^n\}$ and $\mu_n = E\{(\mathbf{x} - \eta)^n\}$ then

$$\mu_n = \sum_{k=0}^{n} \binom{n}{k} m_k(-\eta)^{n-k} \qquad m_n = \sum_{k=0}^{n} \binom{n}{k} \mu_k \eta^{n-k}$$

5-8 Show that if the RV $\mathbf{x}$ is $N(0, \sigma)$, then

$$E\{|\mathbf{x}|^n\} = \begin{cases} 1 \times 3 \cdots (n-1)\sigma^{2k} & n = 2k \\ 2^k k! \sigma^{2k+1}\sqrt{\dfrac{2}{\pi}} & n = 2k+1 \end{cases}$$

Find the mean and the variance of the RV $\mathbf{y} = \mathbf{x}^2$.

5-9 Give the empirical interpretation: **(a)** of the identity $E\{\mathbf{x} + \mathbf{y}\} = E\{\mathbf{x}\} + E\{\mathbf{y}\}$; **(b)** of the fact that, in general, $E\{\mathbf{xy}\} \ne E\{\mathbf{x}\}E\{\mathbf{y}\}$.

5-10 Show that if $\mathbf{z} = a\mathbf{x} + b$, $\mathbf{w} = c\mathbf{y} + d$, and $ac \ne 0$, then $r_{zw}^2 = r_{xy}^2$.

5-11 Show that if the RVs $\mathbf{x}$ and $\mathbf{y}$ are jointly normal with zero mean and equal variance the RVs $\mathbf{z} = \mathbf{x} + \mathbf{y}$ and $\mathbf{w} = \mathbf{x} - \mathbf{y}$ are independent.

5-12 Show that if the RVs $\mathbf{x}_1$ and $\mathbf{x}_2$ are $N(\eta, \sigma)$ and independent, the RVs

$$\bar{\mathbf{x}} = \frac{\mathbf{x}_1 + \mathbf{x}_2}{2} \qquad \mathbf{y} = \frac{(\mathbf{x}_1 - \bar{\mathbf{x}})^2 + (\mathbf{x}_2 - \bar{\mathbf{x}})^2}{2}$$

are independent.

5-13 We denote by $a_1\mathbf{x} + b_1$ the LMS fit of $\mathbf{y}$ on $\mathbf{x}$ and by $a_2\mathbf{y} + b_2$ the LMS fit of $\mathbf{x}$ on $\mathbf{y}$. Show that $a_1a_2 = r_{xy}^2$.

5-14 The RV $\mathbf{x}$ is $N(0, 2)$ and $\mathbf{y} = \mathbf{x}^3$. Find the LMS fit $a + b\mathbf{x}$ of $\mathbf{y}$ on $\mathbf{x}$.

5-15 Using the Taylor series approximation

$$g(x, y) \simeq g(\eta_x, \eta_y) + (x - \eta_x)\frac{\partial g}{\partial x} + (y - \eta_y)\frac{\partial g}{\partial y}$$

show that if $\mathbf{z} = g(\mathbf{x}, \mathbf{y})$, then

$$\sigma_z^2 \simeq \sigma_x^2\left(\frac{\partial g}{\partial x}\right)^2 + \sigma_y^2\left(\frac{\partial g}{\partial y}\right)^2 + 2r\sigma_x\sigma_y\frac{\partial g}{\partial x}\frac{\partial g}{\partial y}$$

5-16 The voltage **v** and the current **i** are two independent RVs with

$$\eta_v = 110\text{V} \qquad \sigma_v = 2\text{V} \qquad \eta_i = 2\text{A} \qquad \sigma_i = 0.1\text{A}$$

Using Problem 5-15 and (5-52) find approximately the mean η_w and the standard deviation σ_w of the power $\mathbf{w} = \mathbf{vi}$.

5-17 The RV **x** is uniform in the interval $(0, c)$. **(a)** Find its moment function $\Phi(s)$. **(b)** Using the moment theorem (5-62), find η_x and σ_x.

5-18 We say that an RV **x** has a Laplace distribution if $f(x) = \frac{c}{2} e^{-c|x|}$. Find the corresponding moment function $\Phi(s)$ and determine η_x and σ_x using (5-62).

5-19 Show that if two RVs **x** and **y** are jointly normal with zero mean, then

$$E\{\mathbf{x}^2\mathbf{y}^2\} = E\{\mathbf{x}^2\}E\{\mathbf{y}^2\} + 2E^2\{\mathbf{xy}\}$$

5-20 The RVs **x** and **y** are jointly normal with zero mean. **(a)** Express the moment function $\Phi_z(s)$ of the RV $\mathbf{z} = a\mathbf{x} + b\mathbf{y}$ in terms of the three parameters σ_x, σ_y, and r_{xy}. **(b)** Show that if two RVs **x** and **y** are such that the sum $\mathbf{z} = a\mathbf{x} + b\mathbf{y}$ is a normal RV for every a and b, then **x** and **y** are jointly normal.

5-21 The logarithm $\Psi(s) = \ln \phi(s)$ of the moment-generating function $\Phi(s)$ is called the *second moment generating* function or the *cumulant generating* function. The derivatives $k_n = \Psi^{(n)}(0)$ of $\Psi(s)$ are called the *cumulants of* **x**. **(a)** Show that

$$k_0 = 0 \qquad k_1 = \eta \qquad k_2 = \sigma^2 \qquad k_3 = m_3 \qquad k_4 = m_4 - 3\sigma^4$$

where $m_n = E\{\mathbf{x}^n\}$. **(b)** Find $\Psi(s)$ if **x** is a Poisson RV with parameter a. Find the mean and the variance of **x** using (a).

5-22 The RVs **x** and **y** are uniform in the interval (0, 1) and independent. Find the density of the RVs $\mathbf{z} = \mathbf{x} + \mathbf{y}$, $\mathbf{w} = \mathbf{x} - \mathbf{y}$, $\mathbf{s} = |\mathbf{x} - \mathbf{y}|$.

5-23 The resistances of two resistors are two independent RVs, and each is uniform in the interval (900Ω, 1,100Ω). Connected in series, they form a resistor with resistance $\mathbf{R} = \mathbf{R}_1 + \mathbf{R}_2$. **(a)** Find the density of **R**. **(b)** Find the probability p that **R** is between 1,900Ω and 2,100Ω.

5-24 The times of arrival of two trains are two independent RVs **x** and **y**, and each is uniform in the interval (0, 20). **(a)** Find the density of the RV $\mathbf{z} = \mathbf{x} - \mathbf{y}$. **(b)** Find the probability p_2 in Example 2-38 using RVs.

5-25 Show that if $\mathbf{z} = \mathbf{xy}$, then

$$f_z(z) = \int_{-\infty}^{\infty} \frac{1}{|x|} f_{xy}\left(x, \frac{z}{x}\right) dx$$

5-26 The RVs **x** and **y** are independent. Express the joint density $f_{zw}(z, w)$ of the RVs $\mathbf{z} = \mathbf{x}^2$, $\mathbf{w} = \mathbf{y}^3$ in terms of $f_x(x)$ and $f_y(y)$. Use the result to show that the RVs **z** and **w** are independent.

5-27 The coordinates of a point (**x**, **y**) on the plane are two RVs **x** and **y** with joint density $f_{xy}(x, y)$. The corresponding polar coordinates are

$$\mathbf{r} = \sqrt{\mathbf{x}^2 + \mathbf{y}^2} \qquad \boldsymbol{\phi} = \arctan \frac{\mathbf{y}}{\mathbf{x}} \qquad -\pi < \phi < \pi$$

Show that their joint density equals

$$f_{r\phi}(r, \phi) = r f_{xy}(r \cos \phi, r \sin \phi)$$

Special case: Show that if the RVs **x** and **y** are $N(0, \sigma)$ and independent, the RVs **r** and $\boldsymbol{\phi}$ are independent, **r** has a Rayleigh distribution, and $\boldsymbol{\phi}$ is uniform in the interval $(-\pi, \pi)$.

5-28 The RVs **x** and **y** are independent and $\mathbf{z} = \mathbf{x} + \mathbf{y}$. **(a)** Find the density of **y** if

$$f_x(x) = ce^{-cx}U(x) \qquad f_z(z) = c^2ze^{-cz}U(z)$$

(b) Show that if **y** is uniform in the interval (0, 1), then

$$f_z(z) = F_x(z) - F_x(z - 1)$$

5-29 Show that if the RVs **x** and **y** are independent with exponential densities $ce^{-cx}U(x)$ and $ce^{-cy}U(y)$, respectively, their difference $\mathbf{z} = \mathbf{x} - \mathbf{y}$ has the Laplace density $\frac{c}{2}\, e^{-c|z|}$.

5-30 Given two normal RVs **x** and **y** with joint density as in (5-99), show that the probability masses m_1, m_2, m_3, m_4 in the four quadrants of the xy plane equal

$$m_1 = m_3 = \frac{1}{4} + \frac{\alpha}{2\pi} \qquad m_2 = m_4 = \frac{1}{4} - \frac{\alpha}{2\pi}$$

where $\alpha = \arcsin r$.

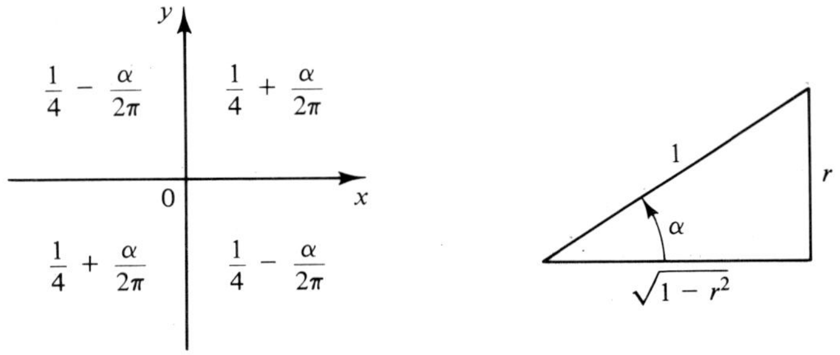

Figure P5.30

5-31 The RVs **x** and **y** are $N(0, \sigma)$ and independent. Show that the RVs $\mathbf{z} = \mathbf{x}/\mathbf{y}$ and $\mathbf{w} = \sqrt{\mathbf{x}^2 + \mathbf{y}^2}$ are independent, **z** has a Cauchy density, and **w** has a Rayleigh density:

$$f_z(z) = \frac{1/\pi}{1 + z^2} \qquad f_w(w) = \frac{w}{\sigma^2}\, e^{-w^2/\sigma^2}U(w)$$

6 Conditional Distributions, Regression, Reliability

In this chapter, we use the concept of the conditional probability of events to define conditional distributions, densities, and expected values. In the first section, we develop various extensions of Bayes' formula, and we introduce the notion of Bayesian estimation in the context of an unknown probability. Later, we treat the nonlinear prediction problem and its relationship to the regression line defined as conditional mean. In the last section, we present a number of basic concepts related to reliability and system failure.

6-1 Conditional Distributions

Recall that the conditional probability of an event $\mathscr{A}$ assuming $\mathscr{M}$ is the ratio

$$P(\mathscr{A}|\mathscr{M}) = \frac{P(\mathscr{A} \cap \mathscr{M})}{P(\mathscr{M})} \tag{6-1}$$

defined for every $\mathscr{M}$ such that $P(\mathscr{M}) > 0$. In the following, we express one or both events $\mathscr{A}$ and $\mathscr{M}$ in terms of various RVs. Unless otherwise stated, it will be assumed that all RVs are of continuous type. The discrete case leads to similar results.

We start with the definition of the conditional distribution $F(x|\mathcal{M})$ of the RV $\mathbf{x}$ assuming $\mathcal{M}$. This is a function defined as in (4-2) where all probabilities are replaced by conditional probabilities. Thus

$$F(x|\mathcal{M}) = P\{\mathbf{x} \leq x|\mathcal{M}\} = \frac{P\{\mathbf{x} \leq x, \mathcal{M}\}}{P(\mathcal{M})} \tag{6-2}$$

Here, $\{\mathbf{x} \leq x, \mathcal{M}\}$ is the intersection of the events $\{\mathbf{x} \leq x\}$ and $\mathcal{M}$; that is, it is an event consisting of all outcomes ζ that are in $\mathcal{M}$ and such that $\mathbf{x}(\zeta) \leq x$.

The derivative

$$f(x|\mathcal{M}) = \frac{dF(x|\mathcal{M})}{dx} \tag{6-3}$$

of $F(x|\mathcal{M})$ is the conditional density of $\mathbf{x}$ assuming $\mathcal{M}$.

It follows from the fundamental note on page 48 that conditional distributions have all the properties of unconditional distributions. For example,

$$F(x_2|\mathcal{M}) - F(x_1|\mathcal{M}) = P\{x_1 < \mathbf{x} \leq x_2|\mathcal{M}\} \tag{6-4}$$

the area of $f(x|\mathcal{M})$ equals 1, and [see (4-32)]

$$f(x|\mathcal{M})\,dx = P\{x < \mathbf{x} < x + dx|\mathcal{M}\} = \frac{P\{x < \mathbf{x} \leq x + \Delta x, \mathcal{M})}{P(\mathcal{M})} \tag{6-5}$$

Example 6.1

In the fair-die experiment, the RV $\mathbf{x}$ is such that $\mathbf{x}(f_i) = 10i$ and its distribution is the staircase function of Fig. 6.1. We shall determine the conditional distribution $F(x|\mathcal{M})$ where $\mathcal{M} = \{f_2, f_4, f_6\}$.

If $x \geq 60$, then $\{\mathbf{x} \leq x\}$ is the certain event and $\{\mathbf{x} \leq x, \mathcal{M}\} = \mathcal{M}$; hence,

$$F(x|\mathcal{M}) = \frac{P(\mathcal{M})}{P(\mathcal{M})} = 1$$

Figure 6.1

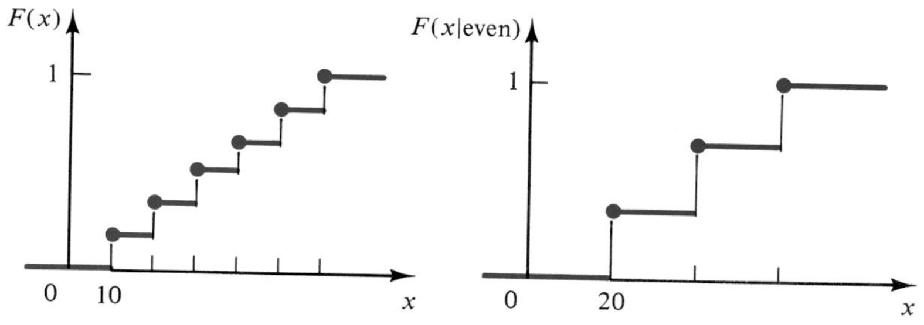

If $40 \le x < 60$, then $\{\mathbf{x} \le x, \mathcal{M}\} = \{f_2, f_4\}$; hence,

$$F(x|\mathcal{M}) = \frac{P\{f_2, f_4\}}{P(\mathcal{M})} = \frac{2/6}{3/6} = \frac{2}{3}$$

If $20 \le x < 40$, then $\{\mathbf{x} \le x, \mathcal{M}\} = \{f_2\}$; hence,

$$F(x|\mathcal{M}) = \frac{P\{f_2\}}{P(\mathcal{M})} = \frac{1/6}{3/6} = \frac{1}{3}$$

Finally, if $x < 20$, then $\{\mathbf{x} \le x, \mathcal{M}\} = \varnothing$; hence, $F(x|\mathcal{M}) = 0$. ■

TOTAL PROBABILITY AND BAYES' FORMULA We have shown [see (2-24)] that if $[\mathcal{A}_1, \ldots, \mathcal{A}_m]$ is a partition of $\mathcal{S}$, then (total probability theorem)

$$P(\mathcal{B}) = P(\mathcal{B}|\mathcal{A}_1)P(\mathcal{A}_1) + \cdots + P(\mathcal{B}|\mathcal{A}_m)P(\mathcal{A}_m) \tag{6-6}$$

for any $\mathcal{B}$. With $\mathcal{B} = \{\mathbf{x} \le x\}$, this yields

$$F(x) = F(x|\mathcal{A}_1)P(\mathcal{A}_1) + \cdots + F(x|\mathcal{A}_m)P(\mathcal{A}_m) \tag{6-7}$$

and by differentiation

$$f(x) = f(x|\mathcal{A}_1)P(\mathcal{A}_1) + \cdots + f(x|\mathcal{A}_m)P(\mathcal{A}_m) \tag{6-8}$$

Example 6.2

Two machines M_1 and M_2 produce cylinders at the rate of 3 units per second and 7 units per second, respectively. The diameters of these cylinders are two RVs with densities $N(\eta_1, \sigma)$ and $N(\eta_2, \sigma)$. The daily outputs are combined into a single lot, and the RV $\mathbf{x}$ equals the diameters of the cylinders. We shall find the density of $\mathbf{x}$.

In this experiment,

$$\mathcal{A}_1 = \{\zeta \text{ came from } M_1\} \qquad \mathcal{A}_2 = \{\zeta \text{ came from } M_2\}$$

are two events consisting of 30% and 70% of the units, respectively. Thus $f(x|\mathcal{A}_1)$ is the conditional density of $\mathbf{x}$ assuming that the unit came from machine M_1; hence, $f(x|\mathcal{A}_1)$ is $N(\eta_1, \sigma)$. Similarly, $f(x|\mathcal{A}_2)$ is $N(\eta_2, \sigma)$. And since the events $\mathcal{A}_1$ and $\mathcal{A}_2$ form a partition, we conclude from (6-8) with $P(\mathcal{A}_1) = .3$ and $P(\mathcal{A}_2) = .7$ that

$$f(x) = \frac{.3}{\sigma\sqrt{2\pi}} e^{-(x-\eta_1)/2\sigma^2} + \frac{.7}{\sigma\sqrt{2\pi}} e^{-(x-\eta_2)^2/2\sigma^2}$$

as in Fig. 6.2 ■

Figure 6.2

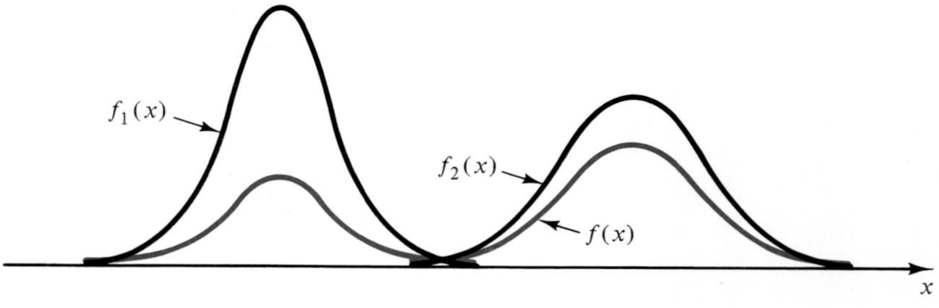

We shall now extend Bayes' formula

$$P(\mathscr{A}|\mathscr{B}) = \frac{P(\mathscr{B}|\mathscr{A})}{P(\mathscr{B})} P(\mathscr{A}) \tag{6-9}$$

to RVs. With $\mathscr{B} = \{\mathbf{x} \le x\}$, (6-9) yields

$$P\{\mathscr{A}|\mathbf{x} \le x\} = \frac{P\{\mathbf{x} \le x|\mathscr{A}\}}{P\{\mathbf{x} \le x\}} P(\mathscr{A}) = \frac{F(x|\mathscr{A})}{F(x)} P(\mathscr{A}) \tag{6-10}$$

Similarly [see (6-4)],

$$P\{\mathscr{A}|x_1 < \mathbf{x} \le x_2\} = \frac{F(x_2|\mathscr{A}) - F(x_1|\mathscr{A})}{F(x_2) - F(x_1)} P(\mathscr{A}) \tag{6-11}$$

Using (6-11), we shall define the conditional probability

$$P\{\mathscr{A}|x) = P\{\mathscr{A}|\mathbf{x} = x\}$$

of an event $\mathscr{A}$ assuming $\mathbf{x} = x$. If $P\{\mathbf{x} = x\} > 0$, we can use (6-9). We cannot do so, however, if $\mathbf{x}$ is of continuous type because then $P\{\mathbf{x} = x\} = 0$. In this case, we define $P\{\mathscr{A}|x)$ as a limit:

$$P\{\mathscr{A}|\mathbf{x} = x\} = \lim_{\Delta x \to 0} P\{\mathscr{A}|x \le \mathbf{x} \le x + \Delta x\}$$

With $x_1 = x$, $x_2 = x + \Delta x$, it follows from this and (6-11) that

$$P(\mathscr{A}|x) = \frac{f(x|\mathscr{A})P(\mathscr{A})}{f(x)} \tag{6-12}$$

We next multiply both sides by $f(x)$ and integrate. Since the area of $f(x|\mathscr{A})$ equals 1, we obtain

$$P(\mathscr{A}) = \int_{-\infty}^{\infty} P(\mathscr{A}|x)f(x)\,dx \tag{6-13}$$

This is another form of the total probability theorem. The corresponding version of Bayes' formula follows from (6-12):

$$f(x|\mathscr{A}) = \frac{P(\mathscr{A}|x)f(x)}{\int_{-\infty}^{\infty} P(\mathscr{A}|x)f(x)\,dx} \tag{6-14}$$

Bayesian Estimation

We are given a coin and we wish to determine the probability p that heads will show. To do so, we toss it n times and observe that heads shows k times. What conclusion can be drawn from this observation about the unknown p?

This problem can be given two interpretations. In the first interpretation, p is viewed as an unknown parameter. In the second, p is the value of an RV $\mathbf{p}$. These interpretations are examined in detail in Part Two. Here we introduce the second interpretation (Bayesian) in the context of Bayes' formula (6-14).

We assume that the probability of heads is an RV $\mathbf{p}$ defined in an experiment $\mathscr{S}_c$. This experiment could be the random selection of a coin from a large supply of coins. The RV $\mathbf{p}$ takes values only in the interval (0, 1); hence, its density $f(p)$ vanishes outside this interval. We toss the selected coin once, and we wish to find the probability $P(\mathcal{H})$ that heads will show, that is, that the event $\mathcal{H} = \{\text{heads}\}$ will occur.

■ ***Theorem.*** In the single toss of a randomly selected coin, the probability that heads will show equals the mean of $\mathbf{p}$:

$$P(\mathcal{H}) = \int_0^1 pf(p)\,dp \tag{6-15}$$

■ ***Proof.*** The experiment of the single toss of a randomly selected coin is a Cartesian product $\mathcal{S}_c \times \mathcal{S}_1$ where $\mathcal{S}_1 = \{h, t\}$. In this experiment, the event $\mathcal{H} = \{\text{heads}\}$ consists of all pairs $\zeta_c h$ where ζ_c is any element of the space $\mathcal{S}_c$. The probability that a particular coin with $\mathbf{p}(\zeta_c) = p$ will show equals p. This is the conditional probability of the event $\mathcal{H}$ assuming $\mathbf{p} = p$. Thus

$$P\{\mathcal{H}|p\} = p \tag{6-16}$$

Inserting into (6-13), we obtain (6-15).

Suppose that the coin is tossed and heads shows. We then have an updated version of the density of $\mathbf{p}$, given by $f(p|\mathcal{H})$. This density is obtained from (6-14) and (6-16):

$$f(p|\mathcal{H}) = \frac{pf(p)}{\int_0^1 pf(p)\,dp} \tag{6-17}$$

REPEATED TRIALS We now consider the toss of a randomly selected coin n times. The space of this experiment is a Cartesian product $\mathcal{S}_c \times \mathcal{S}_n$ where $\mathcal{S}_n$ consists of all sequences of length n formed with h and t. In the space $\mathcal{S}_c \times \mathcal{S}_n$,

$$\mathcal{A} = \{k \text{ heads in a specific order}\}$$

is an event consisting of all outcomes of the form $\zeta_c\, ht \cdots h$ where ζ_c is an element of $\mathcal{S}_c$ and $ht \cdots h$ is an element of $\mathcal{S}_n$. As we know from (3-15),

$$P(\mathcal{A}|p) = p^k q^{n-k} \qquad q = 1 - p \tag{6-18}$$

This is the probability that a particular coin with $\mathbf{p}(\zeta_c) = p$, tossed n times, will show k heads. From this and (6-13) it follows that

$$P(\mathcal{A}) = \int_0^1 p^k q^{n-k} f(p)\,dp \tag{6-19}$$

Equation (6-19) is an extension of (6-15) to repeated trials. The corresponding extension of (6-17) is the conditional density

$$f(p|\mathcal{A}) = \frac{p^k q^{n-k} f(p)}{\int_0^1 p^k q^{n-k} f(p)\,dp} \tag{6-20}$$

of the RV $\mathbf{p}$ assuming that in n trials k heads show. This density is used in the following problem.

We have selected a coin, tossed it n times, and observed k heads. Find the probability that at the next toss heads will show. This is equivalent to the problem of determining the probability of heads of a coin with prior density $f(p|\mathscr{A})$. We can therefore use theorem (6-15). The unknown probability equals

$$\int_0^1 pf(p|\mathscr{A})\,dp = \frac{\int_0^1 pp^k q^{n-k} f(p)\,dp}{\int_0^1 p^k q^{n-k} f(p)\,dp} \tag{6-21}$$

Example 6.3

Suppose that $\mathbf{p}$ is uniform in the interval (0, 1). In this case, $f(p) = 1$; hence, the probability of heads [see (6-15)] equals

$$\int_0^1 p\,dp = \frac{1}{2}$$

We toss the coin n times; if heads shows k times, the updated density of $\mathbf{p}$ is obtained from (6-20). Using the identity

$$\int_0^1 p^k(1-p)^{n-k}\,dp = \frac{k!(n-k)!}{(n+1)!} \tag{6-22}$$

we obtain

$$f(p|\mathscr{A}) = \frac{(n+1)!}{k!(n-k)!} p^k q^{n-k} \tag{6-23}$$

This function is known as *beta density*. Inserting into (6-21), we conclude that the probability of heads at the next toss equals

$$\int_0^1 pf(p|\mathscr{A})\,dp = \frac{(n+1)!}{k!(n-k)!}\int_0^1 p^{k+1}q^{n-k}\,dp = \frac{k+1}{n+2} \tag{6-24}$$

Thus after the observation of k heads, the density of the coin is updated from uniform to beta and the probability of heads from 1/2 to $(k+1)/(n+2)$. This result is known as the *law of succession*. ■

The assumption that the density of the die, prior to any observation, is constant is justified by the subjectivists as a consequence of the principle of insufficient reason (page 17). In the context of our discussion, however, it is only an assumption (and not a very good one, because most dice are fair).

Returning to the general case, we shall call the densities $f(p)$ and $f(p|\mathscr{A})$ in (6-20) *prior* and *posterior*, respectively. The posterior density $f(p|\mathscr{A})$ is an updated version of the prior, and its form depends on the observed number of heads. The factor

$$l(p) = p^k(1-p)^{n-k}$$

in (6-21) is called the *likelihood* function (see Section 9-5). This function is maximum for $p = k/n$.

In Fig. 6.3 we show the functions $f(p)$, $l(p)$, and

$$f(p|\mathscr{A}) \sim l(p)f(p)$$

For moderate values of n, the factor $l(p)$ is smooth, and the product $l(p)f(p)$ exhibits two maxima: one near the maximum k/n of $l(p)$ and one near the

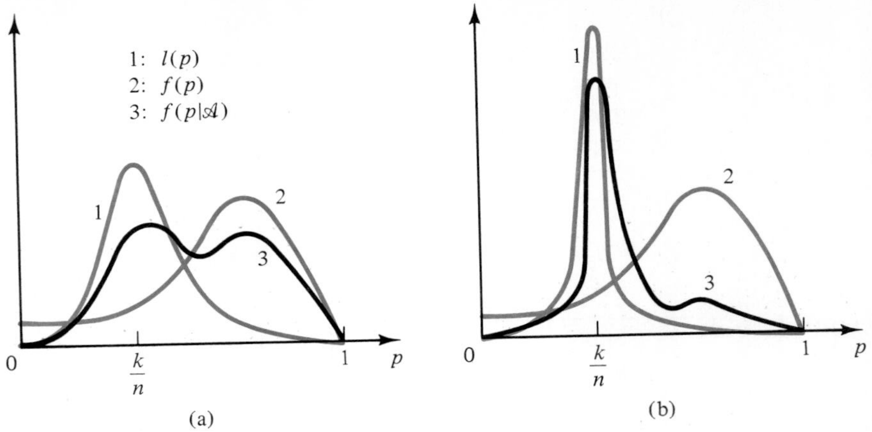

Figure 6.3

maximum of $f(p)$. As n increases, the sharpness of $l(p)$ prevails, and the function $f(p|\mathscr{A})$ approaches $l(p)$ regardless of the form of the prior $f(p)$. Thus as $n \to \infty$, $f(p|\mathscr{A})$ approaches a line at $p = k/n$ and its mean, that is, the probability of heads at the next toss, tends to k/n. This is, in a sense, the model justification of the empirical interpretation k/n of p.

6-2 Bayes' Formulas

The conditional distribution $F(x|\mathscr{M})$ of an RV $\mathbf{x}$ assuming $\mathscr{M}$ involves the event $\{\mathbf{x} \le x, \mathscr{M}\}$. For the determination of $F(x|\mathscr{M})$, knowledge of the underlying experiment is therefore required. However, if $\mathscr{M}$ is an event that can be expressed in terms of the RV $\mathbf{x}$, then $F(x|\mathscr{M})$ can be expressed in terms of the unconditional distribution $F(x)$ of $\mathbf{x}$. Let us look at several illustrations (we assume that all RVs are of continuous type).

Suppose, first, that

$$\mathscr{M} = \{\mathbf{x} \le a\}$$

In this case, $F(x|\mathscr{M})$ is the conditional distribution of $\mathbf{x}$ assuming that $\mathbf{x} \le a$. Thus

$$F(x|\mathbf{x} \le a) = \frac{P\{\mathbf{x} \le x, \mathbf{x} \le a\}}{P\{\mathbf{x} \le a\}} \tag{6-25}$$

where a is a fixed number and x is a variable ranging from $-\infty$ to ∞. The event $\{\mathbf{x} \le x, \mathbf{x} \le a\}$ consists of all outcomes such that $\mathbf{x} \le x$ and $\mathbf{x} \le a$; its probability depends on x.

If $x \ge a$, then $\{\mathbf{x} \le x, \mathbf{x} \le a\} = \{\mathbf{x} \le a\}$; hence,

$$F(x|\mathbf{x} \le a) = \frac{P\{\mathbf{x} \le a\}}{P\{\mathbf{x} \le a\}} = 1$$

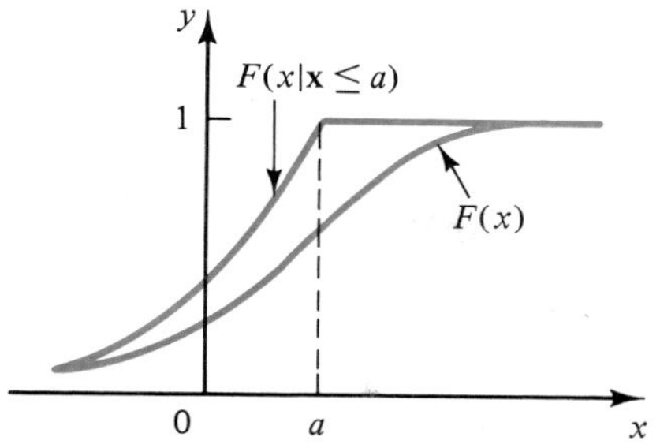

Figure 6.4

If $x < a$, then $\{\mathbf{x} \le x, \mathbf{x} \le a\} = \{\mathbf{x} \le x\}$; hence,

$$F(x|\mathbf{x} \le a) = \frac{P\{\mathbf{x} \le x\}}{P\{\mathbf{x} \le a\}} = \frac{F(x)}{F(a)}$$

Thus $F(x|\mathbf{x} \le a)$ is proportional to $F(x)$ for $x \le a$, and for $x > a$ it equals 1 (Fig. 6.4). The conditional density $f(x|\mathbf{x} \le a)$ is obtained by differentiation.

Suppose, next, that

$$\mathcal{M} = \{a \le \mathbf{x} \le b\}$$

We shall determine $f(x|\mathcal{M})$ directly using (6-5):

$$f(x|a \le \mathbf{x} \le b\}dx = \frac{P\{x < \mathbf{x} \le x + dx, a \le \mathbf{x} \le b\}}{P\{a \le \mathbf{x} \le b\}} \tag{6-26}$$

To do so, we must find the probability of the event

$$\{x < \mathbf{x} \le x + dx, a < \mathbf{x} \le b\} = \begin{cases} \{x < \mathbf{x} \le x + dx\} & a < x < b \\ 0 & \text{otherwise} \end{cases}$$

Since $P\{x < \mathbf{x} \le x + dx\} = f(x)\,dx$, we conclude from (6-5) that

$$f(x|a \le \mathbf{x} \le b) = \frac{f(x)}{F(b) - F(a)} \qquad \text{for } a < x < b \tag{6-27}$$

and zero otherwise.

Example 6.4

The RV $\mathbf{x}$ is $N(\eta, \sigma)$; we shall find its conditional density $f(x|\mathcal{M})$ assuming $\mathcal{M} = \{\eta - \sigma \le \mathbf{x} < \eta + \sigma\}$.

As we know (see Fig. 4.20), the probability that $\mathbf{x}$ is in the interval $(\eta - \sigma, \eta + \sigma)$ equals .683. Setting $F(b) - F(a) = .683$ in (6-27), we obtain the truncated normal density

$$f(x|\,|\mathbf{x} - \eta| < \sigma) = \frac{1}{.683\sigma\sqrt{2\pi}} e^{-(x-\eta)^2/2\sigma^2}$$

for $\eta - \sigma < x < \eta + \sigma$ and zero otherwise (Fig. 6.5). ■

Empirical Interpretation. We wish to determine empirically the conditional distribution $F(x|a \le \mathbf{x} \le b)$. To do so, we repeat the experiment n times, and we reject all outcomes ζ such that $\mathbf{x}(\zeta)$ is outside the interval (a, b). In the subsequence s_1 of the remaining n_1 trials, the function $F(x|a \le \mathbf{x} \le b)$ has the

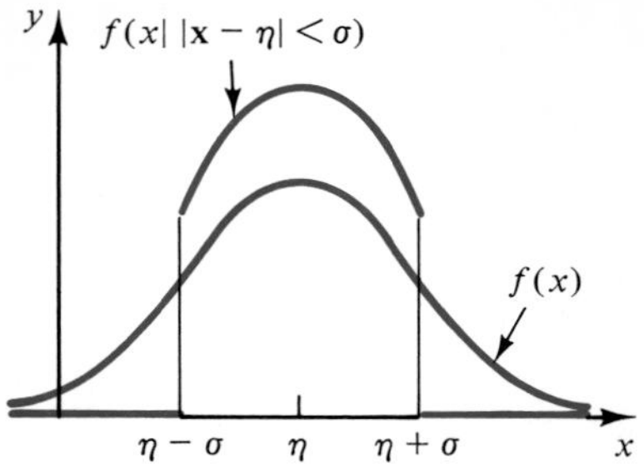

Figure 6.5

empirical interpretation of an unconditional distribution: For a specific x, it equals the ratio n_x/n_1 where n_x is the number of trials of the subsequence s_1 such that $\mathbf{x}(\zeta) \le x$.

Suppose that our experiment is the manufacture of cylinders and that $x_i = \mathbf{x}(\zeta_i)$ is the diameter of the ith unit. To control the quality of the output, we specify a tolerance interval (a, b) and reject all units that fall outside this interval. The function $F(x|a < \mathbf{x} \le b)$ is the distribution of the accepted units.

Note, finally, that if $\mathbf{y} = g(\mathbf{x})$ is a function of the RV $\mathbf{x}$, its conditional density $f_y(y|\mathcal{M})$ is obtained from (4-78), where all densities are replaced by conditional densities. If, in particular, $\mathcal{M}$ is an event that can be expressed in terms of $\mathbf{x}$, then $f_y(y|\mathcal{M})$ can be determined in terms of $f_x(x)$ and the function $g(x)$.

Example 6.5 If $\mathcal{M} = \{\mathbf{x} \ge 0\}$ and

$$\mathbf{y} = \mathbf{x}^2$$

then [see (4-84)]

$$f_y(y|\mathbf{x} \ge 0) = \frac{1}{2\sqrt{y}} \frac{f_x(\sqrt{y})}{1 - F_x(0)} U(y)$$ ■

Joint Distributions

We shall investigate the properties of the conditional probability $P(\mathcal{A}|\mathcal{B})$ when both events $\mathcal{A}$ and $\mathcal{B}$ are specified in terms of the RVs $\mathbf{x}$ and $\mathbf{y}$.

We start with the determination of the function $F_y(y|x_1 \le \mathbf{x} \le x_2)$. With $\mathcal{A} = \{\mathbf{y} \le y\}$ and $\mathcal{B} = \{x_1 \le \mathbf{x} \le x_2\}$, it follows from (6-1) that

$$F_y(y|x_1 \le \mathbf{x} \le x_2) = \frac{P\{x_1 \le \mathbf{x} \le x_2, \mathbf{y} \le y\}}{P\{x_1 \le \mathbf{x} \le x_2\}} = \frac{F(x_2, y) - F(x_1, y)}{F_x(x_2) - F_x(x_1)} \qquad (6\text{-}28)$$

We shall use (6-28) to determine the conditional distribution $F_y(y|x)$ of $\mathbf{y}$ assuming $\mathbf{x} = x$. This function cannot be determined from (6-1) because the event $\{\mathbf{x} = x\}$ has zero probability. It will be defined as a limit:

$$F_y(y|x) = \lim_{\Delta x \to 0} F_y(y|x \le \mathbf{x} \le x + \Delta x)$$

Setting $x_1 = x$ and $x_2 = x + dy$ in (6-28) and dividing numerator and denominator of the right side by Δx, we conclude with $\Delta x \to 0$ that

$$F_y(y|x) = \frac{1}{f_x(x)} \frac{\partial F(x, y)}{\partial x} \tag{6-29}$$

The conditional density $f_y(y|x)$ of **y**, assuming $\mathbf{x} = x$, is the derivative of $F_y(y|x)$ with respect to y. Thus

$$f_y(y|x) = \frac{\partial F_y(y|x)}{\partial y} = \frac{1}{f_x(x)} \frac{\partial^2 F(x, y)}{\partial x \, \partial y}$$

The function $f_x(x|y)$ is defined similarly. Omitting subscripts, we conclude from the foregoing and from (5-8) that

$$f(y|x) = \frac{f(x, y)}{f(x)} \qquad f(x|y) = \frac{f(x, y)}{f(y)} \tag{6-30}$$

If the RVS **x** and **y** are independent,

$$f(x, y) = f(x)f(y) \qquad f(y|x) = f(y) \qquad f(x|y) = f(x)$$

For a specific x, the function $f(x, y)$ is the intersection (*profile*) of the surface $z = f(x, y)$ by the plane x = constant. The conditional density $f(y|x)$, considered as a function of y, is the profile of $f(x, y)$ normalized by the factor $1/f_x(x)$.

From (6-30) and the relationship

$$f(y) = \int_{-\infty}^{\infty} f(x, y)\, dx$$

between marginal and joint densities, it follows that

$$f(y) = \int_{-\infty}^{\infty} f(y|x)f(x)\, dx \tag{6-31}$$

This is another form of the total probability theorem. The corresponding form of Bayes' formula is

$$f(x|y) = \frac{f(y|x)f(x)}{f(y)} = \frac{f(y|x)f(x)}{\int_{-\infty}^{\infty} f(y|x)f(x)\, dx} \tag{6-32}$$

Example 6.6

The RVS **x** and **y** are normal with zero mean and density [see (5-99)]

$$f(x, y) \sim \exp\left\{-\frac{1}{2(1 - r^2)}\left(\frac{x^2}{\sigma_1^2} - 2r\frac{xy}{\sigma_1\sigma_2} + \frac{y^2}{\sigma_2^2}\right)\right\} \tag{6-33}$$

We shall show that

$$f(y|x) = \frac{1}{\sigma_2\sqrt{2\pi(1 - r^2)}} \exp\left\{-\frac{(y - r\sigma_2 x/\sigma_1)^2}{2\sigma_2^2(1 - r^2)}\right\} \tag{6-34}$$

■ ***Proof.*** As we know,

$$f(x) = \frac{1}{\sigma_1\sqrt{2\pi}}\, e^{-x^2/2\sigma_1^2} \tag{6-35}$$

Dividing (6-31) by (6-33), we obtain an exponential with exponent

$$-\frac{1}{2(1-r^2)}\left(\frac{x^2}{\sigma_1^2} - 2r\frac{xy}{\sigma_1\sigma_2} + \frac{y^2}{\sigma_2^2}\right) + \frac{x^2}{2\sigma_2^2} = \frac{1}{2\sigma_2^2(1-r^2)}\left(y - \frac{r\sigma_2}{\sigma_1}x\right)^2$$

and (6-34) results. ■

Conditional Expected Values

The mean $E\{\mathbf{x}\}$ of a continuous type RV $\mathbf{x}$ equals the integral in (4-87). The conditional mean $E\{\mathbf{x}|\mathcal{M}\}$ is given by the same integral where $f(x)$ is replaced by $f(x|\mathcal{M})$. Thus

$$E\{\mathbf{x}|\mathcal{M}\} = \int_{-\infty}^{\infty} xf(x|\mathcal{M})\,dx \tag{6-36}$$

The empirical interpretation of $E\{\mathbf{x}|\mathcal{M}\}$ is the arithmetic mean of the samples x_i of $\mathbf{x}$ in the subsequence of trials in which the event $\mathcal{M}$ occurs.

Example 6.7

Two light bulbs are bought from the same lot. The first is turned on on May 1 and the second on June 1. If the first is still good on June 1, can it, on the average, last longer than the second?

Suppose that the density of the time to failure is the function $f(x)$ of Fig. 6.6. In this case, $E\{\mathbf{x}\} = 3.15$ month. The conditional density $f(x|\mathbf{x} \geq 1)$ is a triangle, and $E\{\mathbf{x}|\mathbf{x} \geq 1\} = 5$ months. Thus the average time to failure after June 1 of the old bulb is $5 - 1 = 4$ months and of the new bulb 3.15 months. Thus the old bulb is better than the new bulb!

This phenomenon is observed in statistics of populations with high infant mortality: The mean life expectancy a year after birth is larger than at birth. ■

If $\mathcal{M}$ is an event that can be expressed in terms of the RV $\mathbf{x}$, the conditional mean $E\{\mathbf{y}|\mathcal{M}\}$ is specified if $f(x, y)$ is known. Of particular interest is

Figure 6.6

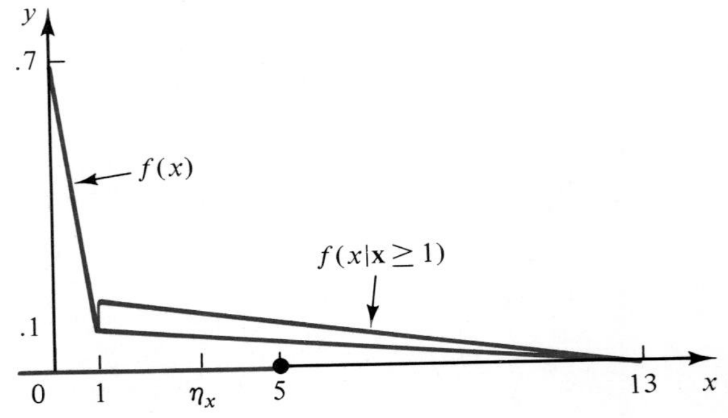

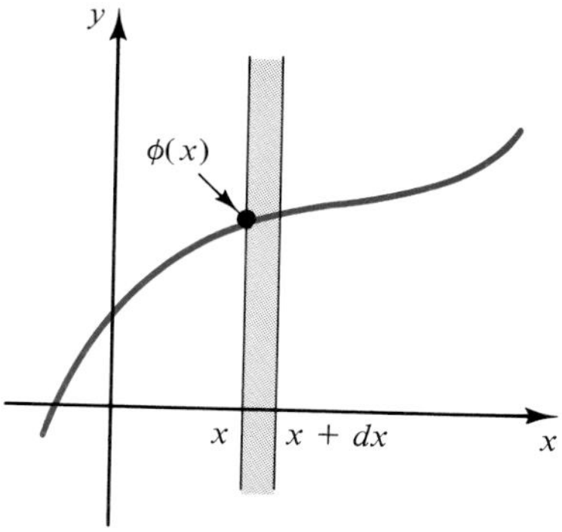

Figure 6.7

the conditional mean $E\{\mathbf{y}|x\}$ of $\mathbf{y}$ assuming $\mathbf{x} = x$. As we shall see in Section 6-3, this concept is important in mean-square estimation.

Setting $\mathcal{M} = \{x \le \mathbf{x} \le x + \Delta x\}$ in (6-36), we conclude with $\Delta x \to 0$ that

$$E\{\mathbf{y}|x\} = \int_{-\infty}^{\infty} y f(y|x)\, dy \tag{6-37}$$

Similarly,

$$E\{g(\mathbf{y})|x\} = \int_{-\infty}^{\infty} g(y) f(y|x)\, dy \tag{6-38}$$

where $f(y|x)$ is given by (6-30). For a given x, the integral in (6-37) is the center of gravity of all masses in the strip $(x, x + dx)$ of the xy plane (Fig. 6.7). The locus of these points is the function

$$\phi(x) = \int_{-\infty}^{\infty} y f(y|x)\, dy \tag{6-39}$$

known as the *regression curve* of $\mathbf{y}$ on $\mathbf{x}$.

Example 6.8

If the RVS $\mathbf{x}$ and $\mathbf{y}$ are jointly normal with zero mean, then [see (6-34)]

$$f(y|x) \sim \exp\left\{-\frac{(y - r\sigma_2 x/\sigma_1)^2}{2\sigma_2^2(1 - r^2)}\right\}$$

For a fixed x, this is a normal density with mean $r\sigma_2 x/\sigma_1$; hence,

$$E\{\mathbf{y}|x\} = \phi(x) = r\frac{\sigma_2}{\sigma_1}x \tag{6-40}$$

If the RVS $\mathbf{x}$ and $\mathbf{y}$ are normal with mean η_x and η_y, respectively, then $f(y|x)$ is given by (6-34) where y and x are replaced by $y - \eta_y$ and $x - \eta_x$, respectively. In this case, $f(y|x)$ is a normal density in y with mean

$$\phi(x) = \eta_y + \frac{r\sigma_2}{\sigma_1}(x - \eta_x) \tag{6-41}$$

Thus the regression curve of normal RVS is a straight line with slope $r\sigma_2/\sigma_1$ passing through the point (η_x, η_y). ■

The mean $E\{\mathbf{y}\}$ of an RV $\mathbf{y}$ is a deterministic number. The conditional mean $E\{\mathbf{y}|x\}$ is a function $\phi(x)$ of the real variable x. Using this function, we form the composite function $\phi(\mathbf{x})$ as in Section 4-4. This function is an RV with domain the space $\mathcal{S}$. Thus starting from the deterministic function $E\{\mathbf{y}|x\}$, we have formed the RV $E\{\mathbf{y}|\mathbf{x}\} = \phi(\mathbf{x})$.

■ ***Theorem.*** The mean of the RV $E\{\mathbf{y}|\mathbf{x}\}$ equals the mean of $\mathbf{y}$

$$E\{E\{\mathbf{y}|\mathbf{x}\}\} = E\{\mathbf{y}\} \tag{6-42}$$

■ ***Proof.*** As we know from (4-94),

$$E\{\mathbf{y}|\mathbf{x}\} = E\{\phi(\mathbf{x})\} = \int_{-\infty}^{\infty} \phi(x)f(x)\,dx$$

Inserting (6-39) into this and using the fact that $f(y|x)f(x) = f(x, y)$, we obtain

$$E\{\phi(\mathbf{x})\} = \int_{-\infty}^{\infty} \left(\int_{-\infty}^{\infty} yf(y|x)\,dy\right) f(x)\,dx = \int_{-\infty}^{\infty}\int_{-\infty}^{\infty} yf(x, y)\,dx\,dy$$

This yields (6-42) because the last integral equals $E\{\mathbf{y}\}$ [see (5-28)].

In certain problems, it is easy to evaluate the function $E\{\mathbf{y}|x\}$. In such cases, (6-42) is used to find $E\{\mathbf{y}\}$. Let us now look at an illustration involving discrete type RVs.

Example 6.9

The number of accidents in a day is a Poisson RV $\mathbf{x}$ with parameter a. The accidents are independent, and the probability that an accident is fatal equals p. Show that the number of fatal accidents in a day is a Poisson RV $\mathbf{y}$ with parameter ap.

To solve this problem, it suffices to show that the moment function of $\mathbf{y}$ [see (5-60)] equals

$$E\{e^{s\mathbf{y}}\} = e^{ap(e^s-1)} \tag{6-43}$$

This involves only expected values; we can therefore apply (6-42).

■ ***Proof.*** The RV $\mathbf{x}$ is Poisson by assumption; hence,

$$P\{\mathbf{x} = n\} = e^{-a}\frac{a^n}{n!} \qquad n = 0, 1, \ldots$$

If $\mathbf{x} = n$, we have n independent accidents during that day, and the probability of each equals p. From this it follows that the conditional distribution of the number $\mathbf{y}$ of fatal accidents assuming $\mathbf{x} = n$ is a binomial distribution:

$$P\{\mathbf{y} = k|\mathbf{x} = n\} = \binom{n}{k} p^k q^{n-k} \qquad k = 0, 1, \ldots, n \tag{6-44}$$

and its moment function [see (5-66)] equals

$$E\{e^{s\mathbf{y}}|\mathbf{x} = n\} = (pe^s + q)^n \tag{6-45}$$

The right side is the value of the RV $(pe^s + q)^{\mathbf{x}}$ for $\mathbf{x} = n$. Since $\mathbf{x}$ has a Poisson distribution, it follows from (4-94) that the expected value of the RV $(pe^s + q)^{\mathbf{x}}$ equals

$$\sum_{n=0}^{\infty} (pe^s + q)^n P\{\mathbf{x} = n\} = \sum_{n=0}^{\infty} (pe^s + q)^n e^{-a} \frac{a^n}{n!} = e^{-a} e^{a(pe^s+q)}$$

Hence [see (6-42)]

$$E\{e^{s\mathbf{y}}\} = E\{E\{e^{s\mathbf{y}}|\mathbf{x}\}\} = E\{(pe^s + q)^{\mathbf{x}}\} = e^{ape^s - a(1-q)}$$

and (6-43) results. ■

We conclude with the specification of the conditional mean $E\{g(\mathbf{x}, \mathbf{y})|x\}$ of the RV $g(\mathbf{x}, \mathbf{y})$ assuming $\mathbf{x} = x$. Clearly, $E\{g(\mathbf{x}, \mathbf{y})|x\}$ is a function of x that can be determined as the limit of the conditional mean $E\{g(\mathbf{x}, \mathbf{y})|x \le \mathbf{x} \le x + dx\}$. We shall, however, specify it using the interpretation of the mean as an empirical average.

The function $E\{g(\mathbf{x}, \mathbf{y})|x\}$ is the average of the samples $g(x_i, y_i)$ in the subsequence of trials in which $x_i = x$. It therefore equals the average of the samples $g(x, y_i)$ of the RV $g(x, \mathbf{y})$. This leads to the conclusion that

$$E\{g(\mathbf{x}, \mathbf{y})|x\} = E\{g(x, \mathbf{y})|x\} \tag{6-46}$$

Note the difference between the RVs $g(\mathbf{x}, \mathbf{y})$ and $g(x, \mathbf{y})$. The first is a function of the RVs $\mathbf{x}$ and $\mathbf{y}$; the second is a function of the RV $\mathbf{y}$ depending on the parameter x. However, as (6-46) shows, both RVs have the same mean, assuming $\mathbf{x} = x$.

Since $g(x, \mathbf{y})$ is a function of $\mathbf{y}$ (depending also on the parameter x), its conditional mean is given [see (6-38)] by

$$E\{g(x, \mathbf{y})|x\} = \int_{-\infty}^{\infty} g(x, y) f(y|x)\, dy \tag{6-47}$$

This integral is a function $\theta(x)$ of x; it therefore defines the RV $\theta(\mathbf{x}) = E\{g(\mathbf{x}, \mathbf{y})|\mathbf{x}\}$. The mean of $\theta(\mathbf{x})$ equals

$$\int_{-\infty}^{\infty} \theta(x) f(x)\, dx = \int_{-\infty}^{\infty}\int_{-\infty}^{\infty} g(x, y) f(y|x) f(x)\, dx\, dy$$

$$= \int_{-\infty}^{\infty}\int_{-\infty}^{\infty} g(x, y) f(x, y)\, dx\, dy$$

But the last integral is the mean of $g(\mathbf{x}, \mathbf{y})$; hence,

$$E\{E\{g(\mathbf{x}, \mathbf{y})|\mathbf{x}\}\} = E\{g(\mathbf{x}, \mathbf{y})\} \tag{6-48}$$

Note the following special cases of (6-46) and (6-48):

$$\begin{aligned} E\{g_1(\mathbf{x})g_2(\mathbf{y})|x\} &= g_1(x)E\{g_2(\mathbf{y})|x\} \\ E\{g_1(\mathbf{x})g_2(\mathbf{y})\} &= E\{g_1(\mathbf{x})E\{g_2(\mathbf{y})|\mathbf{x}\}\} \end{aligned} \tag{6-49}$$

6-3 Nonlinear Regression and Prediction

The RV $\mathbf{y}$ models the values of a physical quantity in a real experiment, and its distribution is a function $F(y)$ determined from past observations. We wish to predict the value $\mathbf{y}(\zeta) = y$ of this RV at the next trial (Fig. 6.8a). The outcome ζ of the trial is an unknown element of the space $\mathscr{S}$; hence, y could

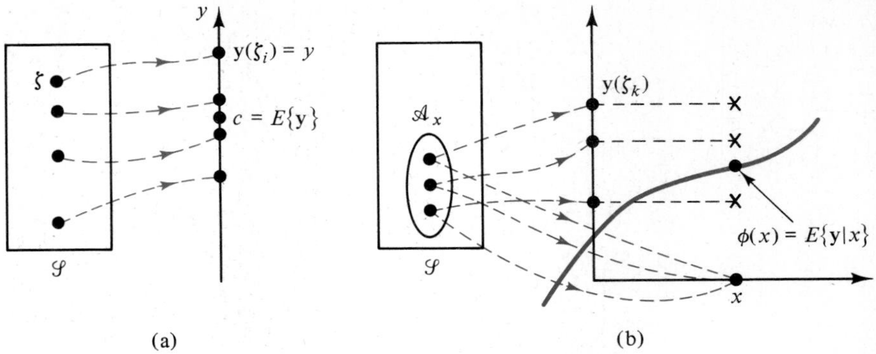

Figure 6.8

be any number in the range of $\mathbf{y}$. We therefore cannot predict y; we can only estimate it. Suppose that we estimate y by a constant c. The estimation error $y - c$ is the value of the difference $\mathbf{y} - c$, and our goal is to choose c so as to minimize in some sense this error. We shall use as our criterion for selecting c the minimization of the mean-square (MS) error

$$e = E\{(\mathbf{y} - c)^2\} = \int_{-\infty}^{\infty} (y - c)^2 f(x)\, dx \tag{6-50}$$

This criterion is reasonable; however, it is used primarily because it leads to a simple solution. At the end of the section, we comment briefly on other criteria.

To find c, we shall use the identity (see Problem 4-28)

$$E\{(\mathbf{y} - c)^2\} = (\eta_y - c)^2 + \sigma_y^2 \tag{6-51}$$

Since η_y and σ_y are given constants, (6-51) is minimum if

$$c = \eta_y = \int_{-\infty}^{\infty} y f_y(y)\, dy \tag{6-52}$$

Thus the least mean square (LMS) estimate of an RV by a constant is its mean.

REGRESSION Suppose now that at the next trial we observe the value $\mathbf{x}(\zeta) = x$ of another RV $\mathbf{x}$. On the basis of this information, we might improve the estimate of $\mathbf{y}$ if we use as its predicted value not a constant but a function $\phi(x)$ of the observed x.

It might be argued that if we know the number $x = \mathbf{x}(\zeta)$, we know also ζ and, hence, the value $y = \mathbf{y}(\zeta)$ of $\mathbf{y}$. This is the case, however, only if $\mathbf{y}$ is a function of $\mathbf{x}$. In general, $\mathbf{x}(\zeta_k) = x$ for every ζ_k in the set $\mathscr{A}_x = \{\mathbf{x} = x\}$, but the corresponding values $\mathbf{y}(\zeta_k)$ of $\mathbf{y}$ might be different (Fig. 6.8b). Thus the observed value x of $\mathbf{x}$ does not determine the unknown value $y = \mathbf{y}(\zeta_k)$ of $\mathbf{y}$. It reduces our uncertainty about y, however, because it tells us that ζ_k is not an arbitrary element of $\mathscr{S}$ but an element of its subset $\mathscr{A}_x$.

For example, suppose that the RV $\mathbf{y}$ represents the height of all boys in a community and the RV $\mathbf{x}$ their weight. We wish to estimate the height y of Jim. The best estimate of y by a number is the mean η_y of $\mathbf{y}$. This is the average of the heights of all boys. Suppose, however, that we weigh Jim and his weight is x. As we shall show, the best estimate of y is now the average $E\{\mathbf{y}|x\}$ of all children that have Jim's weight.

Again using the LMS criterion, we shall determine the function $\phi(x)$ so as to minimize the mean value

$$e = E\{[\mathbf{y} - \phi(\mathbf{x})]^2\} = \int_{-\infty}^{\infty}\int_{-\infty}^{\infty} [y - \phi(x)]^2 f(x, y)\,dx\,dy \tag{6-53}$$

of the square of the estimation error $\mathbf{y} - \phi(\mathbf{x})$.

■ ***Theorem.*** The LMS estimate of the RV $\mathbf{y}$ in terms of the observed value x of the RV $\mathbf{x}$ is the conditional mean

$$\phi(x) = E\{\mathbf{y}|x\} = \int_{-\infty}^{\infty} yf(y|x)\,dy \tag{6-54}$$

■ ***Proof.*** Inserting the identity $f(x, y) = f(y|x)f(x)$ into (6-53), we obtain

$$e = \int_{-\infty}^{\infty}\int_{-\infty}^{\infty} [y - \phi(x)]^2 f(y|x)f(x)\,dx\,dy$$

$$= \int_{-\infty}^{\infty} f(x) \int_{-\infty}^{\infty} [y - \phi(x)]^2 f(y|x)\,dy\,dx$$

All integrands are positive; hence, e is minimum if the inner integral on the right is minimum. This integral is of the form (6-50) if c is changed to $\phi(x)$ and $f(y)$ is changed to $f(y|x)$. Therefore, the integral is minimum if $\phi(x)$ is given by (6-52), *mutatis mutandis*. Changing the function $f_y(y)$ in (6-52) to $f(y|x)$, we obtain (6-54).

We have thus concluded that the LMS estimate of y in terms of x is the ordinate $\phi(x)$ of the regression curve (6-39).

Note that if the RVs $\mathbf{x}$ and $\mathbf{y}$ are normal as in (6-31), the regression curve is a straight line [see (6-40)]. Thus, for normal RVs, linear and nonlinear predictors are identical. Here is another example of RVs with this property.

Example 6.10 Suppose that $\mathbf{x}$ and $\mathbf{y}$ are two RVs with joint density equal to 1 in the parallelogram of Fig. 6-9 and zero elsewhere. In this case, $f(y|x)$ is constant in the segment AB of the line L_x of the figure. Since the center of that segment is at $y = x/2$, we conclude that $E\{\mathbf{y}|x\} = x/2$. ■

Galton's Law The term *regression* has its origin in the following observation by the geneticist and biostatistician Sir Francis Galton (1822–1911): "Population extremes regress toward their mean." In terms of average

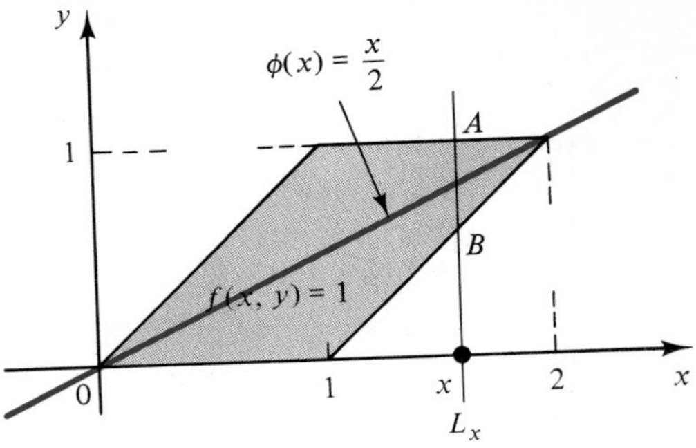

Figure 6.9

heights of parents and their adult children, this can be phrased as follows: Children of tall (short) parents are on the average shorter (taller) than their parents. This observation is based on the fact that the height **y** of children depends not only on the height **x** of their parents but also on other genetic factors. As a result, the conditional mean of children born of tall (short) parents, although larger (smaller) than the population mean, is smaller (larger) than the height of their parents. This process continues until after several generations, the mean height of descendants of tall (short) parents approaches the population mean. This empirical result is called Galton's law.

The statistical interpretation of the law can be expressed in terms of the properties of the regression line $\varphi(x) = E\{\mathbf{y}|x\}$. We observe first, that

$$\text{if } x > \eta, \text{ then } \varphi(x) < x; \text{ if } x < \eta, \text{ then } \varphi(x) > x \tag{6-55}$$

because $\varphi(x)$ is the mean height of all children whose father's height equals x. This shows that $\varphi(x)$ is below the line $y = x$ if $x > \eta$ and above the line $y = x$ if $x < \eta$. The evolution of this process to several generations leads to the following property of $\varphi(x)$. We start with a group of parents with height $x_0 > \eta$ as in Fig. 6.10, and we find the average $y_0 = \varphi(x_0) < x_0$ of the height of their children. We next form a group of parents with height $x_1 = y_0$ and find the average $y_1 = \varphi(x_1) < y_0$ of the height of their children. Continuing this process, we obtain two sequences x_n and y_n such that

$$x_n > \varphi(x_n) = y_n = x_{n+1} > \varphi(x_{n+1}) = y_{n+1} = x_{n+2} \quad \underset{n \to \infty}{\to \eta}$$

Starting with short parents, we obtain similarly two sequences x'_n and y'_n such that

$$x'_n < \varphi(x'_n) = y'_n = x'_{n+1} < \varphi(x'_{n+1}) = y'_{n+1} = x'_{n+2} \quad \underset{n \to \infty}{\to \eta}$$

This completes the properties of a regression curve obeying Galton's law. Today, the term *regression curve* is used to characterize not only a function obeying Galton's law but any conditional mean $E\{\mathbf{y}|x\}$.

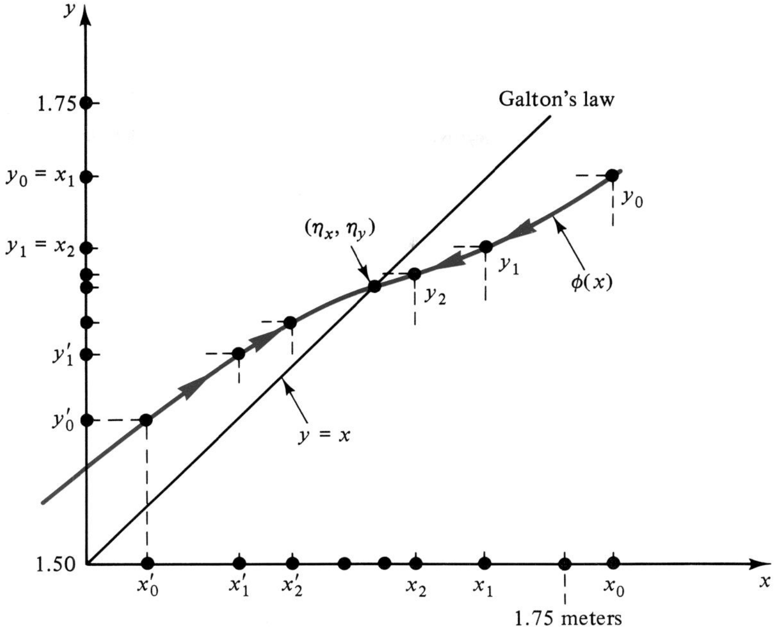

Figure 6.10

THE ORTHOGONALITY PRINCIPLE We have shown in (5-49) that if $a + b\mathbf{x}$ is the best MS fit of $\mathbf{y}$ on $\mathbf{x}$, then

$$E\{[\mathbf{y} - (a + b\mathbf{x})]\mathbf{x}\} = 0 \qquad (6\text{-}56)$$

This result can be phrased as follows: If $a + b\mathbf{x}$ is the linear predictor of $\mathbf{y}$ in terms of $\mathbf{x}$, then the prediction error $\mathbf{y} - (a + b\mathbf{x})$ is orthogonal to $\mathbf{x}$. We show next that if $\phi(\mathbf{x})$ is the nonlinear predictor of $\mathbf{y}$, then the error $\mathbf{y} - \phi(\mathbf{x})$ is orthogonal not only to $\mathbf{x}$ but to any function $q(\mathbf{x})$ of $\mathbf{x}$.

■ ***Theorem.*** If $\phi(x) = E\{\mathbf{y}|x\}$, then

$$E\{[\mathbf{y} - \phi(\mathbf{x})]q(\mathbf{x})\} = 0 \qquad (6\text{-}57)$$

■ ***Proof.*** From the linearity of expected values, it follows that

$$E\{\mathbf{y} - \phi(\mathbf{x})|x\} = E\{\mathbf{y}|x\} - \phi(x) = 0$$

Hence [see (6-49)],

$$E\{[\mathbf{y} - \phi(\mathbf{x})]q(\mathbf{x})\} = E\{q(\mathbf{x})E\{\mathbf{y} - \phi(\mathbf{x})|\mathbf{x}\}\} = 0$$

and (6-57) results.

The Rao-Blackwell Theorem The following corollary of (6-57) is used in parameter estimation (see page 313). We have shown in (6-32) that the mean η_ϕ of $\phi(\mathbf{x})$ equals the mean η_y of $\mathbf{y}$. We show next that the variance of $\phi(\mathbf{x})$

does not exceed the variance of **y**:

$$\eta_\phi = \eta_y \qquad \sigma_\phi^2 \le \sigma_y^2 \tag{6-58}$$

■ ***Proof.*** From the identity $\mathbf{y} - \eta_y = \mathbf{y} - \phi(\mathbf{x}) + \phi(\mathbf{x}) - \eta_\phi$ it follows that

$$(\mathbf{y} - \eta_y)^2 = [\mathbf{y} - \phi(\mathbf{x})]^2 + [\phi(\mathbf{x}) - \eta_\phi]^2 + 2[\mathbf{y} - \phi(\mathbf{x})][\phi(\mathbf{x}) - \eta_\phi]$$

We next take expected values of both sides. With $q(x) = \phi(x) - \eta_\phi$ it follows from (6-57) that the expected value of the last term equals zero. Hence,

$$\sigma_y^2 = E\{[\mathbf{y} - \phi(\mathbf{x})]^2\} + \sigma_\phi^2 \ge \sigma_\phi^2$$

and (6-58) results.

Risk and Loss Returning to the problem of estimating an RV **y** by a constant c, we note that the choice of c is not unique. For each c, we commit an error $\mathbf{y} - c$, and our objective is to reduce some function $L(\mathbf{y} - c)$ of this error. The selection of the form of $L(y - c)$ depends on the applications and is based on factors unrelated to the statistical model. The curve $L(\mathbf{y} - c)$ is called the *loss function*. Its mean value

$$R = E\{L(\mathbf{y} - c)\} = \int_{-\infty}^{\infty} L(y - c)f(y)\,dy$$

is the *average risk*. If $L(\mathbf{y} - c) = (\mathbf{y} - c)^2$, then R is the MS error e in (6-50) and $c = \eta_y$.

Another choice of interest is the loss function $L(\mathbf{y} - c) = |\mathbf{y} - c|$. In this case, our problem is to find c so as to minimize the average risk $R = E\{|\mathbf{y} - c|\}$. Clearly,

$$R = \int_{-\infty}^{\infty} |y - c| f(y)\,dy = \int_{-\infty}^{c} (c - y)f(y)\,dy + \int_{c}^{\infty} (y - c)f(y)\,dy$$

Differentiating with respect to c, we obtain

$$-\frac{dR}{dc} = \int_{-\infty}^{c} f(y)\,dy - \int_{c}^{\infty} f(y)\,dy = F(c) - [1 - F(c)] = 2F(c) - 1$$

Hence, the average risk $E\{|\mathbf{y} - c|\}$ is minimum if $F(c) = 1/2$, that is, if c equals the *median* of **y**.

6-4 System Reliability

A system is an object made to perform a function. The system is good if it performs that function, defective if it does not. In reliability theory, the state of a system is often interpreted statistically. This interpretation has two related forms. The first is time-dependent: The interval of time from the moment the system is put into operation until it fails is a random variable. For example, the life length of a light bulb is the value of a random variable. The second is time-independent: The system is either good with probability

p or defective with probability $1 - p$. In this interpretation, the state of the system is specified in terms of the number p; time is not a factor. For example, for all practical purposes, a bullet is either good or defective. In this section, we deal primarily with time-dependent systems. We introduce the notion of time to failure and explain the meaning of conditional failure rate in the context of conditional probabilities. At the end of the section, we consider the properties of systems formed by the interconnection of components.

■ ***Definition.*** The *time to failure* or *life length* of a system is the time interval from the moment the system is put into operation until it fails. This interval is an RV $\mathbf{x} \geq 0$ with distribution $F(t) = P\{\mathbf{x} \leq t\}$. The difference

$$R(t) = 1 - F(t) = P\{\mathbf{x} > t\} \tag{6-59}$$

is the *system reliability*. Thus $F(t)$ is the probability that the system fails prior to time t, and $R(t)$ is the probability that the system functions at time t.

The mean of $\mathbf{x}$ is called *mean time to failure*. As we see from (4-92),

$$E\{\mathbf{x}\} = \int_0^\infty xf(x)\,dx = \int_0^\infty R(t)\,dt \tag{6-60}$$

because $F(x) = 0$ for $x < 0$.

The conditional distribution

$$F(x|\mathbf{x} > t) = \frac{P\{\mathbf{x} \leq x, \mathbf{x} \geq t\}}{P\{\mathbf{x} \geq t\}} \tag{6-61}$$

is the probability that a system functioning at time t will fail prior to time x. Clearly, $F(x|\mathbf{x} > t) = 0$ if $x < t$ and

$$F(x|\mathbf{x} > t) = \frac{F(x) - F(t)}{1 - F(t)} \qquad x > t \tag{6-62}$$

Differentiating with respect to x, we obtain the conditional density

$$f(x|\mathbf{x} \geq t) = \frac{f(x)}{1 - F(t)} \qquad x > t \tag{6-63}$$

The product $f(x|\mathbf{x} \geq t)\,dx$ is the probability that the system will fail in the time interval $(x, x + dx)$, assuming that it functions at time t.

Example 6.11

Suppose that $\mathbf{x}$ has an exponential distribution

$$F(x) = (1 - e^{-\alpha x})U(x) \qquad f(x) = \alpha e^{-\alpha x}U(x)$$

In this case (Fig. 6.11),

$$f(x|\mathbf{x} \geq t) = \frac{\alpha e^{-\alpha x}}{e^{-\alpha t}} = \alpha e^{-\alpha(x-t)} \qquad x > t \tag{6-64}$$

Thus, in this case, the probability that a system functioning at time t will fail in the interval $(x, x + dx)$ depends only on the difference $x - t$. As we shall see, this is true only if $f(x)$ is an exponential. ■

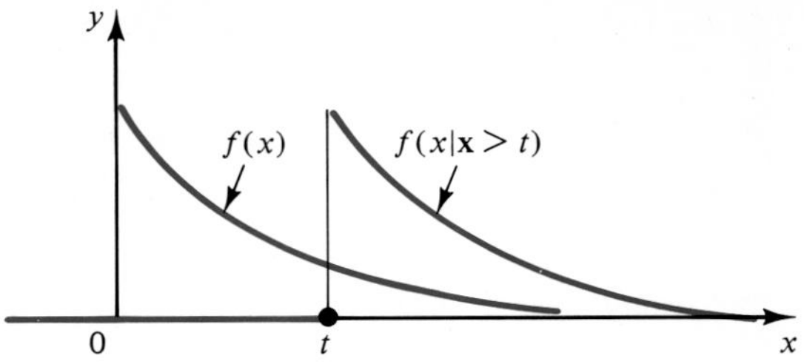

Figure 6.11

Conditional Failure Rate

The conditional density $f(x|\mathbf{x} \geq t)$ is a function of x and t. Its value at $x = t$ is a function of t

$$\beta(t) = f(t|\mathbf{x} \geq t) \tag{6-65}$$

known as the *conditional rate of failure* or *hazard rate*. The product $\beta(t)\,dt$ is the probability that a system functioning at time t will fail in the interval $(t, t + dt)$. Since $f(x) = F'(x) = -R'(x)$, (6-63) yields

$$\beta(t) = \frac{F'(t)}{1 - F(t)} = -\frac{R'(t)}{R(t)} \tag{6-66}$$

Example 6.12

The time to failure is an RV $\mathbf{x}$ uniformly distributed in the interval $(0, T)$ (Fig. 6.12). In this case, $F(x) = x/T$, $R(t) = 1 - x/T$, and (6-65) yields

$$\beta(t) = \frac{1/T}{1 - x/T} = \frac{1}{T - x} \qquad 0 \leq t < T$$ ■

Using (6-66), we shall express $F(x)$ in terms of $\beta(t)$.

■ ***Theorem***

$$1 - F(x) = R(x) = \exp\left\{-\int_0^x \beta(t)\,dt\right\} \tag{6-67}$$

Figure 6.12

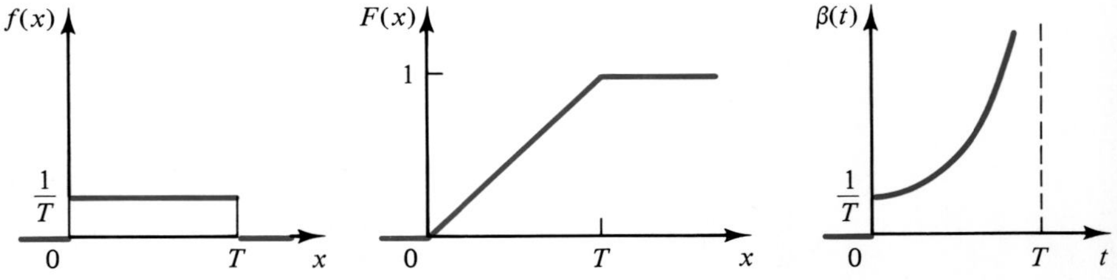

■ ***Proof.*** Integrating (6-66) from 0 to x and using the fact that $F(0) = 1 - R(0) = 0$, we obtain

$$-\int_0^x \beta(t)\,dt = \ln R(x)$$

and (6-67) follows.

To express $f(x)$ in terms of $\beta(t)$, we differentiate (6-67). This yields

$$f(x) = \beta(x) \exp\left\{-\int_0^x \beta(t)\,dt\right\} \qquad (6\text{-}68)$$

Note that the function $\beta(t)$ equals the value of the density $f(x|\mathbf{x} \geq t)$ for $x = t$; however, $\beta(t)$ is not a density. A conditional density $f(x|\mathcal{M})$ has all the properties of a density only if the event $\mathcal{M}$ is fixed (it does not depend on x). Thus $f(x|\ \mathbf{x} \geq t)$ considered as a function of x is a density; however, the function $\beta(t) = f(t|\mathbf{x} \geq t)$ does not have the properties of densities. In fact, its area is infinite for any $F(x)$. This follows from (6-67) because $F(\infty) = 1$.

EXPECTED FAILURE RATE The probability that a given unit functioning at time t will fail in the interval $(t, t + \delta)$ equals $P\{t < \mathbf{x} \leq t + \delta|\mathbf{x} > t\}$. Hence, for small δ,

$$\beta(t)\delta \simeq P\{t < \mathbf{x} \leq t + \delta|\mathbf{x} \geq t\} \qquad (6\text{-}69)$$

This has the following *empirical interpretation:* Suppose, to be concrete, that $\mathbf{x}$ is the time to failure of a light bulb. We turn on n bulbs at $t = 0$ and denote by n_t the number of bulbs that are still good at time t and by Δn_t the number of bulbs that fail in the interval $(t, t + \delta)$. As we know [see (2-54)] the conditional probability $\beta(t)\delta$ equals the relative frequency of failures in the interval $(t, t + \delta)$ in the subsequence of trials involving only bulbs that are still good at time t. Hence

$$\beta(t)\delta \simeq \frac{\Delta n_t}{n_t} \qquad (6\text{-}70)$$

Equation (6-69) is a probabilistic statement involving a single component; (6-70) is an empirical statement involving a large number of components. In the following, we give a probabilistic interpretation of (6-70) in terms of N components, where N is any number large or small.

Suppose that a system consists of N components and that the ith component is modeled by an RV $\mathbf{x}_i$ with $R_i(t) = P\{\mathbf{x}_i > t\}$. The number of units that are still good at time t is an RV $\mathbf{n}(t)$ depending on t. We maintain that its expected value equals

$$\eta(t) = E\{\mathbf{n}(t)\} = R_1(t) + \cdots + R_N(t) \qquad (6\text{-}71)$$

■ ***Proof.*** We denote by $\mathbf{y}_i$ the zero-one RV associated with the event $\{\mathbf{x}_i > t\}$. Thus $\mathbf{y} = 1$ if $\mathbf{x}_i > t$ and 0 if $\mathbf{x}_i \leq t$; hence, $\mathbf{n}(t) = \mathbf{y}_1 + \cdots + \mathbf{y}_N$. This yields

$$E\{\mathbf{n}(t)\} = \sum_{i=1}^{N} E\{\mathbf{y}_i\} = \sum_{i=1}^{N} P\{\mathbf{x}_i > t\}$$

and (6-71) results.

Suppose now that all components are equally reliable. In this case,

$$R_1(t) = \cdots = R_N(t) = R(t) \qquad \eta(t) = NR(t) \tag{6-72}$$

and (6-71) yields [see (6-66)]

$$\beta(t) = -\frac{R'(t)}{R(t)} = -\frac{\eta'(t)}{\eta(t)} \tag{6-73}$$

Thus we have a new interpretation of $\beta(t)$: The product $\beta(t)\,dt$ equals the ratio of the expected number $\eta(t) - \eta(t + dt)$ of failures in the interval $(t, t + dt)$ divided by the expected number $\eta(t)$ of good components at time t. This interpretation justifies the term *expected failure rate* used to characterize the conditional failure rate $\beta(t)$. Note, finally, that (6-73) is the probabilistic interpretation of (6-70).

Weibull Distribution A special case of particular interest in reliability studies is the function

$$\beta(t) = ct^{b-1} \tag{6-74}$$

This is a satisfactory approximation of a failure rate for most applications, at least for values of t near $t = 0$. The density of the corresponding time to failure is given by [see (6-68)]

$$f(x) = cx^{b-1}e^{-cx^b/b}U(x) \tag{6-75}$$

This function is called the *Weibull density*. It depends on the two parameters c and b, and its first two moments equal

$$E\{\mathbf{x}\} = \left(\frac{c}{b}\right)^{-1/b} \Gamma\left(\frac{b+1}{b}\right) \qquad E\{\mathbf{x}^2\} = \left(\frac{c}{b}\right)^{-2/b} \Gamma\left(\frac{b+2}{b}\right)$$

where $\Gamma(x)$ is the gamma function. This follows from (4-54) with $x^b = y$. In Fig. 6.13, we show $f(x)$ and $\beta(t)$ for $b = 1, 2$, and 3. If $b = 1$, then $\beta(t) = c =$ constant and $f(x) = ce^{-cx}U(x)$. This case has the following interesting property.

Figure 6.13

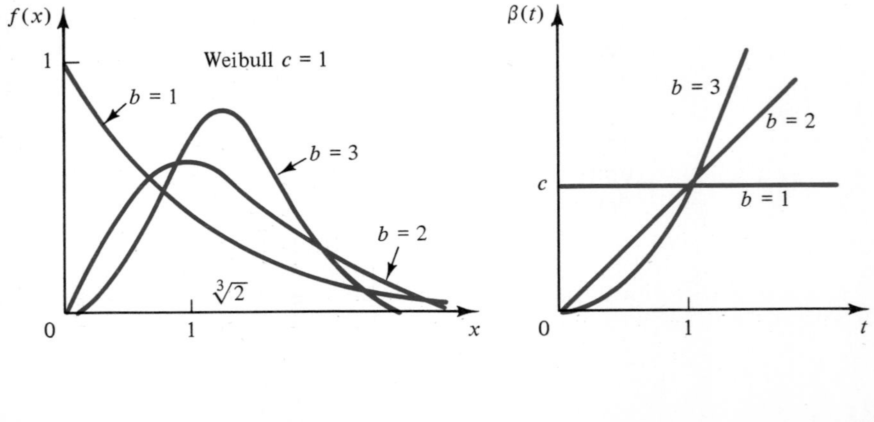

Memoryless Systems We shall say that a system is memoryless if the probability that it fails in an interval (t, x) assuming that it functions at time t, depends only on the length $x - t$ of this interval. This is equivalent to the assumption that

$$f(x|\mathbf{x} \geq t) = f(x - t) \qquad \text{for every } x \geq t \tag{6-76}$$

With $x = t$, it follows from (6-76) and (6-66) that $\beta(t) = f(t|\mathbf{x} \geq t) = f(0) =$ constant. Thus a system is memoryless iff $f(x)$ is an exponential density or, equivalently, iff its conditional failure rate is constant.

Interconnection of Systems

We determine next the reliability of a system S consisting of two or more components.

Parallel Connection We shall say that two components S_1 and S_2 are connected in parallel forming a system S if S functions when at least one of the systems S_1 and S_2 functions (Fig. 6.14a). Denoting by $\mathbf{x}_1$, $\mathbf{x}_2$, and $\mathbf{z}$ the times to failure of the systems S_1, S_2, and S, respectively, we conclude that $\mathbf{z} = t$ if the larger of the numbers $\mathbf{x}_1$ and $\mathbf{x}_2$ equals t; hence,

$$\mathbf{z} = \max(\mathbf{x}_1, \mathbf{x}_2) \tag{6-77}$$

The distribution $F_z(z)$ of $\mathbf{z}$ can be expressed in terms of the joint distribution of $\mathbf{x}_1$ and $\mathbf{x}_2$ as in (5-78). If the systems S_1 and S_2 are independent, that is, if the RVS $\mathbf{x}_1$ and $\mathbf{x}_2$ are independent, then

$$F_z(t) = F_1(t)F_2(t) \tag{6-78}$$

This follows from (5-78); however, we shall establish it directly: The event $\{\mathbf{z} < t\}$ occurs if S fails prior to time t. This is the case if both systems fail prior to time t, that is, if both events $\{\mathbf{x}_1 < t\}$ and $\{\mathbf{x}_2 < t\}$ occur. Thus $\{\mathbf{z} < t\}$ is the intersection of the events $\{\mathbf{x}_1 < t\}$ and $\{\mathbf{x}_2 < t\}$. And since these events are independent, (6-78) follows.

We shall say that n systems S_i are connected in parallel forming a system S if S functions when at least one of the systems S_i functions. Reasoning as in (6-77) and (6-78), we conclude that if the systems S_i are indepen-

Figure 6.14

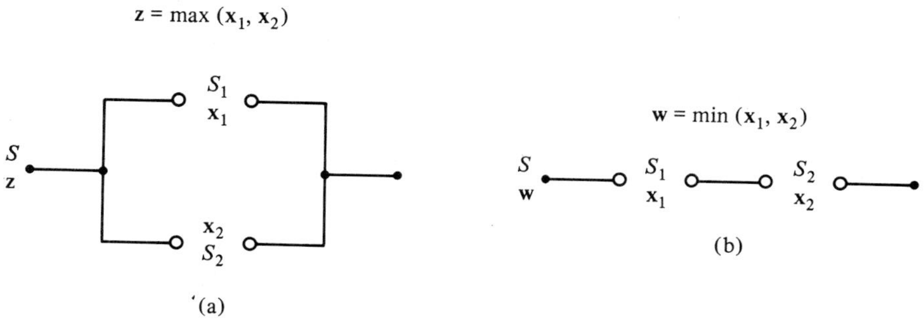

dent, then

$$\mathbf{z} = \max(\mathbf{x}_1, \ldots, \mathbf{x}_n) \qquad F_z(t) = F_1(t) \cdots F_n(t) \tag{6-79}$$

Series Connection Two systems are connected in series forming a system S if S fails when at least one of the systems S_1 and S_2 fails (Fig. 6.14b). Denoting by $\mathbf{w}$ the time to failure of the system S, we conclude that $\mathbf{w} = t$ if the smaller of the numbers $\mathbf{x}_1$ and $\mathbf{x}_2$ equals t; hence,

$$\mathbf{w} = \min(\mathbf{x}_1, \mathbf{x}_2) \tag{6-80}$$

The reliability $R_w(t) = P\{\mathbf{w} > t\}$ of S can be determined from (5-79). If the systems S_1 and S_2 are independent, then

$$R_w(t) = R_1(t)R_2(t) \tag{6-81}$$

Indeed, the event $\{\mathbf{z} > t\}$ occurs if S fails after time t. This is the case if both systems fail after t, that is, if both events $\{\mathbf{x}_1 > t\}$ and $\{\mathbf{x}_2 > t\}$ occur. Thus $\{\mathbf{z} > t\} = \{\mathbf{x}_1 > t\} \cap \{\mathbf{x}_2 > t\}$; and since the two events on the right are independent, (6-81) follows.

Generalizing, we note that if n independent systems are connected in series forming a system with time to failure $\mathbf{w}$, then

$$\mathbf{w} = \min(\mathbf{x}_1, \ldots, \mathbf{x}_n) \qquad R_w(t) = R_1(t) \cdots R_n(t) \tag{6-82}$$

Stand-by Connection We put system S_1 into operation, keeping system S_2 in reserve. When S_1 fails, we put S_2 into operation. When S_2 fails, the system S so formed fails (Fig. 6.14c). Thus if t_1 and t_2 are the times of operation of S_1 and S_2, then $t_1 + t_2$ is the time of operation of S. Denoting by $\mathbf{s}$ its time to failure, we conclude that

$$\mathbf{s} = \mathbf{x}_1 + \mathbf{x}_2 \tag{6-83}$$

The density $f_s(s)$ of $\mathbf{s}$ is obtained from (5-90). If the systems S_1 and S_2 are independent, then [see (5-97)] $f_s(t)$ equals the convolution of the densities $f_1(t)$ and $f_2(t)$

$$f_s(t) = \int_0^t f_1(z)f_2(t - z)\,dz \equiv f_1(t) * f_2(t) \tag{6-84}$$

Note, finally, that if S is formed by the stand-by connection of n independent systems S_i, then

$$\mathbf{s} = \mathbf{x}_1 + \cdots + \mathbf{x}_n \qquad f_s(t) = f_1(t) * \cdots * f_n(t) \tag{6-85}$$

Example 6.13

We connect n identical, independent systems in series. Find the mean time to failure of the system so formed if their conditional failure rate is constant.

In this problem,

$$\beta(t) = c \qquad R_i(t) = e^{-ct} \qquad R(t) = R_1(t) \cdots R_n(t) = e^{-nct}$$

[see (6-82)]. Hence,

$$E\{\mathbf{w}\} = \int_0^\infty e^{-nct}\,dt = \frac{1}{nc}$$

■

TIME-INDEPENDENT SYSTEMS A time-independent system is either good or defective at all times. The probability $p = 1 - q$ that the system is good is called system reliability. Problems involving interconnections of time-independent or time-dependent systems are equivalent if time is only a parameter. This is the case for series-parallel but not for stand-by connections. Thus to find the time-independent form of (6-78) and (6-81), we set $p = R(t)$ and $q = F(t)$.

System interconnections are represented by linear graphs involving links and nodes. A link represents a component and is closed if the component is good, open if it is defective. A system has an input node and an output node. It is good if it contains one or more connected paths linking the input to the output. To find the reliability p of a system, we must trace all paths from the input to the output and the probability that at least one is connected. In general, this is not a simple task. The problem is simplified if we consider only series-parallel connections. Here are the reliabilities $p = 1 - q$ of the four systems of Fig. 6-15:

$$q_a = q_1q_2 \qquad p_b = p_1p_2 \qquad q_c = (1 - p_1p_2)q_3 \qquad p_d = (1 - q_1q_2)p_3$$

Structure Function The state of the system S is specified in terms of an RV $\mathbf{y}$ taking the values 1 and 0 with probabilities

$$P\{\mathbf{y} = 1\} = p \qquad P\{\mathbf{y} = 0\} = q$$

where $p = 1 - q$ is the reliability of the system. This RV will be called the *state variable*. Thus a state variable is the zero-one RV associated with the event {good}. Suppose that S consists of n components S_i with state vari-

Figure 6.15

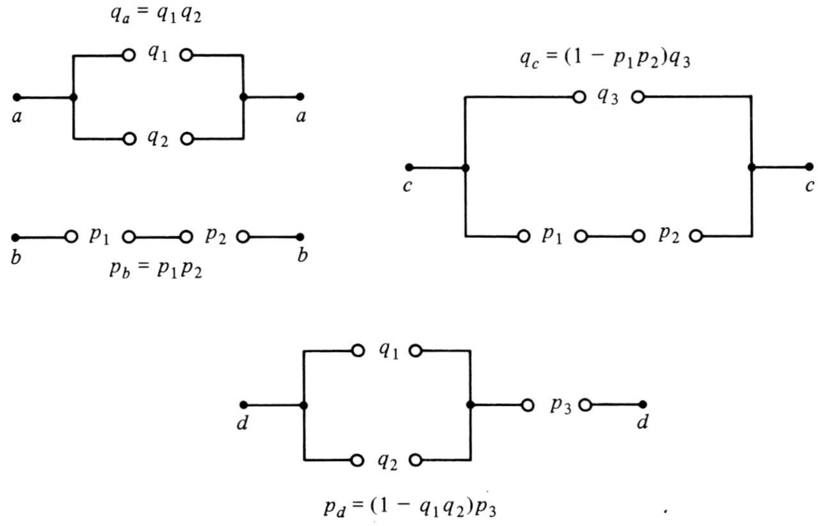

ables $\mathbf{y}_i$. Clearly, $\mathbf{y}$ is a function

$$\mathbf{y} = \psi(\mathbf{y}_1, \cdots, \mathbf{y}_n)$$

of the variables $\mathbf{y}_i$ called the structure function of S. Here are its values for parallel and series connections:

Parallel Connection From (6-79) it follows that

$$\psi(\mathbf{y}_1, \ldots, \mathbf{y}_n) = \max \mathbf{y}_i = 1 - (1 - \mathbf{y}_1) \cdots (1 - \mathbf{y}_n) \qquad (6\text{-}86)$$

Series Connection From (6-82) it follows that

$$\psi(\mathbf{y}_1, \ldots, \mathbf{y}_n) = \min \mathbf{y}_i = \mathbf{y}_1\mathbf{y}_2 \cdots \mathbf{y}_n \qquad (6\text{-}87)$$

To determine the structure function of a system, we identify all paths from the input to the output and use (6-86) and (6-87). We give next an illustration.

Example 6.14

The structure on the left of Fig. 6.16 is called a *bridge*. It consists of four paths forming four subsystems as shown. From (6-87) it follows that the structure functions of these subsystems equal

$$\mathbf{y}_1\mathbf{y}_2 \qquad \mathbf{y}_3\mathbf{y}_4 \qquad \mathbf{y}_1\mathbf{y}_4\mathbf{y}_5 \qquad \mathbf{y}_2\mathbf{y}_3\mathbf{y}_5$$

respectively. The bridge is good if at least one of the substructures is good. Hence [see (6-86)],

$$\begin{aligned}\psi(\mathbf{y}_1, \mathbf{y}_2, \mathbf{y}_3, \mathbf{y}_4) &= \max(\mathbf{y}_1\mathbf{y}_2, \mathbf{y}_3\mathbf{y}_4, \mathbf{y}_1\mathbf{y}_4\mathbf{y}_5, \mathbf{y}_2\mathbf{y}_4\mathbf{y}_5) \\ &= 1 - (1 - \mathbf{y}_1\mathbf{y}_2)(1 - \mathbf{y}_3\mathbf{y}_4)(1 - \mathbf{y}_2\mathbf{y}_3\mathbf{y}_5)\end{aligned}$$

The determination of the reliability of this bridge is discussed in Problem 6-17. ∎

Figure 6.16

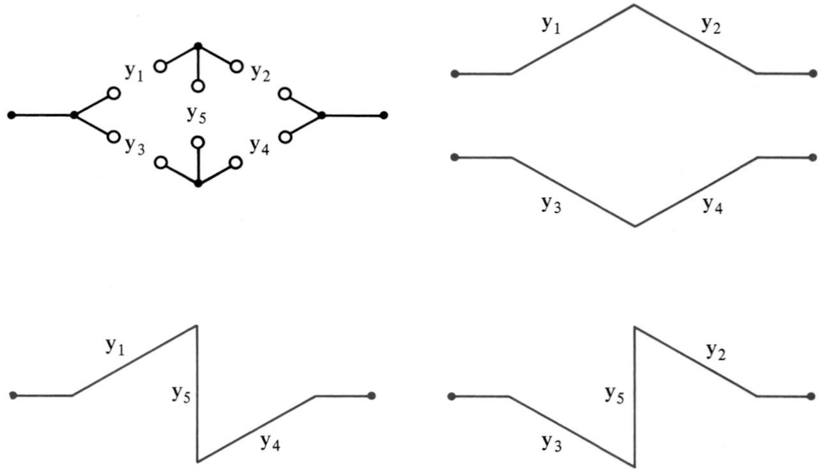

Problems

6-1 We are given 10 coins; one has two heads, and nine are fair. We pick one at random and toss it three times. **(a)** Find the probability p_1 that at the first toss, heads will show. **(b)** We observe that the first three tosses showed heads; find the probability p_2 that at the next toss, heads will show.

6-2 Suppose that the age (in years) of the males in a community is an RV $\mathbf{x}$ with density $.03e^{-.03t}$. **(a)** Find the percentage of males between 20 and 50. **(b)** Find the average age of males between 20 and 50.

6-3 Given two independent $N(0, 2)$ RVs $\mathbf{x}$ and $\mathbf{y}$, we form the RVs $\mathbf{z} = 2\mathbf{x} + \mathbf{y}$, $\mathbf{w} = \mathbf{x} - \mathbf{y}$. Find the conditional density $f(z|w)$ and the conditional mean $E\{\mathbf{z}|\mathbf{w} = 5\}$.

6-4 If $F(x) = (1 - e^{-.2x})U(x)$, find the conditional probabilities

$$p_1 = P\{\mathbf{x} < 5|3 < \mathbf{x} < 6\} \qquad p_2 = P\{\mathbf{x} > 5|3 < \mathbf{x} < 6\}$$

6-5 The RVs $\mathbf{x}$ and $\mathbf{y}$ have a uniform joint density in the shaded region of Fig. P6.5 between the parabola $y = x(2 - x)$ and the x-axis. Find the regression line $\phi(x) = E\{\mathbf{y}|x\}$.

Figure P6.5

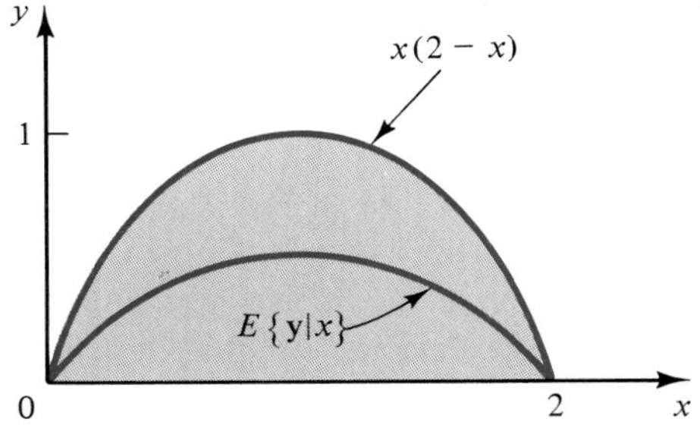

6-6 Suppose that $\mathbf{x}$ and $\mathbf{y}$ are two normal RVs such that $\eta_x = \eta_y = 0$, $\sigma_x = 2$, $\sigma_y = 4$, $r_{xy} = .5$. **(a)** Find the regression line $E\{\mathbf{y}|x\} = \phi(x)$. **(b)** Show that the RVs $\mathbf{x}$ and $\mathbf{y} - \phi(\mathbf{x})$ are independent.

6-7 Using the regression line $\phi(x) = E\{\mathbf{y}|x\}$, we form the RV $\mathbf{z} = \phi(\mathbf{x})$. Show that if $a\mathbf{x}$ is the homogeneous linear MS estimate of the RV $\mathbf{y}$ in terms of $\mathbf{x}$ and $A\mathbf{x}$ is the corresponding estimate of $\mathbf{z}$ in terms of $\mathbf{x}$, then $A = a$.

6-8 The length of rods coming out of a production line is modeled by an RV $\mathbf{c}$ uniform in the interval (10, 12). We measure each rod and obtain the RV $\mathbf{x} = \mathbf{c} + \boldsymbol{\nu}$ where $\boldsymbol{\nu}$ is the error, which we assume uniform in the interval (−0.2, 0.2) and independent of $\mathbf{c}$. **(a)** Find the conditional density $f_x(x|c)$, the joint density $f_{xc}(x, c)$, and the marginal density $f_x(x)$. **(b)** Find the LMS estimate $\hat{c} = E\{\mathbf{c}|x\}$ of the length c of the received rods in terms of its measured value x.

6-9 We toss a coin 18 times, and heads shows 11 times. The probability p of heads is an RV $\mathbf{p}$ with density $f(p)$. Show that the LMS estimate $\hat{p}$ of p equals

$$\hat{p} = \gamma \int_0^1 p^{12}(1-p)^7 f(p)\,dp$$

Find γ if $f(p) = 1$.

6-10 The time to failure of electric bulbs is an RV $\mathbf{x}$ with density $ce^{-cx}U(x)$. A box contains 200 bulbs; of these, 50 are of type A with $c = 4$ per year and 150 are of type B with $c = 6$ per year. A bulb is selected at random. **(a)** Using (6-7), find the probability that it will function at time t. **(b)** The selected bulb has lasted three months; find the probability that it is of type A.

6-11 We are given three systems with reliabilities $p_1 = 1 - q_1$, $p_2 = 1 - q_2$, $p_3 = 1 - q_3$, and we combine them in various ways forming a new system S with reliability $p = 1 - q$. Find the form of S such that: **(a)** $p = p_1p_2p_3$; **(b)** $q = q_1q_2q_3$; **(c)** $p = p_1(1 - q_2q_3)$; **(d)** $q = q_1(1 - p_2p_3)$.

6-12 The hazard rate of a system is $\beta(t) = ct/(1 + ct)$; find its reliability $R(t)$.

6-13 Find and sketch the reliability $R(t)$ of a system if its hazard rate equals $\beta(t) = 6U(t) + 2U(t - t_0)$. Find the mean time to failure of the system.

6-14 **(a)** Show that if $\mathbf{w} = \min(\mathbf{x}, \mathbf{y})$, then

$$P\{\mathbf{x} \geq t \mid \mathbf{w} \leq t\} = \frac{F_y(t) - F_{xy}(t, t)}{F_x(t) + F_y(t) - F_{xy}(t, t)}$$

(b) Two independent systems S_x and S_y are connected in series. The resulting system S_w failed prior to t. Find the probability p that S_x is working at time t. System S_w fails at time t. Find the probability p_1 that S_x is still working.

6-15 We connect n independent systems with hazard rates $\beta_i(t)$ in series forming a system S with hazard rate $\beta(t)$. Show that $\beta_s(t) = \beta_1(t) + \cdots + \beta_n(t)$.

6-16 Given four independent systems with the same reliability $R(t)$, form a new system with reliability $1 - [1 - R^2(t)]^2$.

6-17 Find the reliability of the bridge of Fig. 6-16.

6-18 **(a)** Find the structure function of the systems of Fig. 6-14. **(b)** Given four components with state variables $\mathbf{y}_i$, form a system with structure function

$$\psi(\mathbf{y}_1, \mathbf{y}_2, \mathbf{y}_3, \mathbf{y}_4) = \begin{cases} 1 & \text{if } \mathbf{y}_1 + \mathbf{y}_2 + \mathbf{y}_3 + \mathbf{y}_4 \geq 2 \\ 0 & \text{otherwise} \end{cases}$$

7 Sequences of Random Variables

We extend all concepts introduced in the earlier chapters to an arbitrary number of RVS and develop several applications including sample spaces, measurement errors, and order statistics. In Section 7-3 we introduce the notion of a random sequence, the various interpretations of convergence, and the central limit theorem. In the last section, we develop the chi-square, Student t, and Snedecor distributions. In the appendix, we establish the relationship between the chi-square distribution and various quadratic forms involving normal RVS, and we discuss the noncentral character of the results.

7-1 General Concepts

All concepts developed earlier in the context of one and two RVS can be readily extended to an arbitrary number of RVS. We give here a brief summary. Unless otherwise stated, we assume that all RVS are of continuous type.

Consider n RVS $\mathbf{x}_1, \ldots, \mathbf{x}_n$ defined as in Section 4-1. The joint distribution of these RVS is the function

$$F(x_1, \ldots, x_n) = P\{\mathbf{x}_1 \le x_1, \ldots, \mathbf{x}_n \le x_n\} \qquad (7\text{-}1)$$

specified for every x_i from $-\infty$ to ∞. This function is increasing as x_i increases, and $F(\infty, \ldots, \infty) = 1$. Its derivative

$$f(x_1, \ldots, x_n) = \frac{\partial^n F(x_1, \ldots, x_n)}{\partial x_1 \cdots \partial x_n} \tag{7-2}$$

is the joint density of the RVS $\mathbf{x}_i$.

All probabilistic statements involving the RVS $\mathbf{x}_i$ can be expressed in terms of their joint distribution. Thus the probability that the point $(\mathbf{x}_1, \ldots, \mathbf{x}_n)$ is in a region D of the n-dimensional space equals

$$P\{(\mathbf{x}_1, \ldots, \mathbf{x}_n) \in D\} = \int_D \cdots \int f(x_1, \ldots, x_n)\, dx_1 \cdots dx_n \tag{7-3}$$

If we substitute certain variables in $F(x_1, \ldots, x_n)$ by ∞, we obtain the joint distribution of the remaining variables. If we integrate $f(x_1, \ldots, x_n)$ with respect to certain variables, we obtain the joint density of the remaining variables. For example,

$$\begin{aligned} F(x_1, x_3) &= F(x_1, \infty, x_3, \infty) \\ f(x_1, x_3) &= \int_{-\infty}^{\infty}\int_{-\infty}^{\infty} f(x_1, x_2, x_3, x_4)\, dx_2\, dx_4 \end{aligned} \tag{7-4}$$

The composite functions

$$\mathbf{y}_1 = g_1(\mathbf{x}_1, \ldots, \mathbf{x}_n) \quad \ldots \quad \mathbf{y}_n = g_n(\mathbf{x}_1, \ldots, \mathbf{x}_n)$$

specify the n RVS $\mathbf{y}_1, \ldots, \mathbf{y}_n$. To determine the joint density of these RVS, we proceed as in Section 5-3. We solve the system

$$g_1(x_1, \ldots, x_n) = y_1 \quad \ldots \quad g_n(x_1, \ldots, x_n) = y_n \tag{7-5}$$

for x_i in terms of y_i. If this system has no solution for certain values of y_i, then $f_y(y_1, \ldots, y_n) = 0$ for these values. If it has a single solution $(x_1, \ldots, x_n)$, then

$$f_y(y_1, \ldots, y_n) = \frac{f_x(x_1, \ldots, x_n)}{|J(x_1, \ldots, x_n)|} \tag{7-6}$$

where

$$J(x_1, \ldots, x_n) = \begin{vmatrix} \dfrac{\partial g_1}{\partial x_1} & \cdots & \dfrac{\partial g_1}{\partial x_n} \\ \cdots & \cdots & \cdots \\ \dfrac{\partial g_n}{\partial x_1} & \cdots & \dfrac{\partial g_n}{\partial x_n} \end{vmatrix} \tag{7-7}$$

is the Jacobian of the transformation (7-5). If it has several solutions, we add the corresponding terms as in (4-78).

We can use the preceding result to find the joint density of $r < n$ functions $\mathbf{y}_1, \ldots, \mathbf{y}_r$ of the n RVS $\mathbf{x}_i$. To do so, we introduce $n - r$ auxiliary variables $\mathbf{y}_{r+1} = \mathbf{x}_{r+1}, \ldots, \mathbf{y}_n = \mathbf{x}_n$ find the joint density of the n RVS $\mathbf{y}_1, \ldots, \mathbf{y}_n$ using (7-6), and find the marginal density of $\mathbf{y}_1, \ldots, \mathbf{y}_r$ by integration as in (7-4).

The RVS $\mathbf{x}_i$ are called (mutually) independent if the events $\{\mathbf{x} \le x_i\}$ are independent for every x_i. From this it follows that

$$F(x_1, \ldots, x_n) = F(x_1) \cdots F(x_n)$$
$$f(x_1, \ldots, x_n) = f(x_1) \cdots f(x_n) \tag{7-8}$$

Any subset of a set of independent RVS is itself a set of independent RVS. Suppose that the RVS $\mathbf{x}_1$, $\mathbf{x}_2$, and $\mathbf{x}_3$ are independent. In this case,

$$f(x_1, x_2, x_3) = f(x_1)f(x_2)f(x_3)$$

Integrating with respect to x_3, we obtain $f(x_1, x_2) = f(x_1)f(x_2)$. This shows that the RVS $\mathbf{x}_1$ and $\mathbf{x}_2$ are independent. Note that if the RVS $\mathbf{x}_i$ are independent in pairs, they are not necessarily independent.

Suppose that the RV $\mathbf{y}_i$ is a function $g_i(\mathbf{x}_i)$ depending on the RV $\mathbf{x}_i$ only. Reasoning as in (5-26), we can show that if the RVS $\mathbf{x}_i$ are independent, the RVS $\mathbf{y}_i = g_i(\mathbf{x}_i)$ are also independent.

Note, finally, that the mean of the RV $\mathbf{z} = g(\mathbf{x}_1, \ldots, \mathbf{x}_n)$ equals

$$E\{g(\mathbf{x}_1, \ldots, \mathbf{x}_n)\} = \int_{-\infty}^{\infty} \cdots \int_{-\infty}^{\infty} g(x_1, \ldots, x_n) f(x_1, \ldots, x_n)\, dx_1 \cdots dx_n$$

as in (5-30). From this it follows that

$$E\{\Sigma a_k g_k(\mathbf{x}_1, \ldots, \mathbf{x}_n)\} = \Sigma a_k E\{g_k(\mathbf{x}_1, \ldots, \mathbf{x}_n)\} \tag{7-9}$$

The covariance μ_{ij} of the RVS $\mathbf{x}_i$ and $\mathbf{x}_j$ equals [see (5-34)]

$$\mu_{ij} = E\{(\mathbf{x}_i - \eta_i)(\mathbf{x}_j - \eta_j)\} = E\{\mathbf{x}_i\mathbf{x}_j\} - \eta_i\eta_j \tag{7-10}$$

where $\eta_i = E\{\mathbf{x}_i\}$. The n-by-n matrix

$$\begin{pmatrix} \mu_{11}\mu_{12} & \cdots & \mu_{1n} \\ \mu_{21}\mu_{22} & \cdots & \mu_{2n} \\ \cdots & \cdots & \cdots \\ \mu_{n1}\mu_{n2} & \cdots & \mu_{nn} \end{pmatrix}$$

is called the *covariance matrix* of the n RVS $\mathbf{x}_i$.

We shall say that the RVS $\mathbf{x}_i$ are uncorrelated if $\mu_{ij} = 0$ for every $i \neq j$. In this case, if

$$\mathbf{z} = a_1\mathbf{x}_1 + \cdots + a_n\mathbf{x}_n \qquad \text{then} \qquad \sigma_z^2 = a_1^2\sigma_1^2 + \cdots + a_n^2\sigma_n^2 \tag{7-11}$$

where $\sigma_i^2 = \mu_{ii}$ is the variance of $\mathbf{x}_i$.

Note, finally, that if the RVS $\mathbf{x}_i$ are independent, then

$$\int_{-\infty}^{\infty} \cdots \int_{-\infty}^{\infty} g_1(x_1) \cdots g_n(x_n) f(x_1) \cdots f(x_n)\, dx_1 \cdots dx_n$$
$$= \int_{-\infty}^{\infty} g_1(x_1) f(x_1)\, dx_1 \cdots \int_{-\infty}^{\infty} g_n(x_n) f(x_n)\, dx_n$$

Hence,

$$E\{g_1(\mathbf{x}_1) \cdots g_n(\mathbf{x}_n)\} = E\{g_1(\mathbf{x}_1)\} \cdots E\{g_n(\mathbf{x}_n)] \tag{7-12}$$

As in (5-56), the expression

$$\Phi(s_1, \ldots, s_n) = E\{\exp(s_1\mathbf{x}_1 + \cdots + s_n\mathbf{x}_n)\} \tag{7-13}$$

will be called the joint moment function of the RVS $\mathbf{x}_i$. From (7-12) it follows that if the RVS $\mathbf{x}_i$ are independent, then

$$\Phi(s_1, \ldots, s_n) = E\{e^{s_1\mathbf{x}_1}\} \cdots E\{e^{s_n\mathbf{x}_n}\} = \Phi(s_1) \cdots \Phi(s_n) \tag{7-14}$$

where $\Phi(s_i)$ is the moment function of $\mathbf{x}_i$.

Normal RVS

We shall say that the n RVS $\mathbf{x}_i$ are jointly normal if the RV

$$\mathbf{z} = a_1\mathbf{x} + \cdots + a_n\mathbf{x}_n$$

is normal for any a_i. Introducing a shift if necessary, we shall assume that $E\{\mathbf{x}_i\} = 0$. Reasoning as in (5-98) and (5-99), we can show that the joint density and the joint moment function of the RVS $\mathbf{x}_i$ are exponentials the exponents of which are quadratics in x_i and s_i, respectively. Specifically,

$$\Phi(s_1, \ldots, s_n) = \exp\left\{\frac{1}{2}\sum_{ij=1}^{n} \mu_{ij}s_is_j\right\} \tag{7-15}$$

$$f(x_1, \ldots, x_n) = \frac{1}{\sqrt{(2\pi)^n\Delta}}\exp\left\{-\frac{1}{2}\sum_{ij=1}^{n} \gamma_{ij}x_ix_j\right\} \tag{7-16}$$

where μ_{ij} are the elements of the covariance matrix C of $\mathbf{x}_i$, γ_{ij} are the elements the inverse C^{-1} of C, and Δ is the determinant of C. We shall verify (7-16) for $n = 2$. The proof of the general case will not be given.

If $n = 2$, then $\mu_{11} = \sigma_1^2$, $\mu_{12} = r\sigma_1\sigma_2$, $\mu_{22} = \sigma_2^2$.

$$C = \begin{pmatrix}\sigma_1^2 & r\sigma_1\sigma_2\\ r\sigma_1\sigma_2 & \sigma_2^2\end{pmatrix} \qquad \Delta = \sigma_1^2\sigma_2^2(1 - r^2)$$

$$C^{-1} = \frac{1}{\Delta}\begin{pmatrix}\sigma_2^2 & -r\sigma_1\sigma_2\\ -r\sigma_1\sigma_2 & \sigma_1^2\end{pmatrix} = \begin{pmatrix}\gamma_{11} & \gamma_{12}\\ \gamma_{21} & \gamma_{22}\end{pmatrix}$$

In this case, the sum in (7-16) equals

$$\gamma_{11}x_1^2 + 2\gamma_{12}x_1x_2 + \gamma_{22}x_2^2 = \frac{1}{\Delta}(\sigma_2^2x_1^2 - 2r\sigma_1\sigma_2x_1x_2 + \sigma_1^2x_2^2)$$

in agreement with (5-99).

If the matrix C is diagonal, that is, if $\mu_{ij} = 0$ for $i \neq j$, then $\gamma_{ij} = 0$ for $i \neq j$ and $\gamma_{ii} = 1/\sigma_i^2$; hence,

$$f(x_1, \ldots, x_n) = \frac{1}{\sigma_1 \cdots \sigma_n\sqrt{(2\pi)^n}}\exp\left\{-\frac{1}{2}\left(\frac{x_1^2}{\sigma_1^2} + \cdots + \frac{x_n^2}{\sigma_n^2}\right)\right\} \tag{7-17}$$

Thus if the RVS $\mathbf{x}_i$ are normal and uncorrelated, they are independent.

Suppose, finally, that the RVS $\mathbf{y}_i$ are linearly dependent on the normal RVS $\mathbf{x}_i$. Using (7-6), we conclude that the joint density of the RVS $\mathbf{y}_i$ is an exponential with a quadratic in the exponent as in (7-16); hence, the RVS $\mathbf{y}_i$ are jointly normal.

Conditional Distributions

In Section 6-2 we showed that the conditional density of $\mathbf{y}$ assuming $\mathbf{x} = x$, defined as a limit, equals

$$f(y|x) = \frac{f(x, y)}{f(x)}$$

Proceeding similarly, we can show that the conditional density of the RVS $\mathbf{x}_n, \ldots, \mathbf{x}_{k+1}$ assuming $\mathbf{x}_k = x_k, \ldots, \mathbf{x}_1 = x_1$ equals

$$f(x_n, \ldots, x_{k+1} | x_k, \ldots, x_1) = \frac{f(x_1, \ldots, x_n)}{f(x_1, \ldots, x_k)} \tag{7-18}$$

For example,

$$f(x_3 | x_2, x_1) = \frac{f(x_1, x_2, x_3)}{f(x_1, x_2)}$$

Repeated application of (7-18) leads to the *chain rule*:

$$f(x_1, \ldots, x_n) = f(x_n | x_{n-1}, \ldots, x_1) \cdots f(x_2 | x_1) f(x_1) \tag{7-19}$$

We give next a rule for removing variables on the left or on the right of the conditional line. Consider the identity

$$f(x_3, x_2 | x_1) = \frac{1}{f(x_1)} f(x_1, x_2, x_3)$$

Integrating with respect to x_2, we obtain

$$\int_{-\infty}^{\infty} f(x_3, x_2 | x_1)\, dx_2 = \frac{1}{f(x_1)} \int_{-\infty}^{\infty} f(x_1, x_2, x_3)\, dx_2 = \frac{f(x_1, x_3)}{f(x_1)} = f(x_3 | x_1)$$

Thus to remove one or more variables on the left of the conditional line, we integrate with respect to these variables. This is the relationship between marginal and joint densities [see (7-4)] extended to conditional densities.

We shall next remove the right variables x_2 and x_3 from the conditional density $f(x_4 | x_3, x_2, x_1)$. Clearly,

$$f(x_4 | x_3, x_2, x_1) f(x_3, x_2 | x_1) = f(x_4, x_3, x_2 | x_1)$$

We integrate both sides with respect to x_2 and x_3. The integration of the right side removes the left variables x_2 and x_3, leaving $f(x_4 | x_1)$. Hence,

$$\int_{-\infty}^{\infty}\int_{-\infty}^{\infty} f(x_4 | x_3, x_2, x_1) f(x_3, x_2 | x_1)\, dx_2\, dx_3 = f(x_4 | x_1)$$

Thus to remove one or more variables on the right of the conditional line, we multiply by the conditional density of these variables, assuming the remaining variables, and integrate. The following case, known as the *Chapman-Kolmogoroff equation*, is of special interest:

$$\int_{-\infty}^{\infty} f(x_3 | x_2, x_1) f(x_2 | x_1)\, dx_2 = f(x_3 | x_1)$$

The conditional mean of the RV $\mathbf{y}$ assuming $\mathbf{x}_i = x_i$ equals

$$E\{\mathbf{y} | x_n, \ldots, x_1\} = \int_{\infty}^{\infty} y f(y | x_n, \ldots, x_1)\, dy \tag{7-20}$$

This integral is a function $\phi(x_1, \ldots, x_n)$ of x_i. Using this function, we form the RV

$$\phi(\mathbf{x}_1, \ldots, \mathbf{x}_n) = E\{\mathbf{y}|\mathbf{x}_1, \ldots, \mathbf{x}_n\}$$

The mean of this RV equals

$$\int_{-\infty}^{\infty} \cdots \int_{-\infty}^{\infty} \phi(x_1, \ldots, x_n) f(x_1, \ldots, x_n)\, dx_1 \cdots dx_n$$

Inserting (7-20) into this equation and using the identity

$$f(x_1, \ldots, x_n, y) = f(y|x_n, \ldots, x_1) f(x_1, \ldots, x_n)$$

we conclude that

$$E\{E\{\mathbf{y}|\mathbf{x}_n, \ldots, \mathbf{x}_1\}\} = \int_{-\infty}^{\infty} \cdots \int_{-\infty}^{\infty} y f(x_1, \ldots, x_n, y)\, dx_1 \cdots dx_n\, dy = E\{\mathbf{y}\} \qquad (7\text{-}21)$$

The function $\phi(x_1, \ldots, x_n)$ is the generalization of the regression curve $\phi(x)$ introduced in (6-39), and it is the nonlinear LMS predictor of $\mathbf{y}$ in terms of the RVS $\mathbf{x}_i$ (see Section 11-4).

Sampling

We are given an RV $\mathbf{x}$ with distribution $F(x)$ and density $f(x)$, defined on an experiment $\mathscr{S}$. Using this RV, we shall form n *independent and identically distributed* (i.i.d.) RVS

$$\mathbf{x}_1, \ldots, \mathbf{x}_i, \ldots, \mathbf{x}_n \qquad (7\text{-}22)$$

with distribution equal to the distribution $F(x)$ of the RV $\mathbf{x}$:

$$F_1(x) = \cdots = F_n(x) = F(x) \qquad (7\text{-}23)$$

These RVS are defined not on the original experiment $\mathscr{S}$ but on the product space

$$\mathscr{S}_n = \mathscr{S} \times \cdots \times \mathscr{S}$$

consisting of n independent repetitions of $\mathscr{S}$. As we explained in Section 3-1, the outcomes of this experiment are sequences of the form

$$\xi = \xi_1 \cdots \xi_i \cdots \xi_n \qquad (7\text{-}24)$$

where ξ_i is any one of the elements of $\mathscr{S}$. The ith RV $\mathbf{x}_i$ in (7-22) is so constructed that its value $\mathbf{x}_i(\xi)$ equals the value $\mathbf{x}(\xi_i)$ of the given RV $\mathbf{x}$:

$$\mathbf{x}_i(\xi_1 \cdots \xi_i \cdots \xi_n) = \mathbf{x}(\xi_i) \qquad (7\text{-}25)$$

Thus the values of $\mathbf{x}_i$ depend only on the outcome ξ_i of the ith trial, and its distribution equals

$$F_i(x) = P\{\mathbf{x}_i \le x\} = P\{\mathbf{x} \le x\} = F(x) \qquad (7\text{-}26)$$

This yields

$$E\{\mathbf{x}_i\} = E\{\mathbf{x}\} = \eta \qquad \sigma_{x_i}^2 = \sigma_x^2 = \sigma^2 \qquad (7\text{-}27)$$

Note further that the events $\{\mathbf{x}_i \le x_i\}$ are independent because $\mathscr{S}_n$ is a product space of n independent trials; hence, the n RVS so constructed are i.i.d. with joint distribution

$$F(x_1, \ldots, x_n) = F(x_1) \cdots F(x_n) \qquad (7\text{-}28)$$

Thus, starting from an RV $\mathbf{x}$ defined on an experiment $\mathcal{S}$, we formed the product space $\mathcal{S}_n$ and the n RVS $\mathbf{x}_i$. We shall call the set $\mathcal{S}_n$ *sample space* and the n RVS $\mathbf{x}_i$ a *random sample* of size n. This construction is called a *sampling* of a population. In the context of the theory, "population" is a model concept. This concept might be an abstraction of an existing population or of the repetition of a real experiment.

As we show in Part Two, the concept of sampling is fundamental in statistics. We give next an illustration in the context of a simple problem.

Sample Mean The arithmetic mean

$$\bar{\mathbf{x}} = \frac{\mathbf{x}_1 + \cdots + \mathbf{x}_n}{n} \tag{7-29}$$

of the samples $\mathbf{x}_i$ is called their *sample mean*. The RVS $\mathbf{x}_i$ are independent with the same mean and variance; hence, they are uncorrelated and [see (7-11)]

$$E\{\bar{\mathbf{x}}\} = \frac{\eta + \cdots + \eta}{n} = \eta \qquad \sigma_{\bar{x}}^2 = \frac{\sigma^2 + \cdots + \sigma^2}{n^2} = \frac{\sigma^2}{n} \tag{7-30}$$

Thus $\bar{\mathbf{x}}$ has the same mean as the original RV $\mathbf{x}$, but its variance is σ^2/n. We shall use this observation to estimate η.

The RV $\mathbf{x}$ is defined on the experiment $\mathcal{S}$. To estimate η, we perform the experiment once and observe the number $x = \mathbf{x}(\xi)$. What can we say about η from this observation? As we know from Tchebycheff's inequality (4-115),

$$P\{\eta - 10\sigma \le \mathbf{x} \le \eta + 10\sigma\} \ge .99 \tag{7-31}$$

This shows that the event $\{\eta - 10\sigma \le \mathbf{x} \le \eta \;\; 10\sigma\}$ will almost certainly occur at a single trial, and it leads to the conclusion that the observed value x of the RV $\mathbf{x}$ is between $\eta - 10\sigma$ and $\eta + 10\sigma$ or, equivalently, that

$$x - 10\sigma < \eta < x + 10\sigma \tag{7-32}$$

This conclusion is useful only if $10\sigma \ll \eta$. If this is not the case, a single observation of $\mathbf{x}$ cannot give us an adequate estimate of η. To improve the estimation, we repeat the experiment n times and form the arithmetic average $\bar{x}$ of the samples x_i of $\mathbf{x}$. In the context of a model, $\bar{x}$ is the observed value of the sample mean $\bar{\mathbf{x}}$ of $\mathbf{x}$ obtained by a single performance of the experiment $\mathcal{S}_n$. As we know, the mean of $\bar{\mathbf{x}}$ is η, and its variance equals σ^2/n; hence,

$$P\left\{\eta - \frac{10\sigma}{\sqrt{n}} \le \bar{\mathbf{x}} \le \eta + \frac{10\sigma}{\sqrt{n}}\right\} \ge .99$$

Replacing $\mathbf{x}$ by $\bar{\mathbf{x}}$ and σ by $\sigma/\sqrt{n}$ in (7-31), we conclude with probability .99 that the observed sample mean $\bar{x}$ is between $\eta - 10\sigma/\sqrt{n}$ and $\eta + 10\sigma/\sqrt{n}$ or, equivalently, the unknown η is between $\bar{x} - 10\sigma/\sqrt{n}$ and $\bar{x} + 10\sigma/\sqrt{n}$. Therefore, if n is sufficiently large, we can claim with near certainty that $\eta \simeq \bar{x}$. This topic is developed in detail in Chapter 9.

7-2 Applications

We start with a problem taken from the *theory of measurements*. The distance c between two points is measured at various times with instruments of different accuracies, and the results of the measurements are n numbers x_i. What is our best estimate of c?

One might argue that we should accept the reading obtained with the most accurate instrument, ignoring all other readings. Another choice might be the weighted average of all readings, with the more accurate instruments assigned larger weights. The choice depends on a variety of factors involving the nature of the errors and the optimality criteria. We shall solve this problem according to the following model.

The ith measurement is a sum

$$\mathbf{x}_i = c + \boldsymbol{\nu}_i \tag{7-33}$$

where $\boldsymbol{\nu}_i$ is the measurement error. The RVs $\boldsymbol{\nu}_i$ are independent with mean zero and variance σ_i^2. The assumption that $E\{\boldsymbol{\nu}_i\} = 0$ indicates that the instruments do not introduce systematic errors. We thus have n RVs $\mathbf{x}_i$ with mean c and variance σ_i^2, and our problem is to find the best estimate of c in terms of the n numbers σ_i, which we assume known, and the n observed values x_i of the RVs $\mathbf{x}_i$. If the instruments have the same accuracy, that is, if $\sigma_i = \sigma =$ constant, then, following the reasoning of Section 7-1, we use as the estimate of c the arithmetic mean of x_i. This, however, is not best if the accuracies differ. Guided by (7-30), we shall use as our estimate the value of an RV $\hat{\mathbf{c}}$ with mean the unknown c and variance as small as possible. In the terminololgy of Chapter 9, this RV will be called the *unbiased minimum variance estimator* of c. To simplify the problem, we shall make the additional assumption that $\hat{\mathbf{c}}$ is the weighted average

$$\hat{\mathbf{c}} = \gamma_1\mathbf{x}_1 + \cdots + \gamma_n\mathbf{x}_n \tag{7-34}$$

of the n measurements $\mathbf{x}_i$. Thus our problem is to find the n constants γ_i such that $E\{\hat{\mathbf{c}}\} = c$ and Var $\hat{\mathbf{c}}$ is minimum. Since $E\{\mathbf{x}_i\} = c$, the first condition yields

$$\gamma_1 + \cdots + \gamma_n = 1 \tag{7-35}$$

From (7-11) and the independence of the RVs $\mathbf{x}_i$ it follows that the variance of $\hat{\mathbf{c}}$ equals

$$V = \gamma_1^2\sigma_1^2 + \cdots + \gamma_n^2\sigma_n^2 \tag{7-36}$$

Hence, our objective is to minimize the sum in (7-36) subject to the constraint of (7-35). From those two equations it follows that

$$V = \gamma_1^2\sigma_1^2 + \cdots + \gamma_i^2\sigma_i^2 + \cdots + (1 - \gamma_1 - \cdots - \gamma_{n-1})\sigma_n^2$$

This is minimum if

$$\frac{\partial V}{\partial \gamma_i} = 2\gamma_i\sigma_i^2 - 2(1 - \gamma_1 - \cdots - \gamma_{n-1})\sigma_n^2 = 0 \qquad i = 1, \ldots, n-1$$

And since the expression in parentheses equals γ_n, the equation yields

$$\gamma_i\sigma_i^2 = \gamma_n\sigma_n^2$$

Combining with (7-36), we obtain

$$\gamma_i = \frac{V}{\sigma_i^2} \qquad V = \frac{1}{1/\sigma_1^2 + \cdots + 1/\sigma_n^2} \tag{7-37}$$

We thus reach the reasonable conclusion that the best estimate of c is the weighted average

$$\hat{c} = \frac{x_1/\sigma_1^2 + \cdots + x_n/\sigma_n^2}{1/\sigma_1^2 + \cdots + 1/\sigma_n^2} \tag{7-38}$$

where the weights γ_i are inversely proportional to the variances of the instrument errors.

Example 7.1

The length c of an object is measured with three instruments. The resulting measurements are

$$x_1 = 84 \qquad x_2 = 85 \qquad x_3 = 87$$

and the standard deviations of the errors equal 1, 1.2, and 1.5, respectively. Inserting into (7-38), we conclude that the best estimate of c equals

$$\hat{c} = \frac{x_1 + x_2/1.44 + x_3/2.25}{1 + 1/1.44 + 1/2.25} = 84.95$$ ■

We assumed that the constants σ_i (instrument errors) are known. However, as we see from (7-38), what we need is only their ratios. As the next example suggests, this case is not uncommon.

Example 7.2

A pendulum is set into motion at time $t = 0$ starting from a vertical position as in Fig. 7.1. Its angular motion is a periodic function $\theta(t)$ with period $T = 2c$. We wish to measure c.

To do so, we measure the first 10 zero crossings of $\theta(t)$ using a measuring instrument with variance σ^2. The ith measurement is the sum

$$\mathbf{t}_i = ic + \boldsymbol{\delta}_i \qquad E\{\boldsymbol{\delta}_i\} = 0 \qquad E\{\boldsymbol{\delta}_i^2\} = \sigma^2$$

Figure 7.1

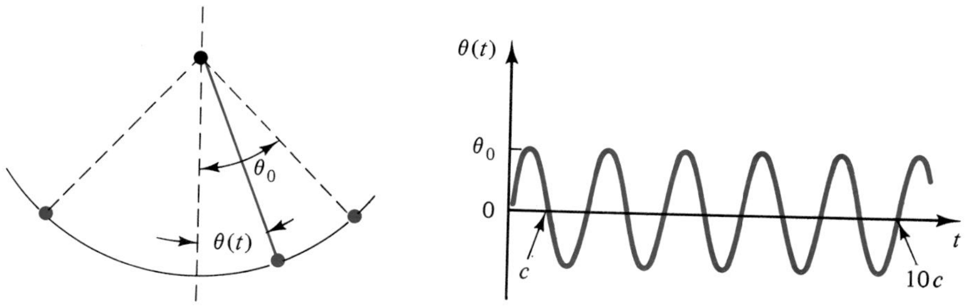

where $\boldsymbol{\delta}_i$ is the measurement error. Thus $\mathbf{t}_i$ is an RV with mean ic and variance an unknown constant σ^2. The results of the measurement are as follows:

$$t_i = 10.5 \quad 20.1 \quad 29.6 \quad 39.8 \quad 50.2 \quad 61 \quad 69.5 \quad 79.1 \quad 89.5 \quad 99.8$$

To reduce this problem to the measurement problem considered earlier, we introduce the RVS

$$\mathbf{x}_i = \frac{\mathbf{t}_i}{i} = c + \boldsymbol{\nu}_i \qquad \boldsymbol{\nu}_i = \frac{\boldsymbol{\delta}_i}{i}$$

Clearly, $E\{\boldsymbol{\nu}_i\} = 0$, $E\{\boldsymbol{\nu}_i^2\} = \sigma^2/i^2$; hence,

$$E\{\mathbf{x}_i\} = c \qquad \sigma_{x_i}^2 = \frac{\sigma^2}{i^2}$$

Finally,

$$x_i = \frac{t_i}{i} = 10.5 \quad 10.05 \quad 9.87 \quad 9.95 \quad 10.04 \quad 10.17 \quad 9.93 \quad 9.89 \quad 9.94 \quad 9.98$$

Inserting into (7-38) and canceling σ^2, we obtain

$$\hat{c} = \frac{x_1 + 4x_2 + \cdots + 100x_{10}}{1 + 4 + \cdots + 100} = 9.97$$

Thus the estimate $\hat{c}$ of c does not depend on σ. ■

Random Sums

Given an RV $\mathbf{n}$ of discrete type taking the values 1, 2, . . . and a sequence of RVS $\mathbf{x}_1, \mathbf{x}_2, \ldots$, we form the sum

$$\mathbf{s} = \sum_{k=1}^{\mathbf{n}} \mathbf{x}_k \tag{7-39}$$

This sum is an RV defined as follows: For a specific ζ, the RV $\mathbf{n}$ takes the value $n = \mathbf{n}(\zeta)$, and the corresponding value $s = \mathbf{s}(\zeta)$ of $\mathbf{s}$ is the sum of the numbers $\mathbf{x}_k(\zeta)$ from 1 to n. Thus the outcomes ζ of the underlying experiment determine not only the values of $\mathbf{x}_k$ but also the number of terms of the sum.

We maintain that if the RVS $\mathbf{x}_k$ have the same mean $E\{\mathbf{x}_k\} = \eta_x$ and they are *independent* of $\mathbf{n}$, then

$$E\{\mathbf{s}\} = \eta_x E\{\mathbf{n}\} \tag{7-40}$$

To prove this, we shall use the identity [see (7-21)]

$$E\{\mathbf{s}\} = E\{E\{\mathbf{s}|\mathbf{n}\}\}$$

If $\mathbf{n} = n$, then $\mathbf{s}$ is a sum of n RVS, and (7-9) yields

$$E\{\mathbf{s}|n\} = E\left\{\sum_{k=1}^{n} \mathbf{x}_k|n\right\} = \sum_{k=1}^{n} E\{\mathbf{x}_k|n\} = \sum_{k=1}^{n} E\{\mathbf{x}_k\}$$

The last equality followed from the assumption that the RVS $\mathbf{x}_k$ are independent of $\mathbf{n}$. Thus

$$E\{\mathbf{s}|n\} = \eta_x n \qquad E\{E\{\mathbf{s}|\mathbf{n}\}\} = E\{\eta_x \mathbf{n}\} = \eta_x E\{\mathbf{n}\}$$

We show next that if the RVS $\mathbf{x}_k$ are *uncorrelated* with variance σ_x^2, then

$$E\{\mathbf{s}^2\} = \eta_x^2 E\{\mathbf{n}^2\} + \sigma_x^2 E\{\mathbf{n}\} \tag{7-41}$$

Clearly,

$$\mathbf{s}^2 = \sum_{k=1}^{\mathbf{n}} \sum_{m=1}^{\mathbf{n}} \mathbf{x}_k \mathbf{x}_m \qquad E\{\mathbf{s}^2|n\} = \sum_{k=1}^{n} \sum_{m=1}^{n} E\{\mathbf{x}_k \mathbf{x}_m\}$$

The last summation contains n terms with $k = m$ and $n^2 - n$ terms with $k \neq m$. And since

$$E\{\mathbf{x}_k\mathbf{x}_m\} = \begin{cases} E\{\mathbf{x}_k^2\} = \sigma_x^2 + \eta_x^2 & k = m \\ E\{\mathbf{x}_k\}E\{\mathbf{x}_m\} = \eta_x^2 & k \neq m \end{cases}$$

we conclude that

$$E\{\mathbf{s}^2|\mathbf{n}\} = (\sigma_x^2 + \eta_x^2)\mathbf{n} + \eta_x^2(\mathbf{n}^2 - \mathbf{n}) = \eta_x^2\mathbf{n}^2 + \sigma_x^2\mathbf{n}$$

Taking expected values of both sides, we obtain (7-41).

Note, finally, that under the stated assumptions, the variance of $\mathbf{s}$ equals

$$\sigma_s^2 = E\{\mathbf{s}^2\} - E^2\{\mathbf{s}\} = \eta_x^2\sigma_n^2 + \sigma_x^2\eta_n \tag{7-42}$$

Example 7.3

The thermal energy $\mathbf{w} = m\mathbf{y}^2/2$ of a particle is an RV having a *gamma density* with parameters

$$b = \frac{3}{2} \qquad c = \frac{1}{kT}$$

where T is the absolute temperature of the particle and $k = 1.37 \times 10^{-23}$ joule degrees is the Boltzmann constant. From this and (5-65) it follows that

$$E\{\mathbf{w}\} = \frac{b}{c} = \frac{3kT}{2} \qquad \sigma_w^2 = \frac{b}{c^2} = \frac{3k^2T^2}{2}$$

The number $\mathbf{n}_t$ of particles emitted from a radioactive substance in t seconds is a Poisson RV with parameter $a = \lambda t$. Find the mean and the variance of the emitted energy

$$\mathbf{s} = \sum_{k=0}^{\mathbf{n}_t} \mathbf{w}_k$$

As we know,

$$E\{\mathbf{n}_t\} = a = \lambda t \qquad \sigma_{n_t}^2 = a = \lambda t$$

Inserting into (7-40) and (7-42), we obtain

$$E\{\mathbf{s}\} = \frac{3kT\lambda t}{2} \qquad \sigma_s^2 = \left(\frac{3kT}{2}\right)^2 \lambda T + \frac{3k^2T^2}{2}\lambda T = \frac{15k^2T^2\lambda T}{4}$$ ■

Order Statistic

We are given a sample of n RVs $\mathbf{x}_i$ defined on the sample space $\mathcal{S}_n$ of repeated trials as in (7-25). For a specific $\xi \in \mathcal{S}_n$, the RVs $\mathbf{x}_i$ take the values $x_i = \mathbf{x}_i(\xi)$. Arranging the n numbers x_i in increasing order, we obtain the sequence

$$x_{r_1} \le x_{r_2} \le \cdots \le x_{r_n} \tag{7-43}$$

We next form the n RVs $\mathbf{y}_i$ such that their values $y_i = \mathbf{y}_i(\xi)$ equal the ordered numbers x_{r_i}

$$y_1 = x_{r_1} \le y_2 = x_{r_2} \le \cdots \le y_n = x_{r_n} \tag{7-44}$$

Note that for a specific i, the values $\mathbf{x}_i(\xi)$ of the ith RV $\mathbf{x}_i$ may occupy different positions in the ordering as ξ ranges over the space $\mathcal{S}_n$. For example, the minimum $\mathbf{y}_1(\xi)$ might equal $\mathbf{x}_3(\xi)$ for some ξ but $\mathbf{x}_8(\xi)$ for some other ξ. The RVs $\mathbf{y}_i$ so constructed are called the *order statistics* of the sample $\mathbf{x}_i$.

As we show in Section 9-4, they are used in nonparametric estimation of percentiles.

The kth RV $\mathbf{y}_k$ in (7-44) is called the *kth order statistic*. We shall determine its density $f_k(y)$ using the identity

$$f_y(y)\,dy = P\{y < \mathbf{y}_k \le y + dy\} \tag{7-45}$$

The event $\mathcal{A} = \{y < \mathbf{y}_k \le y + dy\}$ occurs if $k - 1$ of the RVs $\mathbf{x}_i$ are less than y, $n - k$ are larger than $y + dy$, and, consequently, one is in the internal $(y, y + dy)$. The RVs $\mathbf{x}_i$ are the samples of an RV $\mathbf{x}$ defined on an experiment $\mathcal{S}$. The sets (Fig. 7.2)

$$\mathcal{B}_1 = \{\mathbf{x} \le y\} \qquad \mathcal{B}_2 = \{y < \mathbf{x} \le y + dy\} \qquad \mathcal{B}_3 = \{y + dy < \mathbf{x}\}$$

are events in $\mathcal{S}$, and their probabilities equal

$$P(\mathcal{B}_1) = p_1 = F_x(y) \qquad P(\mathcal{B}_2) = p_2 = f_x(y)\,dy$$
$$P(\mathcal{B}_3) = p_3 = 1 - F_x(y + \Delta y)$$

where $F_x(x)$ is the distribution and $f_x(x)$ is the density of $\mathbf{x}$. The events $\mathcal{B}_1$, $\mathcal{B}_2$, $\mathcal{B}_3$ form a partition of $\mathcal{S}$; therefore, if $\mathcal{S}$ is repeated n times, the probability that the events, $\mathcal{B}_i$ will occur k_i times equals [see (3-41)]

$$\frac{n!}{k_1!k_2!k_3!}\,p_1^{k_1}p_2^{k_2}p_3^{k_3} \qquad k_1 + k_2 + k_3 = n \tag{7-46}$$

Clearly, $\mathcal{A}$ is an event in the sample space $\mathcal{S}_n$, and it occurs if $\mathcal{B}_1$ occurs $k - 1$ times, $\mathcal{B}_2$ occurs once, and $\mathcal{B}_3$ occurs $n - k$ times. With

$$k_1 = k - 1 \qquad k_2 = 1 \qquad k_3 = n - k$$

(7-46) yields

$$P\{y < \mathbf{y}_k \le y + dy\} = \frac{n!}{(k-1)!1!(n-k)!}\,F_x^{k-1}(y)f_x(y)\,dy\,[1 - F_x(y + dy)]^{n-k}$$

Comparing with (7-45), we conclude that

$$f_k(y) = \frac{n!}{(k-1)!(n-k)!}\,F_x^{k-1}(y)[1 - F_x(y)]^{n-k} f_x(y) \tag{7-47}$$

The joint distributions of the order statistics can be determined similarly. We shall carry out the analysis for the maximum $\mathbf{y}_n = \mathbf{x}_{\max}$ and the minimum $\mathbf{y}_1 = \mathbf{x}_{\min}$ of $\mathbf{x}_i$.

Figure 7.2

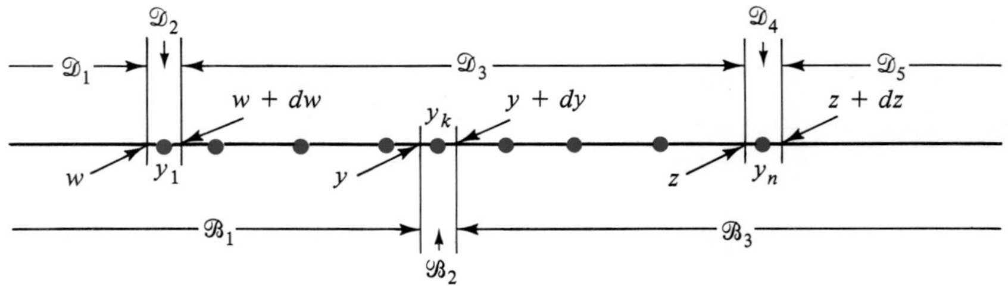

EXTREME ORDER STATISTICS We shall determine the joint density of the RVS

$$\mathbf{z} = \mathbf{y}_n \qquad \mathbf{w} = \mathbf{y}_1$$

using the identity

$$f_{zw}(z, w)\,dz\,dw = P\{z < \mathbf{z} \le z + dz,\ w < \mathbf{w} \le w + dw\} \tag{7-48}$$

The event

$$\mathscr{C} = \{z < \mathbf{z} \le z + dz,\ w < \mathbf{w} \le w + dw\} \qquad z > w \tag{7-49}$$

occurs iff the smallest of the RVS $\mathbf{x}_i$ is in the interval $(w, w + dw)$, the largest is in the interval $(z, z + dz)$, and, consequently, all others are between $w + dw$ and z. To find $P(\mathscr{C})$, we introduce the sets

$$\mathscr{D}_1 = \{\mathbf{x} \le w\} \qquad \mathscr{D}_2 = \{w < \mathbf{x} \le w + dw\} \qquad \mathscr{D}_3 = \{w + dw < \mathbf{x} \le z\}$$
$$\mathscr{D}_4 = \{z < \mathbf{x} \le z + dz\} \qquad \mathscr{D}_5 = \{\mathbf{x} > z + dz\}$$

These sets form a partition of the space $\mathscr{S}$, and their probabilities equal

$$\begin{gathered} p_1 = F_x(w) \qquad p_2 = f_x(w)\,dw \qquad p_3 = F_x(z) - F_x(w + dw) \\ p_4 = f_x(z)\,dz \qquad p_5 = 1 - F_x(z + dz) \end{gathered} \tag{7-50}$$

Clearly, the set $\mathscr{C}$ in (7-49) is an event in the sample space $\mathscr{S}_n$, and it occurs iff $\mathscr{D}_1$ does not occur at all, $\mathscr{D}_2$ occurs once, $\mathscr{D}_3$ occurs $n - 2$ times, $\mathscr{D}_4$ occurs once, and $\mathscr{D}_5$ does not occur at all. With

$$k_1 = 0 \qquad k_2 = 1 \qquad k_3 = n - 2 \qquad k_4 = 1 \qquad k_5 = 0$$

it follows from (3-41) and (7-50) that

$$P\{z < \mathbf{z} \le z + dz,\ w < \mathbf{w} \le w + dw\} = \frac{n!}{(n-2)!}\,p_2 p_3^{n-2} p_4$$
$$= n(n-1) f_x(w)\,dw[F_x(z) - F_x(w + dw)]^{n-2} f_x(z)\,dz$$

for $z > w$ and zero otherwise. Comparing with (7-48), we conclude that

$$f_{zw}(z, w) = \begin{cases} n(n-1) f_x(z) f_x(w)[F_x(z) - F_x(w)]^{n-2} & z > w \\ 0 & z < w \end{cases} \tag{7-51}$$

Integrating with respect to w and z respectively, we obtain the marginal densities $f_z(z)$ and $f_w(w)$. These densities can be found also from (7-47): Setting $k = n$ and $k = 1$ in (7-47), we obtain

$$\begin{aligned} f_z(z) &= n f_x(z) F_x^{n-1}(z) \\ f_w(w) &= n f_x(w)[1 - F_x(w)]^{n-1} \end{aligned} \tag{7-52}$$

Range and Midrange Using (7-51), we shall determine the densities of the RVS

$$\mathbf{r} = \mathbf{z} - \mathbf{w} \qquad \mathbf{s} = \frac{\mathbf{z} + \mathbf{w}}{2}$$

The first is the range of the sample, the second the midpoint between the maximum and the minimum. The Jacobian of this transformation equals 1. Solving for z and w, we obtain $z = s + r/2$, $w = s - r/2$, and (5-88) yields

$$f_{rs}(r, s) = f_{zw}\left(s + \frac{r}{2},\ s - \frac{r}{2}\right) \tag{7-53}$$

Example 7.4 Suppose that $\mathbf{x}$ is uniform in the interval $(0, c)$. In this case, $F_x(x) = x/c$ for $0 < x < c$; hence,

$$F_x(z) - F_x(w) = \frac{z}{c} - \frac{w}{c} \qquad f_x(z) = \frac{1}{c} \qquad f_x(w) = \frac{1}{c}$$

in the shaded area $0 \le w \le z \le c$ of Fig. 7.3 and zero elsewhere. Inserting into (7-51), we obtain

$$f_{zw}(z, w) = \frac{n(n-1)}{c^2}\left(\frac{z-w}{c}\right)^{n-2} \qquad 0 \le w \le z \le c \tag{7-54}$$

and zero elsewhere. This yields

$$f_z(z) = \frac{n(n-1)}{c_n}\int_0^z (z-w)^{n-2}\,dw = \frac{n}{c^n} z^{n-1} \qquad 0 \le z \le c$$

$$f_w(w) = \frac{n(n-1)}{c^n}\int_w^c (z-w)^{n-2}\,dz = \frac{n}{c^n}(c-w)^{n-1} \qquad 0 \le w \le c$$

For $n = 3$ the curves of Fig. 7.3 result. Note that

$$E\{\mathbf{z}\} = \frac{n}{n+1}c \qquad E\{\mathbf{w}\} = \frac{c}{n+1}$$

From (7-53) and (7-54) it follows that

$$f_{rs}(r, s) = \frac{n(n-1)}{c^n} r^{n-2}$$

Figure 7.3

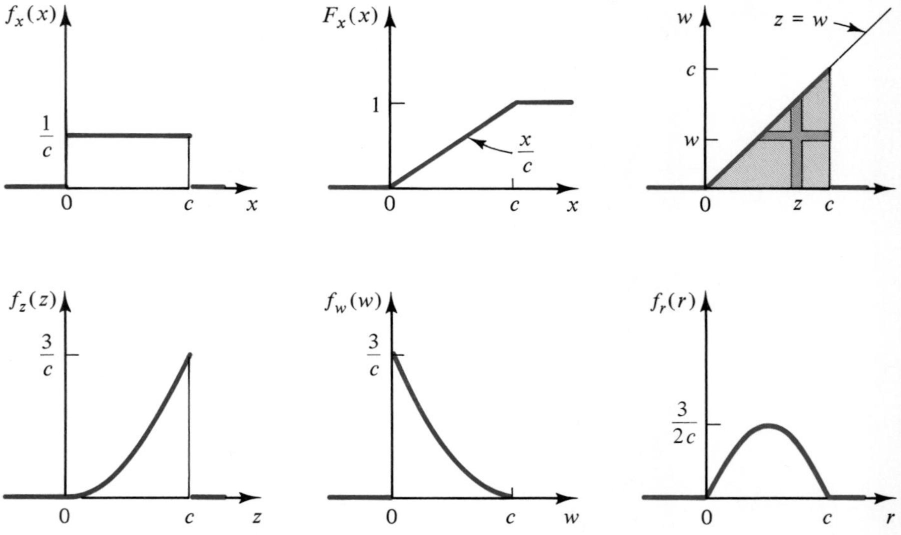

for $0 \le s - r/2 \le s + r/2 \le c$ and zero elsewhere. Hence,

$$f_r(r) = \frac{n(n-1)}{c^n}\int_{r/2}^{c-r/2} r^{n-2}\,ds = \frac{n(n-1)}{c^n} r^{n-2}(c-r) \qquad 0 \le r \le c$$

$$f_s(s) = \begin{cases} \displaystyle\int_0^{2s} r^{n-2}\,dr = \frac{n}{c^n}(2s)^{n-1} & 0 \le s \le \dfrac{c}{2} \\ \displaystyle\int_0^{2c-2s} r^{n-2}\,dr = \frac{n}{c^n}(2c-2s)^{n-1} & \dfrac{c}{2} \le s \le c \end{cases}$$

Note that

$$E\{\mathbf{r}\} = \frac{n-1}{n+1}c \qquad \sigma_r^2 = \frac{2(n-1)c^2}{(n+1)^2(n+2)}$$
$$E\{\mathbf{s}\} = \frac{c}{2} \qquad \sigma_s^2 = \frac{c^2}{2(n+1)(n+2)}. \tag{7-55}$$ ■

Sums of Independent Random Variables

We showed in Section 5-3 that the density of the sum $\mathbf{z} = \mathbf{x} + \mathbf{y}$ of two independent RVS $\mathbf{x}$ and $\mathbf{y}$ is the convolution

$$f_z(z) = \int_{-\infty}^{\infty} f_x(z-y)f_y(y)\,dy \tag{7-56}$$

of their respective densities [see (5-96)]. Convolution is a binary operation between two functions, and it is written symbolically in the form

$$f_z(z) = f_x(z) * f_y(z)$$

From the definition (7-56) it follows that the operation of convolution is commutative and associative. This can also be deduced from the identity [see (5-92)]

$$\Phi_z(s) = \Phi_x(s)\Phi_y(s) \tag{7-57}$$

relating the corresponding moment functions.

Repeated application of (7-57) leads to the conclusion that the density of the sum

$$\mathbf{z} = \mathbf{x}_1 + \cdots + \mathbf{x}_n \tag{7-58}$$

of n independent RVS $\mathbf{x}_i$ equals the convolution

$$f_z(z) = f_1(z) * \cdots * f_n(z) \tag{7-59}$$

of their respective densities $f_i(x)$. From the independence of the RVS $e^{s\mathbf{x}_i}$ it follows as in (7-12) that

$$E\{e^{s\mathbf{z}}\} = E\{e^{s(\mathbf{x}_1+\cdots+\mathbf{x}_n)}\} = E\{e^{s\mathbf{x}_1}\}\cdots E\{e^{s\mathbf{x}_n}\}$$

Hence,

$$\Phi_z(s) = \Phi_1(s)\cdots\Phi_n(s) \tag{7-60}$$

where $\Phi_i(s)$ is the moment function of $\mathbf{x}_i$.

Example 7.5 Using (7-60), we shall show that the convolution of n exponential densities

$$f_1(x) = \cdots = f_n(x) = ce^{-cx}U(x)$$

equals the Erlang density

$$f_z(z) = c^n \frac{z^{n-1}}{(n-1)!} e^{-cz}U(z) \tag{7-61}$$

Indeed, the moment function of an exponential equals

$$\Phi_i(s) = c \int_0^\infty e^{-cx}e^{sx}\,dx = \frac{c}{c-s}$$

Hence,

$$\Phi_z(s) = \Phi_i^n(s) = \frac{c^n}{(c-s)^n} \tag{7-62}$$

and (7-61) follows from (5-63) with $b = n$.

The function $f_z(z)$ in (7-56) involves actually infinitely many integrals, one for each z, and if the integrand consists of several analytic pieces, the computations might be involved. In such cases, a graphical interpretation of the integral is a useful technique for determining the integration limits: To find $f_z(z)$ for a specific z, we form the function $f_x(-y)$ and shift it z units to the right; this yields $f_x(z-y)$. We then form the product $f_x(z-y)f_y(y)$ and integrate. The following example illustrates this.

Example 7.6 The RVS $\mathbf{x}_i$ are uniformly distributed in the interval (0, 1). We shall evaluate the density of the sum $\mathbf{z} = \mathbf{x}_1 + \mathbf{x}_2 + \mathbf{x}_3$. As we showed in Example 5.17, the density of the sum $\mathbf{y} = \mathbf{x}_1 + \mathbf{x}_2$ is a triangle as in Fig. 7.4. Since $\mathbf{z} = \mathbf{y} + \mathbf{x}_3$, the density of $\mathbf{z}$ is the convolution of the triangle $f_y(y)$ with the pulse $f_3(x)$. Clearly, $f_z(z) = 0$ outside the interval (0, 3) because $f_y(y) = 0$ outside the interval (0, 2) and $f_3(x) = 0$ outside the interval (0, 1). To find $f_z(z)$ for $0 \le z \le 3$, we must consider three cases:

If $0 \le z \le 1$, then

$$f_z(z) = \int_0^z y\,dy = \frac{z^2}{2}$$

If $1 \le z \le 2$, then

$$f_z(z) = \int_{z-1}^1 y\,dy + \int_1^z (2-y)\,dy = -z^2 + 3z - \frac{3}{2}$$

Finally, if $2 \le z \le 3$, then

$$f_z(z) = \int_{z-1}^2 (2-y)\,dy = \frac{z^2}{2} - 3z + \frac{9}{2}$$

In all cases, $f_z(z)$ equals the area of the shaded region shown in Fig. 7.4. Thus $f_z(z)$ consists of three parabolic pieces. ■

Binomial Distribution Revisited We shall reestablish the binomial distribution in terms of the sum of n samples $\mathbf{x}_i$ of an RV $\mathbf{x}$. Consider an event $\mathscr{A}$ with probability $P(\mathscr{A}) = p$, defined in an experiment $\mathscr{S}$. The zero-one RV associ-

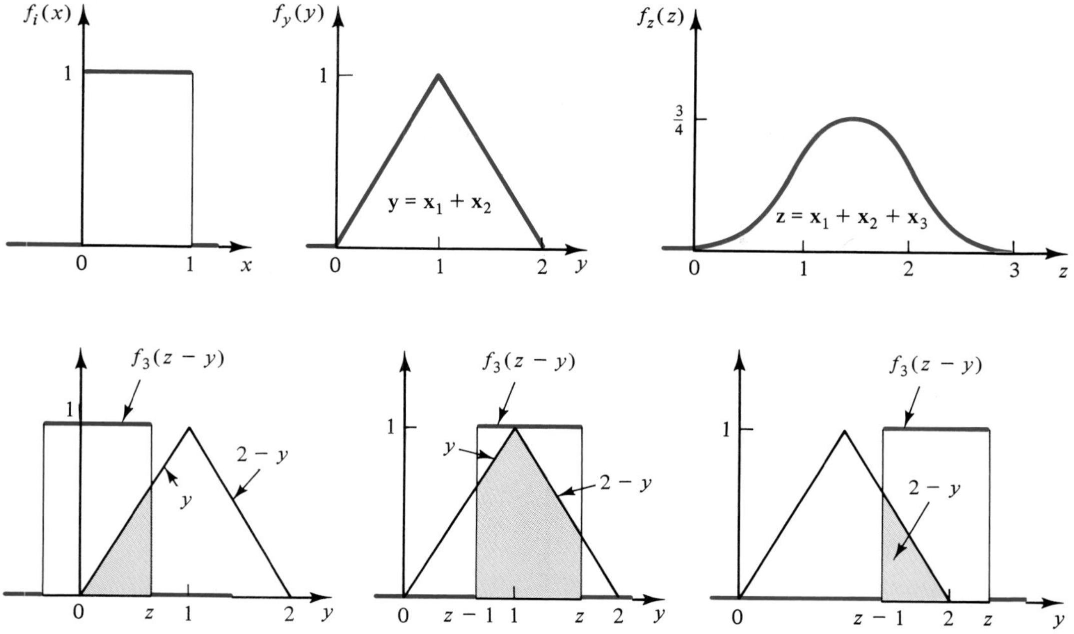

Figure 7.4

ated with this event is by definition an RV $\mathbf{x}$ such that

$$\mathbf{x}(\zeta) = \begin{cases} 1 & \zeta \in \mathscr{A} \\ 0 & \zeta \notin \mathscr{A} \end{cases}$$

Thus $\mathbf{x}$ takes the values 0 and 1 with probabilities p and q, respectively. Repeating the experiment $\mathscr{S}$ n times, we form the sample space $\mathscr{S}_n$ and the samples $\mathbf{x}_i$. We maintain that the sum

$$\mathbf{z} = \mathbf{x}_1 + \cdot \cdot \cdot + \mathbf{x}_n$$

has a binomial distribution. Indeed, the moment function of the RVs $\mathbf{x}_i$ equals

$$\Phi_i(s) = \Phi_x(s) = E\{e^{s\mathbf{x}}\} = pe^s + q$$

Hence,

$$\Phi_z(s) = \Phi_1(s) \cdot \cdot \cdot \Phi_n(s) = (pe^s + q)^n$$

From this and (5-66) it follows that $\mathbf{z}$ has a binomial distribution. We note again that $E\{\mathbf{x}_i\} = p$, $\sigma_i^2 = pq$; hence, $E\{\mathbf{z}\} = np$, $\sigma_z^2 = npq$.

7-3 Central Limit Theorem

Given n independent RVS $\mathbf{x}_i$, we form the sum

$$\mathbf{z} = \mathbf{x}_1 + \cdots + \mathbf{x}_n \tag{7-63}$$

The central limit theorem (CLT) states that under certain general conditions, the distribution of $\mathbf{z}$ approaches a normal distribution:

$$F_z(z) \simeq G\left(\frac{z - \eta_z}{\sigma_z}\right) \tag{7-64}$$

as n increases. Furthermore, if the RVS $\mathbf{x}_i$ are of continuous type, the density of $\mathbf{z}$ approaches a normal density:

$$f_z(z) \simeq \frac{1}{\sigma_z\sqrt{2\pi}} e^{-(z-\eta_z)^2/2\sigma_z^2} \tag{7-65}$$

Under appropriate normalization, the theorem can be stated as a limit: If $\mathbf{z}_0 = (\mathbf{z} - \eta_z)/\sigma_z$, then

$$F_{z_0}(z) \underset{n\to\infty}{\longrightarrow} G(z) \qquad f_{z_0}(z) \underset{n\to\infty}{\longrightarrow} \frac{1}{\sqrt{2\pi}} e^{-z^2/2} \tag{7-66}$$

for the general and for the continuous case, respectively.

No general statement can be made about the required size of n for a satisfactory approximation of $F_z(z)$ by a normal distribution. For a specific n, the nature of the approximation depends on the form of the densities $f_i(x)$. If the RVS $\mathbf{x}_i$ are i.i.d., the value $n = 30$ is adequate for most applications. In fact, if $f_i(x)$ is sufficiently smooth, values of n as low as 5 can be used. The following example illustrates this.

Example 7.7

The RVS $\mathbf{x}_i$ are uniformly distributed in the interval (0, 1), and

$$E\{\mathbf{x}_i\} = \frac{1}{2} \qquad \sigma_i^2 = \frac{1}{12}$$

We shall compare the density of $\mathbf{z}$ with the normal approximation (7-65) for $n = 2$ and $n = 3$.

***n* = 2**

$$\eta_z = \frac{n}{2} = 1 \qquad \sigma_z^2 = \frac{n}{12} = \frac{1}{6} \qquad f_z(z) \simeq \sqrt{\frac{3}{\pi}}\, e^{-3(z-1)^2}$$

***n* = 3**

$$\eta_z = \frac{n}{2} = \frac{3}{2} \qquad \sigma_z^2 = \frac{n}{12} = \frac{1}{4} \qquad f_z(z) \simeq \sqrt{\frac{2}{\pi}}\, e^{-2(z-1.5)^2}$$

As we showed in Example 7.6, $f_z(z)$ is a triangle for $n = 2$ and consists of three

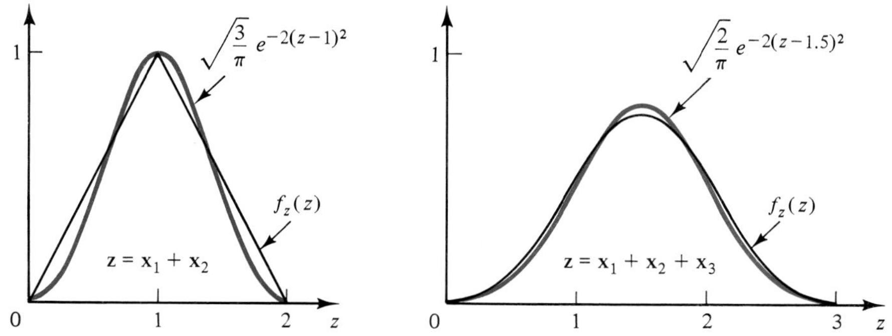

Figure 7.5

parabolic pieces for $n = 3$. In Fig. 7.5, we compare these densities with the corresponding normal approximations. Even for such small values of n, the approximation error is small. ■

The central limit theorem (7-66) can be expressed as a property of convolutions: The convolution of a large number of positive functions is approximately a normal curve [see (7-55)].

The central limit theorem is not always true. We cite next sufficient conditions for its validity:

If the RVs $\mathbf{x}_i$ are i.i.d. and $E\{\mathbf{x}_i^3\}$ is finite, the distribution $F_0(z)$ of their normalized sum $\mathbf{z}_0$ tends to $N(0, 1)$.

If the RVs $\mathbf{x}_i$ are such that $|\mathbf{x}_i| < A < \infty$ and $\sigma_i > a > 0$ for every i, then $F_0(z)$, tends to $N(0, 1)$.

If $E\{\mathbf{x}_i^3\} < B < \infty$ for every i and

$$\sigma_z^2 = \sigma_1^2 + \cdots + \sigma_n^2 \to \infty \qquad \text{as } n \to \infty \tag{7-67}$$

then $F_0(z)$ tends to $N(0, 1)$.

The proofs will not be given.

Example 7.8

If

$$f_i(x) = e^{-x}U(x) \qquad \text{then} \qquad E\{\mathbf{x}_i^n\} = \int_0^\infty x^n e^{-x}\,dx = n!$$

Thus $E\{\mathbf{x}_i^3\} = 6 < \infty$; hence, the central limit theorem applies. In this case, $\eta_z = nE\{\mathbf{x}_i\} = n$, $\sigma_z^2 = n\sigma_i^2 = n$. And since the convolution of n exponential densities equals the Erlang density (7-61), we conclude from (7-65) that

$$\frac{z^{n-1}}{(n-1)!}e^{-z} \simeq \frac{1}{\sqrt{2\pi n}}\, e^{-(z-n)^2/2n}$$

for large n. In Fig. 7.6, we show both sides for $n = 5$ and $n = 10$. ■

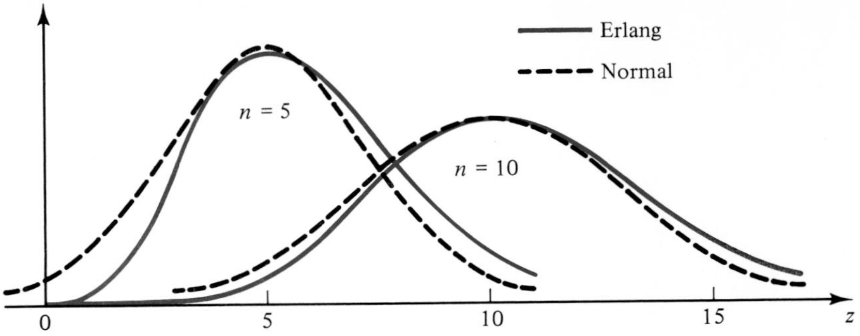

Figure 7.6

The De Moivre–Laplace Theorem The (7-64) form of the central limit theorem holds for continuous and discrete type RVs. If the RVs $\mathbf{x}_i$ are of discrete type, their sum $\mathbf{z}$ is also of discrete type, taking the values z_k, and its density consists of points $f_z(z_k) = P\{\mathbf{z} = z_k\}$. In this case, (7-65) no longer holds because the normal density is a continuous curve. If, however, the numbers z_k form an arithmetic progression, the numbers $f_z(z_k)$ are nearly equal to the values of the normal curve at these points. As we show next, the De Moivre–Laplace theorem (3-27) is a special case.

▪ ***Definition.*** We say that an RV $\mathbf{x}$ is of *lattice* type if it takes the values $x_k = ka$ where k is an integer. If the RVs $\mathbf{x}_i$ are of lattice type with the same a, their sum

$$\mathbf{z} = \mathbf{x}_1 + \cdots + \mathbf{x}_n$$

is also of lattice type. We shall examine the asymptotic form of $f_z(z)$ for the following special case.

Suppose that the RVs $\mathbf{x}_i$ are i.i.d., taking the values 1 and 0 as in (7-60). In this case, their sum $\mathbf{z}$ has a binomial distribution:

$$P\{\mathbf{z} = k\} = \binom{n}{k} p^k q^{n-k} \qquad k = 0, 1, \ldots, n$$

and

$$E\{\mathbf{z}\} = np \qquad \sigma_z^2 = npq \to \infty \qquad \text{as } n \to \infty$$

Hence, we can use the approximation (7-64). This yields

$$F_z(z) = \sum_{k \le z} \binom{n}{k} p^k q^{n-k} \simeq G\left(\frac{z - np}{\sqrt{npq}}\right) \tag{7-68}$$

From this and (4-48) it follows that

$$P\{np - 3\sqrt{npq} \le \mathbf{z} \le np + 3\sqrt{npq}\} = .997 \tag{7-69}$$

In Fig. 7.7a, we show the function $F_n(z)$ and its normal approximation. Clearly, $F_z(z)$ is a staircase function with discontinuities at the points $z = k$. If n is large, then in the interval $(k - 1, k)$ between discontinuity points, the normal density is nearly constant. And since the jump of $F_z(z)$ at $z = k$ equals

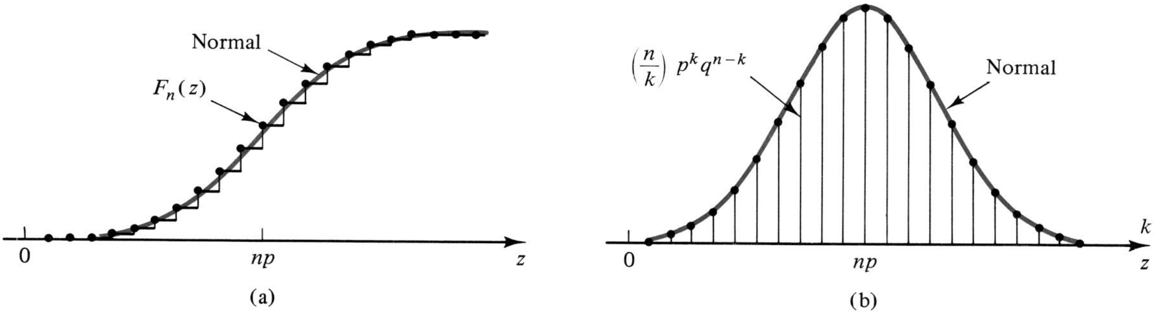

Figure 7.7

$P\{\mathbf{z} = k\}$, we conclude from (7-68) (Fig. 7.7b) that

$$\binom{n}{k} p^k q^{n-k} \simeq \frac{1}{\sqrt{2\pi npq}} e^{-(k-np)^2/2npq} \tag{7-70}$$

as in (3-27).

Multinomial Distribution Suppose that $[\mathcal{A}_1, \ldots, \mathcal{A}_r]$ is a partition of $\mathcal{S}$ with $p_i = P(\mathcal{A}_i)$. We have shown in (3-41) that if the experiment $\mathcal{S}$ is performed n times and the events $\mathcal{A}_i$ occur k_i times, then

$$P\{\mathcal{A}_i \text{ occurs } k_1 \text{ times}\} = \frac{n!}{k_1! \cdots k_r!} p_1^{k_1} \cdots p_r^{k_r}$$

For large n, this can be approximated by

$$\frac{n!}{k_1! \cdots k_r!} p_1^{k_1} \cdots p_r^{k_r} \simeq \frac{\exp\left\{-\frac{1}{2}\left[\frac{(k_1 - np_1)^2}{np_1} + \cdots + \frac{(k_r - np_r)^2}{np_r}\right]\right\}}{\sqrt{(2\pi n)^{r-1} p_1 \cdots p_r}} \tag{7-71}$$

We maintain that for $n = 2$ (7-71) reduces to (7-70). Indeed, in this case,

$$k_1 + k_2 = n \qquad p_1 + p_2 = 1 \qquad |k_1 - np_1| = |k_2 - np_2|$$

Hence,

$$\frac{(k_1 - np_1)^2}{np_1} + \frac{(k_2 - np_2)^2}{np_2} = (k_1 - np_1)^2 \left(\frac{1}{np_1} + \frac{1}{np_2}\right) = \frac{(k_1 - np_1)^2}{np_1 p_2} \tag{7-72}$$

And since $k_1 = k$, $p_1 = p$, $p_2 = 1 - p_1 = q$, (7-70) results.

Sequences and Limits

An infinite sequence

$$\mathbf{x}_1, \ldots, \mathbf{x}_n, \ldots$$

of RVs is called a *random process*. For a specific ζ, the values $\mathbf{x}_n(\zeta)$ of $\mathbf{x}_n$ form a sequence of numbers. If this sequence has a limit, that limit depends in general on ζ. The limit of $\mathbf{x}_n(\zeta)$ might exist for some outcomes but not for others. If the set of outcomes for which it does not exist has zero probability, we say that the sequence $\mathbf{x}_n$ converges *almost everywhere*. The limit of $\mathbf{x}_n$ might be an RV $\mathbf{x}$ or a constant number. We shall consider next sequences converging to a constant c.

Convergence in the MS Sense The mean $E\{(\mathbf{x}_n - c)^2\}$ of $(\mathbf{x}_n - c)^2$ is a sequence of numbers. If this sequence tends to zero

$$E\{(\mathbf{x}_n - c)^2\} \underset{n\to\infty}{\longrightarrow} 0 \tag{7-73}$$

we say that the random sequence $\mathbf{x}_n$ tends to c in the *mean-square sense*. This can be expressed as a limit involving the mean η_n and the variance σ_n^2 of $\mathbf{x}_n$.

■ ***Theorem.*** The sequence $\mathbf{x}_n$ tends to c in the MS sense iff

$$\eta_n \underset{n\to\infty}{\longrightarrow} c \qquad \sigma_n \underset{n\to\infty}{\longrightarrow} 0 \tag{7-74}$$

■ ***Proof.*** This follows readily from the identity

$$E\{(\mathbf{x}_n - c)^2\} = (\eta_n - c)^2 + \sigma_n^2$$

Convergence in Probability Consider the events $\{|\mathbf{x}_n - c| > \varepsilon\}$ where ε is an arbitrary positive number. The probabilities of these events form a sequence of numbers. If this sequence tends to zero

$$P\{|\mathbf{x}_n - c| > \varepsilon\} \underset{n\to\infty}{\longrightarrow} 0 \tag{7-75}$$

for every $\varepsilon > 0$, we say that the random sequence $\mathbf{x}_n$ tends to c *in probability*.

■ ***Theorem.*** If $\mathbf{x}_n$ converges to c in the MS sense, it converges to c in probability.

■ ***Proof.*** Applying (4-116) to the positive RV $|\mathbf{x}_n - c|^2$, we conclude, replacing $\lambda\eta$ by ε^2, that

$$P\{|\mathbf{x}_n - c|^2 \geq \varepsilon^2\} = P\{|\mathbf{x}_n - c| \geq \varepsilon\} \leq \frac{E\{|\mathbf{x}_n - c|^2\}}{\varepsilon^2} \tag{7-76}$$

Hence, if $E\{|\mathbf{x}_n - c|^2\} \to 0$, then $P\{|\mathbf{x}_n - c| > \varepsilon\} \to 0$.

Convergence in Distribution This form of convergence involves the limit of the distributions $F_n(x) = P\{\mathbf{x}_n \leq x\}$ of the RVs $\mathbf{x}_n$. We shall say that the sequence $\mathbf{x}_n$ converges in distribution if

$$F_n(x) \underset{n\to\infty}{\longrightarrow} F(x) \tag{7-77}$$

for every point of continuity of $\mathbf{x}$. In this case, $\mathbf{x}_n(\zeta)$ need not converge for any ζ. If the limit $F(x)$ is the distribution of an RV $\mathbf{x}$, we say that $\mathbf{x}_n$ tends to $\mathbf{x}$ in distribution. From (7-77) it follows that if $\mathbf{x}_n$ tends to $\mathbf{x}$ in distribution, the probability $P\{\mathbf{x}_n \in D\}$ that $\mathbf{x}_n$ is in a region D tends to $P\{\mathbf{x} \in D\}$. The RVs $\mathbf{x}_n$ and $\mathbf{x}$ need not be related in any other way.

The central limit theorem is an example of convergence in distribution: The sum $\mathbf{z}_n = \mathbf{x}_1 + \cdots + \mathbf{x}_n$ in (7-63) is a sequence of RVs converging in distribution to a normal RV.

Law of Large Numbers We showed in Section 3-3 that if an event $\mathscr{A}$ occurs k times in n trials and $P(\mathscr{A}) = p$, the probability that the ratio k/n is in the interval $(p - \varepsilon, p + \varepsilon)$ tends to 1 as $n \to \infty$ for any ε. We shall reestablish this result as a limit in probability of a sequence of RVS.

We form the zero-one RVS $\mathbf{x}_i$ associated with the event $\mathscr{A}$ as in (7-61), and their sample mean

$$\bar{\mathbf{x}}_n = \frac{\mathbf{x}_1 + \cdots + \mathbf{x}_n}{n}$$

Since $E\{\mathbf{x}_i\} = p$ and $\sigma_i^2 = pq$, it follows that

$$E\{\bar{\mathbf{x}}_n\} = p \qquad \sigma_{\bar{x}_n}^2 = \frac{npq}{n^2} = \frac{pq}{n} \underset{n\to\infty}{\longrightarrow} 0$$

Thus the sequence $\bar{\mathbf{x}}_n$ satisfies (7-73); hence, it converges in the MS sense and, therefore, in probability to its mean p. Thus

$$P\{|\bar{\mathbf{x}}_n - p| > \varepsilon\} \underset{n\to\infty}{\longrightarrow} 0$$

And since $\bar{\mathbf{x}}_n$ equals the ratio k/n, we conclude that

$$P\left\{\left|\frac{k}{n} - p\right| > \varepsilon\right\} \underset{n\to\infty}{\longrightarrow} 0$$

for every $\varepsilon > 0$. This is the law of large numbers (3-37) expressed as a limit.

7-4 Special Distributions of Statistics

We discuss next three distributions of particular importance in mathematical statistics.

Chi-Square Distribution

The chi-square density

$$f(x) = \frac{1}{2^{m/2}\,\Gamma(m/2)} x^{m/2-1} e^{-x/2}\, U(x) \qquad (7\text{-}78)$$

is a special case of the gamma density (4-50). The term $\Gamma(m/2)$ is the gamma function given by [see (4-51)]

$$\Gamma\left(\frac{m}{2}\right) = \begin{cases} k! & m = \text{even} = 2k \\ \left(k - \frac{1}{2}\right)\left(k - \frac{3}{2}\right)\cdots\frac{3}{2}\sqrt{\pi} & m = \text{odd} = 2k + 1 \end{cases} \qquad (7\text{-}79)$$

We shall use the notation $\chi^2(m)$ to indicate that an RV $\mathbf{x}$ has a chi-square distribution.* The constant m is called the number of *degrees of*

* For brevity we shall also say: The RV $\mathbf{x}$ is $\chi^2(m)$.

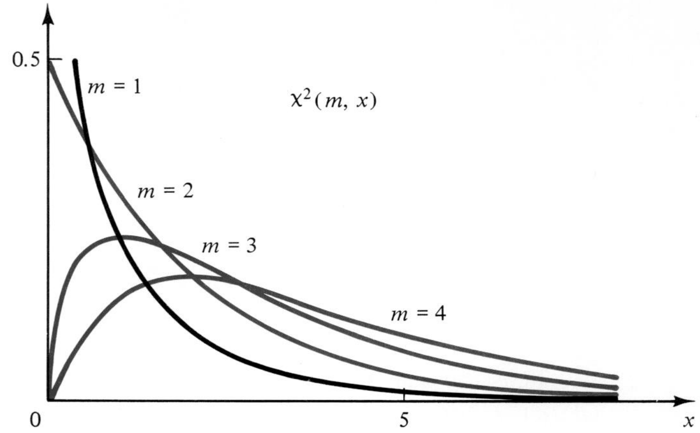

Figure 7.8

freedom. The notation $\chi^2(m, x)$ will mean the function $f(x)$ in (7-78). In Fig. 7.8, we plot this function for $m = 1, 2, 3$, and 4.

With $b = m/2$ and $c = 1/2$, it follows from (5-64) that

$$E\{\mathbf{x}^n\} = m(m+2)\cdots(m+2n-2) \tag{7-80}$$

Hence,

$$E\{\mathbf{x}\} = m \qquad E\{\mathbf{x}^2\} = m(m+2) \qquad \sigma_x^2 = 2m \tag{7-81}$$

Note [see (4-54) and (5-63)] that

$$E\left\{\frac{1}{\mathbf{x}}\right\} = \gamma \int_0^\infty x^{m/2-3} e^{-x/2}\, dx = \frac{1}{m-2} \qquad m > 2 \tag{7-82}$$

$$\Phi(s) = \frac{1}{\sqrt{(1-2s)^m}} \tag{7-83}$$

The χ^2 density with $m = 2$ degrees of freedom is an exponential

$$f(x) = \frac{1}{2} e^{-x/2}\, U(x) \qquad \Phi(s) = \frac{1}{1-2s} \tag{7-84}$$

and for $m = 1$ it equals

$$f(x) = \frac{1}{\sqrt{2\pi x}} e^{-x/2}\, U(x) \qquad \Phi(s) = \frac{1}{\sqrt{1-2s}} \tag{7-85}$$

because $\Gamma(1/2) = \sqrt{\pi}$.

Example 7.9

We shall show that the square of an $N(0, 1)$ RV has a χ^2 density with one degree of freedom. Indeed, if

$$f_x(x) = \frac{1}{\sqrt{2\pi}} e^{-x^2/2} \qquad \text{and} \qquad \mathbf{y} = \mathbf{x}^2$$

then [see (4-84)]

$$f_y(y) = \frac{1}{\sqrt{y}} f_x(\sqrt{y})U(y) = \frac{1}{\sqrt{2\pi y}} e^{-y/2}U(y) \tag{7-86}$$ ■

The following property of χ^2 densities will be used extensively. If the RVS $\mathbf{x}$ and $\mathbf{y}$ are $\chi^2(m)$ and $\chi^2(n)$, respectively, and are independent, their sum

$$\mathbf{z} = \mathbf{x} + \mathbf{y} \qquad \text{is } \chi^2(m+n) \tag{7-87}$$

■ ***Proof.*** As we see from (7-83),

$$\Phi_x(s) = \frac{1}{\sqrt{(1-2s)^m}} \qquad \Phi_y(y) = \frac{1}{\sqrt{(1-2s)^n}}$$

Hence [see (7-14)]

$$\Phi_z(s) = \Phi_x(s)\Phi_y(s) = \frac{1}{\sqrt{(1-2s)^{m+n}}}$$

From this it follows that the convolution of two χ^2 densities is a χ^2 density:

$$\chi^2(m, x) * \chi^2(n, x) = \chi^2(m+n, x) \tag{7-88}$$

Fundamental Property The sum of the squares of n independent $N(0, 1)$ RVS $\mathbf{x}_i$ has a chi-square distribution with n degrees of freedom:

$$\text{If } \mathbf{Q} = \sum_{i=1}^{n} \mathbf{x}_i^2 \qquad \text{then} \qquad \mathbf{Q} \text{ is } \chi^2(n) \tag{7-89}$$

■ ***Proof.*** As we have shown in (7-86), the RVS $\mathbf{x}_i^2$ are $\chi^2(1)$. Furthermore, they are independent because the RVS $\mathbf{x}_i$ are independent. Repeated application of (7-88) therefore yields (7-89).

The RVS $\mathbf{x}_i$ are $N(0, 1)$; hence [see (5-55)],

$$E\{\mathbf{x}_i^2\} = 1 \qquad E\{\mathbf{x}_i^4\} = 3 \qquad \text{Var } \mathbf{x}_i^2 = 2$$

From this it follows that

$$E\{\mathbf{Q}\} = n \qquad \sigma_{\mathbf{Q}}^2 = 2n \tag{7-90}$$

in agreement with (7-81).

As an application, note that if the RVS $\mathbf{w}_i$ are $N(\eta, \sigma)$ and i.i.d., the RVS $\mathbf{x}_i = (\mathbf{w}_i - \eta)/\sigma$ are $N(0, 1)$ and i.i.d., as in (7-89); hence, the sum

$$\mathbf{Q} = \sum_{i=1}^{n} \left(\frac{\mathbf{w}_i - \eta}{\sigma}\right)^2 \qquad \text{is} \qquad \chi^2(n) \tag{7-91}$$

QUADRATIC FORMS The sum

$$\mathbf{Q} = \sum_{i,j=1}^{n} a_{ij}\mathbf{x}_i\,\mathbf{x}_j \tag{7-92}$$

is called a quadratic form of order n in the variables $\mathbf{x}_i$. We shall consider quadratic forms like these where the RVS $\mathbf{x}_i$ are $N(0, 1)$ and independent, as in (7-89). An important problem in statistics is the determination of the conditions that the coefficients a_{ij} must satisfy such that the sum in (7-92) have a χ^2 distribution. The following theorem facilitates in some cases the determination of the χ^2 character of $\mathbf{Q}$.

■ ***Theorem 1.*** Given three quadratic forms $\mathbf{Q}$, $\mathbf{Q}_1$, and $\mathbf{Q}_2$ such that

$$\mathbf{Q} = \mathbf{Q}_1 + \mathbf{Q}_2 \tag{7-93}$$

it can be shown that if

(a) the RV $\mathbf{Q}$ has a $\chi^2(n)$ distribution
(b) the RV $\mathbf{Q}_1$ has a $\chi^2(r)$ distribution
(c) $\mathbf{Q}_2 \geq 0$

then the RVS $\mathbf{Q}_1$ and $\mathbf{Q}_2$ are independent and $\mathbf{Q}_2$ has a χ^2 distribution with $n - r$ degrees of freedom.

The proof of this difficult theorem follows from (7A-6); the details, however, will not be given. We discuss next an important application.

Sample Mean and Sample Variance Suppose that $\mathbf{x}_i$ is a sample from an arbitrary population. The RVS

$$\bar{\mathbf{x}} = \frac{1}{n}\sum_{i=1}^{n} \mathbf{x}_i \qquad \mathbf{s}^2 = \frac{1}{n-1}\sum_{i=1}^{n} (\mathbf{x}_i - \bar{\mathbf{x}})^2 \tag{7-94}$$

are the *sample mean* and the *sample variance*, respectively, of the sample $\mathbf{x}_i$. As we have shown,

$$E\{\bar{\mathbf{x}}\} = \eta \qquad \sigma_{\bar{x}}^2 = \frac{\sigma^2}{n}$$

where η is the mean and σ^2 is the variance of $\mathbf{x}_i$. We shall show that

$$E\{\mathbf{s}^2\} = \sigma^2 \tag{7-95}$$

■ ***Proof.*** Clearly,

$$(\mathbf{x}_i - \eta)^2 = (\mathbf{x}_i - \bar{\mathbf{x}} + \bar{\mathbf{x}} - \eta)^2 = (\mathbf{x}_i - \bar{\mathbf{x}})^2 + 2(\mathbf{x}_i - \bar{\mathbf{x}})(\bar{\mathbf{x}} - \eta) + (\bar{\mathbf{x}} - \eta)^2$$

Summing from 1 to n and using the identity $\sum_{i=1}^{n} (\mathbf{x}_i - \bar{\mathbf{x}}) = 0$, we conclude that

$$\sum_{i=1}^{n} (\mathbf{x}_i - \eta)^2 = \sum_{i=1}^{n} (\mathbf{x}_i - \bar{\mathbf{x}})^2 + n(\bar{\mathbf{x}} - \eta)^2 \tag{7-96}$$

Since the mean of $\bar{\mathbf{x}}$ is η and its variance σ^2/n, we conclude, taking expected values of both sides, that

$$n\sigma^2 = E\{(\mathbf{x}_i - \bar{\mathbf{x}})^2\} + \sigma^2$$

and (7-95) results.

We now introduce the additional assumption that the samples $\mathbf{x}_i$ are normal.

■ ***Theorem 2.*** The sample mean $\bar{\mathbf{x}}$ and the sample variance $\mathbf{s}^2$ of a normal sample are two independent RVS. Furthermore, the RV

$$\boxed{\mathbf{Q}_2 = \frac{(n-1)\mathbf{s}^2}{\sigma^2} \quad \text{is} \quad \chi^2(n-1)} \tag{7-97}$$

■ ***Proof.*** Dividing both sides of (7-96) by σ^2, we obtain

$$\sum_{i=1}^{n}\left(\frac{\mathbf{x}_i-\eta}{\sigma}\right)^2=\left(\frac{\bar{\mathbf{x}}-\eta}{\sigma/\sqrt{n}}\right)^2+\sum_{i=1}^{n}\left(\frac{\mathbf{x}_i-\bar{\mathbf{x}}}{\sigma}\right)^2 \tag{7-98}$$

The RV $\bar{\mathbf{x}}$ is normal with mean η and variance σ^2/n. Hence, the first RV in parentheses on the right side of (7-98) is $N(0, 1)$, and its square is $\chi^2(1)$. Furthermore, the RV $(\mathbf{x}_i - \eta)/\sigma$ is $N(0, 1)$; hence, the left side of (7-98) is $\chi^2(n)$ [see (7-89)]. Theorem 2 follows, therefore, from theorem 1.

Note, finally, that

$$E\{\mathbf{s}^2\}=\sigma^2 \qquad \sigma_{s^2}^2=\frac{2\sigma^4}{n-1} \tag{7-99}$$

This follows from (7-97) and (7-81) because

$$E\{\mathbf{s}^2\}=\frac{\sigma^2}{n-1}E\{\mathbf{Q}_2\} \qquad \sigma_{s^2}^2=\left(\frac{\sigma^2}{n-1}\right)^2\sigma_{Q_2}^2$$

Student t Distribution

We shall say that an RV $\mathbf{x}$ has a Student $t(m)$ distribution with m degrees of freedom if its density equals

$$f(x)=\gamma\frac{1}{\sqrt{(1+x^2/m)^{m+1}}} \qquad \gamma=\frac{\Gamma[(m+1)/2]}{\sqrt{\pi m}\,\Gamma(m/2)} \tag{7-100}$$

This density will be denoted by $t(m, x)$. The importance of the t distribution is based on the following.

■ ***Theorem.*** If the RVs $\mathbf{z}$ and $\mathbf{w}$ are independent, $\mathbf{z}$ is $N(0, 1)$, and $\mathbf{w}$ is $\chi^2(m)$, the ratio

$$\mathbf{x}=\frac{\mathbf{z}}{\sqrt{\mathbf{w}/m}} \qquad \text{is} \qquad t(m) \tag{7-101}$$

■ ***Proof.*** From the independence of $\mathbf{z}$ and $\mathbf{w}$ it follows that

$$f_{zw}(z, w) \sim e^{-z^2/2}(w^{m/2-1}e^{-w/2})U(w)$$

To find the density of $\mathbf{x}$, we introduce the auxiliary variable $\mathbf{y} = \mathbf{w}$ and form the system

$$x = z\sqrt{m/w} \qquad y = w$$

This yields $z = x\sqrt{y/m}$, $w = y$; and since the Jacobian of the transformation equals $\sqrt{m/w}$, we conclude from (5-83) that

$$f_{xy}(x, y)=\sqrt{\frac{y}{m}}f_{zw}\left(x\sqrt{\frac{y}{m}}, y\right)\sim\sqrt{y}e^{-x^2y/m}(y^{m/2-1}e^{-y/2})$$

for $y > 0$ and 0 for $y < 0$. Integrating with respect to y, we obtain

$$f_x(x)\sim\int_0^\infty y^{(m-1)/2}\exp\left\{-\frac{y}{2}\left(1+\frac{x^2}{m}\right)\right\}dy \tag{7-102}$$

With

$$q=\frac{y}{2}\left(1+\frac{x^2}{m}\right) \qquad y=\frac{2q}{1+x^2/m}$$

(7-102) yields

$$f_x(x) \sim \frac{1}{\sqrt{(1 + x^2/m)^{m+1}}} \int_0^\infty q^{(m+1)/2} e^{-q}\, dq \sim \frac{1}{\sqrt{(1 + x^2/m)^{m+1}}}$$

and (7-101) results.

Note that the mean of $\mathbf{x}$ is zero and its variance equals

$$E\{\mathbf{x}\}^2 = mE\{\mathbf{z}^2\}E\left\{\frac{1}{\mathbf{w}}\right\} = \frac{m}{m-2} \tag{7-103}$$

This follows from (7-82) and the independence of $\mathbf{z}$ and $\mathbf{w}$.

■ ***Corollary.*** If $\bar{\mathbf{x}}$ and $\mathbf{s}^2$ are the sample mean and the sample variance, respectively, of an N(η, σ) sample $\mathbf{x}_i$ [see (7-94)], the ratio

$$\mathbf{t} = \frac{\bar{\mathbf{x}} - \eta}{\mathbf{s}/\sqrt{n}} \qquad \text{is } t(n-1) \tag{7-104}$$

■ ***Proof.*** We have shown that the RVs

$$\mathbf{z} = \frac{\bar{\mathbf{x}} - \eta}{\sigma/\sqrt{n}} \qquad \mathbf{w} = (n-1)\frac{\mathbf{s}^2}{\sigma^2}$$

are independent, the first is $N(0, 1)$, and the second is $\chi^2(n-1)$. From this and (7-101) it follows that the ratio

$$\frac{\bar{\mathbf{x}} - \eta}{\sigma/\sqrt{n}} \Big/ \sqrt{\frac{(n-1)\mathbf{s}^2}{\sigma^2(n-1)}} = \frac{\bar{\mathbf{x}} - \eta}{\mathbf{s}/\sqrt{n}}$$

is $t(n-1)$.

Snedecor F Distribution

An RV $\mathbf{x}$ has a Snedecor distribution if its density equals

$$f(x) = \gamma \frac{x^{k/2-1}}{\sqrt{(1 + kx/m)^{k+m}}} U(x) \tag{7-105}$$

This is a two-parameter family of functions denoted by $F(k, m)$. Its importance is based on the following.

■ ***Theorem.*** If the RVs $\mathbf{z}$ and $\mathbf{w}$ are independent, $\mathbf{z}$ is $\chi^2(k)$, and $\mathbf{w}$ is $\chi^2(r)$, the ratio

$$\mathbf{x} = \frac{\mathbf{z}/k}{\mathbf{w}/r} \qquad \text{is} \qquad F(k, r) \tag{7-106}$$

■ ***Proof.*** From the independence of $\mathbf{z}$ and $\mathbf{w}$ it follows that

$$f_{zw}(z, w) \sim (z^{k/2-1}e^{-z/2})(w^{r/2-1}e^{-w/2})\, U(z)U(w)$$

We introduce the auxiliary variable $\mathbf{y} = \mathbf{w}$ and form the system $x = rz/kw$, $y = w$. The solution of this system is $z = kxy/r$, $w = y$, and its Jacobian equals r/kw. Hence [see (5-83)],

$$f_{xy}(x, y) = \frac{kw}{r} f_{zw}\left(\frac{k}{r}xy, y\right) \sim y(xy)^{k/2-1}e^{-kxy/2r}y^{r/2-1}e^{-w/2}$$

for $x > 0$, $y > 0$, and 0 otherwise. Integrating with respect to y, we obtain

$$f_x(x) \sim x^{k/2-1} \int_0^\infty y^{(k+r)/2-1} \exp\left\{-\frac{y}{2}\left(1 + \frac{k}{r}x\right)\right\} dy$$

With

$$q = \frac{y}{2}\left(1 + \frac{k}{r}x\right) \qquad y = \frac{2q}{1 + kx/r}$$

we obtain

$$f_x(x) \sim \frac{x^{k/2-1}}{(1 + kx/r)^{(k+r)/2}} \int_0^\infty q^{(k+r)/2} e^{-q} dq$$

and (7-106) results.

Note that the mean of the RV $\mathbf{x}$ in (7-106) equals [see (7-81) and (7-82)]

$$E\{\mathbf{x}\} = \frac{r}{k} E\{\mathbf{z}\}E\left\{\frac{1}{\mathbf{w}}\right\} = \frac{r}{k}\frac{k}{r-2} = \frac{r}{r-2} \tag{7-107}$$

Furthermore, if

$$\mathbf{x} \text{ is } F(k, r) \qquad \text{then} \qquad \frac{1}{\mathbf{x}} \text{ is } F(r, k) \tag{7-108}$$

Percentiles The u-percentile of an RV $\mathbf{x}$ is by definition a number x_u such that

$$u = P\{\mathbf{x} \le x_u\} = P\{\mathbf{x} > x_{1-u}\} \tag{7-109}$$

The u-percentiles of the $\chi^2(m)$, $t(m)$, and $F(k, r)$ distributions will be denoted by

$$\chi_u^2(m) \qquad t_u(m) \qquad F_u(k, r)$$

respectively.

We maintain that

$$F_u(k, r) = \frac{1}{F_{1-u}(r, k)} \tag{7-110}$$

$$F_{2u-1}(1, r) = t_u^2(r) \tag{7-111}$$

$$F_u(k, r) \underset{r\to\infty}{\longrightarrow} \frac{1}{k}\chi_u^2(k) \tag{7-112}$$

■ ***Proof.*** If the RV $\mathbf{x}$ is $F(k, r)$, then

$$u = P\{\mathbf{x} \le F_u(k, r)\} = P\left\{\frac{1}{\mathbf{x}} \ge \frac{1}{F_u(k, r)}\right\}$$

and (7-110) follows from (7-108) and (7-109).

If $\mathbf{x}$ is $N(0, 1)$, then $\mathbf{x}^2$ is $\chi^2(1)$. From this and (7-101) it follows that if $\mathbf{y}$ is $t(r)$, then $\mathbf{y}^2$ is $F(1, r)$; hence,

$$2u - 1 = P\{|\mathbf{y}| \le t_u(r)\} = P\{\mathbf{y}^2 \le t_u^2(r)\} = P\{\mathbf{y}^2 \le F_{2u-1}(1, r)\}$$

and (7-111) results.

Note, finally, that if $\mathbf{w}$ is $\chi^2(r)$, then [see (7-81)]

$$E\left\{\frac{\mathbf{w}}{r}\right\} = 1 \qquad \operatorname{Var}\frac{\mathbf{w}}{r} = \frac{1}{r}$$

Hence, $\mathbf{w}/r \to 1$ as $r \to \infty$. This leads to the conclusion that the RV $\mathbf{x}$ in (7-106) tends to $\mathbf{z}/k$ as $r \to \infty$, and (7-112) follows because $\mathbf{z}$ is $\chi^2(k)$.

Appendix: Chi-Square Quadratic Forms

Consider the sum

$$\mathbf{Q} = \sum_{i,j=1}^{n} a_{ij}\mathbf{x}_i\mathbf{x}_j \tag{7A-1}$$

where $\mathbf{x}_i$ are n independent $N(0, 1)$ random variables and a_{ij} are n^2 numbers such that $a_{ij} = a_{ji}$. We shall examine the conditions that the numbers a_{ij} must satisfy so that the sum $\mathbf{Q}$ has a χ^2 distribution. For notational simplicity, we shall express our results in terms of the matrices

$$A = \begin{bmatrix} a_{11} & \cdots & a_{1n} \\ \cdots & \cdots & \cdots \\ a_{n1} & \cdots & a_{nn} \end{bmatrix} \qquad \mathbf{X} = \begin{bmatrix} \mathbf{x}_1 \\ \vdots \\ \mathbf{x}_n \end{bmatrix} \qquad \mathbf{X}^* = [\mathbf{x}_1, \ldots, \mathbf{x}_n]$$

Thus A is a symmetrical matrix, $\mathbf{X}$ is a column vector, and $\mathbf{X}^*$ is its transpose. With this notation,

$$\mathbf{Q} = \mathbf{X}^* A\mathbf{X}$$

We define the expected value of a matrix with elements $\mathbf{w}_{ij}$ as the matrix with elements $E\{\mathbf{w}_{ij}\}$. This yields

$$\mathbf{X}\mathbf{X}^* = \begin{bmatrix} \mathbf{x}_1\mathbf{x}_1 & \cdots & \mathbf{x}_1\mathbf{x}_n \\ \cdots & \cdots & \cdots \\ \mathbf{x}_n\mathbf{x}_1 & \cdots & \mathbf{x}_n\mathbf{x}_n \end{bmatrix}$$

$$E\{\mathbf{X}\mathbf{X}^*\} = \begin{bmatrix} 1 & \cdots & 0 \\ \cdots & \cdots & \cdots \\ 0 & \cdots & 1 \end{bmatrix} = 1_n \tag{7A-2}$$

where 1_n is the identity matrix. The last equation follows from the assumption that $E\{\mathbf{x}_i\mathbf{x}_j\} = 1$ if $i = j$ and 0 if $i \neq j$.

Diagonalization It is known from the properties of matrices that if A is a symmetric matrix with eigenvalues λ_i, we can find a matrix T such that the product TAT^* equals a diagonal matrix D:

$$TAT^* = D \qquad D = \begin{bmatrix} \lambda_1 & 0 & \cdots & 0 \\ 0 & \lambda_2 & \cdots & 0 \\ \cdots & \cdots & \cdots & \cdots \\ 0 & 0 & \cdots & \lambda_n \end{bmatrix} \tag{7A-3}$$

From this it follows that the matrix T is unitary; that is, T^* equals its inverse T^{-1}, and hence its determinant equals 1. Using this matrix, we form the vector

$$\mathbf{Z} = T\mathbf{X} = \begin{bmatrix} \mathbf{z}_1 \\ \vdots \\ \mathbf{z}_n \end{bmatrix}$$

We maintain that the components $\mathbf{z}_i$ of $\mathbf{Z}$ are normal, independent, with zero mean, and variance 1.

■ ***Proof.*** The RVS $\mathbf{z}_i$ are linear functions of the normal RVS $\mathbf{x}_i$; hence, they are normal. Furthermore, $E\{\mathbf{z}_i\} = 0$ because $E\{\mathbf{x}_i\} = 0$ by assumption. It suffices, therefore, to show that $E\{\mathbf{z}_i\mathbf{z}_j\} = 1$ for $i = j$ and 0 otherwise. Since

$$\mathbf{Z}^* = \mathbf{X}^*T^* \qquad \mathbf{ZZ}^* = T\mathbf{XX}^*T^* \qquad T^* = T^{-1}$$

we conclude that

$$E\{\mathbf{ZZ}^*\} = T\,E\{\mathbf{XX}^*\}T^* = TT^{-1} = 1_n \tag{7A-4}$$

Hence, the RVS $\mathbf{z}_i$ are i.i.d. and N(0, 1).

The following identity is an important consequence of the diagonalization of the matrix A. The quadratic form $\mathbf{Q}$ in (7A-1) can be written as a sum of squares of normal independent RVS. Indeed, since $\mathbf{X} = T^*\,\mathbf{Z}$ and $\mathbf{X}^* = \mathbf{Z}^*T$, (7A-3) yields

$$\mathbf{Q} = \mathbf{Z}^*\,TAT^*\mathbf{Z} = \mathbf{Z}^*\,D\mathbf{Z} = \sum_{i=1}^{n} \lambda_i \mathbf{z}_i^2 \tag{7A-5}$$

■ ***Fundamental Theorem.*** A quadratic form generated by a matrix A has a χ^2 distribution with r degrees of freedom iff r of the eigenvalues λ_i of A equal 1 and the other $n - r$ equal zero. Rearranging the order of all λ_i as necessary, we can state the theorem as follows:

$$\mathbf{Q} \text{ is } \chi^2(\mathrm{r}) \qquad \text{iff} \qquad \lambda_i = \begin{cases} 1 & 1 \le i \le r \\ 0 & \mathrm{r} < i \le n \end{cases} \tag{7A-6}$$

■ ***Proof.*** From (7A-5) it follows that if eigenvalues λ_i satisfy (7A-6), then

$$\mathbf{Q} = \sum_{i=1}^{r} \mathbf{z}_i^2 \tag{7A-7}$$

Hence [see (7-87)], $\mathbf{Q}$ is $\chi^2(r)$. Conversely, if $\lambda_i \ne 0$ or 1 for some i, then the corresponding term $\lambda_i\mathbf{z}_i^2$ is not χ^2, and the sum $\mathbf{Q}$ in (7A-5) is not χ^2.

The following consequence of (7A-7) justifies the term *degrees of freedom* used to characterize a χ^2 distribution. Suppose that $\mathbf{Q} = \Sigma\,\mathbf{w}_i^2$ where $\mathbf{w}_i$ are n RVS linearly dependent on $\mathbf{x}_i$. If $\mathbf{Q}$ is $\chi^2(r)$, then at most r of these RVS are linearly independent.

Noncentral Distributions

Consider the sum

$$\mathbf{Q}_0 = \mathbf{Q}_0(\mathbf{y}_i) = \sum_{i=1}^{n} \mathbf{y}_i^2 \tag{7A-8}$$

where $\mathbf{y}_i$ are n independent $N(\eta_i, 1)$ RVS. If $\eta_i \ne 0$ for some i, then the RV $\mathbf{Q}_0$ does not have a χ^2 distribution. Its distribution will be denoted by

$$\chi^2(n, e) \qquad \text{where } e = \sum_{i=1}^{n} \eta_i^2 = \mathbf{Q}_0(\eta_i) \tag{7A-9}$$

and will be called noncentral χ^2 with n degrees of freedom and *eccentricity* e. We shall determine its moment function $\Phi(s)$.

■ ***Theorem***

$$\Phi(s) = E\{e^{s\mathbf{Q}_0}\} = \frac{1}{\sqrt{(1-2s)^n}} \exp\left\{\frac{se}{1-2s}\right\} \tag{7A-10}$$

■ ***Proof.*** The moment function of $\mathbf{y}_i^2$ equals

$$E\{e^{s\mathbf{y}_i^2}\} = \frac{1}{\sqrt{2\pi}} \int_{-\infty}^{\infty} e^{sy_i^2} e^{-(y_i-\eta_i)^2/2}\, dy_i$$

The exponent of the integrand equals

$$sy_i^2 - \frac{(y_i-\eta_i)^2}{2} = \left(s - \frac{1}{2}\right)\left(y_i - \frac{\eta_i}{1-2s}\right)^2 + \frac{s\eta_i^2}{1-2s}$$

Hence [see (3A-1)],

$$\Phi_i(s) = E\{e^{s\mathbf{y}_i^2}\} = \frac{1}{\sqrt{1-2s}} \exp\left\{\frac{s\eta_i^2}{1-2s}\right\}$$

Since the RVS $\mathbf{y}_i$ are independent, we conclude from the convolution theorem (7-14) that $\Phi(s)$ is the product of the moment functions $\Phi_i(s)$; this yields (7A-10). From (7A-10) it follows that the $\chi^2(n, e)$ distribution depends not on η_i separately but only on the sum e of their squares.

■ ***Corollary***

$$E\{\mathbf{Q}_0\} = n + e \tag{7A-11}$$

■ ***Proof.*** This follows from the moment theorem (5-62). Differentiating (7A-10) with respect to s and setting $s = 0$, we obtain $\Phi'(0) = n + e$, and (7A-11) results.

Centering In (7A-6), we have established the χ^2 character of certain quadratic forms. In the following, we obtain a similar result for noncentral χ^2 distributions.

Using the RVS $\mathbf{y}_i$ in (7A-8), we form the sum

$$\mathbf{Q}_1(\mathbf{y}_i) = \sum_{i,j=1}^{n} a_{ij}\mathbf{y}_i\mathbf{y}_j$$

where a_{ij} are the elements of the quadratic form $\mathbf{Q}$ in (7A-1). We maintain that if $\mathbf{Q}$ is $\chi^2(r)$, then

$$\mathbf{Q}_1 \text{ is } \chi^2(r, e) \qquad \text{where } e = \mathbf{Q}_1(\eta_i) = \sum_{i,j=1}^{n} a_{ij}\eta_i\eta_j \tag{7A-12}$$

■ ***Proof.*** The RVS $\mathbf{x}_i = \mathbf{y}_i - \eta_i$ are i.i.d. as in (7A-1), and $\mathbf{Q}_1(y_i) = \mathbf{Q}(x_i + \eta_i)$. The RVS $\mathbf{z}_i$ in (7A-6) are linear functions of $\mathbf{x}_i$. A change, therefore, from $\mathbf{x}_i$ to $\mathbf{x}_i + \eta_i$ results in a change from $\mathbf{z}_i$ to $\mathbf{z}_i + \theta_i$ where $\theta_i = E\{\mathbf{z}_i\}$. From this and (7A-7) it follows that

$$\mathbf{Q}_1(\mathbf{y}_i) = \mathbf{Q}(\mathbf{x}_i + \eta_i) = \sum_{i=1}^{r} (\mathbf{z}_i + \theta_i)^2$$

This is a quadratic form as in (7A-8); hence, its distribution is $\chi^2(r, e)$ as in (7A-12).

Application We have shown in (7-97) that if $\bar{\mathbf{x}}$ is the sample mean of the RVS $\mathbf{x}_i$, the sum

$$\mathbf{Q} = \sum_{i=1}^{n} (\mathbf{x}_i - \bar{\mathbf{x}})^2 \text{ is } \chi^2(n-1)$$

With $\mathbf{y}_i = \mathbf{x}_i + \eta_i$, $\bar{\mathbf{y}} = \bar{\mathbf{x}} + \bar{\eta}$, $E\{\bar{\mathbf{y}}\} = \bar{\eta}$, it follows from (7A-12) that the sum

$$\mathbf{Q}_1 = \sum_{i=1}^{n} (\mathbf{y}_i - \bar{\mathbf{y}})^2 \text{ is } \chi^2(n-1, e) \qquad \text{with } e = \sum_{i=1}^{n} (\eta_i - \bar{\eta})^2 \quad \text{(7A-13)}$$

Note, finally, [see (7A-11)] that

$$E\{\mathbf{Q}_1\} = (n-1) + e \tag{7A-14}$$

Noncentral *t* and *F* Distributions Noncentral distributions are used in hypothesis testing to determine the operating characteristic function of various tests. In these applications, the following forms of noncentrality are relevant.

A noncentral t distribution with n degrees of freedom and eccentrity e is the distribution of the ratio

$$\frac{\mathbf{z}}{\sqrt{\mathbf{w}/m}}$$

where $\mathbf{z}$ and $\mathbf{w}$ are two independent RVS, $\mathbf{z}$ is $N(e, 1)$, and $\mathbf{w}$ is $\chi^2(n)$. This is an extension of (7-101).

A noncentral $F(k, r, e)$ distribution is the distribution of the ratio

$$\frac{\mathbf{x}/k}{\mathbf{w}/r}$$

where $\mathbf{z}$ and $\mathbf{w}$ are two independent RVS, $\mathbf{z}$ is noncentral $\chi^2(k, e)$, and $\mathbf{w}$ is $\chi^2(r)$. This is an extension of (7-106).

Problems

7-1 A matrix C with elements μ_{ij} is called nonnegative definite if $\sum_{i,j} c_i c_j \mu_{ij} \geq 0$ for every c_i and c_j. Show that if C is the covariance matrix of n RVS $\mathbf{x}_i$, it is nonnegative definite.

7-2 Show that if the RVS $\mathbf{x}$, $\mathbf{y}$, and $\mathbf{z}$ are such that $r_{xy} = r_{yz} = 1$, then $r_{xz} = 1$.

7-3 The RVS $\mathbf{x}_1$, $\mathbf{x}_2$, $\mathbf{x}_3$ are independent with densities $f_1(x_1)$, $f_2(x_2)$, $f_3(x_3)$. Show that the joint density of the RVS

$$\mathbf{y}_1 = \mathbf{x}_1 \qquad \mathbf{y}_2 = \mathbf{x}_2 - \mathbf{x}_1 \qquad \mathbf{y}_3 = \mathbf{x}_3 - \mathbf{x}_2$$

equals the product $f_1(y_1)f_2(y_2 - y_1)f_3(y_3 - y_2)$.

7-4 **(a)** Using (7-13), show that

$$\frac{\partial^4 \Phi(s_1, s_2, s_3, s_4)}{\partial s_1 \partial s_2 \partial s_3 \partial s_4} = E\{\mathbf{x}_1\mathbf{x}_2\mathbf{x}_3\mathbf{x}_4\}$$

(b) Show that if the RVS $\mathbf{x}_i$ are normal with zero mean and $E\{\mathbf{x}_i\mathbf{x}_j\} = \mu_{ij}$, then

$$E\{\mathbf{x}_1\mathbf{x}_2\mathbf{x}_3\mathbf{x}_4\} = \mu_{12}\mu_{34} + \mu_{13}\mu_{24} + \mu_{14}\mu_{23}$$

7-5 Show that if the RVs $\mathbf{x}_i$ are i.i.d. with sample mean $\bar{\mathbf{x}}$, then $E\{(\mathbf{x}_i - \bar{\mathbf{x}})^2\} = (n-1)\sigma^2/n$.

7-6 The length c of an object is measured three times. The first two measurements, made with instrument A, are 6.19 cm and 6.25 cm. The third measurement, made with instrument B, is 6.21 cm. The standard derivations of the instruments are 0.5 cm and 0.2 cm, respectively. Find the unbiased linear minimum variance estimate $\hat{c}$ of c.

7-7 Consider the sum [see also (7-39)]

$$\mathbf{z} = \sum_{k=1}^{\mathbf{n}} \mathbf{x}_k \qquad \text{where} \qquad P\{\mathbf{n} = k\} = p_k \qquad k = 1, 2, \cdots$$

and the RVs $\mathbf{x}_i$ are i.i.d. with density $f_x(x)$. Using moment functions, show that if $\mathbf{n}$ is independent of $\mathbf{x}_i$, then

$$f_z(z) = \sum_{k=1}^{\infty} p_k f_x^{(k)}(z)$$

where $f_x^{(k)}(x)$ is the convolution of $f_x(x)$ with itself k times.

7-8 A plumber services N customers. The probability that in a given month a particular customer needs service equals p. The duration of each service is an RV $\mathbf{x}_i$ with density $ce^{-cx}U(x)$. The total service time during the month is thus the random sum $\mathbf{z} = \mathbf{x}_1 + \cdots + \mathbf{x}_\mathbf{n}$ where $\mathbf{n}$ is an RV with binomial distribution. **(a)** Show that

$$f_z(z) = \sum_{n=0}^{N} \binom{N}{n} p^n q^{N-n} \frac{c^n}{(n-1)!} z^{n-1} e^{-cz}$$

(b) The plumber charges \$40 per hour. Find his mean monthly income if $c = 1/2$, $N = 1{,}000$ and $p = .1$.

7-9 The RVs $\mathbf{x}_i$ are i.i.d., and each is uniform in the interval (9, 11). Find and sketch the density of their sample mean $\bar{\mathbf{x}}$ for $n = 2$ and $n = 3$.

7-10 The RVs $\mathbf{x}_i$ are i.i.d., and $N(\eta, \sigma)$. Show that

$$\textbf{(a)}\ \text{if } \mathbf{y} = \frac{1}{n}\sqrt{\frac{\pi}{2}} \sum_{i=1}^{n} |\mathbf{x}_i - \eta| \qquad \text{then} \qquad E\{\mathbf{y}\} = \sigma \qquad \sigma_y^2 = \left(\frac{\pi}{2} - 1\right)\frac{\sigma^2}{n}$$

$$\textbf{(b)}\ \text{if } \mathbf{z} = \frac{1}{2(n-1)} \sum_{i=2}^{n} (\mathbf{x}_i - \mathbf{x}_{i-1})^2 \qquad \text{then} \qquad E\{\mathbf{z}\} = \sigma^2$$

7-11 The n RVs $\mathbf{x}_i$ are i.i.d. with distribution $F(x) = 1 - R(x)$ and $\mathbf{z} = \max \mathbf{x}_i$, $\mathbf{w} = \min \mathbf{x}_i$.

(a) Show that $\qquad F_z(z) = F^n(z) \qquad R_w(w) = R^n(w)$

(b) Show that if

$$f(x) = e^{-(x-\theta)}U(x-\theta) \qquad \text{then} \qquad F_w(w) = e^{-n(w-\theta)}U(w-\theta)$$

(c) Find $F_z(z)$ and $F_w(w)$ if $f(x) = ce^{-cx}U(x)$. **(d)** Find $F_z(z)$ and $F_w(w)$ if $\mathbf{x}$ has a geometric distribution as in (4-66).

7-12 The n RVs $\mathbf{x}_i$ are i.i.d. and uniformly distributed in the interval $c - 0.5 < x < c + 0.5$. Show that if $\mathbf{z} = \max \mathbf{x}_i$, $\mathbf{w} = \min \mathbf{x}_i$, then **(a)** $P\{\mathbf{w} < c < \mathbf{z}\} = 1 - 1/2^n$;

$$\textbf{(b)} \qquad f_{zw}(z, w) = \begin{cases} n(n-1)(z-w)^{n-2} & c - 0.5 < w < z < c + 0.5 \\ 0 & \text{otherwise} \end{cases}$$

7-13 Show that if $\mathbf{y}_k$ is the kth-order statistic of a sample of size n of $\mathbf{x}$ and $x_{.5}$ is the median of $\mathbf{x}$, then

$$P\{\mathbf{y}_k < x_{.5}\} = \frac{1}{2^n} \sum_{m=k}^{n} \binom{n}{m}$$

7-14 The height of children in a certain grade is an RV $\mathbf{x}$ uniformly distributed between 55 and 65 inches. We pick at random five students and line them up according to height. Find the distribution of the height $\mathbf{t}$ of the child in the center.

7-15 A well pipe consists of four sections. Each section is an RV $\mathbf{x}_i$ uniformly distributed between 9.9 and 10.1 feet. Using (7-64), find the probability that the length $\mathbf{x} = \mathbf{x}_1 + \mathbf{x}_2 + \mathbf{x}_3 + \mathbf{x}_4$ of the pipe is between 39.8 and 40.2 feet.

7-16 The RVs $\mathbf{x}_i$ are $N(0, \sigma)$ and independent. Using the CLT, show that if $\mathbf{y} = \mathbf{x}_1^2 + \cdots + \mathbf{x}_n^2$, then for large n,

$$f_y(y) = \frac{1}{\sigma^2\sqrt{4\pi n}} \exp\left\{-\frac{1}{4n\sigma^4}(y - n\sigma^2)^2\right\}$$

7-17 (Random walk) We toss a fair coin and take a step of size c to the right if heads shows, to the left if tails shows. At the nth toss, our position is an RV $\mathbf{x}_n$ taking the values mc where $m = n, n-2, \ldots, -n$.

(a) Show that $\quad P\{\mathbf{x}_n = mc\} = \binom{n}{k}\frac{1}{2^n} \qquad k = \frac{m+n}{2}$

(b) Show that for large n,

$$P\{\mathbf{x}_n = mc\} \simeq \frac{1}{\sqrt{n\pi/2}}\, e^{-m^2/2n}$$

(c) Find the probability $\{\mathbf{x}_{50} > 6c\}$ that at the 50th step, $\mathbf{x}_n$ will exceed $6c$.

7-18 (Proof of the CLT) Given n i.i.d. RVs $\mathbf{x}_i$ with zero mean and variance σ^2, we form their moment functions $\Phi(s) = E\{e^{s\mathbf{x}_i}\}$, $\Psi(s) = \ln \Phi(s)$. Show that (Taylor series)

$$\Psi(s) = \frac{\sigma^2}{2}s^2 + \text{higher power of } s$$

If $\mathbf{z} = (1/\sqrt{n})(\mathbf{x}_1 + \cdots + \mathbf{x}_n)$, then $\Phi_z(s) = \Phi^n(s/\sqrt{n})$

$$\Psi_z(s) = \frac{\sigma^2}{2}s^2 + \text{powers of } \frac{1}{\sqrt{n}}$$

Using the foregoing, show that $f_z(z)$ tends to an $N(0, \sigma)$ density as $n \to \infty$.

7-19 The n RVs $\mathbf{x}_i$ are i.i.d. with mean η and variance σ^2. Show that if $\bar{\mathbf{x}}$ is their sample mean and $g(x)$ is a differentiable function, then for large n, the RV $\mathbf{y} = g(\bar{\mathbf{x}})$ is nearly normal with $\eta_y = g'(\eta)$ and $\sigma_y = |g'(\eta)|\sigma/\sqrt{n}$.

7-20 The RVs $\mathbf{x}_i$ are positive, independent, and $\mathbf{y}$ equals their product:

$$\mathbf{y} = \mathbf{x}_1 \cdots \mathbf{x}_n$$

Show that for large n, the density of $\mathbf{y}$ is approximately *lognormal* (see Problem 4-22)

$$f_y(y) \simeq \frac{1}{cy\sqrt{2\pi}} \exp\left\{-\frac{1}{2c^2}(\ln y - b)\right\} U(y)$$

where b is the mean and c^2 the variance of the RV

$$\mathbf{z} = \ln \mathbf{y} = \sum \ln \mathbf{x}_i$$

This result leads to the following form of the central limit theorem: The distribution of the product $\mathbf{y}$ of n independent positive RVs $\mathbf{x}_i$ tends to a lognormal distribution as $n \to \infty$.

7-21 Show that if the RV $\mathbf{x}$ has a $\chi^2(7)$ distribution, then $E\{1/\mathbf{x}\} = 1/5$.

7-22 (Chi distribution) Show that if $\mathbf{x}$ has a $\chi^2(m)$ distribution and $\mathbf{y} = \sqrt{\mathbf{x}}$, then

$$f_y(y) = \gamma y^{n-1} e^{-y^2/2} U(y) \qquad \gamma = \frac{2}{2^{n/2}\Gamma(n/2)}$$

7-23 Show that if the RVs $\mathbf{x}_i$ are independent $N(\eta, \sigma)$ and

$$\mathbf{s}^2 = \frac{1}{n-1}\sum_{i=1}^{n}\left(\frac{x_i - \bar{x}}{\sigma}\right)^2 \qquad \text{then} \qquad E\{\mathbf{s}\} = \sqrt{\frac{2}{n-1}}\,\frac{\Gamma(n/2)\sigma}{\Gamma[(n-1)/2]}$$

7-24 The n RVs $\mathbf{x}_i$ i.i.d. with density $f(x) = ce^{-cx}U(x)$. Find the density of their sample mean $\bar{\mathbf{x}}$, and show that

$$E\left\{\frac{1}{\bar{\mathbf{x}}}\right\} = \frac{nc}{n-1} \qquad E\left\{\frac{1}{\bar{\mathbf{x}}^2}\right\} = \frac{n^2c^2}{(n-1)(n-2)}$$

7-25 The RVs $\mathbf{x}_i$ are $N(0, \sigma)$ and independent. Find the density of the sum $\mathbf{z} = \mathbf{x}_1^2 + \mathbf{x}_2^2 + \mathbf{x}_3^2$.

7-26 The components $\mathbf{v}_x$, $\mathbf{v}_y$, and $\mathbf{v}_z$ of the velocity $\mathbf{v} = \sqrt{\mathbf{v}_x^2 + \mathbf{v}_y^2 + \mathbf{v}_z^2}$ of a particle are three independent RVs with zero mean and variance $\sigma^2 = kT/m$. Show that $\mathbf{v}$ has a *Maxwell* distribution

$$f_v(v) = \frac{1}{\sigma^3}\sqrt{\frac{2}{\pi}}\, v^2 e^{-v^2/2\sigma^2} U(v)$$

7-27 Show that if the RVs $\mathbf{x}$ and $\mathbf{y}$ are independent, $\mathbf{x}$ is $N(0, 1/2)$ and $\mathbf{y}$ is $\chi^2(5)$, then the RV $\mathbf{z} = 4\mathbf{x}^2 + \mathbf{y}$ is $\chi^2(6)$.

7-28 The RVs $\mathbf{z}$ and $\mathbf{w}$ are independent, $\mathbf{z}$ is $N(0, \sigma)$, and $\mathbf{w}$ is $\chi^2(9)$. Show that if $\mathbf{x} = 3\mathbf{z}/\sqrt{\mathbf{w}}$, then $f_x(x) \sim 1/(9\sigma^2 + x^2)^5$.

7-29 The RVs $\mathbf{z}$ and $\mathbf{w}$ are independent, $\mathbf{z}$ is $\chi^2(4)$, and $\mathbf{w}$ is $\chi^2(6)$. Show that if $\mathbf{x} = \mathbf{z}/2\mathbf{w}$, then $f_x(x) \sim [x/(1 + 2x)^5]U(x)$.

7-30 The RVs $\mathbf{x}$ and $\mathbf{y}$ are independent, $\mathbf{x}$ is $N(\eta_x, \sigma)$, and $\mathbf{y}$ is $N(\eta_y, \sigma)$. We form their sample means $\bar{\mathbf{x}}$, $\bar{\mathbf{y}}$ and variances $\mathbf{s}_x^2$ and $\mathbf{s}_y^2$ using samples of length n and m, respectively. **(a)** Show that if $\hat{\boldsymbol{\sigma}}^2 = a\mathbf{s}_x^2 + b\mathbf{s}_y^2$ is the LMS unbiased estimator of σ^2, then

$$a = \frac{v_2}{v_1 + v_2} \qquad b = \frac{v_1}{v_1 + v_2} \qquad \text{where } v_1 = \frac{2\sigma^4}{n-1} \qquad v_2 = \frac{2\sigma^4}{m-1}$$

are the variances of $\mathbf{s}_x^2$ and $\mathbf{s}_y^2$, respectively. *Hint:* Use (7-38) and (7-97). **(b)** Show that if $\bar{\mathbf{w}} = \bar{\mathbf{x}} - \bar{\mathbf{y}}$, $\eta_{\bar{w}} = \eta_x - \eta_y$, and

$$\hat{\boldsymbol{\sigma}}_w^2 = (a\mathbf{s}_x^2 + b\mathbf{s}_y^2)\left(\frac{1}{n} + \frac{1}{m}\right)$$

then the RV $(\bar{\mathbf{w}} - \eta_{\bar{w}})/\hat{\boldsymbol{\sigma}}_{\bar{w}}$ has a t distribution with $n + m - 2$ degrees of freedom.

7-31 Show that the quadratic form $\mathbf{Q}$ in (7A-1) has a χ^2 distribution if the matrix A is *idempotent*, that is, if $A^2 = A$.

7-32 The RVs $\mathbf{x}$ and $\mathbf{y}$ are independent, and their distributions are noncentral $\chi^2(m, e_1)$ and $\chi^2(n, e_2)$, respectively. Show that their sum $\mathbf{z} = \mathbf{x} + \mathbf{y}$ has a noncentral $\chi^2(m + n, e_1 + e_2)$ distribution.

PART TWO

STATISTICS

8 The Meaning of Statistics

Statistics is part of the theory of probability relating theoretical concepts to reality. As in all probabilistic predictions, statistical statements concerning the real world are only inferences. However, statistical inferences can be accepted as near certainties because they are based on probabilities that are close to 1. This involves a change from statements dealing with an arbitrary space $\mathcal{S}$ to statements involving the space $\mathcal{S}_n$ of repeated trials. In this chapter, we present the underlying reasoning. In Section 8-2, we introduce the basic areas of statistics, including estimation, hypothesis testing, Bayesian statistics, and entropy. In the final section, 8-3, we comment on the interaction of computers and statistics. We explain the meaning of random numbers and their computer generation and conclude with a brief comment on the significance of simulation in Monte Carlo methods.

8-1 Introduction

Probability is a mathematical discipline developed on the basis of an abstract model, and its conclusions are *deductions* based on the axioms. Statistics deals with the applications of the theory to real problems, and its conclusions are *inferences* based on observations. Statistics consists of two parts: analysis and design. *Analysis*, or mathematical statistics, is part of the theory of probability involving RVs generated mainly by repeated trials. A major task of analysis is the construction of events the probability of which is close to 0 or to 1. As we shall see, this leads to inferences that can be accepted as near certainties. *Design*, or applied statistics, deals with the selection of analytical methods that are best suited to particular problems and with the construction of experiments that can be adequately described by theoretical models. This book covers only mathematical statistics.

The connection between probabilistic concepts and reality is based on the empirical formula

$$n_{\mathscr{A}} \simeq np \tag{8-1}$$

relating the probability $p = P(\mathscr{A})$ of an event $\mathscr{A}$ to the number $n_{\mathscr{A}}$ of successes of $\mathscr{A}$ in n trials of the underlying physical experiment $\mathscr{S}$. This formula can be used to *estimate* the model parameter p in terms of the observed number $n_{\mathscr{A}}$. If p is known, it can be used to *predict* the number $n_{\mathscr{A}}$ of successes of $\mathscr{A}$ in n future trials. Thus (8-1) can be viewed as a rudimentary form of statistical analysis: The ratio

$$\hat{p} = \frac{n_{\mathscr{A}}}{n} \tag{8-2}$$

is the point estimate of the parameter p.

Suppose that $\mathscr{S}$ is the polling experiment and $\mathscr{A}$ is the event {Republican}. We question 620 voters and find that 279 voted Republican. We then conclude that $\hat{p} = 279/620 = .45$. Using this estimate, we predict that 45% of all voters are Republican.

Formula (8-1) is only an approximation. A major objective of statistics is to replace it by an exact statement about the value $n_{\mathscr{A}} - np$ of the approximation error. Since p is a model parameter, to find such a statement we must interpret also the numbers n and $n_{\mathscr{A}}$ as model parameters. For this purpose, we form the product space

$$\mathscr{S}_n = \mathscr{S} \times \cdots \times \mathscr{S}$$

consisting of the n repetitions of the experiment $\mathscr{S}$ (see Section 3-1), and we denote by $n_{\mathscr{A}}$ the number of successes of $\mathscr{A}$. We next form the set

$$\mathscr{B} = \{np - 3\sqrt{npq} < n_{\mathscr{A}} < np + 3\sqrt{npq}\} \tag{8-3}$$

This set is an event in the space $\mathscr{S}_n$, and its probability equals [see (7-69)]

$$P\{np - 3\sqrt{npq} < n_{\mathscr{A}} < np + 3\sqrt{npq}\} = .997 \tag{8-4}$$

We can thus claim with probability .997 that $n_{\mathscr{A}}$ will be in the interval $np \pm 3\sqrt{npq}$. This is an interval estimate of $n_{\mathscr{A}}$.

We shall use (8-4) to obtain an interval estimate of p in terms of its point estimate $\hat{p} = n_{\mathscr{A}}/n$. From (8-4) it follows with simple algebra that

$$P\{(p - \hat{p})^2 < \frac{9}{n} p(1 - p)\} = .997$$

Denoting by p_1 and p_2 the roots of the quadratic

$$(p - \hat{p})^2 = \frac{9}{n} p(1 - p) \tag{8-5}$$

we conclude from (8-4) that

$$P\{p_1 < \hat{p} < p_2\} = .997 \tag{8-6}$$

We can thus claim with probability .997 that the unknown parameter p is in the interval (p_1, p_2). Thus using statistical analysis, we replaced the empirical *point* estimate (8-2) of p by the precise *interval* estimate (8-6).

In the polling experiment, $n = 620$, $\hat{p} = .45$, and (8-5) yields

$$(p - .45)^2 = \frac{9}{620} p(1 - p) \qquad p_1 = .39 \qquad p_2 = .51$$

Note In news broadcasts, this result is phrased as follows: "A poll showed that forty-five percent of all voters are Republican; the margin of error is ±6%." The number ±6 is the difference from 45 to the endpoints 39 and 51 of the interval (39, 51). The fact that the result is correct with probability .997 is not mentioned.

The change from the empirical formula (8-1) to the precise formula (8-4) did not solve the problem of relating p to real quantities. Since $P(\mathcal{B})$ is also a model concept, its relationship to the real world must again be based on (8-1). This relationship now takes the following form. If we repeat the experiment $\mathcal{S}_n$ a large number of times, in 99.7% of these cases, the number $n_{\mathcal{A}}$ of success of $\mathcal{A}$ will be in the interval $np \pm 3\sqrt{npq}$. There is, however, a basic difference between this and (8-1). Unlike $P(\mathcal{A})$, which could be any number between 0 and 1, the probability of the event $\mathcal{B}$ is almost 1. We can therefore expect with near certainty that the event $\mathcal{B}$ will occur in a *single* performance of the experiment $\mathcal{S}_n$. Thus the change from the event $\mathcal{A}$ with an arbitrary probability to an event $\mathcal{B}$ with $P(\mathcal{B}) \simeq 1$ leads to a conclusion that can be accepted as near certainty.

The foregoing observations are relevant, although not explicitly stated, in most applications of statistics. We rephrase them next in the context of statistical inference.

Statistics and Induction

Suppose that we know from past observations the probability $P(\mathcal{M})$ of an event $\mathcal{M}$. What conclusion can we draw about the occurrence of this event in a *single performance* of the underlying experiment? We shall answer this question in two ways, depending on the size of $P(\mathcal{M})$. We shall give one answer if $P(\mathcal{M})$ is a number distinctly different from 0 or 1—for example, .6—and a different answer if $P(\mathcal{M})$ is close to 0 or 1—for example, .997. Although the boundary between the two probabilities is not sharply defined (.9 or .9999?), the answers are fundamentally different.

Case 1 We assume, first, that $P(\mathcal{M}) = .6$. In this case, the number .6 gives us only some degree of confidence that the event $\mathcal{M}$ will occur. Thus the known probability is used merely as a measure of our state of knowledge about the possible occurrence of $\mathcal{M}$. This interpretation of $P(\mathcal{M})$ is subjective and cannot be verified experimentally. At the next trial, the event $\mathcal{M}$ will either occur or not occur. If it does not, we will not question the validity of the assumption that $P(\mathcal{M}) = .6$.

Case 2 Suppose, next, that $P(\mathcal{M}) = .997$. Since .997 is close to 1, we expect with near certainty that the event $\mathcal{M}$ will occur. If it does not occur, we shall question the assumption that $P(\mathcal{M}) = .997$.

Mathematical statistics can be used to change case 1 to case 2. Suppose that $\mathcal{A}$ is an event with $P(\mathcal{A}) = .45$. As we have noted, no reliable prediction about $\mathcal{A}$ in a single performance of $\mathcal{S}$ is possible (case 1). In the space $\mathcal{S}_n$ of 620 repetitions of $\mathcal{S}$, the set $\mathcal{B} = \{242 < n_{\mathcal{A}} < 316\}$ is an event with $P(\mathcal{B}) = .997 \simeq 1$. Hence (case 2), we can predict with near certainty that if we perform $\mathcal{S}_n$ once, that is, if $\mathcal{S}$ is repeated 620 times, the number $n_{\mathcal{A}}$ of successes of $\mathcal{A}$ will be between 242 and 316. We have thus changed subjective knowledge about the occurrence of $\mathcal{A}$ based on the *given* information that $P(\mathcal{A}) = .45$ to an objective prediction about $\mathcal{B}$ based on the *derived* probability that $P(\mathcal{B}) \simeq .997$. Note that both conclusions are inductive inferences. The difference between the two, although significant, is only quantitative. As in case 1, the conclusion that $\mathcal{B}$ will occur at a single performance of the experiment $\mathcal{S}_n$ is not a logical certainty but only an inference. In the last analysis, no prediction about the future can be accepted as logical necessity.

8-2 The Major Areas of Statistics

We introduce next the major areas of statistics stressing basic concepts. We shall make frequent use of the following:

Percentiles The u-percentile of an RV $\mathbf{x}$ is a number x_u such that $u = F(x_u)$. The percentiles of the normal, the χ^2, the t, and the F distributions are listed in the tables in the appendix. The u-percentile of the $N(0, 1)$ distribution is denoted by z_u. If $\mathbf{x}$ is $N(\eta, \sigma)$, then

$$x_u = \eta + z_u\sigma \qquad \text{because } F(x) = G\left(\frac{x - \eta}{\sigma}\right)$$

With $\gamma = 1 - \delta$ a given constant, it follows that (see Fig. 8.1)

$$P\{\eta - z_{1-\delta/2}\sigma < \mathbf{x} < \eta + z_{1-\delta/2}\sigma\} = \gamma \tag{8-7}$$

Sample Mean Consider an RV $\mathbf{x}$ with density $f(x)$, defined on an experiment $\mathcal{S}$. A *sample* of $\mathbf{x}$ of length n is a sequence of n independent and

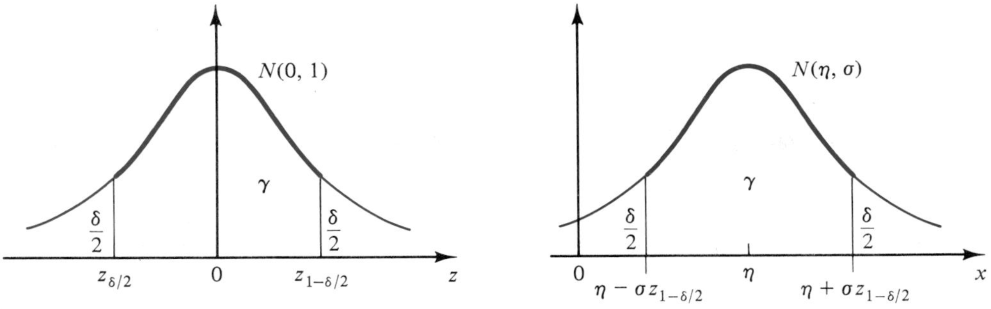

Figure 8.1

identically distributed (i.i.d.) RVS $\mathbf{x}_1, \ldots, \mathbf{x}_n$ with density $f(x)$ defined on the sample space $\mathcal{S}_n = \mathcal{S} \times \cdots \times \mathcal{S}$ as in Section 7-1. The arithmetic mean $\bar{\mathbf{x}}$ of $\mathbf{x}_i$ is called the *sample mean* of $\mathbf{x}$. Thus [see (7-30)]

$$\bar{\mathbf{x}} = \frac{1}{n}\sum_{i=1}^{n}\mathbf{x}_i \qquad \eta_{\bar{x}} = \eta_x \qquad \sigma_{\bar{x}}^2 = \frac{\sigma_x^2}{n} \tag{8-8}$$

If we perform the corresponding physical experiment n times, we obtain the n numbers $x_1, \ldots, x_n$. These numbers are the values $x_i = \mathbf{x}_i(\zeta)$ of the samples $\mathbf{x}_i$. They will be called *observations*; their arithmetic mean $\bar{x} = \bar{\mathbf{x}}(\zeta)$ is the *observed sample mean*.

From the CLT theorem (7-64) it follows that for large n, the RV $\bar{\mathbf{x}}$ is approximately normal. In fact, if $f(x)$ is smooth, this is true for n as small as 10 or even less. Applying (8-7) to the RV $\bar{\mathbf{x}}$, we obtain the basic formula

$$P\left\{\eta - \frac{z_{1-\delta/2}\sigma}{\sqrt{n}} < \bar{\mathbf{x}} < \eta + \frac{z_{1-\delta/2}\sigma}{\sqrt{n}}\right\} = \gamma = 1 - \delta \tag{8-9}$$

Estimation

Estimation is the most important topic in statistics. In fact, the underlying ideas form the basis of most statistical investigations. We introduce this central topic by means of specific illustrations.

We wish to measure the diameter θ of a rod. The results of the measurements are the values of the RV $\mathbf{x} = \theta + \boldsymbol{\nu}$ where $\boldsymbol{\nu}$ is the measurement error. We know from past observations that $\boldsymbol{\nu}$ is a normal RV with 0 mean and known variance. Thus $\mathbf{x}$ is an $N(\theta, \sigma)$ RV with known σ, and our task is to find θ. This is a problem in *parameter estimation*: The distribution of the RV $\mathbf{x}$ is a function $F(x, \theta)$ of known form depending on an unknown parameter θ. The problem is to estimate θ in terms of one or more observations (measurements) of the RV $\mathbf{x}$.

We buy an electric motor. We are told that its life length is a normal RV $\mathbf{x}$ with known η and σ. On the basis of this information, we wish to estimate the life length of our motor. This is a problem in *prediction*. The distribution $F(x)$ of the RV $\mathbf{x}$ is completely known, and our problem is to predict its value $x = \mathbf{x}(\zeta)$ at a single performance of the underlying experiment (the life length of a specific motor).

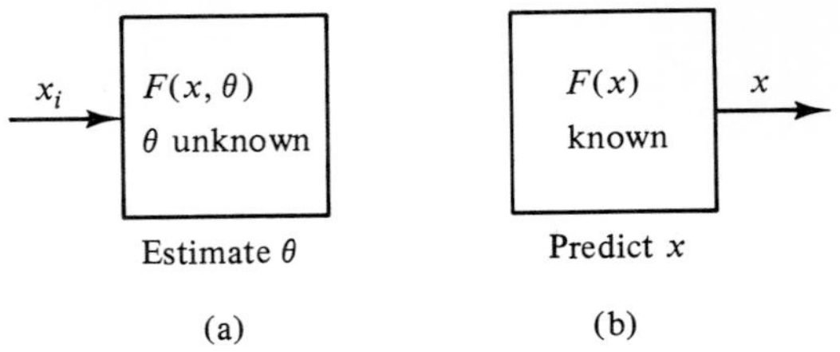

Figure 8.2

In both examples, we deal with a *classical* esimation problem. In the first example, the unknown θ is a model parameter, and the data are observations of real experiments. We thus proceed from the observations to the model (Fig. 8.2a). In the second example, the model is completely known, and the number to be estimated is the value x of a real quantity. In this case, we proceed from the model to the observation (Fig. 8.2b). We continue with the analysis.

PREDICTION We are given an RV $\mathbf{x}$ with known density $f(x)$, and we wish to predict its value $x = \mathbf{x}(\zeta)$ at the next trial. Clearly, x can be any number in the range of $\mathbf{x}$, hence, it cannot be predicted; it can only be estimated. Thus our problem is to find a constant c such as to minimze in a sense some function $L(\mathbf{x} - c)$ of the estimation error $\mathbf{x} - c$. This problem was considered in Section 6-3. We have shown that if our criterion is the minimization of the MS error $E\{(\mathbf{x} - c)^2\}$, then c equals the mean η_x of $\mathbf{x}$; if it is the minimization of $E\{|\mathbf{x} - c|\}$, then c equals the median $x_{.5}$ of $\mathbf{x}$. The constant c so obtained is a *point estimate* of x. A point estimate is used if we wish to find a number c that is close to x on the average. In certain applications, we are interested only in a single value of the RV $\mathbf{x}$. Our problem then is to find two tolerance limits c_1 and c_2 for the unknown x. For example, we buy a door from a factory and wish to know with reasonable certainty that its length x is between c_1 and c_2. For our purpose, the minimization of the average prediction error is not a relevant objective. Our objective is to find not a point estimate but an *interval estimate* of x.

To solve this problem, we select a number γ close to 1 and determine the constants c_1 and c_2 such as to minimize the length $c_2 - c_1$ of the interval (c_1, c_2) subject to the condition that

$$P\{c_1 < \mathbf{x} < c_2\} = \gamma \tag{8-10}$$

If this condition is satisfied and we predict that x will be in the interval (c_1, c_2), then our prediction is correct in $100\gamma\%$ of the cases. We shall find c_1 and c_2 under the assumption that the density $f(x)$ is unimodal (has a single maximum). Suppose, first, that $f(x)$ is also symmetrical about its mode $x_{\max}$ as in Fig. 8.3. In this case, $x_{\max} = \eta$ and $c_2 - c_1$ is minimum if $c_1 = \eta - a$, $c_2 = \eta + a$ where a is a constant that can be determined from (8-10). As we see from Fig. 8.3a,

$$a = \eta - x_{\delta/2} = x_{1-\delta/2} - \eta \qquad \delta = 1 - \gamma \tag{8-11}$$

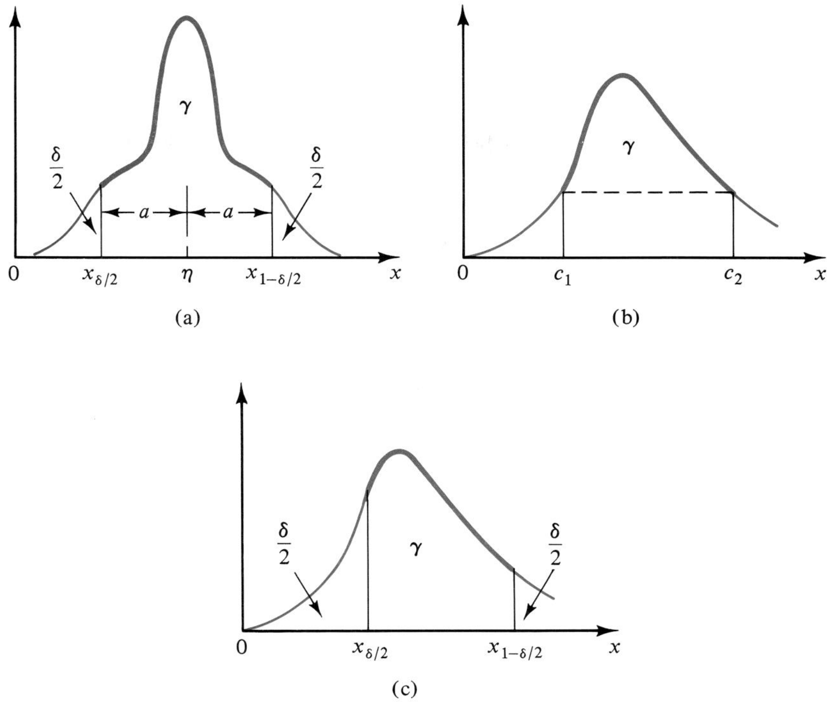

Figure 8.3

For an arbitrary unimodal density, $c_2 - c_1$ is minimum if $f(c_1) = f(c_2)$ as in Fig. 8.3b. (Problem 9-9) This condition, combined with (8-10), leads by trial and error to a unique determination of the constants c_1 and c_2. For computational simplicity we shall use the constants (Fig. 8.3c)

$$c_1 = x_{\delta/2} \qquad c_2 = x_{1-\delta/2} \tag{8-12}$$

For asymmetrical densities, the length of the resulting interval (c_1, c_2) is no longer minimum. The constant $\gamma = 1 - \delta$ is called the *confidence coefficient* and the tolerance interval (c_1, c_2) the γ *confidence interval* of the prediction.

The value of γ is dictated by two conflicting requirements: If γ is close to 1, the estimate is reliable but the size $c_2 - c_1$ of the confidence interval is large; if γ is reduced, $c_2 - c_1$ is reduced but the estimate is less reliable. The commonly used values of γ are .9, .95, .99, and .999.

Example 8.1

The life length of tires of a certain type is a normal RV with $\eta = 25{,}000$ miles and $\sigma = 3{,}000$ miles. We buy a set of such tires and wish to find the .95-confidence interval of their life.

From the normality assumption it follows that

$$P\{\eta - a < \mathbf{x} < \eta + a\} = 2G\left(\frac{a}{\sigma}\right) - 1 = .95$$

This yields $G(a/\sigma) = .975$, $a/\sigma = z_{.975}$, $a = z_{.975}\sigma \simeq 2\sigma$; hence,

$$P\{19{,}000 < \mathbf{x} < 31{,}000\} = .95$$

Since .95 is close to 1, we expect with reasonable confidence that in a single performance of the experiment, the event $\{\eta - a < \mathbf{x} < \eta + a\}$ will occur; that is, the life length of our tires will be between 19,000 and 31,000 miles. ■

PARAMETER ESTIMATION In the foregoing discussion, we used statistics to change empirical statements to exact formulas involving events with probabilities close to 1. We show next that the underlying ideas can be applied to parameter estimation. To be concrete, we shall consider the problem of estimating the mean η of an RV with known variance σ^2. An example is the measurement problem.

The empirical estimate of the mean $\eta = E\{\mathbf{x}\}$ of an RV $\mathbf{x}$ is its observed sample mean $\bar{x}$ [see (4-90)]:

$$\eta \simeq \bar{x} = \frac{1}{n}\sum_{i=1}^{n} x_i \tag{8-13}$$

This is an approximate formula relating the model concept η to the real observations x_i. We shall replace it by a precise probabilistic statement involving the approximation error $\bar{x} - \eta$. For this purpose, we form the sample mean $\bar{\mathbf{x}}$ of $\mathbf{x}$ and the event $\mathcal{B} = \{\eta - a < \bar{\mathbf{x}} < \eta + a\}$ where a is a constant that can be expressed in terms of the probability of $\mathcal{B}$. Assuming that $\bar{\mathbf{x}}$ is normal, we conclude as in (8-9) that if $P(\mathcal{B}) = \gamma = 1 - \delta$, then $a = z_{1-\delta/2}\sigma/\sqrt{n}$. This yields

$$P\left\{\eta - \frac{z_{1-\delta/2}\sigma}{\sqrt{n}} < \bar{\mathbf{x}} < \eta + \frac{z_{1-\delta/2}\sigma}{\sqrt{n}}\right\} = \gamma \tag{8-14}$$

or, equivalently,

$$P\left\{\bar{\mathbf{x}} - \frac{z_{1-\delta/2}\sigma}{\sqrt{n}} < \eta < \bar{\mathbf{x}} + \frac{z_{1-\delta/2}\sigma}{\sqrt{n}}\right\} = \gamma \tag{8-15}$$

We have thus replaced the empirical statement (8-13) by the exact probabilistic equation (8-15). This equation leads to the following *interval estimate* of the parameter η: The probability that η is in the interval

$$\bar{x} \pm \frac{z_u\sigma}{\sqrt{n}} \qquad u = 1 - \frac{\delta}{2}$$

equals $\gamma = 1 - \delta$. As in the prediction problem, γ is the confidence coefficient of the estimate.

Note that equations (8-14) and (8-15) are equivalent; however, their statistical interpretations are different. The first is a *prediction:* It states that if we predict that $\bar{\mathbf{x}}$ will be in the *fixed* interval $\eta \pm z_u\sigma/\sqrt{n}$, our prediction will be correct in $100\gamma\%$ of the cases. The second is an *estimation:* It states that if we estimate that the unknown number η is in the *random* interval $\bar{\mathbf{x}} \pm$

$z_u\sigma/\sqrt{n}$, our estimation will be correct in $100\gamma\%$ of the cases. We thus conclude that if γ is close to 1, we can claim with near certainty that η is in the interval $\bar{x} \pm z_u\sigma/\sqrt{n}$. This claim involves the average $\bar{x}$ of the n observations x_i in a single performance of the experiment $\mathcal{S}_n$. Using statistics, we have as in (8-4) reached a conclusion that can be accepted as near certainty.

Example 8.2

We wish to estimate the diameter η of a rod using a measuring instrument with zero mean error and standard deviation $\sigma = 1$ mm. We measure the rod 64 times and find that the average of the measurements equals 40.3 mm. Find the .95-confidence interval of η.

In this problem

$$\gamma = .95 \qquad \frac{1-\delta}{2} = .975 \qquad z_{.975} \simeq 2 \qquad n = 64 \qquad \sigma = 1$$

Inserting into (8-15), we obtain the confidence interval 40.3 ± 0.25 mm. ■

Hypothesis Testing

Hypothesis testing is an important area of *decision theory* based on statistical considerations. Does smoking decrease life expectancy? Do IQ scores depend on parental education? Is drug A more effective than drug B? The investigation starts with an assumption about the values of one or more parameters of a probabilistic model $\mathcal{S}_0$. Various factors of the underlying physical experiment are modified, and the problem is to decide whether these modifications cause changes in the model parameters, thereby generating a new model $\mathcal{S}_1$. Suppose that the mean blood pressure of patients treated with drug A is η_0. We change from drug A to drug B and wish to decide whether this results in a decrease of the mean blood pressure.

We shall introduce the underlying ideas in the context of the following problem. The mean cholesterol count of patients in a certain group is 240 and the standard deviation equals 32. A manufacturer introduces a new drug with the claim that it decreases the mean count from 240 to 228. To test the claim, we treat 64 patients with the new drug and observe that the resulting average count is reduced to 230. Should we accept the claim?

In terms of RVs, this problem can be phrased as follows: We assume that the distribution of an RV $\mathbf{x}$ is either a function $F_0(x)$ with $\eta_0 = 240$ and $\sigma_0 = \sigma = 32$ or a function $F_1(x)$ with $\eta_1 = 228$ and $\sigma_1 = \sigma = 32$. The first assumption is denoted by H_0 and is called the *null hypothesis*; the second is denoted by H_1 and is called the *alternative hypothesis*. In the drug problem, H_0 is the hypothesis that the treatment has no effect on the count. Our task is to establish whether the evidence supports the alternative hypothesis H_1. As we have noted, the sample mean $\bar{\mathbf{x}}$ of $\mathbf{x}$ is a normal RV with variance $\sigma/\sqrt{n} = 4$. Its mean equals $\eta_0 = 240$ if H_0 is true and $\eta_1 = 228$ if H_1 is true. In Fig. 8.4, we show the density of $\bar{\mathbf{x}}$ for each case. The values of $\bar{\mathbf{x}}$ are concentrated near its mean. It is reasonable, therefore, to reject H_0 iff $\bar{x}$ is to the left of some constant c. This leads to the following test: Reject the null hypothesis

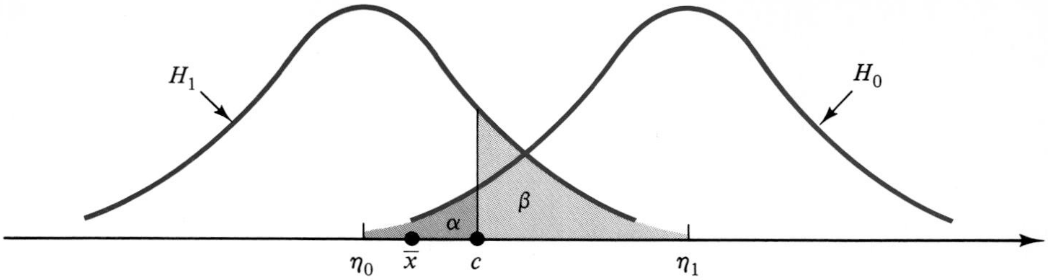

Figure 8.4

iff $\bar{x} < c$. The region $x < c$ of rejection of H_0 is denoted by R_c and is called the *critical region* of the test. To complete the test, we must specify the constant c. To do so, we shall examine the nature of the resulting errors.

Suppose, first, that H_0 is true. If $\bar{x}$ is in the critical region, that is, if $\bar{x} < c$, we reject H_0 even though it is true. Our decision is thus wrong. We then say that we committed a *Type I error*. The probability for such an error is denoted by α and is called Type I error probability or *significance level* of the test. Thus

$$\alpha = P\{\bar{\mathbf{x}} < c | H_0\} = \int_{-\infty}^{c} f_{\bar{x}}(x, \eta_0)\, dx \qquad \text{(8-16)}†$$

Suppose next that H_1 is true. If $\bar{x} > c$, we do not reject H_0 even though H_1 is true. Our decision is wrong. In this case, we say that we committed a *Type II error*. The probability for such an error is denoted by β and is called Type II error probability. The difference $P = 1 - \beta$ is called the *power* of the test. Thus

$$\beta = P\{\bar{\mathbf{x}} > c | H_1\} = \int_{c}^{\infty} f_{\bar{x}}(x, \eta_1)\, dx \qquad \text{(8-17)}$$

For a satisfactory test, it is desirable to keep both errors small. This is not, however, possible because if c moves to the left, α decreases but β increases; if c moves to the right, β decreases but α increases. Of the two errors, the control of α is more important.

Note In general, the purpose of a test is to examine whether the available evidence supports the rejection of the null hypothesis; it is not to establish whether H_0 is true. If $\bar{x}$ is in the critical region R_c, we reject H_0. However, if $\bar{x}$ is not in R_c, we do not conclude that H_0 is true. We conclude merely that the evidence does not justify rejection of H_0. Let us clarify with a simple example. We wish to examine whether a coin is loaded. To do so, we toss it 100 times and observe that heads shows k times. If $k = 15$, we reject the null hypothesis because 15 is much smaller than 50. If $k = 49$, we conclude that the evidence does not support the rejection of the fair-coin hypothesis. The evidence, how-

† The expression $P\{\bar{\mathbf{x}} < c | H_0\}$ is not a conditional probability. It is the probability that $\bar{\mathbf{x}} < c$ under the assumption that H_0 is true.

ever, does not lead to the conclusion that the coin is fair. We could have as well concluded that $p = .49$.

To carry out a test, we select α and determine c from (8-16). This yields

$$\alpha = F_{\bar{x}}(c, \eta_0) = G\left(\frac{c - \eta_0}{\sigma/\sqrt{n}}\right) \qquad c = \eta_0 + \frac{z_\alpha \sigma}{\sqrt{n}} \tag{8-18}$$

The resulting β is obtained from (8-17):

$$\beta = 1 - F_{\bar{x}}(c, \eta_1) = 1 - G\left(\frac{c - \eta_1}{\sigma/\sqrt{n}}\right) \tag{8-19}$$

Example 8.3

In the drug problem,

$$\eta_0 = 240 \qquad \eta_1 = 228 \qquad \sigma = 32 \qquad n = 64 \qquad \bar{x} = 230$$

We wish to test the hypothesis H_0 that the new drug is not effective, with significance level .05. In this case,

$$\alpha = .05 \qquad z_\alpha = -1.645 \qquad c = 233.4 \qquad \beta = 1 - G(1.35) = .089$$

Since $\bar{x} = 230 < c$, we reject the null hypothesis; the new drug is recommended. The power of the test is $P = 1 - \beta \simeq .911$ ■

Fundamental Note There is a basic conceptual difference between parameter estimation and hypothesis testing, although the underlying analysis is the same. In parameter estimation, we have a *single model* and use the observations to estimate its parameters. Our estimate involves only precise statistical considerations. In hypothesis testing, we have *two models:* A model $\mathcal{S}_0$ representing the null hypothesis and a model $\mathcal{S}_1$ representing the alternative hypothesis (Fig. 8.5). We start with the assumption that $\mathcal{S}_0$ is the correct model and use the observations to *decide* whether this assumption must be rejected. Our decision is not based on statistical considerations alone. Mathematical statistics leads only to the following statements:

$$\begin{aligned} &\text{If } \mathcal{S}_0 \text{ is the true model, then } P\{\bar{\mathbf{x}} > c\} = \alpha \\ &\text{If } \mathcal{S}_1 \text{ is the true model, then } P\{\bar{\mathbf{x}} < c\} = \beta \end{aligned} \tag{8-20}$$

These statements do not—indeed, cannot—lead to a decision. A decision involves the selection of the critical region and is based on other considerations, often subjective, that are outside the scope of mathematical statistics.

Figure 8.5

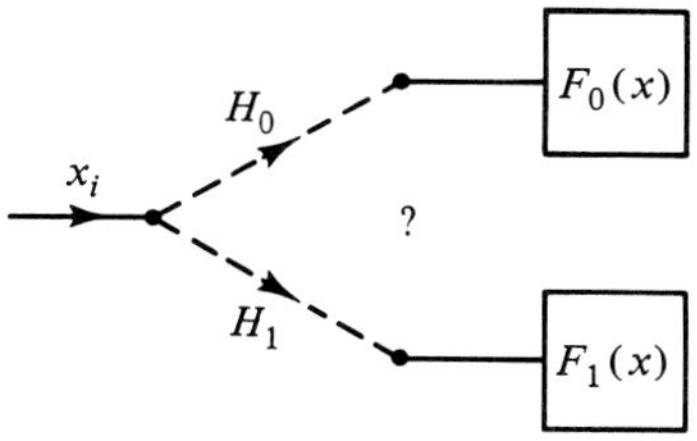

Hypothesis testing

Bayesian Statistics

In the *classical* approach to the estimation problem, the parameter θ of a distribution $F(x, \theta)$ is viewed as an unknown *constant*. In this approach, the estimate of θ is based solely on the observed values x_i of the RV $\mathbf{x}$. In certain applications, θ is not totally unknown. If, for example, θ is the probability of heads in the coin experiment, we expect that its possible values are close to .5 because most coins are fair. In *Bayesian* statistics, the available prior information about θ is used in the estimation process. In this approach, the unknown parameter θ is viewed as the value of a *random variable* $\boldsymbol{\theta}$, and the distribution of $\mathbf{x}$ is interpreted as the conditional distribution $F_x(x|\theta)$ of $\mathbf{x}$ assuming $\boldsymbol{\theta} = \theta$. The prior information is used to assign somehow a density $f_\theta(\theta)$ to the RV $\boldsymbol{\theta}$, and the problem is to estimate the value θ of $\boldsymbol{\theta}$ in terms of the observed value x of $\mathbf{x}$ and the density $f_\theta(\theta)$ of $\boldsymbol{\theta}$. The problem of estimating the unknown parameter θ is thus changed to the problem of estimating the value θ of an RV $\boldsymbol{\theta}$. In other words, in Bayesian statistics, estimation is changed to prediction.

We shall illustrate with the measurement problem. We measure a rod of diameter θ; the results are the values $x_i = \theta + \nu_i$ of the sum $\theta + \boldsymbol{\nu}$ where $\boldsymbol{\nu}$ is the measurement error. We wish to estimate θ. If we interpret θ as an unknown number, we have a classical estimation problem. Suppose, however, that the rod is picked from a production line. In this case, its diameter θ can be interpreted as the value of an RV $\boldsymbol{\theta}$ modeling the diameters of all rods. This is now a problem in Bayesian estimation involving the RVs $\boldsymbol{\theta}$ and $\mathbf{x} = \boldsymbol{\theta} + \boldsymbol{\nu}$. As we have seen in Section 6-3, the LMS estimate $\hat{\theta}$ of θ is the regression line

$$\hat{\theta} = E\{\boldsymbol{\theta}|x\} = \int_{-\infty}^{\infty} \theta f_\theta(\theta|x)\, d\theta \tag{8-21}$$

Here $\hat{\theta}$ is the point estimate of our rod and x is its measured value. To find $\hat{\theta}$, it suffices to find the conditional density $f_\theta(\theta|x)$ of $\boldsymbol{\theta}$ assuming $\mathbf{x} = x$. As we know [see (6-32)],

$$f_\theta(\theta|x) = \frac{f_x(x|\theta)}{f_x(x)} f_\theta(\theta) \tag{8-22}$$

The function $f_\theta(\theta)$ is the unconditional density of $\boldsymbol{\theta}$, which we assume known. This function is called *prior* (before the measurement). The conditional density $f_\theta(\theta|x)$ is called *posterior* (after the measurement). The conditional density $f_x(x|\theta)$ is assumed known. In the measurement problem, it can be expressed in terms of the density $f_\nu(\nu)$ of the error. Indeed, if $\boldsymbol{\theta} = \theta$, then $\mathbf{x} = \theta + \boldsymbol{\nu}$; hence,

$$f_x(x|\theta) = f_\nu(x - \theta)$$

Finally, $f_x(x)$ can be obtained from the total probability theorem (6-31):

$$f_x(x) = \int_{-\infty}^{\infty} f_x(x|\theta) f_\theta(\theta)\, d\theta \tag{8-23}$$

The conditional density $f_x(x|\theta)$ considered as a function of θ is called the *likelihood function*. Omitting factors that do not depend on x, we can write (8-22) in the following form:

$$\text{Posterior} \sim \text{likelihood} \times \text{prior} \tag{8-24}$$

This is the basis of Bayesian estimation.

We conclude with the model interpretation of the densities: The prior density $f_{\boldsymbol{\theta}}(\theta)$ models the diameters of all rods coming out of the production line. The posterior density $f_\theta(\theta|x)$ models the diameters of all rods of measured diameter x. The conditional density $f_x(x|\theta)$ models all measurements of a particular rod of true diameter θ. The unconditional density $f_x(x)$ models all measurements of all rods.

Note In Bayesian estimation, the underlying model is a product space $\mathcal{S} = \mathcal{S}_\theta \times \mathcal{S}_x$ where $\mathcal{S}_\theta$ is the space of the RV $\boldsymbol{\theta}$ and $\mathcal{S}_x$ is the space of the RV $\mathbf{x}$ (Fig. 8.6). In the measurement problem, $\mathcal{S}_\theta$ is the space of all rods and $\mathcal{S}_x$ is the space of all measurements of a particular rod. The product space $\mathcal{S}$ is the space of all measurements of all rods. The number θ has two meanings: It is the value of the RV $\boldsymbol{\theta}$ in the space $\mathcal{S}_\theta$; it is also a parameter specifying the density $f_x(x|\theta) = f_\nu(x - \theta)$ in the space $\mathcal{S}_x$.

The Controversy. Bayesian statistics is a topic of continuing controversy between those who interpret probability "objectively," as a measure of averages, and those who interpret it "subjectively," as a measure of belief. The controversy centers on the meaning of the prior distribution $F_\theta(\theta)$. For the objectivists, $F_\theta(\theta)$ is interpreted in terms of averages as in (4-25); for the subjectives, $F_\theta(\theta)$ is a measure of our state of knowledge concerning the unknown parameter θ.

According to the objectivists, parameter estimation can be classical or Bayesian, depending on the nature of the problem. The classical approach is used if θ is a single number (the diameter of a single rod, for example). The Bayesian approach is used if θ is one of the values of an RV $\boldsymbol{\theta}$ (the diameters of all rods in the production line). The subjectivists use the Bayesian approach in all cases. They assign somehow a prior to the unknown θ even if little or nothing is known about θ. A variety of methods have been proposed for doing so; however, they are of limited interest.

Figure 8.6

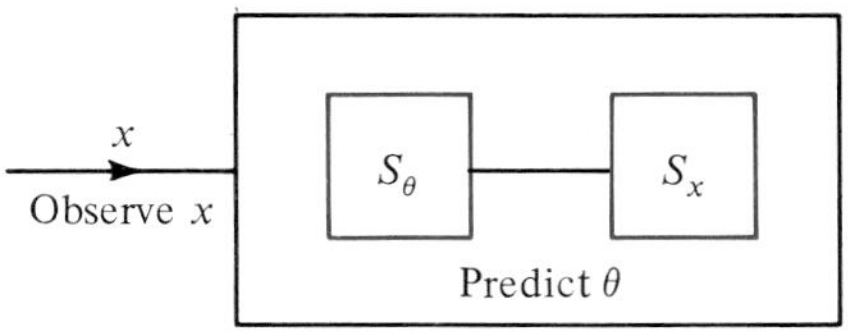

Bayesian estimation

The practical difference between the two approaches depends on the number n of available samples. If n is small, the two methods can lead to very different results; however, the results are not very reliable for either method. As n increases, the role of the prior decreases, and for large n it has no effect on the result. The following example is an illustration.

Example 8.4

We toss a coin n times, and heads shows k times. Estimate the probability p of heads.

In the classical approach to estimation, p is an unknown parameter, and its empirical estimate is the ratio k/n.

In the Bayesian approach, p is the value of an RV $\mathbf{p}$ with density $f_p(p)$, and the LMS estimate $\hat{p}$ of p equals

$$\hat{p} = \gamma \int_0^1 pp^k(1-p)^{n-k} f_p(p)\, dp \tag{8-25}$$

This follows from (6-21) and (6-54) (see Problem 6-9). For small values of n, the estimate depends on $f_p(p)$. For large n, the term $p^k(1-p)^{n-k}$ is a sharp narrow function centered at the point k/n (Fig. 6.3). This shows that the right side of (8-25) approaches k/n regardless of the form of $f_p(p)$. Note that if $f_p(p)$ is constant, then [see (6-24)]

$$\hat{p} = \frac{k+1}{n+1} \xrightarrow[n\to\infty]{} \frac{k}{n}$$

Thus, for large n, the Bayesian estimate of p equals its classical estimate. ■

Entropy

Given a partition $A = [\mathcal{A}_1, \ldots, \mathcal{A}_N]$ consisting of N events $\mathcal{A}_i$, we form the sum

$$H(A) = -\sum_{i=1}^{N} p_i \ln p_i \qquad p_i = P(\mathcal{A}_i) \tag{8-26}$$

This sum is called the *entropy* of the partition A. Thus entropy is a number associated to a partition as probability is a number associated to an event. Entropy is a fundamental concept in statistics. It is used to complete the specification of a partially known model in terms not of observations but of a *principle*. The following problem is an illustration.

The average number of faces up of a given die equals 4.5. On the basis of this information, we wish to estimate the probabilities $p_i = P\{f_i\}$ of the six faces $\{f_i\}$. This leads to the following problem: Find six numbers p_i such that

$$p_1 + p_2 + \cdots + p_6 = 1 \qquad p_1 + 2p_2 + \cdots + 6p_6 = 4.5 \tag{8-27}$$

This problem is ill-posed; that is, it does not have a unique solution because there are six unknowns and only two equations. Suppose, however, that we wish to find one solution. What do we do? Among the infinitely many solutions, is there one that is better in some sense than the others? To answer this question, we shall invoke the following principle: Among all possible solutions of (8-27), choose the one that maximizes the entropy $H(A) = -(p_1 \ln p_1 + \cdots + p_6 \ln p_6)$ of the partition A consisting of the six events $\{f_i\}$. This is called the *principle of maximum entropy*.

What is the justification of this principle? The pragmatic answer is that it leads to results that agree with observations. Conceptually, the principle is often justified by the interpretation of entropy as a *measure of uncertainty*: As we know, the probability of an event $\mathcal{A}$ is used as a measure of our uncertainty about its occurrence. If $P(\mathcal{A}) = .999$, we are practically certain that $\mathcal{A}$ will occur; if $P(\mathcal{A}) = .2$, we are reasonably certain that it will not occur; our uncertainty is maximum if $P(\mathcal{A}) = .5$. Guided by this, we interpret the entropy $H(A)$ of a partition A as a measure of uncertainty not about a single event but about the occurrence of any event of A. This is supported by the following properties of entropy (see Section 12-1): $H(A)$ is a nonnegative number. If $P(\mathcal{A}_k) = 1$ for some k, then $P(\mathcal{A}_i) = 0$ for every $i \neq k$ and $H(A) = 0$; in this case, our uncertainty is zero because at the next trial, we are certain that only the event $\mathcal{A}_k$ will occur. Finally, $H(A)$ is maximum if all events of A have the same probability; our uncertainty is then maximum.

The notion of entropy as a measure of uncertainty is subjective. In Section 12-3, we give a different interpretation to $H(A)$ based on the concept of typical sequences. This concept is related to the relative frequency interpretation of probability, and it leads to an objective meaning of the notion of entropy. We introduce it next in the context of a partition consisting of two events.

TYPICAL SEQUENCES AND RELATIVE FREQUENCY Consider the partition $A = [\mathcal{A}, \overline{\mathcal{A}}]$ consisting of an event $\mathcal{A}$ and its complement $\overline{\mathcal{A}}$. If we repeat the experiment n times, we obtain sequences s_j of the form

$$s_j = \mathcal{A}\overline{\mathcal{A}}\,\overline{\mathcal{A}} \cdots \mathcal{A} \qquad j = 1, \ldots, 2^n \tag{8-28}$$

Each sequence s_j is an event in the space $\mathcal{S}_n$, and its probability equals

$$P(s_j) = p^k q^{n-k} \tag{8-29}$$

where k is the number of successes of $\mathcal{A}$. The total number of sequences of the form (8-28) equals 2^n. We shall show that if p is not .5 and n is large, only a small number of sequences of the forms (8-28) is likely to occur. Our reasoning is based on the familiar approximation $k \simeq np$. This says that of all 2^n sequences, only the ones for which k is close to np are likely to occur. Such sequences are denoted by t_j and are called *typical*; all other sequences are called *atypical*. If t_j is a typical sequence, then $k \simeq np$, $n - k \simeq n - np = nq$. Inserting into (8-29), we obtain

$$P(t_j) = p^k q^{n-k} \simeq p^{np} q^{nq} \tag{8-30}$$

Since $p = e^{\ln p}$ and $q = e^{\ln q}$, this yields

$$P(t_j) = e^{np \ln p + nq \ln q} = e^{-nH(A)} \tag{8-31}$$

where $H(A) = -(p \ln p + q \ln q)$ is the entropy of the partition $A = [\mathcal{A}, \overline{\mathcal{A}}]$.

We denote by $\mathcal{T}$ the set $\{k \simeq np\}$ consisting of all typical sequences. As we noted, it is almost certain that $k \simeq np$; hence, we expect that almost every observed sequence is typical. From this it follows that $P(\mathcal{T}) \simeq 1$.

Denoting by n_t the number of typical sequences, we conclude that

$$P(\mathcal{T}) = n_t P\{t_j\} \simeq 1 \qquad n_t \simeq e^{nH(A)} \tag{8-32}$$

This fundamental formula relates the number of typical sequences to the entropy of the partition $[\mathcal{A}, \overline{\mathcal{A}}]$, and it shows the connection between entropy and relative frequency. We show next that it leads to an empirical justification of the principle of maximum entropy.

If $p = .5$, then $H(A) = -(.5 \ln .5 + .5 \ln .5) = \ln 2$ and $n_t \simeq e^{n \ln 2} = 2^n$. In this case, all sequences of the form (8-28) are typical. For any other value of p, $H(A)$ is less than ln 2. From this it follows that if n is large,

$$n_t \simeq e^{nH(A)} \ll e^{n \ln 2} = 2^n \tag{8-33}$$

Thus if n is large, the number 2^n of all possible sequences is vastly larger than the number $e^{nH(A)}$ of typical sequences. This result is fundamental in coding theory. Equation (8-33) shows that $H(A)$ is maximum iff n_t is maximum. The principle of maximum entropy can thus be stated in terms of the observable number n_t. If we choose p_i so as to maximize $H(A)$, the resulting number of typical sequences is maximum. As we explain in Chapter 12, this leads to the empirical interpretation of the maximum entropy principle.

Concluding Remarks

All statistical statements are probabilistic based on the assumption that the data are samples of independent and identically distributed (i.i.d.) RVs. The theoretical results lead to useful practical inferences only if the underlying physical experiment meets this assumption. The i.i.d. assumption is often phrased as follows:

1. The trials must be independent.
2. They must be performed under essentially equivalent conditions.

These conditions are empirical and cannot be verified precisely. Nevertheless, the experiment must be so designed that they are somehow met. For certain applications, this is a simple task (coin tossing) or requires minimum effort (card games). In other applications, it involves the use of special techniques (polling). In physical sciences, the i.i.d. condition follows from theoretical considerations supported by experimental evidence (statistical mechanics). In the following chapters, we develop various techniques that can be used to establish the validity of the i.i.d. conditions. This, however, requires an infinite number of tests: We must show that the events $\{\mathbf{x}_1 \le x_1\}$, . . . , $\{\mathbf{x}_n \le x_n\}$ are independent for every n and for every x_i. In specific applications, we select only a small number of tests. In many cases, the validity of the i.i.d. conditions is based on our experience.

Numerical data obtained from repeated trials of a physical experiment form a sequence of *numbers*. Such a sequence will be called *random* if the experiment satisfies the i.i.d. condition. The concept of a random sequence of physically generated numbers is empirical because the i.i.d. condition applied to real experiments is an empirical concept. In the next section, we examine the problem of generating random numbers using computers.

8-3 Random Numbers and Computer Simulation

Computer simulation of experimental data is an important discipline based on computer generation of sequences of random numbers (RN). It has applications in many fields, including: use of statistical methods in the numerical solution of deterministic problems; analysis of random physical phenomena by simulation; and use of random numbers in the design of random experiments, in tests of computer algorithms, in decision theory, and in other areas. In this section, we introduce the basic concepts stressing the meaning and generation of random numbers. As a motivation, we start with the explanation of the Monte Carlo method in the evaluation of definite integrals.

Suppose that a physical quantity is modeled by an RV $\mathbf{u}$ uniformly distributed in the interval (0, 1) and that $\mathbf{x} = g(\mathbf{u})$ is a function of $\mathbf{u}$. Since $f_u(u) = 1$ for u in the interval (0, 1) and 0 elsewhere, the mean of $\mathbf{x}$ equals

$$\eta_x = E\{g(\mathbf{u})\} = \int_0^1 g(u) f_u(u)\, du = \int_0^1 g(u)\, du \tag{8-34}$$

As we know [see (4-90)], the mean of $\mathbf{x}$ can be expressed in terms of its empirical average $\bar{x}$. Inserting this approximation into (8-34), we obtain

$$\int_0^1 g(u)\, du \simeq \frac{1}{n} \sum_{i=1}^{n} x_i = \frac{1}{n} \sum_{i=1}^{n} g(u_i) \tag{8-35}$$

where u_i are the observed values of $\mathbf{u}$ in n repetitions of the underlying physical experiment. The approximation (8-35) is an empirical statement relating the model parameter η_x to the experimental data u_i. It is based on the empirical interpretation (1-1) of probability and is valid if n is large and the data u_i satisfy the i.i.d. condition. This suggests the following method for evaluating statistically a deterministic integral. The data u_i, no matter how they are obtained, are RNs; that is, they are numbers having certain properties. If, therefore, we can develop a method for generating such numbers, we have a method for evaluating the integral in (8-35).

Random Numbers

"What are RNs? Can they be generated by a computer? Are there truly random sequences of numbers?" Such questions were raised from the early years of computer simulation, and to this day they do not have a generally accepted answer. The reason is that the term *random sequence* has two very different meanings. The first is empirical: RNS are real (physical) sequences of numbers generated either as observations of a physical quantity, or by a computer. The second is conceptual: RNS are mental constructs.

Consider, for example, the following, extensively quoted, definitions.+

D. H. Lehmer (1951): A random sequence is a vague notion embodying the ideas of a sequence in which each term is unpredictable to the uninitiated and whose digits pass a certain number of tests, traditional with statisticians and depending somewhat on the uses to which the sequence is to be put.

J. M. Franklin (1962): A sequence of numbers is random if it has every property that is shared by all infinite sequences of independent samples of random variables from the uniform distribution.

These definitions are fundamentally different. Lehmer's is empirical: The terms *vague, unpredictable to the uninitiated,* and *depending somewhat on uses* are heuristic characterizations of sequences of real numbers. Franklin's is conceptual: *infinite sequences, independent samples,* and *uniform distribution* are model concepts. Nevertheless, although so different, both are used to define random number sequences.

To overcome this conceptual ambiguity, we shall proceed as in the interpretation of probability (Chapter 1). We shall make a clear distinction between RNS as physically generated numbers, and RNS as a theoretical concept.

THE DUAL INTERPRETATION OF RNS Statistics is a discipline dealing with averages of real quantities. Computer-generated random numbers are used to apply statistical methods to the solution of various problems. Results obtained with such numbers are given the same interpretation as statistical results involving real data. This is based on the assumption that computer-generated RNS have the same properties as numbers obtained from real experiments. It suggests therefore that we relate the empirical and the theoretical properties of computer-generated RNS to the corresponding properties of random data.

Empirical Definition A sequence of real numbers will be called *random* if it has the same properties as a sequence of numerical data obtained from a random experiment satisfying the i.d.d. condition.

As we have repeated noted, the i.d.d. condition applied to real data is a heuristic statement that can be claimed only as an approximation. This vagueness cannot, however, be avoided no matter how a sequence of random numbers is specified. The above definition also has the advantage that it is phrased in terms of concepts with which we are already familiar. We can

+ Knuth, D. E., *The Art of Computer Programming* (Reading, MA: Addison Wesley, 1969).

thus draw directly on our experience with random experiments and use the well-established tests of randomness for testing the i.d.d. condition.

These considerations are relevant also in the conceptual definition of RNs: All statistical results are based on model concepts developed in the context of a probabilistic space. It is natural therefore to define RNs in terms of RVs.

Conceptual Definition A sequence of numbers x_i is called *random* if it equals the samples $x_i = \mathbf{x}_i(\zeta)$ of a sequence of i.d.d. RVs $\mathbf{x}_i$.

This is essentially Franklin's definition expressed directly in terms of RVs. It follows from this definition that in applications involving computer-generated numbers we can rely solely on the theory of probability. We conclude with the following note.

In computer simulation of random phenomena we use a *single* sequence of numbers. In the theory of probability we use a *family* of sequences $\mathbf{x}_i(\zeta)$ forming a sequence $\mathbf{x}_i$ of RVs. From the early years of simulation, various attempts were made to express the theoretical properties of RNs in terms of the properties of a single sequence of numbers. This is in principle possible; however, to be used as the theoretical foundation of the statistical applications of RNs, it must be based on the interpretation of probability as a limit. This approach was introduced by Von Mises early in the century [see (1-7)] but has not been generally accepted. As the developments of the last 50 years have shown, Kolmogoroff's definition is preferable. In the study of the properties of RNs, a new theory is not needed.

GENERATION OF RNs A good source of random numbers is a physical experiment properly selected. Repeated tosses of a coin generate a random sequence of zeros (heads) and ones (tails). We expect from our past experience that this sequence satisfies the i.i.d. requirements for randomness. Tables of random numbers generated by random experiments are available; however, they are not suitable for computer applications. They require excessive memory, access is slow, and implementation is too involved. An efficient source of RNs is a simple algorithm. The problem of designing a good algorithm for generating RNs is old. Most algorithms used today are based on the following solution proposed in 1948.

Lehmer's Algorithm Select a large prime number m and an integer a between 2 and $m - 1$. Form the sequence

$$z_n = az_{n-1} \bmod m \qquad n \geq 1 \tag{8-36)*}$$

Starting with a number $z_0 \neq 0$ we obtain a sequence of numbers z_n such that

$$1 \leq z_n \leq m - 1$$

From this it follows that at least two of the first m numbers of the sequence z_n

* The notation $A = B \bmod m$ means that A equals the remainder of the division of B by m. For example, 20 mod 13 = 7; 9 mod 13 = 9.

will be equal. Therefore, for $n \geq m$, the sequence will be periodic with period $m_0 \leq m - 1$:

$$z_{n+m_0} = z_n \qquad m_0 \leq m - 1$$

Example 8.5

We shall illustrate with $m = 13$. Suppose first that $a = 5$. If $z_0 = 1$, then z_n equals 1, 5, 12, 8, 1, . . . ; if $z_0 = 2$, then z_n equals 2, 10, 11, 3, 2, . . . The sequences so generated are periodic with period $m_0 = 4$.

Suppose next that $a = 6$. In this case, the sequence 1, 6, 10, 8, 9, 2, 12, 7, 3, 5, 4, 11, 1, . . . with the maximum period $m_0 = m - 1 = 12$ results. ■

A periodic sequence is not random. However, if in the problems for which it is intended the required number of samples is smaller than m, periodicity is irrelevant. It is therefore desirable to select a such that the period m_0 of z_n is as large as possible. We discuss next the properties that the multiplier a must satisfy so that the resulting sequence z_n has the maximum period $m_0 = m - 1$.

■ ***Theorem.*** If $m_0 = m - 1$, then a is a *primitive root* of m; that is,

$$a^{m-1} = 1 \bmod m \qquad a^n \neq 1 \bmod m \text{ for } 1 < n < m - 1 \tag{8-37}$$

■ ***Proof.*** From (8-36) it follows by a simple induction that

$$z_n = z_0 a^n \bmod m$$

Since z_n has the maximum period $m - 1$, it takes all values between 1 and $m - 1$; we can therefore assume that $z_0 = 1$. If $a^{m_0} = 1 \bmod m$ and $m_0 < m - 1$, then $z_{m_0} = a^{m_0} = 1 \bmod m = z_0$. Thus m_0 is a period; this, however, is impossible; hence, (8-37) is true.

Note that if m is a prime number, (8-37) is also a sufficient condition for maximum period. Thus the sequence z_n generated by the recursion equation (8-36) has maximum period iff a is a primitive root of m.

Suppose that $m = 13$. In this case, 6 is a primitive root of m because the smallest integer n such that $6^n = 1 \bmod m$ is $12 = 13 - 1$. However, 5 is not a primitive root because $5^4 = 1 \bmod 13$.

To complete the specification of the RN generator (8-36), we must select the integers m and a. A value for m, suggested many years ago[+] and used extensively today, is the number

$$m = 2^{31} - 1 = 2{,}147{,}483{,}647$$

This number is prime and is large enough for most applications. In the selection of the constant a, our first requirement is that the resulting sequence z_n have maximum period. For this to occur, a must satisfy (8-37). Over half a billion numbers satisfy (8-37), but most of them yield poor RNs. To arrive at a satisfactory choice, we subject the multipliers that yield maximum period to a variety of tests. Each test reduces the potential choices

[+] D. H. Lehmer, "Mathematical Methods in Large Scale Computing Units," *Annu. Comput. Lab. Harvard Univ.* 26 (1951).

until all standard tests are passed. We are thus left with a relatively small number of choices. Some of those are then subjected to special tests and are used in particular applications. Through a combination of additional tests and experience in solving problems, we arrive at a small number of multipliers. Such a multiplier is the constant

$$a = 2^7 - 1 = 16{,}807$$

The resulting RN generator is*

$$z_n = 16807 z_{n-1} \bmod 2147483647 \tag{8-38}$$

General Algorithms We describe next a number of more complex algorithms for generating RNs. We should stress that complexity does not necessarily improve the quality of a generator. The algorithm in (8-38) is very simple, but the quality of the resulting sequence is very high.

Equation (8-36) is a first-order linear congruential recursion; that is, z_n depends only on z_{n-1}, and the dependence is linear. A general nonlinear congruential recursion is an equation of the form

$$z_n = f(z_{n-1}, \ldots, z_{n-r}) \bmod m$$

The recursion is linear if

$$z_n = (a_1 z_{n-1} + \cdots + a_r z_{n-r} + c) \bmod m$$

The special case

$$z_n = (z_{n-1} + z_{n-r}) \bmod m \tag{8-39}$$

is of particular interest. It is simple and efficient, and if r is large, it might generate good sequences with period larger than m.

Note, finally, that the assumption that m is prime complicates the evaluation of the product az^{n-1} mod m. The computation is simplified if $m = 2^r$ where r is the available word length. This, however, weakens the randomness of the resulting sequence. In such cases, algorithms of the form

$$z_n = (a z_{m-1} + c) \bmod m \tag{8-40}$$

are used.

TESTS OF RANDOMNESS The sequence z_i in (8-38) is modeled by a sequence of discrete type RVS $\mathbf{z}_i$ taking all integer values from 1 to $m - 1$. The sequence

$$u_i = \frac{z_i}{m} \tag{8-41}$$

is modeled essentially by a sequence of continuous type RVS $\mathbf{u}_i$ taking all values from 0 to 1. If the RVS $\mathbf{u}_i$ are i.i.d. and uniformly distributed, we shall say that the numbers u_i are random and uniformly distributed or, simply, random. The objective of testing is to establish whether a given sequence of RNs is random. The i.i.d. condition requires an infinite number of tests. We

* S. K. Park and K. W. Miller, "Random Number Generations: Good Ones Are Hard to Find," *Communications of the ACM* 31, no. 10 (October 1988).

must show that

$$
\begin{aligned}
P\{\mathbf{u}_i \le u\} &= u \qquad 0 < u < 1 \\
P\{\mathbf{u}_1 \le u_1, \mathbf{u}_2 \le u_2\} &= P\{\mathbf{u}_1 \le u_1\}P\{\mathbf{u}_2 \le u_2\} \qquad (8\text{-}42) \\
P\{\mathbf{u}_1 \le u_1, \mathbf{u}_2 \le u_2, \mathbf{u}_3 \le u_3\} &= P\{\mathbf{u}_1 \le u_1\}P\{\mathbf{u}_2 \le u_2\}P\{\mathbf{u}_3 \le u_3\}
\end{aligned}
$$

and so on. In real problems, we can have only a finite number of tests. We cannot therefore claim that a particular sequence is random but another is not. Tests lead merely to plausible inferences. We might conclude, for example, that a given sequence is reasonably random for certain applications or that this sequence is more random than that.

There are two kinds of tests: theoretical and statistical. We shall explain with an example. We wish to test whether the RNs z_i generated by the algorithm (8-38) are the samples of an RV with mean $m/2$. In a theoretical test, we reason using the properties of prime numbers that in a sequence of $m - 1$ samples, each integer will appear once. This yields the average

$$\frac{1}{m-1}\sum_{i=1}^{m-1} z_i = \frac{1 + 2 + \cdots + m - 1}{m - 1} = \frac{m}{2}$$

In a statistical test, we generate n numbers and form their average $\bar{z} = \Sigma z_i/n$. If our assumption is correct, then $\bar{z} \simeq m/2$.

In a theoretical test, no samples are used. All conclusions are exact statements based on mathematical reasoning; however, for most tests the analysis is difficult. Furthermore, all results involve averages over the entire period. Thus they might not hold for various subsequences of z_i. For example, the fact that each integer appears once does not guarantee that the sequence is uniform.

In an empirical test, we generate n numbers z_i where n is reasonably large but much smaller than m, and we use the numbers z_i to form various averages. The underlying theory is simple for most tests. However, the computations are time-consuming, and the conclusions are probabilistic, leading to subjective decisions.

All empirical tests are tests of statistical hypotheses. The theory is developed in Chapter 10. Let us look at some illustrations.

Distribution We wish to test whether the RNs u_i are the samples of an RV $\mathbf{u}$ with uniform distribution. To do so, we use either the χ^2 test (10-85) or the Kolmogoroff-Smirnov test (10-44).

Independence We wish to test whether the subsequences

$$x_i = u_{2i} \qquad y_i = u_{2i+1}$$

are the samples of two independent RVs $\mathbf{x}$ and $\mathbf{y}$. To do so, we form two partitions involving the events $\{\alpha_i \le \mathbf{x} \le \beta_i\}$ and $\{\gamma_i \le \mathbf{y} \le \delta_i\}$ and apply the χ^2 test (10-78).

Tests based directly on (8-42) are, in general, complex. Most standard tests use (8-42) indirectly. For example, to test the hypothesis that the RVs $\mathbf{x}$ and $\mathbf{y}$ are independent, we test the weaker hypothesis that they are uncorrelated using as an estimate of their correlation coefficient r the empirical ratio $\hat{r}$ in (10-40). If the test fails, we reject the independence hypothesis. We give next two rather special illustrations of indirect testing.

Gap Test Given a uniform RV $\mathbf{u}$ and two constants α and β such that $0 < \alpha < \beta < 1$, we form the event

$$\mathscr{A} = \{\alpha < \mathbf{u} < \beta\} \qquad P(\mathscr{A}) = \beta - \alpha = p$$

We repeat the underlying experiment and observe sequences of the form

$$\mathscr{A}\overline{\mathscr{A}}\,\overline{\mathscr{A}}\mathscr{A}\mathscr{A}\mathscr{A}\overline{\mathscr{A}}\mathscr{A}\overline{\mathscr{A}}\,\overline{\mathscr{A}}\mathscr{A} \tag{8-43}$$

where $\mathscr{A}$ appears in the ith position if the event $\mathscr{A}$ occurs at the ith trial. We next form an RV $\mathbf{x}$ the values of which equal the gap lengths in (8-43), that is, the number of times $\overline{\mathscr{A}}$ appears between successive $\mathscr{A}$'s. The RV $\mathbf{x}$ so constructed has a *geometric distribution* as in (4-66):

$$P\{\mathbf{x} = r\} = p_r = (1 - p)^r p \qquad r = 0, 1, \ldots \tag{8-44}$$

We shall use (8-44) to test whether a given sequence u_i is random. If the numbers u_i are the samples of $\mathbf{u}$, then $\alpha < u_i < \beta$ iff the event $\mathscr{A}$ occurs. Denoting by n_r the number of gaps of length r of the resulting sequence (8-43), we expect with near certainty that $n_r \simeq p_r n$ where n is the length of the sequence. This empirical statement is rephrased in Section 10-4 as a goodness-of-fit problem: To examine whether the numbers n_r fit the probabilities p_r in (8-44), we apply the χ^2 test (10-61). If the test fails, we conclude that u_i is not a good RN sequence.

Spectral Test Here is a simplified development of this important test, limited to its statistical version. Given an RV $\mathbf{u}$ with uniform distribution, we form its moment-generating function

$$\Phi(s) = E\{e^{s\mathbf{u}}\} = \int_0^1 e^{su}\, du = \frac{e^s - 1}{s}$$

With $s = j\omega$, this yields (Fig. 8.7)

$$|\Phi(j\omega)| = \frac{2|\sin \omega/2|}{|\omega|}$$

Figure 8.7

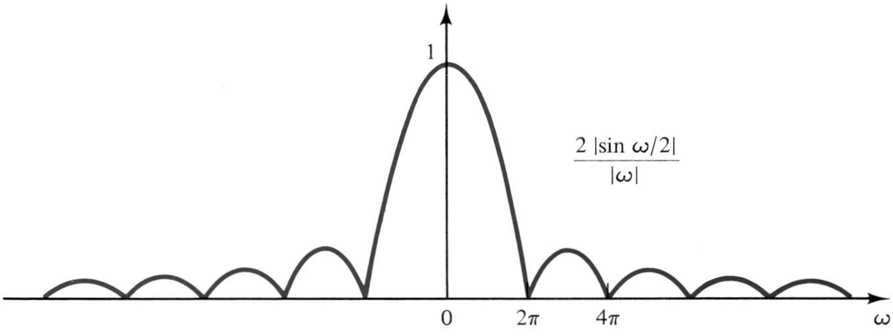

Hence, for any integer r,

$$\Phi(j2\pi r) = \begin{cases} 1 & r = 0 \\ 0 & r \neq 0 \end{cases} \tag{8-45}$$

We shall use this result to test whether a given sequence $u_i = z_i/m$ is uniform. For this purpose, we approximate the mean of the function $e^{j2\pi r\mathbf{u}}$ by its empirical average:

$$\Phi(j2\pi r) = E\{e^{j2\pi r\mathbf{u}}\} \simeq \frac{1}{n}\sum_{i=1}^{n} e^{j2\pi r z_i/m} \tag{8-46}$$

If the uniformity assumption is correct, the right side is small compared to 1 for every integer $r \neq 0$.

We shall now test for the independence and uniformity of the sequences

$$u_{2i} = \frac{z_{2i}}{m} \qquad u_{2i+1} = \frac{z_{2i+1}}{m}$$

If two RVs $\mathbf{u}$ and $\mathbf{v}$ are uniform and independent, their joint moment function equals

$$\Phi_{uv}(s_1, s_2) = E\{e^{s_1\mathbf{u}+s_2\mathbf{v}}\} = \Phi(s_1)\Phi(s_2)$$

From this and (8-46) it follows that

$$\Phi(j2\pi r_1, j2\pi r_2) = \begin{cases} 1 & r_1 = r_2 = 0 \\ 0 & \text{otherwise} \end{cases}$$

Proceeding as in (8-45), we conclude that if the subsequences z_{2i} and z_{2i+1} of an RN sequence z_i are independent, then

$$\frac{1}{n}\sum_{i=1}^{n} e^{j2\pi(r_1 z_{2i} + r_2 z_{2i+1})/m} \simeq \begin{cases} 1 & r_1 = r_2 = 0 \\ 0 & \text{otherwise} \end{cases} \tag{8-47}$$

The method can be extended to an arbitrary number of subsequences.

Note Hypothesis testing is based on a number of untested assumptions. The theoretical results are probabilistic statements, and the conclusions involve the subjective choice of various parameters. Applied to RN sequences, testing leads, therefore, only to plausible inferences. In the final analysis, specific RN generators are adopted for general use not only because they pass standard tests but also because they have been successfully applied to many problems.

RNs *with Arbitrary Distributions*

All RN sequences z_n generated by congruential algorithms are integers with uniform distribution between 1 and $m - 1$. The sequence

$$u_n = \frac{z_n}{m} \tag{8-48}$$

is essentially of continuous type, and the corresponding RV $\mathbf{u}$ is uniform in the interval (0, 1). The RV $a + b\mathbf{u}$ is uniform in the interval $(a, a + b)$, and the RV $1 - \mathbf{u}$ is uniform in the interval (0, 1). Their samples are the sequences $a + bu_i$ and $1 - u_i$, respectively.

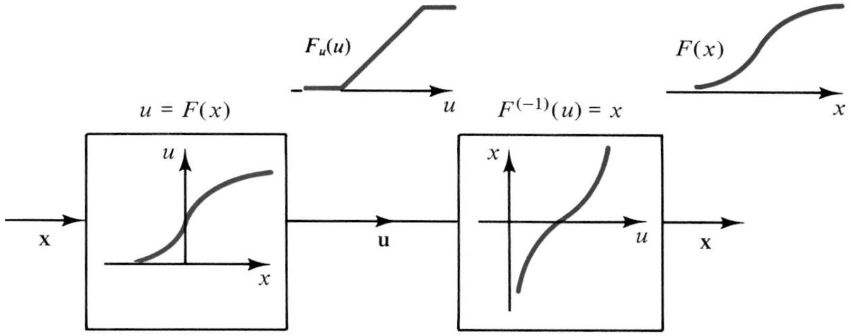

Figure 8.8

We shall use the sequence u_i or, equivalently, $1 - u_i$ to generate RN sequences with arbitrary distributions. Algorithms generating nonuniform RNs directly are not available. We shall discuss two general methods and several special techniques.

PERCENTILE-TRANSFORMATION METHOD If z_i and w_i are the samples of two RVs $\mathbf{z}$ and $\mathbf{w}$ and if $\mathbf{w} = g(\mathbf{z})$, then $w_i = g(z_i)$. Using this observation, we shall generate an RN sequence x_i with distribution a given function $F(x)$, in terms of the RN sequence u_i in (8-48). The proposed method is based on the following theorem.

■ ***Theorem.*** If $\mathbf{x}$ is an RV with distribution $F(x)$ and

$$\mathbf{u} = F(\mathbf{x}) \tag{8-49}$$

then $\mathbf{u}$ is uniformly distributed in the interval (0, 1); that is, $F_u(u) = u$ for $0 \le u \le 1$.

■ ***Proof.*** From (8-49) and the monotonicity of $F(x)$ it follows that the events $\{\mathbf{u} \le u\}$ and $\{\mathbf{x} \le x\}$ are equal. Hence (Fig. 8.8),

$$F_u(u) = P\{\mathbf{u} \le u\} = P\{\mathbf{x} \le x\} = F(x) = u$$

and the proof is complete.

Denoting by $F^{(-1)}(u)$ the inverse of the function $u = F(x)$, we conclude from (8-49) that

$$\mathbf{x} = F^{(-1)}(\mathbf{u}) \tag{8-50}$$

From this it follows that if u_i are the samples of $\mathbf{u}$, then

$$x_i = F^{(-1)}(u_i) \tag{8-51}$$

is an RN sequence with distribution $F(x)$. Thus, to form a sequence x_i with distribution $F(x)$, it suffices to form the inverse $F^{(-1)}(u)$ of $F(x)$ and to compute its values for $u = u_i$.

Example 8.6

We wish to generate a sequence of RNs with distribution

$$F(x) = 1 - e^{-x/\lambda} \qquad x > 0 \tag{8-52}$$

The inverse of $F(x)$ is the function $x = -\lambda \ln(1 - u)$. If $\mathbf{u}$ is uniform in (0, 1), then $1 -$

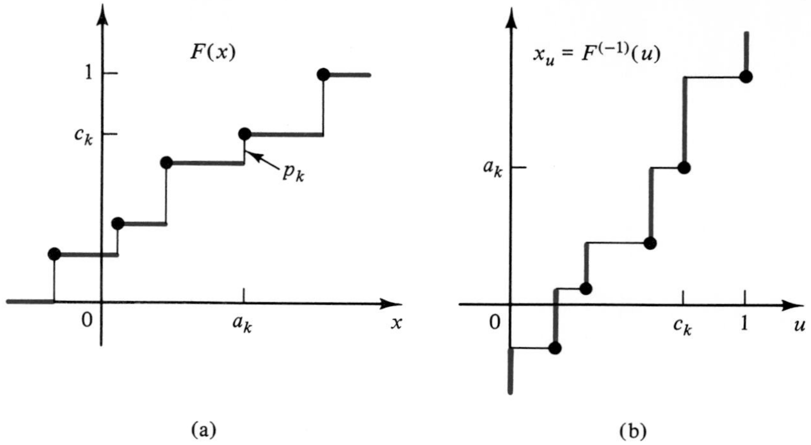

Figure 8.9

u is also uniform in (0, 1). Hence, the sequence

$$x_i = -\lambda \ln u_i \tag{8-53}$$

has an exponential distribution as in (8-52). ■

Discrete Type RNs We wish to generate an RN sequence x_i taking the m values a_k with probabilities p_k. The corresponding distribution is a staircase function with discontinuities at the points $a_1 < \cdots < a_m$ (Fig. 8.9a); its inverse is a staircase function with discontinuities at the points $c_1 < \cdots < c_m = 1$ where (Fig. 8.9b)

$$c_k = F(a_k) = p_1 + \cdots p_k \qquad k = 1, \ldots, m$$

Applying (8-51) we obtain the following rule for generating x_i:

$$\text{Set} \quad x_i = a_k \quad \text{iff} \quad c_{k-1} \le u_i < c_k \qquad c_0 = 0 \tag{8-54}$$

Example 8.7

(a) *Binary* RNs. The sequence

$$x_i = \begin{cases} 0 & 0 < u_i < p \\ 1 & p < u_i < 1 \end{cases}$$

takes the values 0 and 1 with probabilities p and $1 - p$, respectively.

(b) *Decimal* RNs. The sequence

$$x_i = k \quad \text{iff} \quad \frac{k}{10} < u_i < \frac{k+1}{10} \qquad k = 0, 1, \ldots, 9$$

takes the values 0, 1, . . . , 9 with equal probability.

(c) *Bernoulli* RNs. If

$$a_k = k \qquad p_k = \binom{n}{k} p^k q^{n-k} \qquad k = 0, 1, \ldots, n$$

then the sequence x_i in (8-54) has a Bernoulli distribution with parameters n and p.

(d) *Poisson* RNs. If

$$a_k = k \qquad p_k = e^{-\lambda}\frac{\lambda^k}{k!} \qquad k = 0, 1, \ldots$$

then the sequence x_i in (8-54) has a Poisson distribution with parameter λ. ■

From $F_x(x)$ to $F_y(y)$ We have an RN sequence x_i with distribution $F_x(x)$ and we wish to generate another RN sequence y_i with distribution $F_y(y)$. As we know [see (8-40)], the sequence $u_i = F_x(x_i)$ is uniform. Applying (8-51), we conclude that the numbers

$$y_i = F_y^{-1}(u_i) = F_y^{-1}[F_x(x_i)] \tag{8-55}$$

are the values of an RN sequence with distribution $F_y(y)$ (Fig. 8.10).

REJECTION METHOD The determination of the inverse $x = F^{-1}(u)$ of a function $F(x)$ by a computer is a difficult task involving the solution of the equation $F(x) = u$ for every u. It is therefore desirable to avoid using the transformation method if the function $F^{-1}(u)$ is not known or cannot be computed efficiently. We shall now develop a method for realizing an arbitrary $F(x)$ that avoids the inversion problem. The method is based on the properties of conditional distributions and their empirical interpretation as averages of subsequences.

Conditional Distributions Given an RV $\mathbf{x}$ and an event $\mathcal{M}$, we form the conditional distribution

$$F_x(x|\mathcal{M}) = P\{\mathbf{x} \le x|\mathcal{M}\} = \frac{P\{\mathbf{x} \le x, \mathcal{M}\}}{P(\mathcal{M})} \tag{8-56}$$

[see (6-8)]. The empirical interpretation of $F_x(x|\mathcal{M})$ is the relative frequency of the occurrence of the event $\{\mathbf{x} \le x\}$ in the subsequence

$$y_i = x_{k_1}$$

of trials in which the event $\mathcal{M}$ occurs. From this it follows that y_i is a sequence of RNs with distribution

$$F_y\{x\} = P\{\mathbf{y} \le x\} = F_x(x|\mathcal{M}) \tag{8-57}$$

We shall use this result to generate a sequence y_i of RNs with distribution a given function.

Figure 8.10

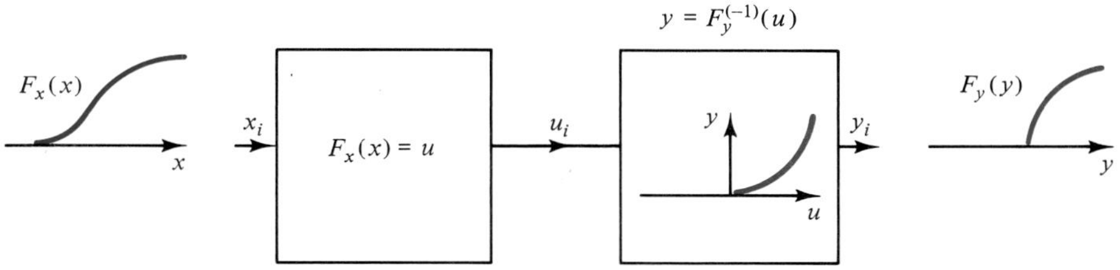

■ ***Rejection Theorem.*** Given an RV $\mathbf{u}$ uniform in the interval (0, 1) and independent of $\mathbf{x}$ and a function $r(x)$ such that

$$0 \le r(x) \le 1$$

we form the event $\mathcal{M} = \{\mathbf{u} \le r(\mathbf{x})\}$. We maintain that

$$f_x(x|\mathcal{M}) = \frac{1}{c} f_x(x) r(x) \qquad c = \int_{-\infty}^{\infty} f_x(x) r(x)\, dx \tag{8-58}$$

■ ***Proof.*** The density of $\mathbf{u}$ equals 1 in the interval (0, 1) by assumption; hence,

$$f_{xu}(x, u) = f_x(x) f_u(u) = f_x(x) \qquad 0 \le u \le 1$$

As we know

$$f_x(x|\mathcal{M})\, dx = P\{x < \mathbf{x} \le x + dx|\mathcal{M}\} = \frac{P\{x < \mathbf{x} \le x + dx, \mathcal{M}\}}{P(\mathcal{M})} \tag{8-59}$$

The set of points on the xu plane such that $u \le r(x)$ is the shaded region of Fig. 8.11. The event $\{x < \mathbf{x} \le x + dx\}$ consists of all outcomes such that $\mathbf{x}$ is the vertical strip $(x, x + dx)$. The intersection of these two regions is the part of the vertical strip in the shaded region $u \le r(x)$. In this region, $f_x(x)$ is constant; hence,

$$P\{x \le \mathbf{x} < x + dx, \mathcal{M}\} = f_x(x) r(x)\, dx \qquad P(\mathcal{M}) = \int_{-\infty}^{\infty} f_x(x) r(x)\, dx = c$$

Inserting into (8-59), we obtain (8-58).

We shall use the rejection theorem to construct an RN sequence y_i with a specified density $f_y(y)$. The function $f_y(y)$ is arbitrary, subject only to the mild condition that $f_y(x) = 0$ for every x for which $f_x(x) = 0$. We form the function

$$r(x) = a \frac{f_y(x)}{f_x(x)}$$

where a is a constant such that $0 \le r(x) \le 1$. Clearly,

$$\int_{-\infty}^{\infty} r(x) f_x(x)\, dx = a \int_{-\infty}^{\infty} f_y(x)\, dx = a$$

Figure 8.11

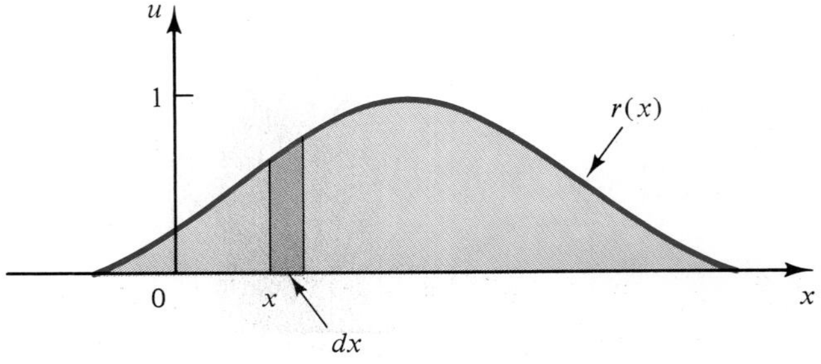

and (8-58) yields

$$f_x(x|\mathcal{M}) = f_y(x) \qquad \mathcal{M} = \left\{\mathbf{u} \le \frac{af_y(\mathbf{x})}{f_x(\mathbf{x})}\right\} \tag{8-60}$$

From (8-60) it follows that if x_i and u_i are the samples of $\mathbf{x}$ and $\mathbf{u}$, respectively, then the desired sequence y_i is formed according to the following rejection rule:

$$\text{Set } y_i = x_i \quad \text{if} \quad u_i \le a\frac{f_y(x_i)}{f_x(x_i)} \qquad \text{Reject } x_i \text{ otherwise} \tag{8-61}$$

Example 8.8

We wish to generate a sequence y_i with a truncated normal distribution starting from a sequence x_i with expontential distribution:

$$f_y(y) = \frac{2}{\sqrt{2\pi}}e^{-y^2/2}U(y) \qquad f_x(x) = e^{-x}U(x)$$

In this problem,

$$\frac{f_y(x)}{f_x(x)} = \sqrt{\frac{2}{\pi}}e^{x-x^2/2} = \sqrt{\frac{2e}{\pi}}e^{-(x-1)^2/2}$$

With $a = \sqrt{\pi/2e}$, (8-61) yields the following rejection rule:

$$\text{Set } y_i = x_i \quad \text{if} \quad u_i < e^{-(x_i-1)^2/2}$$
$$\text{Reject } x_i \text{ otherwise}$$

■

SPECIAL METHODS The preceding methods are general. For specific distributions, faster and more accurate methods are available, some of them tricky. Here are several illustrations, starting with some simple observations.

We shall use superscripts to identify several RVs. Thus $\mathbf{x}^1, \ldots, \mathbf{x}^m$ are m RVs and $x_i^1, \ldots, x_i^m$ are their samples. If x_i is an RN sequence with distribution $f(x)$, then any subsequent of x_i is an RN sequence with distribution $f(x)$. From this it follows that the RNs

$$x_i^1 = x_{mi-m+1} \qquad x_i^2 = x_{mi-m+2} \cdots x_i^m = x_{mi} \tag{8-62}$$

are the samples of the m i.i.d. RVs $\mathbf{x}^1, \ldots, \mathbf{x}^m$; their distribution equals the function $f(x)$.

Transformations If $\mathbf{z}$ is a function

$$\mathbf{z} = g(\mathbf{x}^1, \ldots, \mathbf{x}^m) \tag{8-63}$$

of the m RVs $\mathbf{x}^k$, then the numbers $z_i = g(x_i^1, \ldots, x_i^m)$ form an RN sequence with distribution $f_z(z)$. To generate an RN sequence z_i with distribution a given function $f_z(z)$ it suffices, therefore, to find g such that the distribution of the resulting $\mathbf{z}$ equals $f_z(z)$.

Example 8.9

We wish to generate an RN sequence z_i with Gamma distribution:

$$f_z(z) \sim z^{m-1}e^{-z/\lambda}U(z) \tag{8-64}$$

We know (see Example 7.5) that if the RVs $\mathbf{x}^1, \ldots, \mathbf{x}^m$ are i.i.d. with exponential distribution, their sum

$$\mathbf{z} = \mathbf{x}^1 + \cdots + \mathbf{x}^m$$

has a gamma distribution. From this it follows (see Example 8.6) that

$$z_i = \sum_{k=1}^{m} x_i^k = -\lambda \sum_{k=1}^{m} \ln u_i^k \qquad u_i^k = u_{mi-m+k} \tag{8-65}$$ ■

Example 8.10

(a) *Chi-square* RNs. We wish to generate an RN sequence z_i with distribution $\chi^2(n)$:

$$f_z(z) \sim z^{n/2-1}e^{-z/2}$$

If $n = 2m$, then this is a special case of (8-64) with $\lambda = 2$; hence,

$$z_i = -2\sum_{k=1}^{m} \ln u_i^k$$

To realize z_i for $n = 2m + 1$, we observe that if $\mathbf{y}$ is $\chi^2(2m)$, $\mathbf{w}$ is $N(0, 1)$, and $\mathbf{y}$ and $\mathbf{w}$ are independent, then [see (7-87)] the sum $\mathbf{z} = \mathbf{y} + \mathbf{w}^2$ is $\chi^2(2m + 1)$; hence,

$$z_i = -2\sum_{k=1}^{m} \ln u_i^k + (w_i)^2$$

where u_i^k are uniform RNs and w_i are RNs with normal distribution.

(b) *Student t* RNs. If the RN sequences z_i and w_i are independent with distributions $N(0, 1)$ and $\chi^2(n)$, respectively, then [see (7-101)] the ratio

$$x_i = \frac{z_i}{\sqrt{w_i/n}}$$

is an RN sequence with distribution $t(n)$.

(c) *Snedecor* RNs. If the RN sequences z_i and w_i are independent with distributions $\chi^2(m)$ and $\chi^2(n)$, respectively, then [see (7-106)] the ratio

$$x_i = \frac{z_i/m}{w_i/n}$$

is an RN sequence with distribution $F(m, n)$. ■

Example 8.11

If $\mathbf{x}^1, \ldots, \mathbf{x}^m$ are m RVs taking the values 0 and 1 with probability p and q, respectively, their sum has a binomial distribution. From this it follows that if x_i is a binary sequence as in Example 8.7, then the sequence

$$z_1 = x_1 + \cdots + x_m \qquad z_2 = x_{m+1} + \cdots + x_{2m} \qquad \cdots$$

has a binomial distribution. ■

Mixing We wish to realize an RN sequence x_i with density a given function $f(x)$. We assume that $f(x)$ can be written as a sum

$$f(x) = p_1 f_1(x) + \cdots + p_m f_m(x) \tag{8-66}$$

where $f_k(x)$ are m densities and p_i are m positive numbers such that $p_1 + \cdots + p_m = 1$. We develop a method based on the assumption that we know the RN realizations x_i^k of the densities $f_k(x)$.

We introduce the constants

$$c_0 = 0 \qquad c_k = p_1 + \cdots + p_k \qquad k = 1, \ldots, m$$

and form the events

$$\mathscr{A}_k = \{c_{k-1} \le \mathbf{u} < c_k\} \qquad k = 1, \ldots, m$$

where $\mathbf{u}$ is an RV uniform in the interval (0, 1) with samples u_i. We maintain that the sequence x_i can be realized by mixing the sequences x_i^k according to the following rule:

$$\text{Set} \quad x_i = x_i^k \qquad \text{if} \qquad c_{k-1} \le u_i < c_k \tag{8-67}$$

■ ***Proof.*** We must show that the density $f_x(x)$ of the RN sequence x_i so constructed equals the function $f(x)$ in (8-66). The sequence x_i^k is a subsequence of the sequence x_i conditioned by the event $\mathscr{A}_k$; hence, its density equals $f_x(x|\mathscr{A}_k)$. This yields

$$f_x(x|\mathscr{A}_k) = f_k(x)$$

because the distribution of x_i^k equals $f_k(x)$ by assumption. From the total probability theorem (6-8) it follows that

$$f_x(x) = f_x(x|\mathscr{A}_1)P(\mathscr{A}_1) + \cdots + f_x(x|\mathscr{A}_m)P(\mathscr{A}_m)$$

But $P(\mathscr{A}_k) = c_k - c_{k-1} = p_k$; hence, $f_x(x) = f(x)$.

Example 8.12

We wish to construct an RN sequence x_i with a Laplace distribution

$$f(x) = \frac{1}{2} e^{-|x|} = \frac{1}{2} e^{-x}U(x) + \frac{1}{2} e^{x}U(-x)$$

(Fig. 8.12). This is a special case of (8-66) with

$$f_1(x) = e^{-x}U(x) \qquad f_2(x) = f_1(-x) \qquad p_1 = p_2 = .5 \quad .$$

The functions $f_1(x)$ and $f_2(x)$ are the densities of the RVs $-\ln \mathbf{v}$ and $\ln \mathbf{v}$ respectively, where $\mathbf{v}$ is uniform in the interval (0, 1). Inserting into (8-67), we obtain the following rule:

$$\text{Set } x_i = -\ln v_i \qquad \text{if} \qquad 0 \le u_i < .5$$
$$\text{Set } x_i = \ln v_i \qquad \text{if} \qquad .5 \le u_i < 1.0$$

■

Figure 8.12

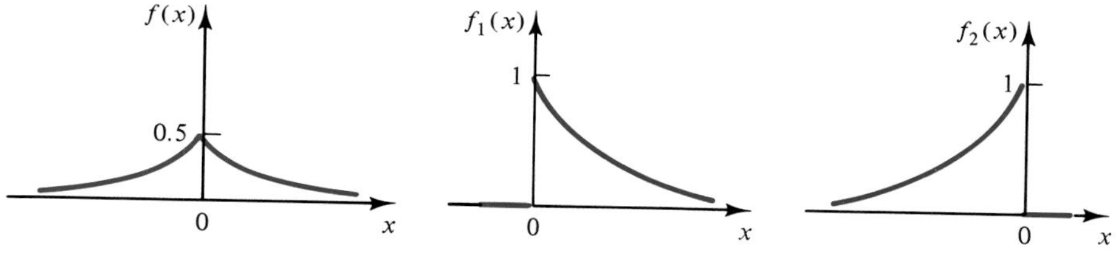

Normal RNs We discuss next some of the many methods for realizing a normal RN sequence z_i. The percentile-transformation method cannot be used because the normal distribution cannot be efficiently inverted.

Rejection and Mixing. In Example 8.8 we constructed a sequence x_i with density the truncated normal curve

$$f_1(x) = \frac{2}{\sqrt{2\pi}} e^{-x^2/2} U(x)$$

The normal density $f(z)$ can be written as a sum

$$f(z) = \frac{1}{2} f_1(z) + \frac{1}{2} f_1(-z)$$

Applying (8-67), we obtain the following rule for realizing z_i:

$$\text{Set } z_i = x_i \quad \text{if} \quad 0 \le u_i < .5$$
$$\text{Set } z_i = -x_i \quad \text{if} \quad .5 \le u_i < 1.0$$

Polar Coordinates If the RVs $\mathbf{z}$ and $\mathbf{w}$ are $N(0, 1)$ and independent and

$$\mathbf{z} = \mathbf{r} \cos \boldsymbol{\phi} \qquad \mathbf{w} = \mathbf{r} \sin \boldsymbol{\phi}$$

then (see Problem 5-27) the RVs $\mathbf{r}$ and $\boldsymbol{\phi}$ are independent, $\boldsymbol{\phi}$ is uniform in the interval $(-\pi, \pi)$, and $\mathbf{r}$ has a Rayleigh density

$$f_r(r) = re^{-r^2/2}$$

Thus $\boldsymbol{\phi} = \pi(2 - \mathbf{u})$ where $\mathbf{u}$ is uniform in the interval (0, 1). From (4-78) it follows (see also Problem 5-27) that if $\mathbf{x}$ has an exponential density $e^{-x}U(x)$ and $\mathbf{r} = \sqrt{2\mathbf{x}}$, then $\mathbf{r}$ has a Rayleigh density. Combining with (8-63), we conclude that if the sequences u_i and v_i are uniform and independent, the RN sequences

$$z_i = \sqrt{-2 \ln v_i} \cos \pi(2 - u_i) \qquad w_i = \sqrt{-2 \ln v_i} \sin \pi(2 - u_i)$$

are normal and independent.

Central Limit Theorem If $\mathbf{u}^1, \ldots, \mathbf{u}^m$ are m independent uniform RVs and $m \gg 1$, then the sum

$$\mathbf{z} = \mathbf{u}^1 + \cdots + \mathbf{u}^m$$

is approximately normal [see (7-66)]. In fact, the approximation is very good even if m is as small as 10. From this it follows that if

$$u_i^k = u_{mi-m+k}$$

are m subsequences of a uniform RN sequence u_i as in (8-62), then the sequence

$$z_1 = u_1 + \cdots + u_m \qquad z_2 = u_{m+1} + \cdots + u_{2m} \qquad \cdots$$

is normal.

Mixing In the computer literature, a number of elaborate algorithms have been proposed for the generation of normal RNs, based on the expansion of $f(z)$ into a sum of simpler functions, as in (8-67). In Fig. 8.13, we show such an expansion. The major part of $f(z)$ consists of r rectangles $f_1, \ldots, f_r$. These functions, properly normalized, are densities that can be

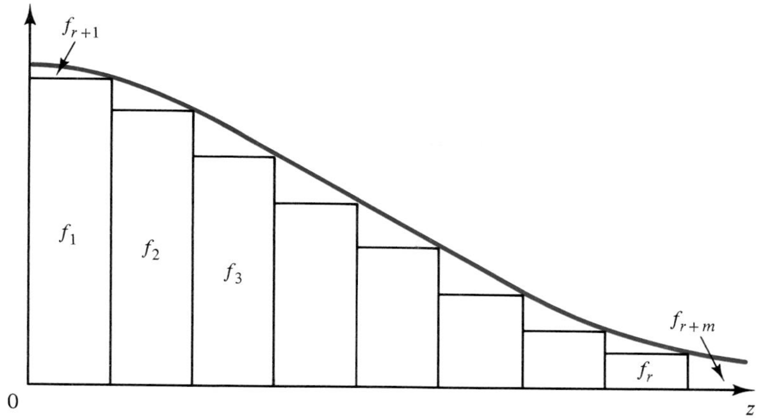

Figure 8.13

easily realized in terms of a uniform RN sequence. The remaining functions $f_{r+1}, \ldots, f_{r+m}$ are approximated by simpler curves; however, since their contribution is small, the approximation need not be very accurate.

Pairs of RNs We conclude with a brief note on the problem of generating pairs of RNs (x_i, y_i) with a specified joint distribution. This problem can, in principle, be reduced to the problem of generating one-dimensional RNs if we use the identity

$$f(x, y) = f(x)f(y|x)$$

The implementation, however, is not simple. The normal case is an exception. The joint normal density is specified in terms of five parameters: η_x, η_y, σ_x, σ_y, and μ. This leads to the following method for generating pairs of normal RN sequences.

Suppose that the RVs $\mathbf{z}$ and $\mathbf{w}$ are independent $N(0, 1)$ with samples z_i and w_i and

$$\mathbf{x} = a_1\mathbf{z} + b_1\mathbf{w} + c_1 \qquad \mathbf{y} = a_2\mathbf{z} + b_2\mathbf{w} + c_2 \tag{8-68}$$

As we know, the RVs $\mathbf{x}$ and $\mathbf{y}$ are jointly normal, and

$$\eta_x = c_1 \qquad \eta_y = c_2 \qquad \sigma_x^2 = a_1^2 + b_1^2 \qquad \sigma_y^2 = a_2^2 + b_2^2 \qquad \mu = a_1a_2 + b_1b_2$$

By a proper choice of the coefficients in (8-68), we can thus obtain two RVs $\mathbf{x}$ and $\mathbf{y}$ with an arbitrary normal distribution. The corresponding RN sequences are

$$x_i = a_1z_i + b_1w_i + c_1 \qquad y_i = a_2z_i + b_2w_i + c_2$$

At the end of this section we discuss the generation of RNs with multinomial distributions.

The Monte Carlo Method

A major application of the Monte Carlo method is the evaluation of multidimensional integrals using statistical methods. We shall discuss this important topic in the context of a one-dimensional integral. We assume—introducing suitable scaling and shifting, if necessary—that the interval of integration is (0, 1) and the integrand is between 0 and 1. Thus our problem is to estimate the integral

$$I = \int_0^1 g(u)du \qquad \text{where } 0 \le g(u) \le 1 \tag{8-69}$$

Method 1 Suppose that $\mathbf{u}$ is an RV uniform in the interval (0, 1) and $\mathbf{x} = g(\mathbf{u})$ is a function of $\mathbf{u}$. As we know, I is the mean of the RV $\mathbf{x} = g(\mathbf{u})$:

$$I = E\{g(\mathbf{u})\} = \eta_x$$

Hence, our problem is to estimate the mean of $\mathbf{x}$. To do so, we apply the estimation techniques introduced in Section 8-2 (which will be developed in detail in Chapter 9). However, instead of real data, we use as samples of $\mathbf{u}$ the computer-generated RNs u_i. This yields the empirical estimate

$$I \simeq \frac{1}{n}\sum_{i=1}^{n} x_i = \frac{1}{n}\sum_{i=1}^{n} g(u_i) \tag{8-70}$$

To evaluate the quality of this approximation, we shall interpret the RNs $x_i = g(u_i)$ as the values of the samples $\mathbf{x}_i = g(\mathbf{u}_i)$ of the RV $\mathbf{x} = g(\mathbf{u})$. With

$$\bar{\mathbf{x}} = \frac{1}{n}\sum_{i=1}^{n} \mathbf{x}_i$$

we conclude from (8-8) that

$$E\{\bar{\mathbf{x}}\} = \eta_x = I \qquad \sigma_{\bar{x}}^2 = \frac{\sigma_x^2}{n} = \sigma_1^2$$

where

$$\sigma_x^2 = E\{\mathbf{x}^2\} - \eta_x^2 = \int_0^1 g^2(u)\,du - \left[\int_0^1 g(u)\,du\right]^2 \tag{8-71}$$

Thus the average $\bar{x}$ in (8-70) is the point estimate of the unknown integral I. The corresponding interval estimate is obtained from (8-15), and it leads to the following conclusion: If we compute the average in (8-70) a large number of times, using a different set of RNs u_i each time, in $100(1 - \alpha)\%$ of the cases, the correct value of I will be between $\bar{x} + z_{\alpha/2}\sigma_x/\sqrt{n}$ and $\bar{x} - z_{\alpha/2}\sigma_x/\sqrt{n}$.

Method 2 Consider two independent RVs $\mathbf{u}$ and $\mathbf{v}$, uniform in the interval (0, 1). Their joint density equals 1 in the square $0 \le u < 1, 0 \le v < 1$; hence, the probability masses in the region $v \le g(u)$ (shaded in Fig. 8.14) equals I. From this it follows that the probability $p = P(\mathscr{A})$ of the event $\mathscr{A} = \{\mathbf{v} \le g(\mathbf{u})\}$ equals

$$p = P\{\mathbf{v} \le g(\mathbf{u})\} = I$$

This leads to the following estimate of I. Form n independent pairs (u_i, v_i) of uniform RNs. Count the number $n_{\mathscr{A}}$ of times that $y_i \le g(u_i)$. Use the approximation

$$p = \int_0^1 g(u)\,du \simeq \frac{n_{\mathscr{A}}}{n}$$

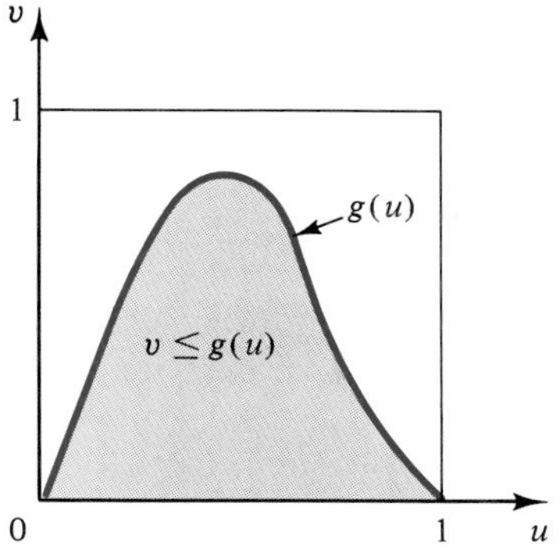

Figure 8.14

To find the variance of the estimate, we observe that the RV $\mathbf{n}_{\mathscr{A}}$ has a binomial distribution; hence, the ratio $\mathbf{n}_{\mathscr{A}}/n$ is an RV with mean $p = I$ and variance $\sigma_2^2 = I(1 - I)/n$.

Note, finally, that

$$\sigma_2^2 - \sigma_1^2 = \frac{1}{n}\int_0^1 g(u)\,du\,[1 - g(u)]\,du > 0$$

Thus the first method is more accurate. Furthermore, it requires only one RN sequence u_i.

COMPUTERS IN STATISTICS A computer is used in statistics in two fundamentally different ways. It is used to perform various computations and to store and analyze statistical data. This function involves mostly standard computer programs, and the fact that the problems originate in statistics is incidental. The second use entails the numerical solution of various problems that originate in statistics but are actually deterministic, by means of statistical methods; they are thus statistical applications of the Monte Carlo method. The underlying principle is a direct consequence of the empirical interpretation of probability: All parameters of a probability space are deterministic, they can be estimated in terms of data obtained from real experiments, and the same estimates can be used if the data are replaced by computer-generated RNs.

Next we shall apply the Monte Carlo method to the problem of estimating the distribution $F(x)$ and the percentiles x_u of an RV $\mathbf{x}$.

Distributions To estimate the distribution

$$F(x) = P\{\mathbf{x} \le x\}$$

of $\mathbf{x}$ for a specific x, we generate n RNs x_i with distribution $F(x)$ and count the number n_x of entries such that $x_i \le x$. The desired estimate of $F(x)$ [see (4-24)] is

$$F(x) \simeq \frac{n_x}{n} \tag{8-72}$$

This method is based on the assumption that we can generate the RNs x_i. We can do so if $\mathbf{x}$ is a function of other RVs with known distributions. The method is used even if $F(x)$ is of known form but its evaluation is complicated or not tabulated or if access to existing tables is not convenient.

Example 8.13

We wish to estimate the values of a chi-square distribution with m degrees of freedom. As we know [see (7-89)], if $\mathbf{z}^1, \ldots, \mathbf{z}^m$ are m independent $N(0, 1)$ RVs, then the RV

$$\mathbf{x} = (\mathbf{z}^1)^2 + \cdots + (\mathbf{z}^m)^2$$

is $\chi^2(m)$. It therefore suffices to form m normal RN sequences z_i^k. To do so, we generate a single normal sequence z_i and form m subsequences $z_i^k = z_{mi-m+k}$ as in (8-62). The sum

$$x_i = (z_i^1)^2 + \cdots + (z_i^m)^2$$

is an RN sequence with distribution $\chi^2(m)$. Using the sequence x_i so generated, we estimate the chi-square distribution from (8-72). Another method for generating x_i is discussed in Example 8.10. ■

Percentiles The u-percentile of a function $F(u)$ is a number x_u such that

$$F(x_u) = u$$

Whereas $F(x)$ is a probability, x_u is not a probability; therefore, it cannot be estimated empirically. As in the inversion problem, it can be only approximated by trial and error: Select x and determine $F(x)$; if $F(x) < u$, try a larger value for x; if $F(x) > u$, try a smaller value; continue until $F(x)$ is close enough to u.

In certain applications, the problem is not to find x_u but to establish whether a given number x is larger or smaller than the unknown x_u. In such problems, we need not find x_u. Indeed, from the monotonicity of $F(x)$ it follows that

$$x > x_u \text{ iff } F(x) > u \quad ; \quad x < x_u \text{ iff } F(x) < u \tag{8-73}$$

It thus suffices to find $F(x)$ and to compare it to the given value of u. We discuss next an important application.

Computer Simulation in Hypothesis Testing Suppose that

$$\mathbf{q} = g(\mathbf{x}^1, \ldots, \mathbf{x}^m)$$

is a function of m RVs $\mathbf{x}^r$. We observe the values x^r of these RVs and form the corresponding value $q = g(x^1, \ldots, x^m)$ of the RV $\mathbf{q}$. We wish to establish whether the number q so obtained is between the percentiles q_α and q_β of $\mathbf{q}$ where α and β are two given numbers:

$$q_\alpha < q < q_\beta \qquad \alpha = F_q(q_\alpha) \qquad \beta = F_q(q_\beta) \tag{8-74}$$

In hypothesis testing, $\mathbf{q}$ is called a test statistic and (8-74) is the null hypothesis.

To solve this problem we must determine q_α and q_β. This, however, can be avoided if we use (8-73): From the monotonicity of $F_q(q)$ it follows that (8-74) is true iff

$$\alpha < F_q(q) < \beta \tag{8-75}$$

To establish (8-74), it suffices therefore to establish (8-75). This involves the determination of $F_q(q)$ where q is the value of $\mathbf{q}$ obtained from the experimental data x^r. The function $F_q(q)$ can be determined in principle in terms of the distribution of the RVs $\mathbf{x}^r$. This, however, might be a difficult task particularly if many parameters are involved. In such cases, a computer simulation is used based on (8-72):

We start with m RN sequences x_i^r simulating the samples of the RVs $\mathbf{x}^r$ and form the RN sequence

$$q_i = g(x_i^1, \ldots, x_i^m) \qquad i = 1, \ldots, n \tag{8-76}$$

The numbers q_i are the computer-generated samples of the RV $\mathbf{q}$. Hence, their distribution is the unknown function $F_q(q)$. We next count the number n_q of samples such that $q_i \le q$ and form the ratio n_q/q. This ratio is the desired estimate of $F_q(q)$. Thus, (8-75) is true iff

$$\alpha < n_q/n < \beta \tag{8-77}$$

Note that we have here two kinds of random numbers. The first consists of the data x^r obtained from a physical experiment. These data are used to form the value

$$q = g(x^1, \ldots, x^r, \ldots, x^m)$$

of the test statistic $\mathbf{q}$. The second consists of the computer-generated sequences x_i^r. These sequences are used to determine the sequence q_i in (8-76) and the value $F_q(q)$ of the distribution of $\mathbf{q}$ from (8-72).

Example 8.14

We have a partition $A = [\mathcal{A}_1, \ldots, \mathcal{A}_m]$ consisting of the m events $\mathcal{A}_r$, and we wish to test the hypothesis that the probabilities $P(\mathcal{A}_r)$ of these events equal m given numbers p_r. To do so, we perform the underlying physical experiment N times and we observe that the event $\mathcal{A}_r$ occurs k^r times where

$$k^1 + \cdots + k^m = N \qquad p_1 + \cdots + p^m = 1$$

Using the data k^r, we form the sum

$$q = \sum_{r=1}^{m} \frac{(k^r - Np_r)^2}{Np_r} \tag{8-78}$$

Our objective is to establish whether the number q so obtained is smaller than the u-percentile q_u of the RV

$$\mathbf{q} = \sum_{r=1}^{m} \frac{(\mathbf{k}^r - Np_r)^2}{Np_r} \tag{8-79}$$

This RV is called *Pearson's test statistic*, and the resulting test *chi-square test* (see Section 10-4).

As we have explained,

$$q < q_u \text{ iff } F_q(q) < u \tag{8-80}$$

To solve the problem, it suffices therefore, to find $F_q(q)$ and compare it to u. For large N, the RV $\mathbf{q}$ has approximately a $\chi^2(m - 1)$ distribution [see (10-63)]. For moderate values of N, however, its determination is difficult. We shall find it using computer simulation.

The RVs $\mathbf{k}^r$ have a multinomial distribution as in (3-41). It suffices therefore to generate m RN sequences

$$k_i^1, \ldots, k_i^m \qquad k_i^1 + \cdots + k_i^m = N$$

with such a distribution. We do so at the end of this section.

Using the sequences k_i^r so generated, we form the samples

$$q_i = \sum_{i=1}^{m} \frac{(k_i^r - Np_r)^2}{Np_r} \qquad i = 1, 2, \ldots, n \tag{8-81}$$

of Pearson's test statistic $\mathbf{q}$, and we denote by n_q the number of entries q_i such that $q_i < q$. The ratio n_q/n is the desired estimate of $F_q(q)$. Thus (8-80) is true iff $n_q/q < u$.

Note that q is a number determined from (8-78) in terms of the experimental data k^r, and q_i is a sequence determined numerically from (8-81) in terms of the computer-generated RNs k_i^r. ■

RNs with Multinomial Distribution In the test of Example 8.13 we made use of the multinomial vector sequence

$$K_i = [k_i^1, \ldots, k_i^m]$$

This sequence forms the samples of m multinomially distributed RVs

$$\mathbf{K} = [\mathbf{k}^1, \ldots, \mathbf{k}^m]$$

of order N. To carry out the test, we must generate such a sequence. This can be done as follows:

Starting with a sequence u_i of RNs uniformly distributed in the interval (0, 1), we form N subsequences

$$U_i = [u_i^1, \ldots, u_i^j, \ldots, u_i^N] \qquad i = 1, 2, \ldots \tag{8-82}$$

as in (8-62). These sequences are the samples of the i.i.d. RVs

$$\mathbf{u}^1, \ldots, \mathbf{u}^j, \ldots, \mathbf{u}^N$$

From this it follows that

$$P\{p_1 + \cdots + p_{r-1} \le \mathbf{u}^j < p_1 + \cdots + p_r\} = p_r \tag{8-84}$$

The vector U_i in (8-82) consists of N components.

We denote by k_i^r the number of components u_i^j such that for a specific i

$$p_1 + \cdots + p_{r-1} < u_i^j < p_1 + \cdots + p_{r-1} + p_r \tag{8-85}$$

Comparing with (8-84), we conclude after some thought, that the sequence

$$k_i^1, \ldots, k_i^r, \ldots, k_i^m$$

so generated has a multinomial distribution of order N as in (3-41).

9 Estimation

Estimation is a fundamental discipline dealing with the specification of a probabilistic model in terms of observations of real data. The underlying theory is used not only in parameter estimation but also in most areas of statistics, including hypothesis testing. The development in this chapter consists of two parts. In the first part (Sections 9-1 to 9-4), we introduce the notion of estimation and develop various techniques involving the commonly used parameters. This includes estimates of means, variances, probabilities, and distributions. In the last part (Sections 9-5 and 9-6), we develop general methods of estimation. We establish the Rao-Cramer bound, and we introduce the notions of efficiency, sufficiency, and completeness.

9-1 General Concepts

Suppose that the distribution of an RV $\mathbf{x}$ is a function $F(x, \theta)$ of known form depending on an unknown parameter θ, scalar or vector. Parameter estimation is the problem of estimating θ. To solve this problem, we repeat the underlying physical experiment n times and denote by x_i the observed values of the RV $\mathbf{x}$. We shall find a point estimate and an interval estimate of θ in terms of these observations.

A *(point) estimate* is a function $\theta = g(X)$ of the observation vector $X = [x_1, \ldots, x_n]$. Denoting by $\mathbf{X} = [\mathbf{x}_1, \ldots, \mathbf{x}_n]$ the sample vector of $\mathbf{x}$, we form the RV $\boldsymbol{\theta} = g(\mathbf{X})$. This RV is called the *(point) estimator* of θ. A *statistic* is a function of the sample vector $\mathbf{X}$. Thus an estimator is a statistic.

We shall say that $\boldsymbol{\theta}$ is an *unbiased* estimator of θ if $E\{\boldsymbol{\theta}\} = \theta$; otherwise, $\boldsymbol{\theta}$ is called *biased*, and the difference $E\{\boldsymbol{\theta}\} - \theta$ is *bias*. In general, the estimation error $\boldsymbol{\theta} - \theta$ decreases as n increases. If it tends to 0 in probability (see Section 7-3) as $n \to \infty$, then $\boldsymbol{\theta}$ is a *consistent* estimator. The sample mean $\bar{\mathbf{x}}$ of $\mathbf{x}$ is an unbiased estimator of its mean $\eta = E\{\mathbf{x}\}$ and its variance equals σ^2/n; hence, $E\{(\bar{\mathbf{x}} - \eta)^2\} = \sigma^2/n \to 0$. From this it follows that $\bar{\mathbf{x}}$ tends to η in the MS sense, therefore, also in probability. In other words, $\bar{\mathbf{x}}$ is a consistent estimator of η.

In parameter estimation, it is desirable to keep the error $\boldsymbol{\theta} - \theta$ small in some sense. This requirement leads to a search for a statistic $g(\mathbf{X})$ having as density a function centered near the unknown θ. The optimum choice of $g(\mathbf{X})$ depends on the error criterion. If we use the LMS criterion, the optimum $\boldsymbol{\theta}$ is called best.

■ ***Definition.*** A statistic $\boldsymbol{\theta} = g(\mathbf{X})$ is the *best* estimator of θ if the function $g(X)$ is so chosen as to minimize the MS error

$$e = E\{[\theta - g(\mathbf{X})]^2\} = \int_{-\infty}^{\infty} \cdots \int_{-\infty}^{\infty} [\theta - g(X)]^2 f(x_1, \theta) \cdots f(x_n, \theta)\, dx_1 \cdots dx_n \tag{9-1}$$

Unlike the nonlinear prediction problem (see Section 6-3), the problem of determining the best estimator does not have a simple solution. The reason is that the unknown in (9-1) is not only the function $g(X)$ but also the parameter θ. In Section 9-6, we determine best estimators for certain classes of distributions. The results, however, are primarily of theoretical interest. For most applications, θ is expressed as the empirical estimate of the mean of some function of $\mathbf{x}$. This approach is simple and in many cases leads to best estimates. For example, we show in Section 9-6 that if $\mathbf{x}$ is normal, then its sample mean $\bar{\mathbf{x}}$ is the best estimator of η.

An *interval estimate* of a parameter θ is an interval of the form (θ_1, θ_2) where $\theta_1 = g_1(X)$ and $\theta_2 = g_2(X)$ are functions of the observation vector X. The corresponding interval $(\boldsymbol{\theta}_1, \boldsymbol{\theta}_2)$ is the *interval estimator* of θ. Thus an interval estimator is a *random interval*, that is, an interval the endpoints of which are two statistics $\boldsymbol{\theta}_1 = g_1(\mathbf{X})$ and $\boldsymbol{\theta}_2 = g_2(\mathbf{X})$.

■ ***Definition.*** We shall say that $(\boldsymbol{\theta}_1, \boldsymbol{\theta}_2)$ is a γ-confidence interval of θ if

$$P\{\boldsymbol{\theta}_1 < \theta < \boldsymbol{\theta}_2\} = \gamma \tag{9-2}$$

where γ is a given constant. This constant is called the *confidence coefficient*. The difference $\delta = 1 - \gamma$ is the *confidence level* of the estimate. The statistics $\boldsymbol{\theta}_1$ and $\boldsymbol{\theta}_2$ are called *confidence limits*.

If γ is a number close to 1, we can expect with near certainty that the unknown θ is in the interval (θ_1, θ_2). This expectation is correct in $100\gamma\%$ of the cases.

Parameter Transformation Suppose that (θ_1, θ_2) is a γ confidence interval of a parameter θ and $q(\theta)$ is a function of θ. This function specifies the parameter $\tau = q(\theta)$. We maintain that if $q(\theta)$ is a monotonically increasing function of θ, the statistics $\hat{\boldsymbol{\tau}}_1 = q(\boldsymbol{\theta}_1)$ and $\hat{\boldsymbol{\tau}}_2 = q(\boldsymbol{\theta}_2)$ are the γ confidence limits of τ. Indeed, the events $\{\boldsymbol{\theta}_1 < \theta < \boldsymbol{\theta}_2\}$ and $\{\hat{\boldsymbol{\tau}}_1 < \tau < \hat{\boldsymbol{\tau}}_2\}$ are equal; hence,

$$P\{\hat{\boldsymbol{\tau}}_1 < \tau < \hat{\boldsymbol{\tau}}_2\} = P\{\boldsymbol{\theta}_1 < \theta < \boldsymbol{\theta}_2\} = \gamma \qquad (9\text{-}3)$$

If $q(\theta)$ is monotonically decreasing, the corresponding interval is $(\hat{\tau}_2, \hat{\tau}_1)$.

The objective of interval estimation is to find two statistics $\boldsymbol{\theta}_1$ and $\boldsymbol{\theta}_2$ such as to minimize in some sense the length $\boldsymbol{\theta}_2 - \boldsymbol{\theta}_1$ of the estimation interval subject to the constraint (9-2). This problem does not have a simple solution. In the applications of this chapter, the statistics $\boldsymbol{\theta}_1$ and $\boldsymbol{\theta}_2$ are expressed in terms of various point estimators with known distributions. The results involve percentiles of the normal, the chi-square, the Student t and the Snedecor F distributions introduced in Section 7-4 and tabulated at the back of the book.

9-2 Expected Values

We start with the estimation of the mean η of an RV $\mathbf{x}$. We shall use as its point estimate the average

$$\bar{x} = \frac{1}{n}\sum_{i=1}^{n} x_i$$

of the observations x_i and as interval estimate the interval $(\bar{x} - a, \bar{x} + a)$. To find a, we need to know the distribution of the sample mean $\bar{\mathbf{x}}$ of $\mathbf{x}$. We shall assume that $\bar{\mathbf{x}}$ is a normal RV. This assumption is true if $\mathbf{x}$ is normal; it is approximately true in general if n is sufficiently large (central limit theorem).

Suppose first that the variance of $\mathbf{x}$ is known. From the normality assumption it follows as in (8-15) that

$$P\left\{\bar{\mathbf{x}} - \frac{z_u\sigma}{\sqrt{n}} < \eta < \bar{\mathbf{x}} + \frac{z_u\sigma}{\sqrt{n}}\right\} = \gamma \qquad (9\text{-}4)$$

where (Fig. 9.1)

$$u = \frac{1+\gamma}{2} = 1 - \frac{\delta}{2}$$

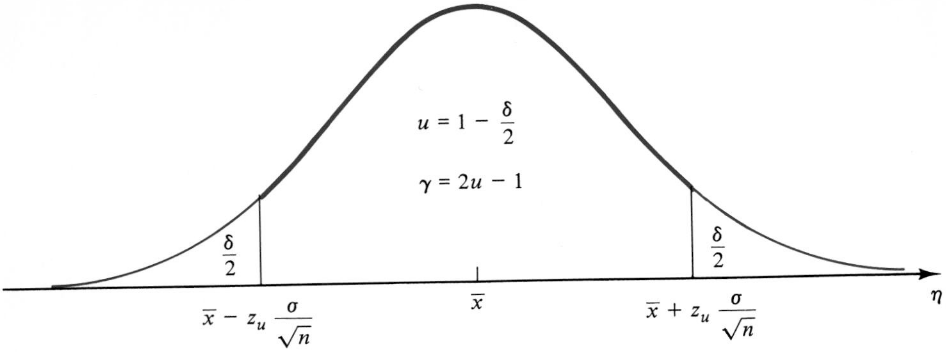

Figure 9.1

Unless otherwise stated, it will be assumed that u and γ are so related.

Equation (9-4) shows that the interval

$$\bar{x} - \frac{z_u\sigma}{\sqrt{n}} < \eta < \bar{x} + \frac{z_u\sigma}{\sqrt{n}} \tag{9-5}$$

is a γ confidence interval of η; in fact, it is the smallest such interval. Thus to find the γ confidence interval of the mean η of an RV $\mathbf{x}$, we proceed as follows:

1. Observe the samples x_i of $\mathbf{x}$ and form their average $\bar{x}$.
2. Select a number $\gamma = 2u - 1$ close to 1.
3. Find the percentile z_u of the normal distribution.
4. Form the interval $\bar{x} \pm z_u\sigma/\sqrt{n}$.

As in the prediction problem, the choice of the confidence coefficient γ is dictated by two conflicting requirements: If γ is close to 1, the estimate is reliable, but the size $2z_u\sigma/\sqrt{n}$ of the confidence interval is large; if γ is reduced, z_u is reduced, but the estimate is less reliable. The final choice is a compromise based on the applications. The commonly used values of γ are .9, .95, .99, and .999. The corresponding values of u are .925, .975, .995, and .9995 yielding the percentiles (Table 1)

$$z_{.925} = 1.440 \qquad z_{.975} = 1.967 \qquad z_{.995} = 2.576 \qquad z_{.9995} = 3.291$$

Note that $z_{.975} \simeq 2$. This leads to the slightly conservative estimate

$$\bar{x} \pm \frac{2\sigma}{\sqrt{n}} \qquad \text{for } \gamma = .95 \tag{9-6}$$

Example 9.1

We wish to estimate the weight w of a given object. The error of the available scale is an $N(0, \sigma)$ RV $\boldsymbol{\nu}$ with $\sigma = 0.06$ oz. Thus the scale readings are the samples of the RV $\mathbf{x} = w + \boldsymbol{\nu}$.

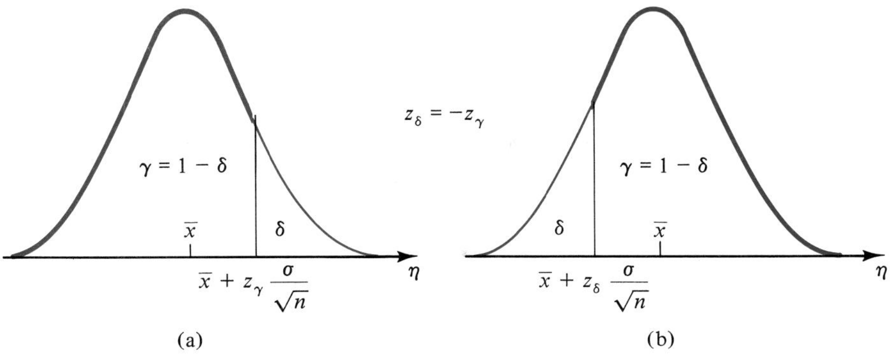

Figure 9.2

(a) We weigh the object four times, and the results are 16.02, 16.09, 16.13, and 16.16 oz. Their average $\bar{x} = 16.10$ is the point estimate of w. The .95 confidence interval is obtained from (9-6):

$$\bar{x} \pm \frac{2\sigma}{\sqrt{n}} = 16.10 \pm 0.06$$

(b) We wish to obtain the confidence interval $\bar{x} \pm 0.02$. How many times must we weigh the object? Again using (9-6), we obtain $2\sigma/\sqrt{n} = 0.02$; hence, $n = 36$. ■

One-Sided Intervals In certain applications, the objective is to establish whether the unknown η is larger or smaller than some constant. The resulting confidence intervals are called *one-sided*.

As we know (Fig. 9.2a),

$$P\{\mathbf{\bar{x}} < \eta + a\} = G\left(\frac{a}{\sigma/\sqrt{n}}\right) = \gamma \qquad \text{if } a = \frac{z_\gamma \sigma}{\sqrt{n}}$$

Hence,

$$P\left\{\eta > \mathbf{\bar{x}} - \frac{z_\gamma \sigma}{\sqrt{n}}\right\} = \gamma \tag{9-7}$$

This leads to the right γ confidence interval $\eta > \bar{x} - z_\gamma\sigma/\sqrt{n}$.

Similarly, the formula (Fig. 9.2b)

$$P\left\{\eta < \mathbf{\bar{x}} - \frac{z_\delta \sigma}{\sqrt{n}}\right\} = \gamma = 1 - \delta \tag{9-8}$$

leads to the left γ confidence interval $\eta < \bar{x} + z_\gamma\sigma/\sqrt{n}$ because $z_\delta = -z_\gamma$.

For these estimates we use the percentiles

$$z_{.9} = 1.282 \qquad z_{.95} = 1.645 \qquad z_{.99} = 2.326 \qquad z_{.999} = 3.090$$

Example 9.2

A drug contains a harmful substance H. We analyze four samples and find that the amount of H per gallon equals .41, .46, .50 and .54 oz. The analysis error is an $N(0, \sigma)$ RV with $\sigma = 0.05$ oz. On the basis of these observations, we wish to state with confidence coefficient .99 that the amount of H per gallon does not exceed c. Find c.

In this problem,

$$\bar{x} \simeq .478 \qquad \sigma = 0.05 \qquad n = 4 \qquad z_{.99} = 2.326$$

and (9-8) yields $c = \bar{x} + z_\gamma \sigma/\sqrt{n} = 0.536$ oz. Thus we can state with confidence coefficient .99 that on the basis of the measurements, the amount of H does not exceed 0.54 oz. ■

Tchebycheff Inequality If the RV $\mathbf{x}$ is not normal and the number n of observations is not large, we cannot use (9-4) because then $\bar{\mathbf{x}}$ is not normal. To find a confidence interval for η, we must find first the distribution of $\bar{\mathbf{x}}$. This involves the evaluation of $n - 1$ convolutions. To avoid this difficulty, we use Tchebycheff's inequality. Setting $k = 1/\sqrt{\delta}$ in (4-115) and replacing $\mathbf{x}$ by $\bar{\mathbf{x}}$ and σ by $\sigma/\sqrt{n}$, we obtain

$$P\left\{\bar{\mathbf{x}} - \frac{\sigma}{\sqrt{n\delta}} < \eta < \bar{\mathbf{x}} + \frac{\sigma}{\sqrt{n\delta}}\right\} \ge 1 - \delta = \gamma \qquad (9\text{-}9)$$

Thus the interval $\bar{x} \pm \sigma/\sqrt{n\delta}$ contains the γ confidence interval of η. If we therefore claim that η is in this interval, the probability that our claim is wrong will not exceed δ regardless of the form of $F(x)$ or the size n of the given sample.

Note that if $\gamma = .95$, then $1/\sqrt{\delta} = 4.47$ and (9-9) yields the interval $\bar{x} \pm 4.47/\sqrt{n}$. Under the normality assumption, the corresponding interval is $\bar{x} \pm 2/\sqrt{n}$.

UNKNOWN VARIANCE If σ is unknown, we cannot use (9-4). To find an interval estimate of η, we introduce the *sample variance*:

$$s^2 = \frac{1}{n-1}\sum_{i=1}^{n}(x_i - \bar{x})^2$$

As we know [see (7-99)], the RV $\mathbf{s}^2$ is an unbiased estimator of σ^2 and its variance tends to 0 as $n \to \infty$; hence, $\mathbf{s} \simeq \sigma$ for large n. Inserting this approximation into (9-5), we obtain the estimate

$$\bar{x} - \frac{z_u s}{\sqrt{n}} < \eta < \bar{x} + \frac{z_u s}{\sqrt{n}} \qquad (9\text{-}10)$$

This is satisfactory for $n > 30$. For smaller values of n, the probability that η is in the interval is somewhat smaller than γ. In other words, the exact γ confidence interval is larger than (9-10).

To determine the exact interval estimate of η, we assume that the RV $\mathbf{x}$ is normal and form the ratio

$$\frac{\bar{\mathbf{x}} - \eta}{\mathbf{s}/\sqrt{n}}$$

Under the normality assumption, this ratio has a Student t distribution with

$n - 1$ degrees of freedom [see (7-104)]. Denoting by t_u its percentile, we obtain

$$P\left\{-t_u < \frac{\bar{\mathbf{x}} - \eta}{\mathbf{s}/\sqrt{n}} < t_u\right\} = 2u - 1$$

With $2u - 1 = \gamma = 1 - \delta$, this yields

$$P\left\{\bar{\mathbf{x}} - \frac{t_u \mathbf{s}}{\sqrt{n}} < \eta < \bar{\mathbf{x}} + \frac{t_u \mathbf{s}}{\sqrt{n}}\right\} = \gamma \tag{9-11}$$

Hence, the exact γ confidence interval of η (Fig. 9.3) is

$$\bar{x} - \frac{t_u s}{\sqrt{n}} < \eta < \bar{x} + \frac{t_u s}{\sqrt{n}} \tag{9-12}$$

For $n > 20$, the $t(n)$ distribution is approximately $N(\eta, \sigma)$ with variance $\sigma^2 = n/(n - 2)$ [see (7-103)]. This yields

$$t_u(n) \simeq z_u \sqrt{\frac{n}{n-2}} \qquad \text{for } n > 20$$

The determination of the interval estimate of η when σ is unknown involves the following steps.

1. Observe the samples x_i of $\mathbf{x}$, and form the sample mean $\bar{x}$ and the sample variance s^2.
2. Select a number $\gamma = 2u - 1$ close to 1. Find the percentile t_u of the $t(n - 1)$ distribution.
3. Form the interval $\bar{x} \pm t_u s/\sqrt{n}$.

Figure 9.3

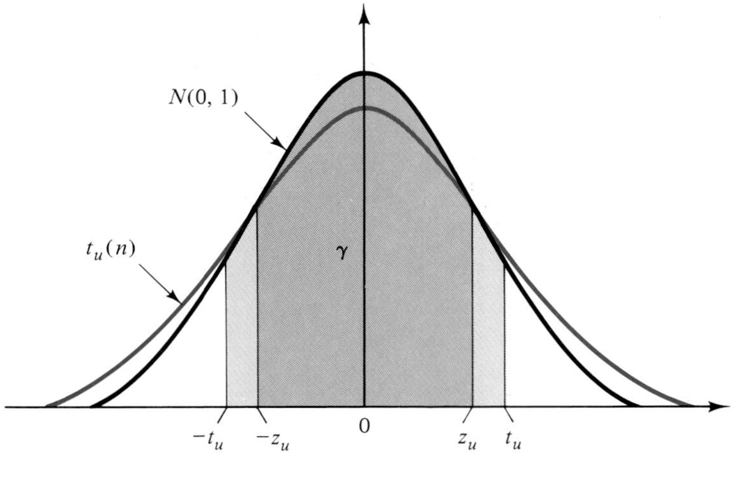

Example 9.3 We wish to estimate the mean η of the diameter of rods coming out of a production line. We measure 10 units and obtain the following readings in millimeters:

10.23	10.22	10.15	10.23	10.26
10.15	10.26	10.19	10.14	10.17

Assuming normality, find the .99 confidence interval of η.

In this problem, $n = 10$, $t_{.995}(9) = 3.25$

$$\bar{x} = \frac{1}{10}\sum_{i=1}^{10} x_i = 10.2 \qquad s^2 = \frac{1}{9}\sum_{i=1}^{10} (x_i - \bar{x})^2 = 0.0021$$

Inserting into (9-11), we obtain the interval 10.2 ± 0.05 mm. ■

Reasoning similarly, we conclude that the one-sided confidence intervals are given by (9-8) provided that we replace z_γ by t_γ and σ by s. This yields

$$P\left\{\eta > \bar{\mathbf{x}} - \frac{t_\gamma \mathbf{s}}{\sqrt{n}}\right\} = \gamma \qquad P\left\{\eta < \bar{\mathbf{x}} + \frac{t_\gamma \mathbf{s}}{\sqrt{n}}\right\} = \gamma \tag{9-13}$$

Example 9.4 A manufacturer introduces a new type of electric bulb. He wishes to claim with confidence coefficient .95 that the mean time to failure $\eta = \bar{x}$ exceeds c days. For this purpose, he tests 20 bulbs and observes the following times to failure:

78	82	84	75	80	86	74	87	82	77
83	70	79	85	81	84	73	91	73	77

Assuming that the time to failure of the bulbs is normal, find c.

In this problem,

$$\bar{x} = \frac{1}{20}\sum_{i=1}^{20} x_i = 80.05 \qquad s^2 = \frac{1}{19}\sum_{i=1}^{20} (x_i - \bar{x})^2 = 29.74$$

Thus

$$s = 5.454 \qquad n = 20 \qquad t_{.95}(19) = 2.09$$

and (9-13) yields $c = \bar{x} - t_{.95}s/\sqrt{n} = 77.94$. ■

MEAN-DEPENDENT VARIANCE In a number of applications, the distribution of $\mathbf{x}$ is specified in terms of a single parameter θ. In such cases, the mean η and the variance σ^2 of $\mathbf{x}$ can be expressed in terms of θ; hence, they are functionally related. Thus we cannot use the preceding results. To estimate η, we must develop special techniques for each case. We illustrate with two examples. We assume in both cases that n is large. This leads to the assumption that the sample mean $\bar{\mathbf{x}}$ is normal.

Exponential Distribution Suppose first that

$$f(x) = \frac{1}{\lambda} e^{-x/\lambda}\, U(x)$$

(Fig. 9.4). As we know, $\eta = \lambda$ and $\sigma = \lambda = \eta$. From this and the normality assumption it follows that the RV $\bar{\mathbf{x}}$ is normal with mean λ and variance λ^2/n.

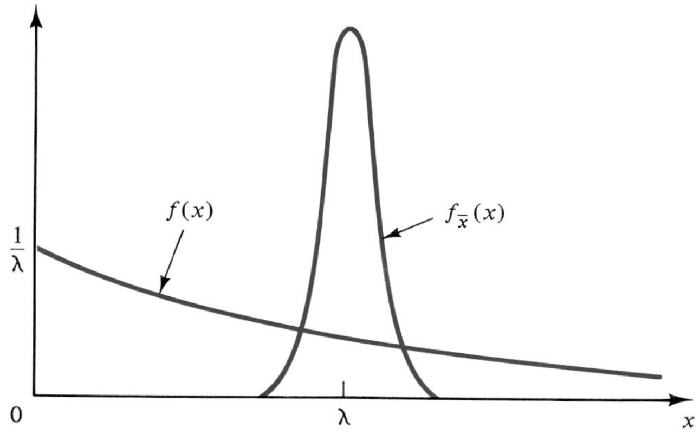

Figure 9.4

We shall use this observation to find the γ confidence interval of λ. From (9-4) it follows with $\eta = \sigma = \lambda$ that

$$P\left\{\bar{\mathbf{x}} - \frac{z_u\lambda}{\sqrt{n}} < \lambda < \bar{\mathbf{x}} + \frac{z_u\lambda}{\sqrt{n}}\right\} = \gamma$$

Rearranging terms, we obtain

$$P\left\{\frac{\bar{\mathbf{x}}}{1 + z_u/\sqrt{n}} < \lambda < \frac{\bar{\mathbf{x}}}{1 - z_u/\sqrt{n}}\right\} = \gamma \qquad (9\text{-}14)$$

and the interval

$$\frac{\bar{x}}{1 \pm z_u\sqrt{n}} \qquad \gamma = 2u - 1$$

results.

Example 9.5

Suppose that the duration of telephone calls in a certain area is an exponentially distributed RV with mean η. We monitor 100 calls and find that the average duration is 1.8 minutes.

(a) Find the .95 confidence interval of η.
With $\bar{x} = 1.8$ and $z_u \simeq 2$, (9-14) yields

$$\frac{\bar{x}}{1 \pm 2/\sqrt{n}} \simeq (1.5, 2.25)$$

(b) In this problem, $F(x) = 1 - e^{-x/\eta}$; hence, the probability p that a telephone call lasts more than 2.5 minutes equals

$$p = 1 - F(2.5) = e^{-2.5/\eta}$$

Find the .95 confidence interval of p.

The number p is a monotonically increasing function of the parameter η. We can therefore use (9-3). In our case, the confidence limits of η are 1.5 and 2.25;

hence, the corresponding limits of p are

$$p_1 = e^{-2.5/1.5} = .19 \qquad p_2 = e^{-2.5/2.25} = .33$$

We can thus claim with confidence coefficient .95 that the percentage of calls lasting more than 2.5 minutes is between 19% and 33%. ■

Poisson Distribution Consider next a Poisson-distributed RV with parameter λ

$$P\{\mathbf{x} = k\} = e^{-\lambda}\frac{\lambda^k}{k!} \qquad k = 0, 1, \ldots$$

In this case [see (4-110)], $\eta = \lambda$ and $\sigma^2 = \lambda = \eta$; hence, the sample mean $\bar{\mathbf{x}}$ of $\mathbf{x}$ is approximately $N(\lambda, \sqrt{\lambda})$. This approximation holds, of course, only for the distribution of $\mathbf{x}$. Since $\mathbf{x}$ is of the discrete type, its density consists of points.

With $\sigma = \sqrt{\lambda}$, (9-4) yields

$$P\left\{\bar{\mathbf{x}} - z_u\sqrt{\frac{\lambda}{n}} < \lambda < \bar{\mathbf{x}} + z_u\sqrt{\frac{\lambda}{n}}\right\} = \gamma \tag{9-15}$$

Unlike (9-14), this does not lead readily to an interval estimate. To find such an estimate, we note that (9-15) can be written in the form

$$P\left\{(\lambda - \bar{\mathbf{x}})^2 < \frac{z_u^2}{n}\lambda\right\} = \gamma = 2u - 1 \tag{9-16}$$

The points $(\bar{x}, \lambda)$ that satisfy the inequality are in the interior of the parabola

$$(\lambda - \bar{x})^2 = \frac{z_u^2}{n}\lambda \tag{9-17}$$

From this it follows that the γ confidence interval of λ is the vertical segment (λ_1, λ_2) of Fig. 9.5. The endpoints of this interval are the roots of the quadratic (9-17).

Figure 9.5

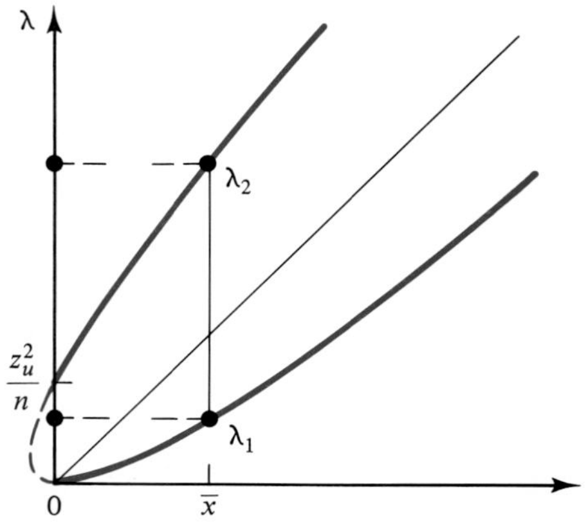

Note that in this problem, the normal approximation holds even for moderate values of n provided that $n\lambda > 25$. The reason is that the RV $n\bar{\mathbf{x}} = \mathbf{x}_1 + \cdots + \mathbf{x}_n$ is Poisson-distributed with parameter $n\lambda$; hence, the normality approximation is based on the size of $n\lambda$.

Example 9.6

The number of monthly fatal accidents in a region is a Poisson RV with parameter λ. In a 12-month period, the reported accidents per month were

4 2 5 2 7 3 1 6 3 8 4 5

Find the .95 confidence interval of λ.

In this problem, $\bar{x} = 4.25$, $n = 12$, and $z_u \simeq 2$. Inserting into (9-17), we obtain the equation

$$(\lambda - 4.25)^2 = \frac{4}{12}\lambda$$

The roots of this equation are $\lambda_1 = 3.23$, $\lambda_2 = 5.59$. We can therefore claim with confidence coefficient .95 that the mean number of monthly accidents is between 3.23 and 5.59. ■

Probabilities

We wish to estimate the probability $p = P(\mathscr{A})$ of an event $\mathscr{A}$. To do so, we repeat the experiment n times and denote by k the number of successes of $\mathscr{A}$. The ratio $\hat{p} = k/n$ is the point estimate of p. This estimate tends to p in probability as $n \to \infty$ (law of large numbers). The problem of estimating an interval estimate of p is equivalent to the problem of estimating the mean of an RV $\mathbf{x}$ with mean-dependent variance.

We introduce the zero-one RV associated with the event $\mathscr{A}$. This RV takes the values 1 and 0 with

$$P\{\mathbf{x} = 1\} = p \qquad P\{\mathbf{x} = 0\} = q = 1 - p$$

Hence,

$$\eta_x = p \qquad \sigma_x^2 = pq$$

The sample mean $\bar{x}$ or $\mathbf{x}$ equals k/n. Furthermore, $\eta_{\bar{x}} = p$, $\sigma_{\bar{x}}^2 = pq/n$, as in (8-8).

Large n The RV $\bar{\mathbf{x}}$ is of discrete type, taking the values k/n. For large n, its distribution approaches a normal distribution with mean p and variance pq/n. It therefore follows from (9-4) that

$$P\left\{\bar{\mathbf{x}} - z_u\sqrt{\frac{pq}{n}} < p < \bar{\mathbf{x}} + z_u\sqrt{\frac{pq}{n}}\right\} = \gamma \tag{9-18}$$

We cannot use this expression directly to find an interval estimate of the unknown p because p appears in the variance term pq/n. To avoid this difficulty, we introduce the approximation $pq \simeq 1/4$. This yields the interval

$$\bar{x} \pm \frac{z_u}{2\sqrt{n}} \qquad \bar{x} = \frac{k}{n} \tag{9-19}$$

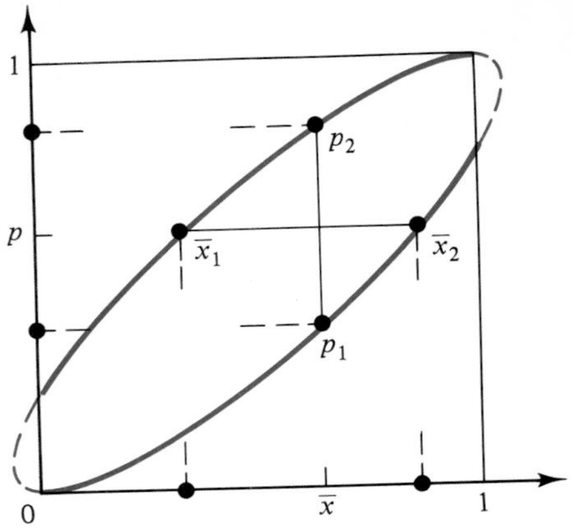

Figure 9.6

This approximation is too conservative because $p(1-p) \le 1/4$, and it is tolerable only if p is close to 1/2. We mention it because it is used sometimes.

The unknown p is close to k/n; therefore, a better approximation results if we set $p \simeq \bar{x}$ in the variance term of (9-18). This yields the interval

$$\bar{x} \pm z_u \sqrt{\frac{\bar{x}(1-\bar{x})}{n}} \qquad \gamma = 2u - 1 \tag{9-20}$$

for the unknown parameter p.

We shall now find an exact interval. To do so, we write (9-18) in the form

$$P\left\{(p - \bar{\mathbf{x}})^2 < \frac{z_u^2}{n} p(1-p)\right\} = \gamma \tag{9-21}$$

The points $(\bar{x}, p)$ that satisfy the inequality are in the interior of the ellipse

$$(p - \bar{x})^2 = \frac{z_u^2}{n} p(1-p) \qquad \bar{x} = \frac{k}{n} \tag{9-22}$$

From this it follows that the γ-confidence interval of p is the vertical segment (p_1, p_2) of Fig. 9.6. The endpoints of this segment are the roots of the quadratic (9-22).

Example 9.7

In a local poll, 400 women were asked whether they favor abortion; 240 said yes. Find the .95 confidence interval of the probability p that women favor abortion.

In this problem, $\bar{x} = .6$, $n = 400$, and $z_u \simeq 2$. The exact confidence limits p_1 and p_2 are the roots of the quadratic

$$(p - .6)^2 = \frac{1}{100} p(1-p)$$

Solving, we obtain $p_1 = .550$, $p_2 = .647$. This result is usually phrased as follows: "Sixty percent of all women favor abortion. The margin of error is $\pm 5\%$." The approximations (9-19) and (9-20) yield the intervals

$$\bar{x} \pm \frac{2}{2\sqrt{n}} = .6 \pm .1 \qquad \bar{x} \pm \frac{1}{\sqrt{n}} \sqrt{\bar{x}(1-\bar{x})} = .6 \pm .049$$ ■

Note that (9-18) can be used to *predict* the number k of successes of $\mathcal{A}$ if p is known. It leads to the conclusion that, with confidence coefficient γ, the ratio $\bar{x} = k/n$ is the interval

$$p - z_u \sqrt{\frac{p(1-p)}{n}} < \frac{k}{n} < p + z_u \sqrt{\frac{p(1-p)}{n}} \tag{9-23}$$

This interval is the horizontal segment $(\bar{x}_1, \bar{x}_2)$ of Fig. 9.6.

Example 9.8

We receive a box of 100 fuses. We know that the probability that a fuse is defective equals .2. Find the .95 confidence interval of the number $k = n\bar{x}$ of the good fuses.

In this problem, $p = .8$, $z_u \simeq 2$, $n = 100$, and (9-23) yields the interval

$$np \pm z_u \sqrt{np(1-p)} = 80 \pm 8 \tag{9-24}$$

Thus the number of good fuses is between 72 and 88. ■

Small *n* For small n, the determination of the confidence interval of p is conceptually and computationally more difficult. We shall first solve the prediction problem. We assume that p is known, and we wish to find the smallest interval (k_1, k_2) such that if we predict that the number k of successes of $\mathcal{A}$ is between k_1 and k_2, our prediction will be correct in $100\gamma\%$ of the cases. The number k is the value of the RV

$$\mathbf{k} = n\bar{\mathbf{x}} = \mathbf{x}_1 + \cdots \mathbf{x}_n \tag{9-25}$$

This RV has a binomial distribution; hence, our problem is to find an interval (k_1, k_2) of minimum size such that

$$\gamma = P\{k_1 \le \mathbf{k} \le k_2\} = \sum_{k=k_1}^{k_2} \binom{k}{n} p^k q^{n-k} = 1 - \delta \tag{9-26}$$

This problem can be solved by trial and error. A simpler solution is obtained if we search for the largest integer k_1 and the smallest integer k_2 such that

$$\sum_{k=0}^{k_1} \binom{n}{k} p^k q^{n-k} < \frac{\delta}{2} \qquad \sum_{k=k_2}^{n} \binom{n}{k} p^k q^{n-k} < \frac{\delta}{2} \tag{9-27}$$

To find k_1 for a specific p, we add terms on the left side starting from $k = 0$ until we reach the largest k_1 for which the sum is less than $\delta/2$. Repeating this for every p from 0 to 1, we obtain a staircase function $k_1(p)$, shown in Fig. 9.7 as a smooth curve. The function $k_2(p)$ is determined similarly. The functions $k_1(p)$ and $k_2(p)$ depend on n and on the confidence coefficient γ. In Fig. 9.8, we show them for $n = 10, 20, 50, 100$ and for $\gamma = .95$ and $.99$ using as horizontal variable the ratio k/n. The curves $k_1(p)/n$ and $k_2(p)/n$ approach the two branches of the ellipse of Fig. 9.6.

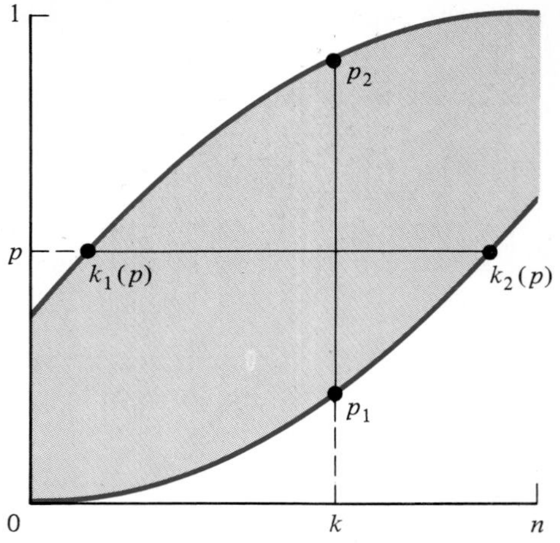

Figure 9.7

Example 9.9

A fair coin is tossed 20 times. Find the .95 confidence interval of the number k of heads.

The intersection points of the line $p = .55$ with the $n = 20$ curves of Fig. 9.8 are $k_1/n = .26$ and $k_2/n = .74$. This yields the interval $6 \le k \le 14$. ∎

We turn now to the problem of estimating the confidence interval of p. From the foregoing discussion it follows that

$$P\{k_1(p) \le \mathbf{k} \le k_2(p)\} = \gamma \tag{9-28}$$

Figure 9.8

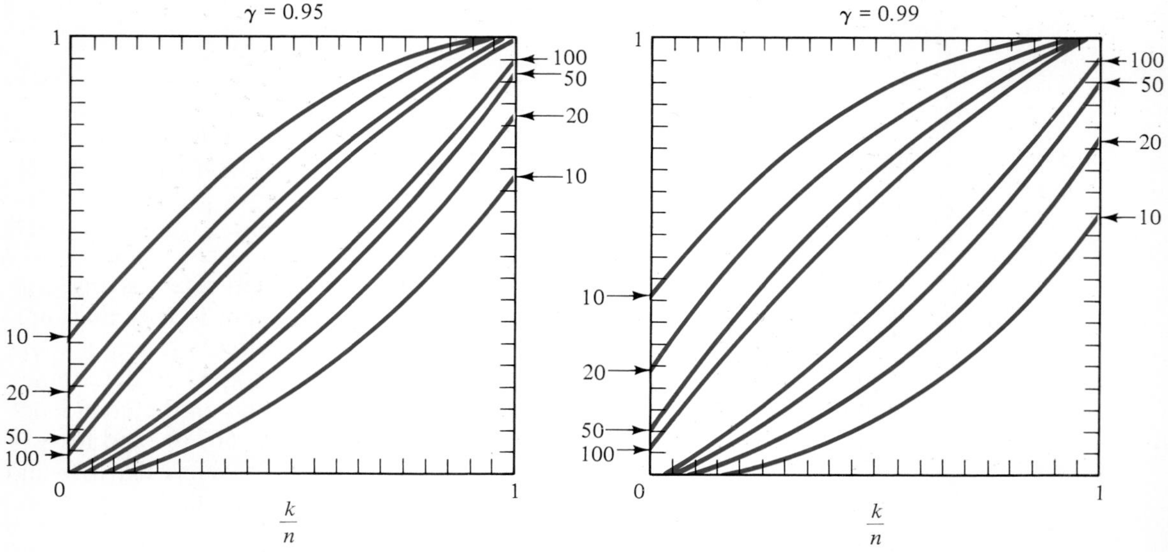

This shows that the set of points (k, p) that satisfy the inequality is the shaded region of Fig. 9.7 between the curves $k_1(p)$ and $k_2(p)$; hence, for a specific k, the γ confidence interval of p is the vertical segment (p_1, p_2) between these curves.

Example 9.10

We examine $n = 50$ units out of a production line and find that $k = 10$ are defective. Find the .95 confidence interval of the probability p that a unit is defective.

The intersection of the line $\bar{x} = k/n = .2$ with the $n = 50$ curves of Fig. 9.8 yields $p_1 = .1$ and $p_2 = .33$; hence, $.1 < p < .33$. If we use the approximation (9-20), we obtain the interval $.12 < p < .28$. ■

Bayesian Estimation

In Bayesian estimation, the unknown parameter is the value θ of an RV $\boldsymbol{\theta}$ with known density $f_\theta(\theta)$. The available information about the RV $\mathbf{x}$ is its conditional density $f_x(x|\theta)$. This is a function of known form depending on θ, and the problem is to predict the RV $\boldsymbol{\theta}$, that is, to find a point and an interval estimate of θ in terms of the observed values x_i of x. Thus the problem of Bayesian estimation is equivalent to the prediction problem considered in Sections 6-3 and 8-2. We review the results.

In the absence of any observations, the LMS estimate $\hat{\theta}$ of $\boldsymbol{\theta}$ is its mean $E\{\boldsymbol{\theta}\}$. To improve the estimate, we form the average $\bar{x}$ of the n observations x_i. Our problem now is to find a function $\phi(\bar{x})$ so as to minimize the MS error $E\{[\boldsymbol{\theta} - \phi(\bar{\mathbf{x}})]^2\}$. We have shown in (6-54) that the optimum $\phi(\bar{x})$ is the conditional mean

$$\hat{\theta} = E\{\boldsymbol{\theta} \mid \bar{x}\} = \int_{-\infty}^{\infty} \theta f_\theta(\theta \mid \bar{x})\, d\theta \qquad \text{(9-29)}$$

The function $f_\theta(\theta|\bar{x})$ is the conditional density of θ assuming $\bar{\mathbf{x}} = \bar{x}$, called *posterior density*. From Bayes' formula (6-32) it follows (Fig. 9.9) that

$$f_\theta(\theta \mid \bar{x}) = \gamma f_{\bar{x}}(\bar{x}|\theta) f_\theta(\theta) \qquad \gamma = \frac{1}{f_{\bar{x}}(x)} \qquad \text{(9-30)}$$

The unconditional density $f_{\bar{x}}(x)$ equals the integral of the product $f_{\bar{x}}(\bar{x}|\theta)f_\theta(\theta)$. Thus to find the LMS Bayesian point estimate of θ, it suffices to determine the posterior density of $\boldsymbol{\theta}$ from (9-30) and to evaluate the integral in (9-29). The function $f_{\bar{x}}(\bar{x}|\theta)$ can be expressed in terms of $f_x(x|\theta)$. For large n, it is approximately normal.

The γ confidence interval of θ is an interval $(\hat{\theta} - a_1, \hat{\theta} + a_2)$ of minimum length $a_1 + a_2$ such that the area of $f_\theta(\theta|\bar{x})$ in this interval equals γ

$$P\{\hat{\theta} - a_1 < \boldsymbol{\theta} < \hat{\theta} + a_2 | \bar{x}\} = \gamma \qquad \text{(9-31)}$$

These results hold also for discrete type RVs provided that the relevant densities are replaced by point densities and the corresponding integrals by sums; example 9-12 is an illustration.

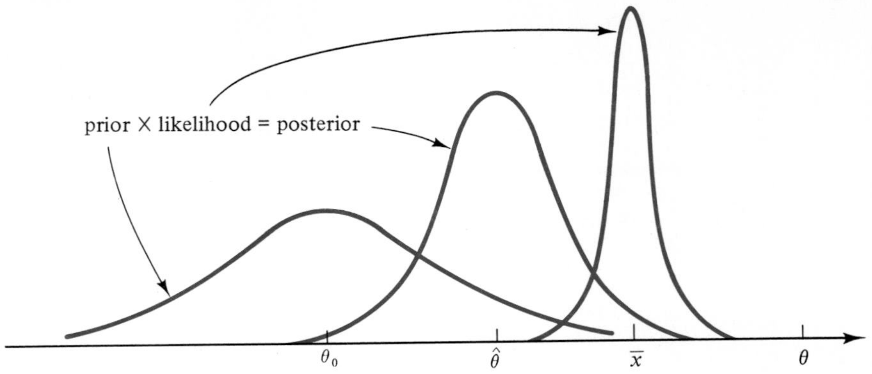

Figure 9.9

Example 9.11

The diameter of rods coming out of a production line is a normal RV with density

$$f_\theta(\theta) = \frac{1}{\sigma_0\sqrt{2\pi}} e^{-(\theta-\theta_0)^2/2\sigma_0^2} \tag{9-32}$$

We receive one rod and wish to estimate its diameter θ. In the absence of any measurements, the best estimate of θ is the mean θ_0 of $\boldsymbol{\theta}$. To improve the estimate, we measure the rod n times and obtain the samples $x_i = \theta + \nu_i$ of the RV $\mathbf{x} = \theta + \boldsymbol{\nu}$. We assume that the measurement error $\boldsymbol{\nu}$ is an $N(0, \sigma)$ RV. From this it follows that the conditional density of $\mathbf{x}$ assuming $\boldsymbol{\theta} = \theta$ is $N(\theta, \sigma)$ and the density of its sample mean $\bar{\mathbf{x}}$ is $N(\theta, \sigma/\sqrt{n})$. Thus

$$f_{\bar{x}}(\bar{x}|\theta) = \frac{1}{\sigma\sqrt{2\pi/n}} e^{-n(\bar{x}-\theta)^2/2\sigma^2} \tag{9-33}$$

Inserting (2-28) and (2-29) into (2-26) we conclude omitting the fussy details that the posterior density of $f_\theta(\theta|\bar{x})$ of $\boldsymbol{\theta}$ is a normal curve with mean $\hat{\theta}$ and standard deviation $\hat{\sigma}$ (see Fig. 9.9) where

$$\frac{n\hat{\sigma}^2}{\sigma^2} = \frac{\sigma_o^2}{\sigma_o^2 + \sigma^2/n} \xrightarrow[n\to\infty]{} 1$$
$$\hat{\theta} = \frac{\hat{\sigma}^2}{\sigma^2}\theta_o + \frac{n\hat{\sigma}^2}{\sigma^2}\bar{x} \xrightarrow[n\to\infty]{} \bar{x} \tag{9-34}$$

This shows that the Bayesian estimate $\hat{\theta}$ of θ is the weighted average of the prior estimate θ_o and the classical estimate $\bar{x}$. Furthermore, as n increases, $\hat{\theta}$ tends to $\bar{x}$. Thus, as the number of measurements increases, the effect of the prior becomes negligible.

From (8-9) and the normality of the posterior density $f_\theta(\theta|\bar{x})$ it follows that

$$P\{\hat{\theta} - \hat{\sigma}z_u < \boldsymbol{\theta} < \hat{\theta} + \hat{\sigma}z_u|\bar{x}\} = \gamma = 2u - 1 \tag{9-35}$$

This shows that the Bayesian γ confidence interval of θ is the interval

$$\hat{\theta} \pm z_u\hat{\sigma}$$

The constants $\hat{\theta}$ and $\hat{\sigma}$ are obtained from (9-34). This interval tends to the classical interval $\bar{x} \pm z_u\sigma/\sqrt{n}$ as $n \to \infty$. ■

Bayesian Estimation of Probabilities In Bayesian estimation, the unknown probability of an event $\mathscr{A}$ is the value p of an RV $\mathbf{p}$ with known density $f_p(p)$. The available information is the observed value k of the number $\mathbf{k}$ of successes of the event $\mathscr{A}$ in n trials, and the problem is to find the estimate of p in terms of k. This problem was discussed in Section 6-1. We reexamine it here in the context of RVs.

In the absence of any observation, the LMS estimate $\hat{p}$ of $\mathbf{p}$ is its mean

$$E\{\mathbf{p}\} = \int_0^1 p f_p(p)\, dp \tag{9-36}$$

in agreement with (6-15). The updated estimate $\hat{p}$ of p is the conditional mean

$$\hat{p} = E\{\mathbf{p}|k\} = \int_0^1 p f_p(p|k)\, dp \tag{9-37}$$

as in (6-21). To find $\hat{p}$, it thus suffices to find the posterior density $f_p(p|k)$. With $\mathscr{A} = \{\mathbf{k} = k\}$ it follows from (6-14) that

$$f_p(p|k) = \gamma P\{\mathbf{k} = k|p\} f_p(p) \qquad \gamma = \frac{1}{P\{\mathbf{k} = k\}} \tag{9-38}$$

where $P\{\mathbf{k} = k|p\} = \binom{n}{k} p^k q^{n-k}$ and

$$P\{\mathbf{k} = k\} = \int_0^1 P\{\mathbf{k} = k|p\} f_p(p)\, dp = \int_0^1 \binom{n}{k} p^k q^{n-k} f_p(p)\, dp$$

Thus to find the Bayesian estimate $\hat{p}$ of p, we determine $f_p(p|k)$ from (9-38) and insert it into (9-37). The result agrees with (6-21) because the term $\binom{n}{k}$ cancels.

Example 9.12

We have two coins. The first is fair, and the second is loaded with $P\{h\} = .35$. We pick one of the coins at random and toss it 10 times. Heads shows 6 times. Find the estimate $\hat{p}$ of the probability p of heads.

In this problem, p takes the values $p_1 = .5$ and $p_2 = .35$ with probability 1/2; hence, the prior density $f_p(p)$ consists of two points as in Fig. 9.10a, and the prior

Figure 9.10

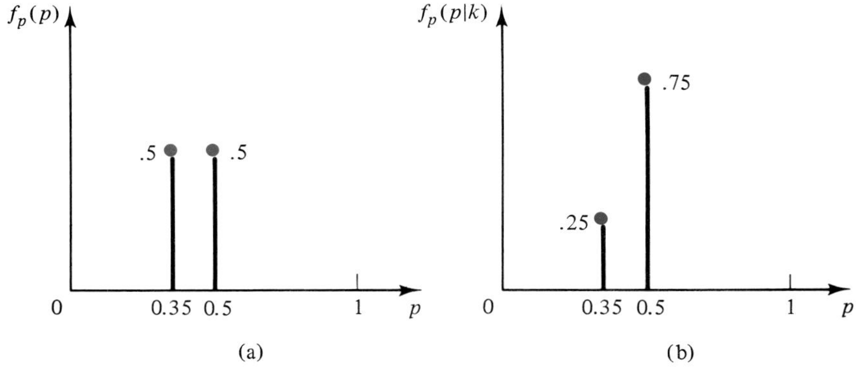

estimate of p equals $p_1/2 + p_2/2 = .425$. To find the posterior estimate $\hat{p}$, we observe that

$$P\{\mathbf{k} = 6\} = \binom{10}{6}\left(\frac{1}{2} \times \frac{1}{2^{10}} + \frac{1}{2} \times .35^6 \times \frac{1}{2^4}\right) = \binom{10}{6} \times .00135$$

Inserting into (9-38) and canceling the factor $\binom{10}{6}$ we obtain

$$f_p(p_i|k) = \frac{p_i^6 q_i^4}{.00135} \simeq \begin{cases} .75 & i = 1 \\ .25 & i = 2 \end{cases}$$

Thus the posterior density of $\mathbf{p}$ consists of two points as in Fig. 9.10b, and

$$\hat{p} = .75 \times .5 + .25 \times .35 = .4625$$

■

Difference of Two Means

We consider, finally, the problem of estimating the difference $\eta_x - \eta_y$ of the means η_x and η_y of two RVs $\mathbf{x}$ and $\mathbf{y}$ defined on an experiment $\mathcal{S}$. It appears that this problem is equivalent to the problem of estimating the mean of the RV

$$\mathbf{w} = \mathbf{x} - \mathbf{y}$$

considered earlier. The problem can be so interpreted only if the experiment $\mathcal{S}$ is repeated n times and at each trial both RVs are observed yielding the pairs (Fig. 9.11a)

$$(x_1, y_1), (x_2, y_2), \ldots, (x_n, y_n) \tag{9-39}$$

The corresponding samples of $\mathbf{w}$ are the n numbers $w_i = x_i - y_i$. The point estimate of the mean $\eta_w = \eta_x - \eta_y$ of $\mathbf{w}$ is the difference $\bar{w} = \bar{x} - \bar{y}$, and the corresponding γ confidence interval equals [see (9-5)]

$$\bar{x} - \bar{y} - \frac{z_u \sigma_w}{\sqrt{n}} < \eta_x - \eta_y < \bar{x} - \bar{y} + \frac{z_u \sigma_w}{\sqrt{n}} \tag{9-40}$$

where $\sigma_w^2 = \sigma_x^2 + \sigma_y^2 - 2\mu_{xy}$ is the variance of $\mathbf{w}$.

This approach can be used only if the available observations are *paired samples* as in (9-39). In a number of applications, the two RVs must be sampled sequentially. At a particular trial, we can observe either the RV $\mathbf{x}$ or the RV $\mathbf{y}$, not both. Observing $\mathbf{x}$ n times and $\mathbf{y}$ m times, we obtain the values (Fig. 9.11b)

$$x_1, \ldots, x_n, y_{n+1}, \ldots, y_{n+m} \tag{9-41}$$

of the n samples $\mathbf{x}_i$ of $\mathbf{x}$ and of the m samples $\mathbf{y}_i$ of $\mathbf{y}$. In this interpretation, the $n + m$ RVs $\mathbf{x}_i$ and $\mathbf{y}_k$ are independent. Let us look at two illustrations of the need for sequential sampling.

Figure 9.11

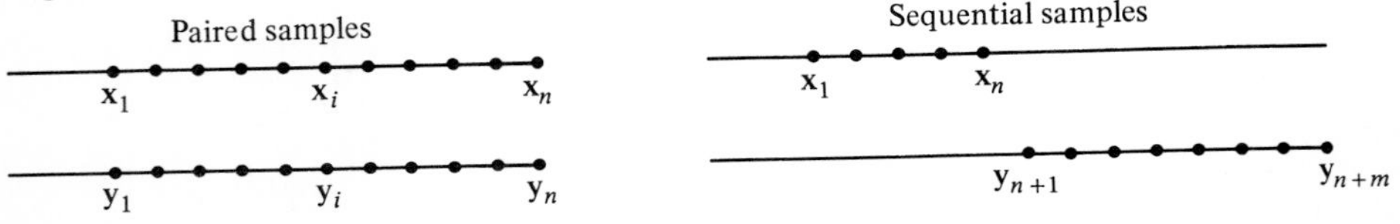

1. Suppose that **x** is an RV with mean η_x. We introduce various changes in the underlying physical experiment and wish to establish whether these changes have any effect on η_x. To do so, we repeat the original experiment n times and the modified experiment m times. We thus obtain the $n + m$ samples in (9-41). The first n numbers are the samples x_i of the RV **x** representing the original experiment, and the next m numbers y_i are the m samples of the RV **y** representing the modified experiment. This problem is basic in hypothesis testing.
2. A system consists of two components. The time to failure of the first component is the RV **x** and of the second, the RV **y.** To estimate the difference $\eta_x - \eta_y$, we test a number of systems. If at each test we can determine the times to failure x_i and y_i of both components, we can obtain the paired samples (x_i, y_1). Suppose, however, that the components are connected in series. In this case, we cannot observe both failure times at a single trial. To estimate the difference $\eta_x - \eta_y$, we must use the independent samples in (9-41).

We introduce the RV

$$\bar{\mathbf{w}} = \bar{\mathbf{x}} - \bar{\mathbf{y}} \qquad \text{where } \bar{\mathbf{x}} = \frac{1}{n}\sum_{i=1}^{n} \mathbf{x}_i \qquad \bar{\mathbf{y}} = \frac{1}{m}\sum_{i=1}^{m} \mathbf{y}_i \tag{9-42}$$

The RV $\bar{\mathbf{w}}$ is *defined* by (9-42); it is not a sample mean. Clearly,

$$\eta_{\bar{w}} = \eta_{\bar{x}} - \eta_{\bar{y}} = \eta_x - \eta_y$$

Hence, $\bar{\mathbf{w}}$ is an unbiased point estimator of $\eta_x - \eta_y$. To find an interval estimate, we must determine the variance of $\bar{\mathbf{w}}$. From the independence of the samples $\mathbf{x}_i$ and $\mathbf{y}_i$ it follows that

$$\sigma_{\bar{x}}^2 = \frac{\sigma_x^2}{n} \qquad \sigma_{\bar{y}} = \frac{\sigma_y^2}{m} \qquad \sigma_{\bar{w}}^2 = \frac{\sigma_x^2}{n} + \frac{\sigma_y^2}{m} \tag{9-43}$$

Under the normality assumption, this leads to the γ confidence interval

$$\bar{x} - \bar{y} - \sigma_{\bar{w}} z_u < \eta_x - \eta_y < \bar{x} - \bar{y} + \sigma_{\bar{w}} z_u \tag{9-44}$$

This estimate is used of σ_x and σ_y are known. If they are unknown and n is large, we use the approximations

$$\sigma_x^2 \simeq s_x^2 = \frac{1}{n-1}\sum_{i=1}^{n}(x_i - \bar{x})^2 \qquad \sigma_y^2 \simeq s_y^2 = \frac{1}{m-1}\sum_{i=1}^{m}(y_i - \bar{y})^2$$

in the evaluation of $\sigma_{\bar{w}}$.

Example 9.13

Suppose that **x** models the math grades of boys and **y** the math grades of girls in a senior class. We examine the grades of 50 boys and 100 girls and obtain $\bar{x} = 80$, $\bar{y} = 82$, $s_x^2 = 32$, $s_y^2 = 36$. Find the .99 confidence interval of the difference $\eta_x - \eta_y$ of the grade means η_x and η_y.

In this problem, $n = 50$, $m = 100$, $z_u = z_{.995} \simeq 2.58$. Inserting the approximation

$$\sigma_{\bar{w}}^2 \simeq \frac{s_x^2}{50} + \frac{s_y^2}{100} = 1$$

into (9-44), we obtain the interval -2 ± 2.58. ■

The estimation of the difference $\eta_x - \eta_y$ when σ_x and σ_y are unknown and the numbers n and m are small is more difficult. If $\sigma_x \neq \sigma_y$, the problem does not have a simple solution because we cannot find an estimator the density of which does not depend on the unknown parameters. We shall solve the problem only for

$$\sigma_x = \sigma_y = \sigma$$

The unknown σ^2 can be estimated either by s_x^2 or by s_y^2. The estimation error is reduced, however, if we use some function of s_x and s_y. We shall select as the estimate $\hat{\sigma}^2$ of σ^2 the sum $as_x^2 + bs_y^2$ where the constants a and b are so chosen as to minimize the MS error. Proceeding as in (7-38), we obtain (see Problem 7-29)

$$\hat{\sigma}^2 = \frac{(n-1)s_x^2 + (m-1)s_y^2}{n+m-2} \tag{9-45}$$

The variance of $\bar{\mathbf{w}}$ equals $\sigma_{\bar{w}}^2 = \sigma^2(1/n + 1/m)$. To find its estimate, we replace σ^2 by $\hat{\sigma}^2$. This yields the estimate

$$\hat{\sigma}_{\bar{w}}^2 = \frac{(n-1)s_x^2 + (m-1)s_y^2}{n+m-2}\left(\frac{1}{n} + \frac{1}{m}\right) \tag{9-46}$$

We next form the RV

$$\frac{\bar{\mathbf{w}} - \eta_{\bar{w}}}{\hat{\boldsymbol{\sigma}}_{\bar{w}}} \tag{9-47}$$

As we show in Problem 7-29, this RV has a t distribution with $n + m - 2$ degrees of freedom. Denoting by t_u its u-percentile, we conclude that

$$P\left\{-t_u < \frac{\bar{\mathbf{w}} - \eta_{\bar{w}}}{\hat{\boldsymbol{\sigma}}_{\bar{w}}} < t_u\right\} = \gamma = 2u - 1$$

Hence,

$$P\{\hat{\mathbf{w}} - \hat{\boldsymbol{\sigma}}_{\bar{w}} t_u < \eta_{\bar{w}} < \hat{\mathbf{w}} + \hat{\boldsymbol{\sigma}}_{\bar{w}} t_u\} = \gamma \tag{9-48}$$

Thus to estimate the difference $\eta_x - \eta_y$ of two RVs $\mathbf{x}$ and $\mathbf{y}$ sequentially sampled, we proceed as follows:

1. Observe the $n + m$ samples x_i and y_i and compute their sample means $\bar{x}$, $\bar{y}$ and variances s_x^2, s_y^2.
2. Determine $\hat{\sigma}$ from (9-45) and $\hat{\sigma}_{\bar{w}}$ from (9-46).
3. Select the coefficient $\gamma = 2u - 1$ and find the percentile $t_u(n + m - 2)$ from Table 3.
4. Form the interval $\bar{x} - \bar{y} \pm t_u\sigma_{\bar{w}}$.

Example 9.14 Two machines produce cylindrical rods with diameters modeled by two normal RVs **x** and **y**. We wish to estimate the difference of their means. To do so, we measure 14 rods from the first machine and 18 rods from the second. The resulting sample means and standard deviations, in millimeters, are as follows:

$$\bar{x} = 85 \qquad \bar{y} = 85.05 \qquad s_x = 0.5 \qquad s_y = 0.3$$

Inserting into (9-46), we obtain

$$\hat{\sigma}^2_{\bar{w}} = \frac{13s_x^2 + 17s_y^2}{30}\left(\frac{1}{14} + \frac{1}{18}\right) = 0.02$$

Thus $\hat{\sigma}_{\bar{w}} = 0.14$, $n + m - 2 = 30$, $t_{.9}(30) = 1.31$, and (9-48) yields the estimate

$$-0.24 < \eta_x - \eta_y < 0.14$$

■

9-3 Variance and Correlation

We shall estimate the variance $v = \sigma^2$ of a normal RV **x**. We assume first that its mean η is known, and we use as point estimate of v the sum

$$\hat{v} = \frac{1}{n}\sum_{i=1}^{n}(x_i - \eta)^2 \tag{9-49}$$

where x_i are the observed samples of **x**. As we know, the corresponding estimator $\hat{\mathbf{v}}$ is unbiased, and its variance tends to 0 as $n \to \infty$; hence, it is consistent.

To find an interval estimate of v, we observe that the RV $n\hat{\mathbf{v}}/\sigma^2$ has a χ^2 distribution with n degrees of freedom [see (7-92)]. Denoting by $\chi^2_u(n)$ its u-percentiles, we conclude (Fig. 9.12) that

$$P\left\{\chi^2_{\delta/2}(n) < \frac{n\hat{\mathbf{v}}}{\sigma^2} < \chi^2_{1-\delta/2}(n)\right\} = \gamma = 1 - \delta \tag{9-50}$$

This yields the γ confidence interval

$$\frac{n\hat{v}}{\chi^2_{\delta/2}(n)} > \sigma^2 > \frac{n\hat{v}}{\chi^2_{1-\delta/2}(n)} \tag{9-51}$$

The percentiles of the χ^2 distribution are listed in Table 2.

Figure 9.12

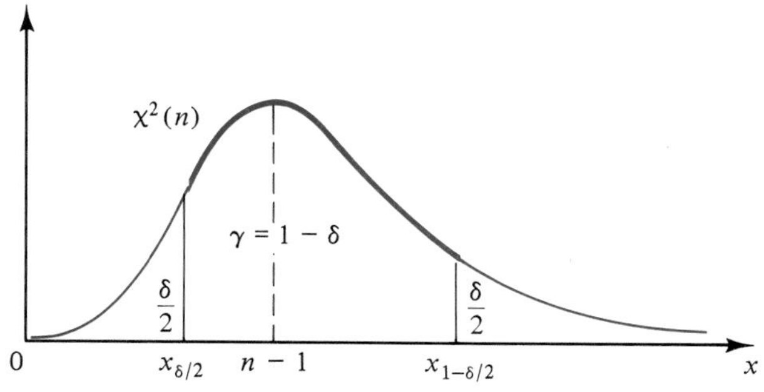

We note that the interval so obtained does not have minimum length because the χ^2 density is not symmetrical; as noted in (8-12), it is used for convenience.

If η is unknown, we use as point estimate of v the sample variance

$$s^2 = \frac{1}{n-1}\sum_{i=1}^{n}(x_i - \bar{x})^2$$

Again, this is a consistent estimate of v.

To find an interval estimate, we observe that the RV $(n-1)\mathbf{s}^2/\sigma^2$ has a $\chi^2(n-1)$ distribution; hence,

$$P\left\{\chi^2_{\delta/2}(n-1) < \frac{(n-1)\mathbf{s}^2}{\sigma^2} < \chi^2_{1-\delta/2}(n-1)\right\} = \gamma \tag{9-52}$$

This yields the interval estimate

$$\frac{(n-1)s^2}{\chi^2_{1-\delta/2}(n-1)} < \sigma^2 < \frac{(n-1)s^2}{\chi^2_{\delta/2}(n-1)} \tag{9-53}$$

Example 9.15

The errors of a given scale are modeled by a normal RV $\boldsymbol{\nu}$ with zero mean and variance σ^2. We wish to find the .95 confidence interval of σ^2. To do so, we weigh a standard object 11 times and obtain the following readings in grams:

98.92	98.68	98.85	97.07	98.98	99.36
98.70	99.33	99.31	98.84	99.20	

These numbers are the samples of the RV $\mathbf{x} = w + \boldsymbol{\nu}$ where w is the true weight of the given object.

(a) Suppose, first, that $w = 99$ g. In this case, the mean w of $\mathbf{x}$ is known; hence, we can use (9-51) with

$$\hat{v} = \frac{1}{11}\sum_{i=1}^{11}(x_i - 99)^2 = 0.626$$

Inserting the percentiles $\chi^2_{.025}(11) = 3.82$ and $\chi^2_{.975}(11) = 21.92$ of $\chi^2(11)$, into (9-51) we obtain the interval

$$\frac{11\hat{v}}{21.92} = 0.30 < \sigma^2 < \frac{11\hat{v}}{3.82} = 1.8$$

Note that the corresponding interval for the standard deviation σ [see (9-3)] is

$$\sqrt{0.30} = .55 < \sigma < \sqrt{1.8} = 1.34$$

(b) We now assume that w is unknown. From the given data, we obtain

$$\bar{x} = \frac{1}{11}\sum_{i=1}^{11} x_i = 99.02 \qquad s^2 = \frac{1}{10}\sum_{i=1}^{11}(x_i - 99.02)^2 = 0.622$$

In this case, $\chi^2_{.025}(10) = 3.25$ and $\chi^2_{.975}(10) = 20.48$. Inserting into (9-53), we obtain the estimate

$$0.303 < \sigma^2 < 1.91 \qquad \blacksquare$$

Covariance and Correlation

A basic problem in many scientific investigations is the determination of the existence of a causal relationship between observable variables: smoking and cancer, poverty and crime, blood pressure and salt. The methods used to establish whether two quantities X and Y are causally related are often statistical. We model X and Y by two RVS $\mathbf{x}$ and $\mathbf{y}$ and draw inferences about the *causal* dependence of X and Y from the *statistical* dependence of $\mathbf{x}$ and $\mathbf{y}$. Such methods are useful, however, they might lead to wrong conclusions if they are interpreted literally: Roosters crow every morning and then the sun rises; but the sun does not rise because roosters crow every morning.

To establish the independence of two RVS, we must show that their joint distribution equals the product of their marginal distributions. This involves the estimate of a function of two variables. A simpler problem is the estimation of the covariance μ_{11} of the RVS $\mathbf{x}$ and $\mathbf{y}$. If $\mu_{11} = 0$, $\mathbf{x}$ and $\mathbf{y}$ are uncorrelated but not necessarily independent. However, if $\mu_{11} \neq 0$, $\mathbf{x}$ and $\mathbf{y}$ are not independent. The size of the covariance of two RVS is a measure of their linear dependence, but this measure is scale-dependent (is 2 a high degree of correlation or 200?). A normalized measure is the correlation coefficient $r = \mu_{11}/\sigma_x\sigma_y$ of $\mathbf{x}$ and $\mathbf{y}$. As we know, $|r| \le 1$, and if $|r| = 1$, the RVS $\mathbf{x}$ and $\mathbf{y}$ are linearly dependent—that is, $\mathbf{y} = a\mathbf{x} + b$; if $r = 0$, they are uncorrelated.

Since $\mu_{11} = E\{(\mathbf{x} - \eta_x)(\mathbf{y} - \eta_y)\}$, the empirical estimate of μ_{11} is the sum $(1/n)\Sigma(x_i - \bar{x})(y_i - \bar{y})$. This estimate is used if the means η_x and η_y are known. If they are unknown, we use as estimate of μ_{11} the *sample covariance*

$$\hat{\mu}_{11} = \frac{1}{n-1}\sum_{i=1}^{n}(x_i - \bar{x})(y_i - \bar{y}) \tag{9-54}$$

The resulting estimator $\boldsymbol{\mu}_{11}$ is unbiased (see Problem 9-26) and consistent.

We should stress that since r is a parameter of the joint density of the RVS $\mathbf{x}$ and $\mathbf{y}$, the observations x_i and y_i used for its estimate must be *paired samples,* as in (9-39). They cannot be obtained sequentially, as in (9-41).

We estimate next the correlation coefficient r using as its point estimate the ratio

$$\hat{r} = \frac{\hat{\mu}_{11}}{s_x s_y} = \frac{\Sigma(x_i - \bar{x})(y_i - \bar{y})}{\sqrt{\Sigma(x_i - \bar{x})^2\Sigma(y_i - \bar{y})^2}} \tag{9-55}$$

This ratio is called the *sample correlation coefficient* of the RVS $\mathbf{x}$ and $\mathbf{y}$. To find an interval estimate of r, we must determine the density $f_{\hat{r}}(r)$ of the RV $\hat{\mathbf{r}}$. The function $f_{\hat{r}}(r)$ vanishes outside the interval $(-1, 1)$ (Fig. 9.13a) because $|\mathbf{r}| \le 1$ (Problem 9-27). However, its exact form cannot be found easily. We give next a large sample approximation.

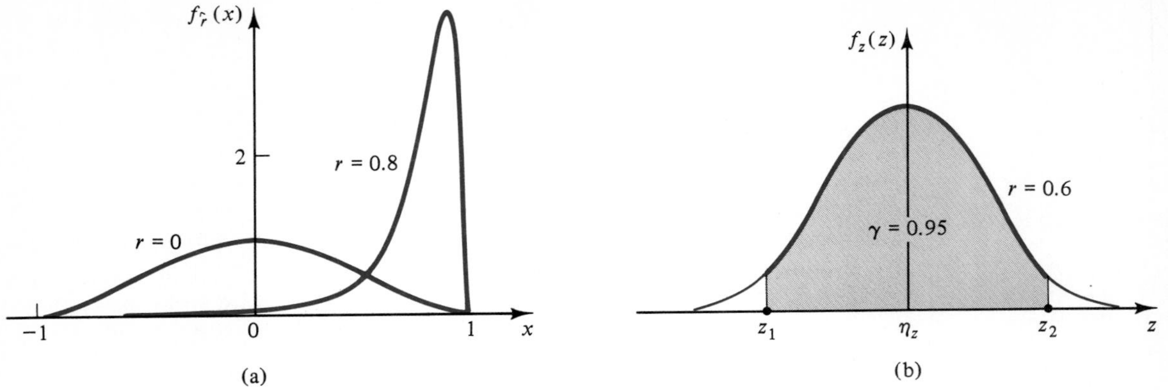

Figure 9.13

Fisher's Auxiliary Variable We introduce the transformations

$$\mathbf{z} = \frac{1}{2} \ln \frac{1 + \hat{\mathbf{r}}}{1 - \hat{\mathbf{r}}} \qquad \hat{\mathbf{r}} = \frac{e^{2\mathbf{z}} - 1}{e^{2\mathbf{z}} + 1} \tag{9-56}$$

It can be shown that for large n, the distribution of the RV $\mathbf{z}$ so constructed is approximately normal (Fig. 9.13b) with mean and variance

$$\eta_z \simeq \frac{1}{2} \ln \frac{1 + r}{1 - r} \qquad \sigma_z^2 \simeq \frac{1}{n - 3} \tag{9-57}$$

The proof of this theorem will not be given.

From (9-57) and (8-7) it follows that

$$P\left\{\eta_z - \frac{z_u}{\sqrt{n-3}} < \mathbf{z} < \eta_z + \frac{z_u}{\sqrt{n-3}}\right\} = \gamma = 2u - 1$$

In this expression, z_u is not the u-percentile of the RV $\mathbf{z}$ in (9-56); it is the normal percentile.

To continue the analysis, we shall replace in (9-57) the unknown r by its empirical estimate (9-55). This yields

$$\hat{\eta}_z \simeq \frac{1}{2} \ln \frac{1 + \hat{r}}{1 - \hat{r}} \tag{9-58}$$

From (9-58) and the monotonicity of the transformations (9-56) it follows [see also (9-49)] that

$$P\{r_1 < \hat{\mathbf{r}} < r_2\} = \gamma$$

where

$$r_1 = \frac{e^{2z_1} - 1}{e^{2z_1} + 1} \qquad r_2 = \frac{e^{2z_2} - 1}{e^{2z_2} + 1} \qquad \begin{matrix} z_2 \\ z_1 \end{matrix} = \hat{\eta}_z \pm \frac{z_u}{\sqrt{n-3}} \tag{9-59}$$

Thus to find the γ confidence interval (r_1, r_2) of the correlation coefficient r, we compute $\hat{r}$ from (9-55), $\hat{\eta}_z$ from (9-58), and r_1 and r_2 from (9-59).

Example 9.16

We wish to estimate the correlation coefficient r of SAT scores, modeled by the RV $\mathbf{x}$, and freshman rankings, modeled by the RV $\mathbf{y}$. For this purpose, we examine the relevant records of 52 students and obtain the 52 paired samples (x_i, y_i). We then compute the fraction in (9-55) and find $\hat{r} = .6$. This is the empirical point estimate of r. We shall find its .95 confidence interval. Inserting the number $\hat{r} = .6$ into (9-58), we obtain $\hat{\eta}_z \simeq .693$, and with $z_u \simeq 2$, (9-59) yields

$$z_1 = .41 \qquad z_2 = .98 \qquad r_1 = .39 \qquad r_2 = .75$$

Hence, $.39 < r < .75$ with confidence coefficient .95. ■

9-4 Percentiles and Distributions

Consider an RV $\mathbf{x}$ with distribution $F(x)$. The u-percentile of $\mathbf{x}$ is the inverse of $F(x)$; that is, it is a number x_u such that $F(x_u) = u$ (Fig. 9.14a). In this section, we estimate the functions x_u and $F(x)$ in terms of the n observations x_i of the RV $\mathbf{x}$. In both cases, we assume that nothing is known about $F(x)$. In this sense, the estimates are *distribution-free*.

Percentiles

We write the observations x_i in ascending order and denote by y_i the ith number so obtained. The resulting RVS $\mathbf{y}_i$ are the *ordered statistics*

$$\mathbf{y}_1 \le \mathbf{y}_2 \le \cdots \le \mathbf{y}_n$$

introduced in (7-44). In particular, $\mathbf{y}_1$ equals the minimum and $\mathbf{y}_n$ the maximum of $\mathbf{x}_i$. The determination of the interval estimate of x_u is based on the following theorem.

■ ***Theorem.*** For any k and r,

$$P\{\mathbf{y}_k < x_u < \mathbf{y}_{k+r}\} = \sum_{m=k}^{k+r-1} \binom{n}{m} u^m(1-u)^{n-m} \tag{9-60}$$

Figure 9.14

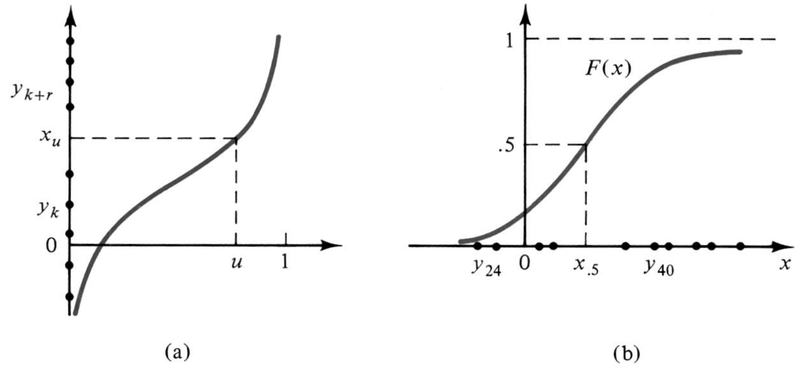

■ ***Proof.*** From the definition of the order statistics it follows that $y_k < x_u$ iff at least k of the samples x_i are less than x_u; similarly, $y_{k+r} > x_u$ iff at least $k + r$ of the samples x_i are greater than x_u. Therefore, $y_k < x_u < y_{k+r}$ iff at least k and at most $k + r - 1$ of the samples x_i are less than x_u. In other words, in the sample space $\mathscr{S}_n$, the event $\{\mathbf{y}_k < x_u < \mathbf{y}_{k+r}\}$ occurs iff the number of successes of the event $\{\mathbf{x} \le x_u\}$ is at least k and at most $k + r - 1$. This yields (9-60) because $P\{\mathbf{x} \le x_u\} = u$.

This theorem leads to the following γ confidence interval of x_u: We select k and r such as to minimize the length $y_{k+r} - y_k$ of the interval (y_k, y_{k+r}) subject to the condition that the sum in (9-60) equals γ. The solution is obtained by trial and error involving the determination of γ for various values of k and r.

Example 9.17

We have a sample of size 4 and use as estimate of x_u the interval (y_1, y_4). In this case, γ equals the sum in (9-60) for m from 1 to 3. Hence,

$$\gamma = P\{\mathbf{y}_1 < x_u < \mathbf{y}_4\} = 4u(1-u)^3 + 6u^2(1-u)^2 + 4u^3(1-u)$$

For $u = .5, .4, .3$, we obtain $\gamma = .875, .845, .752$. ■

If n is large, we can use the approximation (3-35) in the evaluation of γ. This yields

$$\gamma = P\{\mathbf{y}_k < x_u < \mathbf{y}_{k+r}\} \simeq G\left(\frac{k + r - 0.5 - nu}{\sqrt{nu(1-u)}}\right) - G\left(\frac{k + 0.5 - nu}{\sqrt{nu(1-u)}}\right) \qquad (9\text{-}61)$$

For a given γ, the length of the interval (y_k, y_{k+r}) is minimum if its center is y_{n_u} where n_u is the integer closest to nu. Setting $k + 0.5 \simeq n_u - m$ and $k + r - 0.5 \simeq n_u + m$ in (9-61), we obtain

$$\gamma = P\{\mathbf{y}_{n_u - m} < x_u < \mathbf{y}_{n_u + m}\} \simeq 2G\left(\frac{m}{\sqrt{nu(1-u)}}\right) - 1 \qquad (9\text{-}62)$$

This yields the $\gamma = 1 - \delta$ confidence interval

$$y_{n_u - m} < x_u < y_{n_u + m} \qquad m \simeq z_{1-\delta/2}\sqrt{nu(1-u)} \qquad (9\text{-}63)$$

Example 9.18

We have 64 samples of $\mathbf{x}$, and we wish to find the .95 confidence interval of its median $x_{.5}$. In this problem,

$$n = 64 \qquad u = .5 \qquad n_u = nu = 32 \qquad z_{1-\delta/2} \simeq 2 \qquad m = 8$$

and (9-63) yields the estimate $y_{24} < x_{.5} < y_{40}$. We can thus claim with probability .95 that the unknown median is between the 24th and the 40th ordered observation of $\mathbf{x}$ (Fig. 9.14b). ■

Distribution

The point estimate of $F(x)$ is its empirical estimate

$$\hat{F}(x) = \frac{k_x}{n} \qquad (9\text{-}64)$$

where k_x is the number of samples x_i that do not exceed x [see (4-25)]. The function $\hat{F}(x)$ has a staircase form with discontinuities at the points x_i. If $\mathbf{x}$ is of continuous type, almost certainly the samples x_i are distinct and the jump of $\hat{F}(x)$ at x_i equals $1/n$. It is convenient, however, if n_i samples are close, to bunch them together into a single sample x_i of multiplicity n_i. The corresponding jump at x_i is then n_i/n.

We shall find two interval estimates of $F(x)$. The first will hold for a specific x, the second for any x.

Variable-length estimate For a specific x, the function $F(x)$ is the probability $p = P\{\mathbf{x} \le x\}$ of the event $\mathcal{A} = \{\mathbf{x} \le x\}$. We can therefore apply the earlier results involving estimates of probabilities. For large n, we shall use the approximation (9-20) with p replaced by $F(x)$ and $\bar{x}$ by $\hat{F}(x)$. This yields the γ confidence interval

$$\hat{F}(x) \pm a \qquad a = \frac{z_u}{\sqrt{n}} \sqrt{\hat{F}(x)(1 - \hat{F}(x))} \tag{9-65}$$

for the unknown $F(x)$. In this estimate, the length $2a$ of the confidence interval *depends* on x.

Kolmogoroff estimate The empirical estimate of $F(x)$ is a function $\hat{F}(x)$ of x depending on the samples x_i of the RV $\mathbf{x}$. It specifies therefore a random family of functions $\hat{\mathbf{F}}(x)$, one for each set of samples x_i. We wish to find a number c, independent of x, such that

$$P\{|\hat{\mathbf{F}}(x) - F(x)| \le c\} \le \gamma$$

for every x. The constant γ is the confidence coefficient of the desired estimate $\hat{\mathrm{F}}(x) \pm c$ of $F(x)$.

The difference $|\hat{F}(x) - F(x)|$ is a function of x and its maximum (or least upper bound) is a number w (Fig. 9.15) depending on the samples x_i. It specifies therefore the RV

$$\mathbf{w} = \max_{-\infty<x<\infty} |\hat{\mathbf{F}}(x) - F(x)| \tag{9-66}$$

Figure 9.15

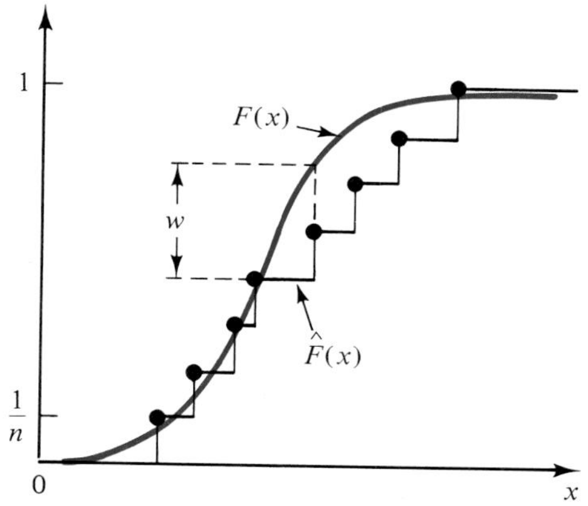

From the definition of w it follows that $w < c$ iff $|\hat{F}(x) - F(x)| < c$ for every x, hence,

$$P\{\max_x |\hat{\mathbf{F}}(x) - F(x)| \le c\} = P\{\mathbf{w} \le c\} = F_w(c) = \gamma \qquad (9.67)$$

where $F_x(w)$ is the distribution of the RV $\mathbf{w}$. To find c, it suffices therefore to find $F_w(w)$.

The function $F_w(w)$ does not depend on the distribution $F(x)$ of $\mathbf{x}$. This unexpected property of the maximum distance w between the curves $\hat{F}(x)$ and $F(x)$ can be explained as follows: The difference $\hat{F}(x) - F(x)$ does not change if the x axis is subjected to a nonlinear transformation or, equivalently, if the RV $\mathbf{x}$ is replaced by any other RV $\mathbf{y} = g(\mathbf{x})$. To determine $F_w(w)$, we can assume therefore without loss of generality that $F(x) = x$ for $0 \le x \le 1$ [see (4-68)]. Even with this simplification, however, the exact form of $F_w(w)$ is complicated. For most purposes, it is sufficient to use the following approximation due to Kolmogoroff:

$$F_w(w) \simeq 1 - 2e^{-2nw^2} \text{ for } w > 1/\sqrt{n} \qquad (9\text{-}68)$$

This yields

$$\gamma = 1 - \delta = F_w(c) \simeq 1 - 2e^{-2nc^2} \qquad \delta \simeq 2e^{-2nc^2}$$

Hence, the γ confidence interval of $F(x)$ is

$$\hat{F}(x) \pm c \qquad c = \sqrt{-\frac{1}{2n}\ln\frac{\delta}{2}} \qquad (9\text{-}69)$$

Thus we can state with confidence coefficient γ that the unknown distribution is a function $F(x)$ located in the zone bounded by the curves $\hat{F}(x) + c$ and $\hat{F}(x) - c$.

Example 9.19

The IQ scores of 40 students rounded off to multiples of 5 are as follows:

x_i	75	80	85	90	95	100	105	110	115	120	125
n_i	1	2	3	5	6	8	6	4	2	2	1

Find the .95 confidence interval of the distribution $F(x)$ of the scores.

In this problem, we have 11 samples x_i with multiplicity n_i, and

$\hat{F}(x) = .025$ .075 .150 .275 .425 .625 .775 .875 .925 .975 1.000

for $x_i \le x < x_{i+1}$, $i = 1, \ldots, 11$ where $x_{12} = \infty$. With $\delta = .05$, equation (9-69) yields the interval

$$\hat{F}(x) \pm c \qquad c = \sqrt{-\frac{1}{80} \ln \frac{.05}{2}} = .217$$

For example, for $90 \le x < 95$, the interval $.425 \pm .217$ results. ■

Kolmogoroff's test leads to reasonable estimates only if n is large. As the last example suggests, n should be at least of the order of 100 (see also Problem 9-31). If this is true, the approximation error in (9-68) is negligible.

9-5 Moments and Maximum Likelihood

The general parameter estimation problem can be phrased as follows: The density of an RV $\mathbf{x}$ is a function $f(x, \theta_1, \ldots, \theta_r)$ depending on $r \ge 1$ parameters θ_i taking values in a region Θ called the *parameter space*. Find a point in that space that is close in some sense to the unknown parameter vector $(\theta_1, \ldots, \theta_r)$. In the earlier sections of this chapter, we developed special techniques involving the commonly used parameters. The results were based on the following form of (4-89): If a parameter θ equals the mean of some function $q(\mathbf{x})$ of $\mathbf{x}$, then we use as its estimate $\hat{\theta}$ the empirical estimate of the mean of $q(\mathbf{x})$:

$$\theta = E\{q(\mathbf{x})\} \qquad \hat{\theta} = \frac{1}{n} \sum q(x_i) \tag{9-70}$$

In this section, we develop two *general methods* for estimating arbitrary parameters. The first is based on (9-70).

Method of Moments

The moment m_k of an RV $\mathbf{x}$ is the mean of $\mathbf{x}^k$. It can therefore be estimated from (9-70) with $q(x) = x^k$. Thus

$$m_k = E\{\mathbf{x}^k\} \qquad \hat{m}_k = \frac{1}{n} \sum x_i^k \tag{9-71}$$

The parameters θ_i of the distribution of $\mathbf{x}$ are functions of m_k because

$$m_k = \int_{-\infty}^{\infty} x^k f(x, \theta_1, \ldots, \theta_r)\, dx \tag{9-72}$$

To find these functions, we assign to the index k the values 1 to r or some other set of r integers. This yields a system of r equations relating the unknown parameters θ_i to the moments m_k. The solution of this system expresses θ_i as functions $\theta_i = y_i(m_1, \ldots, m_r)$ of m_k. Replacing m_k by their estimates $\hat{m}_k$, we obtain

$$\hat{\theta}_i = y_i(\hat{m}_1, \ldots, \hat{m}_r) \tag{9-73}$$

These are the estimates of θ_i obtained with the method of moments.

Note that if $\hat{\theta}$ is the estimate of a parameter θ so obtained, the estimate of a function $\tau(\theta)$ of θ is $\tau(\hat{\theta})$.

Example 9.20

Suppose that **x** has the Rayleigh density

$$f(x, \theta) = \frac{x}{\theta^2} e^{-x^2/2\theta^2} U(x)$$

To estimate θ, we set $k = 1$ in (9-72). This yields

$$m_1 = \frac{1}{\theta^2} \int_0^\infty x^2 e^{-x^2/2\theta^2} dx = \theta \sqrt{\frac{\pi}{2}}$$

Hence,

$$\theta = m_1 \sqrt{\frac{2}{\pi}} \qquad \hat{\theta} = \hat{m}_1 \sqrt{\frac{2}{\pi}} = \bar{x} \sqrt{\frac{2}{\pi}}$$ ■

Example 9.21

The RV **x** is normal with mean 2. We shall estimate its standard deviation σ. Since $m_2 = \eta^2 + \sigma^2 = 4 + \sigma^2$, we conclude that

$$\hat{\sigma} = \sqrt{\hat{m}_2 - 4} \qquad \hat{m}_2 = \frac{1}{n} \sum x_i^2$$ ■

Example 9.22

We wish to estimate the mean η and the variance σ^2 of a normal RV. In this problem, $m_1 = \eta$ and $\sigma^2 = m_2 - m_1^2$; hence,

$$\hat{\eta} = \hat{m}_1 = \bar{x} \qquad \hat{\sigma}^2 = \hat{m}_2 - (\bar{x})^2.$$ ■

Method of Maximum Likelihood

We shall introduce the method of maximum likelihood (ML) starting with the following prediction problem: Suppose that the RV **x** has a known density $f(x, \theta)$. We wish to predict its value x at the next trial; that is, we wish to find a number $\hat{x}$ close in some sense to x. The probability that **x** is the interval $(x, x + dx)$ equals $f(x, \theta)dx$. If we decide to select x such as to maximize this probability, we must set $\hat{x} = x_{max}$ where x_{max} is the *mode* of x. In this problem, θ is specified and x_{max} is the value of x for which the density $f(x, \theta)$, plotted as a function of x, is maximum (Fig. 9.16a).

In the estimation problem, θ is unknown. We observe the value x of **x**, and we wish to find a number $\hat{\theta}$ close in some sense to the unknown θ. To do so, we plot the density $f(x, \theta)$ as a function of θ where x is the observed value of **x**. The curve so obtained will be called the *likelihood function* of θ. The value $\hat{\theta} = \theta_{max}$ of θ for which this curve is maximum is the ML estimate of θ. Again, the probability that **x** is in the interval $(x, x + dx)$ equals $f(x, \theta)dx$. For a given x, this is maximum if $\theta = \theta_{max}$.

We repeat: In the prediction problem, θ is known and x_{max} is the value of x for which the density $f(x, \theta)$ is maximum. In the estimation problem, x is

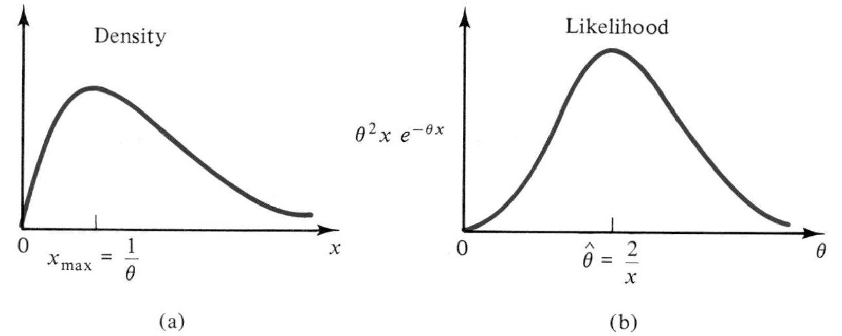

Figure 9.16

known and θ_{max} is the value of θ for which the likelihood function $f(x, \theta)$ is maximum.

Note Suppose that $\theta = \theta(\tau)$ and $f(x, \theta(\tau))$ is maximum for $\tau = \hat{\tau}$. It then follows that $f(x, \theta)$ is maximum for $\theta = \theta(\hat{\tau})$. From this we conclude that if $\hat{\tau}$ is the ML estimate of a parameter τ, the ML estimate of a function $\theta \equiv \theta(\hat{\tau})$ of τ equals $\theta(\hat{\tau})$.

Example 9.23

The time to failure of type A bulbs is an RV **x** with density

$$f(x, \theta) = \theta^2 x e^{-\theta x} U(x)$$

In this example, the density of **x** (Fig. 9.16a) is maximum for $x = 1/\theta$, and its likelihood function (Fig. 9.16b) is maximum for $\theta = 2/x$. This leads to the following estimates:

Prediction: The ML estimate $\hat{x}$ of the life length x of a particular bulb equals $1/\theta$.

Estimation: The ML estimate $\hat{\theta}$ of the parameter θ in terms of the observed life length x of a particular bulb equals $2/x$. ■

We shall now estimate θ in terms of the n observations x_i. The joint density of the corresponding samples $\mathbf{x}_i$ is the product

$$f(X, \theta) = f(x_1, \theta) \cdot \cdot \cdot f(x_n, \theta) \qquad X = [x_1, \ldots x_n]$$

For a given θ, $f(X, \theta)$ is a function of the n variables x_i. For a given X, $f(X, \theta)$ is a function of the single variable θ. If X is the vector of the observations x_i, then $f(X, \theta)$ is called the *likelihood function* of θ. Its logarithm

$$L(X, \theta) = \ln f(X, \theta) = \Sigma \ln f(x_i, \theta) \tag{9-74}$$

is the *log-likelihood* function of θ. The ML estimate $\hat{\theta}$ of θ is the value of θ for which the likelihood function is maximum. If the maximum of $f(X, \theta)$ is in the interior of the domain Θ of θ, then $\hat{\theta}$ is a root of the equation

$$\frac{\partial f(X, \theta)}{\partial \theta} = 0 \tag{9-75}$$

In most cases, it is also a root of the equation

$$\frac{\partial L(X, \theta)}{\partial \theta} = \sum \frac{1}{f(x_i, \theta)} \frac{\partial f(x_i, \theta)}{\partial \theta} = 0 \qquad (9\text{-}76)$$

Thus to find $\hat{\theta}$, we solve either (9-75) or (9-76). The resulting solution depends on the observations x_i.

Example 9.24

The RV $\mathbf{x}$ has an exponential density $\theta e^{-\theta x} U(x)$. We shall find the ML estimate $\hat{\theta}$ of θ. In this example,

$$f(X, \theta) = \theta^n e^{-\theta(x_1 + \cdots + x_n)} = \theta^n e^{-\theta n \bar{x}} \qquad L(X, \theta) = n \ln \theta - \theta n \bar{x}$$

Hence,

$$\frac{\partial L}{\partial \theta} = \frac{n}{\theta} - n\bar{x} = 0 \qquad \hat{\theta} = \frac{1}{\bar{x}}$$

Thus the ML estimator of θ equals $1/\bar{\mathbf{x}}$. This estimator is biased because $E\{\bar{\mathbf{x}}\} = \eta_x = 1/\theta$ and $E\{1/\bar{\mathbf{x}}\} \neq 1/\eta_x$. ■

In Examples 9.23 and 9.24, the likelihood function was differentiable at $\theta = \hat{\theta}$, and its maximum was determined from (9-76) by differentiation. In the next example, the maximum is a corner point and $\hat{\theta}$ is determined by inspection.

Example 9.25

The RV $\mathbf{x}$ is uniform in the interval $(0, \theta)$, as in Fig. 9.17a. We shall find the ML estimate $\hat{\theta}$ of θ.

The joint density of the sample $\mathbf{x}_i$ equals

$$f(X, \theta) = \frac{1}{\theta^n} \qquad \text{for } 0 < x_1, \ldots, x_n < \theta$$

and it is 0 otherwise. The corresponding likelihood function is shown in Fig. 9.17b, where z is the maximum of x_i. As we see from the figure, $f(X, \theta)$ is maximum at the corner point $z = \max x_i$ of the sample space; hence, the ML estimate of θ equals the maximum of the observations x_i. This estimate is biased (see Example 7.4) because $E\{\mathbf{x}_{\max}\} = (n + 1)\theta/n$. However, the estimate is consistent. ■

Figure 9.17

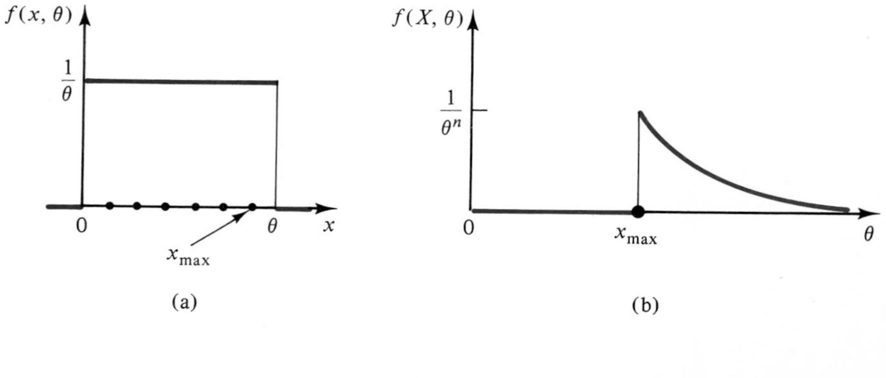

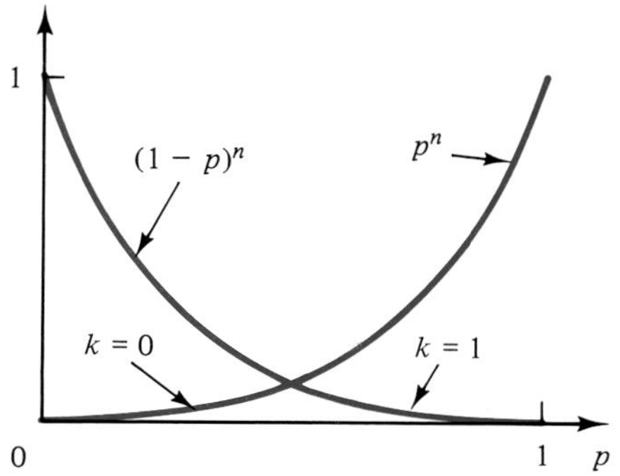

Figure 9.18

Example 9.26

We shall find the ML estimate $\hat{p}$ of the probability p of an event $\mathscr{A}$. For this purpose, we form the zero-one RV $\mathbf{x}$ associated with the event $\mathscr{A}$. The joint density of the corresponding samples $\mathbf{x}_i$ equals $p^k q^{n-k}$ where k is the number of successes of $\mathscr{A}$. This yields

$$L(X, p) = k \ln p + (n - k)\ln q \qquad \frac{\partial L}{\partial p} = \frac{k}{p} - \frac{n-k}{q} = 0$$

Solving for p, we obtain

$$k(1 - p) - (n - k)p = 0 \qquad \hat{p} = \frac{k}{n}$$

This holds if $\hat{p}$ is an interior point of the parameter space $0 \le p \le 1$. If $p = 0$ or 1, the maximum is an endpoint (Fig. 9.18). However, even in this case, $\hat{p} = k/n$. ■

The determination of the ML estimates $\hat{\theta}_i$ of several parameters θ_i proceeds similarly. We form their likelihood function $f(X, \theta_1, \ldots, \theta_r)$, and we determine its maxima or, equivalently, the maxima of the log-likelihood

$$L(X, \theta_1, \ldots, \theta_r) = \ln f(X, \theta_1, \ldots, \theta_r)$$

Example 9.27

We shall find the ML estimates $\hat{\eta}$ and $\hat{v}$ of the mean η and the variance $v = \sigma^2$ of a normal RV.

In this case,

$$f(X, \eta, v) = \frac{1}{(\sqrt{2\pi v})^n} \exp\left\{-\frac{1}{2v}\sum (x_i - \eta)^2\right\}$$

$$L(X, \eta, v) = -\frac{n}{2}\ln(2\pi v) - \frac{1}{2v}\sum (x_i - \eta)^2 \tag{9-77}$$

$$\frac{\partial L}{\partial \eta} = \frac{1}{v}\sum (x_i - \eta) = 0 \qquad \frac{\partial L}{\partial v} = -\frac{n}{2v} + \frac{1}{2v^2}\sum (x_i - \eta)^2 = 0$$

Solving the system, we obtain

$$\hat{\eta} = \bar{x} \qquad \hat{v} = \frac{1}{n}\sum (x_i - \bar{x})^2$$

The estimate $\hat{\eta}$ is unbiased, but the estimate $\hat{v}$ is biased because $E\{\hat{\mathbf{v}}\} = (n-1)\sigma^2/n$. However, both estimates are consistent. ■

Asymptotic Properties of ML Estimators The ML method can be used to estimate any parameter. For moderate values of n, the estimate is not particularly good: It is generally biased, and its variance might be large. Furthermore, the determination of the distribution of $\hat{\boldsymbol{\theta}}$ is not simple. As n increases, the estimate improves, and for large n, $\hat{\boldsymbol{\theta}}$ is nearly normal with mean θ and minimum variance. This is based on the following important theorem.

■ ***Theorem.*** For large n, the distribution of the ML estimator $\hat{\boldsymbol{\theta}}$ of a parameter θ approaches a normal curve with mean θ and variance $1/nI$ where

$$I = E\left\{\left|\frac{\partial}{\partial\theta} L(\mathbf{x}, \theta)\right|^2\right\} \qquad L(x, \theta) = \ln f(x, \theta) \tag{9-78}$$

In Section 9-6, we show that the variance of any estimator of θ cannot be smaller than $1/nI$. From this and the theorem it follows that the ML estimator of a parameter θ is asymptotically the best estimator of θ. The number I in (9-78) is called the *information* about θ contained in $\mathbf{x}$. This concept is important in information theory. Using integration by parts, we can show (Problem 9-36) that

$$I = -E\left\{\frac{\partial^2}{\partial\theta^2} L(\mathbf{x}, \theta)\right\} \tag{9-79}$$

In many cases, it is simpler to evaluate I from (9-79).

The theorem is not always true. For its validity, the likelihood function must be differentiable. This condition is not too restrictive, however. The proof of the theorem is based on the central limit theorem but it is rather difficult and will be omitted. We shall merely demonstrate its validity with an example.

Example 9.28

Given an RV $\mathbf{x}$ with known mean η, we wish to find the ML estimate $\hat{v}$ of its variance $v = \sigma^2$. As in (9-77),

$$\hat{v} = \frac{1}{n}\sum (x_i - \eta)^2$$

The RV $n\hat{\mathbf{v}}/\sigma^2$ has a $\chi^2(n)$ distribution; hence,

$$E\{\hat{\mathbf{v}}\} = \sigma^2 \qquad \sigma_{\hat{v}}^2 = \frac{2\sigma^4}{n}$$

Furthermore, for large $\hat{n}$, $\hat{\mathbf{v}}$ is normal because it is the sum of the independent RVs $(\mathbf{x}_i - \eta)^2$. Thus, asymptotically, the ML estimate $\hat{\mathbf{v}}$ of σ^2 is normal with mean σ^2 and variance $2\sigma^4/n$. We shall show that this agrees with the theorem. In this problem,

$$\frac{\partial L(x, v)}{\partial v} = \frac{-1}{2v} + \frac{(x-\eta)^2}{2v^2} \qquad \frac{\partial^2 L(x, v)}{\partial v^2} = \frac{1}{2v^2} - \frac{(x-\eta)^2}{v^3}$$

and (9-79) yields

$$I = E\left\{\frac{-1}{2v^2} + \frac{(\mathbf{x}-\eta)^2}{v^3}\right\} = \frac{1}{2\sigma^4} \qquad \frac{1}{nI} = \frac{2\sigma^4}{n}$$

in agreement with the theorem. ■

9-6

Best Estimators and the Rao-Cramèr Bound

All estimators considered earlier were more or less empirical. In this section, we examine the problem of determining *best* estimators. The best estimator of a parameter θ is a statistic $\hat{\boldsymbol{\theta}} = g(\mathbf{X})$ minimizing the MS error

$$e = E\{(\hat{\boldsymbol{\theta}} - \theta)^2\} = \int_R (g(X) - \theta)^2 f(X, \theta)\, dX \tag{9-80}$$

In this equation, θ is the parameter to be estimated, $f(X, \theta)$ is the joint density of the samples $\mathbf{x}_i$, and $g(X)$ is the function to be determined. To simplify the problem somewhat, we shall impose the condition that the estimator $\hat{\boldsymbol{\theta}}$ be unbiased:

$$E\{\hat{\boldsymbol{\theta}}\} = \theta \tag{9-81}$$

This condition is only mildly restrictive. As we shall see, in many cases, the best estimators are also unbiased. For unbiased estimators, the LMS error e equals the variance of $\hat{\boldsymbol{\theta}}$. Hence, best unbiased estimators are also minimum-variance unbiased estimators.

The problem of determining an unbiased best estimator is difficult because not only the function $g(X)$ but also the parameter θ is unknown. In fact, it is even difficult to establish whether a solution exists. We show next that if a solution does exist, it is unique.

▪ ***Theorem.*** If $\hat{\boldsymbol{\theta}}_1$ and $\hat{\boldsymbol{\theta}}_2$ are two unbiased minimum-variance estimators of a parameter θ, then $\hat{\boldsymbol{\theta}}_1 = \hat{\boldsymbol{\theta}}_2$.

▪ ***Proof.*** The variances of $\hat{\boldsymbol{\theta}}_1$ and $\hat{\boldsymbol{\theta}}_2$ must be equal because otherwise the one or the other would not be best. Denoting by σ^2 their common variance, we conclude that the statistic

$$\hat{\boldsymbol{\theta}} = \frac{1}{2}(\hat{\boldsymbol{\theta}}_1 + \hat{\boldsymbol{\theta}}_2)$$

is an unbiased estimator of θ and its variance equals

$$\sigma^2_{\hat{\theta}} = \frac{1}{4}(\sigma^2 + \sigma^2 + 2r\sigma^2) = \frac{1}{2}\sigma^2(1 + r)$$

where r is the correlation coefficient of $\hat{\boldsymbol{\theta}}_1$ and $\hat{\boldsymbol{\theta}}_2$. This shows that if $r < 1$, then $\sigma_{\hat{\theta}} < \sigma$, which is impossible because $\hat{\boldsymbol{\theta}}_1$ is best; hence, $r = 1$. And since the RVs $\hat{\boldsymbol{\theta}}_1$ and $\hat{\boldsymbol{\theta}}_2$ have the same mean and variance, we conclude as in (5-46) that $\hat{\boldsymbol{\theta}}_1 = \hat{\boldsymbol{\theta}}_2$.

We continue with the search for best estimators. We establish a lower bound for the MS error of all estimators and develop the class of distributions for which this bound is reached. This material is primarily of theoretical interest.

Regularity The density of an RV $\mathbf{x}$ satisfies the area condition

$$\int_{-\infty}^{\infty} f(x, \theta)\, dx = 1$$

The limits of integration may be finite, but in most cases they do not depend on the parameter θ. We can then deduce, differentiating with respect to θ, that

$$\int_{-\infty}^{\infty} \frac{\partial f(x, \theta)}{\partial \theta}\, dx = 0 \qquad \text{(9-82)}$$

We shall say that the density $f(x, \theta)$ is *regular* if it satisfies (9-82). Thus $f(x, \theta)$ is regular if it is differentiable with respect to θ and the limits of integration in (9-82) do not depend on θ. Most densities of interest are regular; there are, however, exceptions. If, for example, $\mathbf{x}$ is uniform in the interval $(0, \theta)$, its density is not regular because θ is a boundary point of the range $0 \le x \le \theta$ of $\mathbf{x}$. In the following, we consider only regular densities.

Information The log-likelihood of the RV $\mathbf{x}$ is by definition the function $L(x, \theta) = \ln f(x, \theta)$. From (9-82) it follows that

$$\int_{-\infty}^{\infty} \frac{\partial L(x, \theta)}{\partial \theta} f(x, \theta)\, dx = \int_{-\infty}^{\infty} \frac{1}{f(x, \theta)} \frac{\partial f(x, \theta)}{\partial \theta} f(x, \theta)\, dx = 0$$

This shows that the mean of the RV $\partial L(\mathbf{x}, \theta)/\partial\theta$ equals 0. Denoting its variance by I, we conclude that

$$I = E\left\{\left|\frac{\partial L(\mathbf{x}, \theta)}{\partial \theta}\right|^2\right\} \qquad \text{(9-83)}$$

The number I is the information about θ contained in $\mathbf{x}$ [see also (9-78)].

Consider, next, the likelihood function $L(X, \theta) = \ln f(X, \theta)$ of the sample $\mathbf{X} = [\mathbf{x}_1, \ldots, \mathbf{x}_n]$. Its derivative equals

$$\frac{\partial L(X, \theta)}{\partial \theta} = \frac{1}{f(X, \theta)} \frac{\partial f(X, \theta)}{\partial \theta} = \sum \frac{\partial L(x_i, \theta)}{\partial \theta}$$

The RVs $\partial L(\mathbf{x}_i, \theta)/\partial\theta$ have zero mean and variance I. Furthermore, they are independent because they are functions of the independent RVs $\mathbf{x}_i$. This leads to the conclusion that

$$E\left\{\frac{\partial L(\mathbf{X}, \theta)}{\partial \theta}\right\} = 0 \qquad E\left\{\left|\frac{\partial L(\mathbf{X}, \theta)}{\partial \theta}\right|^2\right\} = nI \qquad \text{(9-84)}$$

The number nI is the information about θ contained in $\mathbf{X}$.

We turn now to our main objective, the determination of the greatest lower bound of the variance of any estimator $\hat{\boldsymbol{\theta}} = g(\mathbf{X})$ of θ. We assume first that $\hat{\boldsymbol{\theta}}$ is an unbiased estimator. From this assumption it follows that

$$E\{\hat{\boldsymbol{\theta}}\} = \int_R g(X) f(X, \theta)\, dX = \theta$$

Differentiating with respect to θ, we obtain

$$\int_R g(X) \frac{\partial f(X, \theta)}{\partial \theta}\, dX = 1$$

This yields the identity

$$E\left\{g(\mathbf{X}) \frac{\partial L(\mathbf{X}, \theta)}{\partial \theta}\right\} = 1 \qquad \text{(9-85)}$$

The relationships just established will be used in the proof of the following fundamental theorem. The proof is based on Schwarz's inequality [see (5-42)]: For any $\mathbf{z}$ and $\mathbf{w}$,

$$E^2\{\mathbf{zw}\} \le E\{\mathbf{z}^2\}E\{\mathbf{w}^2\} \tag{9-86}$$

Equality holds iff $\mathbf{z} = c\mathbf{w}$.

THE RAO-CRAMÈR BOUND The variance of an *unbiased* estimator $\hat{\boldsymbol{\theta}} = g(\mathbf{X})$ of a parameter θ cannot be smaller than the inverse $1/nI$ of the information nI contained in the sample $\mathbf{X}$:

$$\sigma_{\hat{\theta}}^2 = E\{[g(\mathbf{X}) - \theta]^2\} \ge \frac{1}{nI} \tag{9-87}$$

Equality holds iff

$$\frac{\partial L(X, \theta)}{\partial \theta} = nI\,[g(X) - \theta] \tag{9-88}$$

■ ***Proof.*** Multiplying the first equation in (9-84) by θ and subtracting from (9-85), we obtain

$$1 = E\left\{[g(\mathbf{X}) - \theta]\,\frac{\partial L(\mathbf{X}, \theta)}{\partial \theta}\right\}$$

We square both sides and apply (9-86) to the RVs $g(\mathbf{X}) - \theta$ and $\partial L(\mathbf{X}, \theta)/\partial\theta$. This yields [see (9-84)]

$$1 \le E\{[g(\mathbf{X}) - \theta]^2\}E\left\{\left|\frac{\partial L(\mathbf{X}, \theta)}{\partial \theta}\right|^2\right\} = \sigma_{\hat{\theta}}^2 nI$$

and (9-87) results.

To prove (9-88), we observe that (9-87) is an equality iff $g(X) - \theta$ equals $c\partial L(X, \theta)/\partial\theta$. This yields

$$1 = E\left\{c\left|\frac{\partial L(\mathbf{X}, \theta)}{\partial \theta}\right|^2\right\} = cnI$$

Hence, $c = 1/nI$, and (9-88) results.

Suppose now that $\hat{\boldsymbol{\theta}}$ is a biased estimator of the parameter θ with mean $E\{\hat{\boldsymbol{\theta}}\} = \tau(\theta)$. If we interpret $\hat{\boldsymbol{\theta}}$ as the estimator of $\tau(\theta)$, our estimator is unbiased. We can therefore apply (9-87) subject to the following modifications: We replace the function $\partial L(X, \theta)/\partial\theta$ by the function

$$\frac{\partial L[X, \theta(\tau)]}{\partial \tau} = \frac{\partial L(X, \theta)}{\partial \theta}\,\frac{1}{\tau'(\theta)}$$

and the information nI about θ contained in $\mathbf{X}$ by the information

$$E\left\{\left|\frac{\partial L[\mathbf{X}, \theta(\tau)]}{\partial \tau}\right|^2\right\} = \frac{1}{[\tau'(\theta)]^2}\,E\left\{\left|\frac{\partial L(\mathbf{X},\theta)}{\partial \theta}\right|^2\right\} = \frac{nI}{[\tau'(\theta)]^2}$$

about $\tau(\theta)$ contained in **X**. This yields the following generalization of the Rao-Cramèr bound.

■ ***Corollary.*** If $\hat{\boldsymbol{\theta}} = g(\mathbf{X})$ is a *biased* estimator of a parameter θ and $E\{\hat{\boldsymbol{\theta}}\} = \tau(\theta)$, then

$$\sigma_{\hat{\theta}}^2 = E\{[g(\mathbf{X}) - \tau(\theta)]^2\} \geq \frac{[\tau'(\theta)]^2}{nI} \tag{9-89}$$

Equality holds iff

$$\frac{\partial L(X, \theta)}{\partial \theta} = \frac{nI}{\tau'(\theta)} [g(X) - \theta] \tag{9-90}$$

EFFICIENT ESTIMATORS AND DENSITIES OF EXPONENTIAL TYPE We shall say that $\hat{\boldsymbol{\theta}}$ is the *most efficient* estimator of a parameter θ if it is unbiased and its variance equals the bound $1/nI$ in (9-87). If $\hat{\boldsymbol{\theta}}$ is biased with mean $\tau(\theta)$ and its variance equals the bound in (6-89), it is the most efficient estimator of the parameter $\tau(\theta)$.

The Rao-Cramèr bound applies only to regular densities. If $f(x, \theta)$ is regular, the most efficient estimator of θ is also the best estimator.

The class of distributions that lead to most efficient estimators satisfy the equality condition (9-88) or (9-90). This condition leads to the following class of densities.

■ ***Definition.*** We shall say that a density $f(x, \theta)$ is of the exponential type with respect to the parameter θ if it is of the form

$$f(x, \theta) = h(x) \exp\{a(\theta)q(x) - b(\theta)\} \tag{9-91}$$

where the functions $a(\theta)$ and $b(\theta)$ depend only on θ and the functions $h(x)$ and $q(x)$ depend only on x.

We shall show that the class of exponential type distributions generates the class of most efficient estimators.

■ ***Theorem.*** If the density $f(x, \theta)$ is of the form (9-91), the statistic

$$\hat{\boldsymbol{\theta}} = g(\mathbf{X}) = \frac{1}{n} \sum q(\mathbf{x}_i) \tag{9-92}$$

is the most efficient estimator of the parameter

$$\tau(\theta) = \frac{b'(\theta)}{a'(\theta)} \tag{9-93}$$

The corresponding Rao-Cramèr bound equals

$$\sigma_{\hat{\theta}}^2 = \frac{[\tau'(\theta)]^2}{nI} = \frac{\tau'(\theta)}{na'(\theta)} \tag{9-94}$$

■ ***Proof.*** From (9-91) it follows that

$$\ln f(x, \theta) = \ln h(x) + a(\theta)q(x) - b(\theta)$$

and (9-74) yields

$$\frac{\partial L(X, \theta)}{\partial \theta} = a'(\theta) \sum q(x_i) - nb'(\theta) = na'(\theta)[g(X) - \tau(\theta)]$$

This function satisfies (9-90) with

$$\frac{nI}{\tau'(\theta)} = na'(\theta)$$

Hence, $I = a'(\theta)\tau'(\theta)$. Inserting into (9-89), we obtain (9-94). Note that the converse is also true: If $L(X, \theta)$ satisfies (9-90), then $f(x, \theta)$ is of the exponential type. We give next several illustrations of exponential type distributions.

Normal **(a)** The normal density is of the exponential type with respect to its mean because

$$f(x, \eta) = \frac{1}{\sqrt{2\pi v}} \exp\left\{-\frac{1}{2v}(x^2 - 2x\eta + \eta^2)\right\}$$

This density is of the form (9-91) with

$$a(\eta) = \frac{\eta}{v} \qquad b(\eta) = \frac{\eta^2}{2v} \qquad q(x) = x$$

In this case,

$$\hat{\eta} = g(X) = \frac{1}{n}\sum x_i = \bar{x} \qquad \tau(\eta) = \frac{b'(\eta)}{a'(\eta)} = \eta$$

Hence, the sample mean $\bar{\mathbf{x}}$ is the most efficient estimator of η.

(b) The normal density is also of the exponential type with respect to its variance because

$$f(x, v) = \frac{1}{\sqrt{2\pi}} \exp\left\{-\ln\sqrt{v} - \frac{1}{2v}(x - \eta)^2\right\}$$

This is also of the form (9-91) with

$$a(v) = -\frac{1}{2v} \qquad b(v) = \ln\sqrt{v} \qquad q(x) = (x - \eta)^2$$

In this case, $\tau(v) = v$; hence, the statistic $\hat{v} = \Sigma(x_i - \eta)^2/n$ is the best estimator of v. The variance of the estimation [see (9-94)] equals

$$\sigma_{\hat{v}}^2 = \frac{\tau'(v)}{na'(v)} = \frac{2\sigma^4}{n}$$

Exponential The exponential density

$$f(x, \theta) = \theta e^{-\theta x} = \exp\{-\theta x + \ln\theta\}$$

is of the exponential type with $a(\theta) = -\theta$, $b(\theta) = -\ln\theta$, and $q(x) = x$. Hence, the statistic $\Sigma x_i/n = \bar{x}$ is the most efficient estimator of the parameter

$$\tau(\theta) = \frac{b'(\theta)}{a'(\theta)} = \frac{1}{\theta}$$

The Rao-Cramèr bound also holds for discrete type RVs. Next we give two illustrations of most efficient estimators and point densities of the exponential type.

Poisson The Poisson density

$$f(x, \theta) = e^{-\theta}\frac{\theta^x}{x!} = \frac{1}{x!}\exp\{x\ln\theta - \theta\} \qquad x = 0, 1, \ldots$$

is of the exponential type with

$$a(\theta) = \ln \theta \qquad b(\theta) = \theta \qquad q(x) = x \qquad \tau(\theta) = \frac{b'(\theta)}{a'(\theta)} = \theta$$

Hence, the statistic $\Sigma\, q(x_i)/n = \bar{x}$ is the most efficient estimator of θ, and its variance equals $1/na'(\theta) = \theta/n$. Note that the corresponding measure of information about θ contained in the sample X equals $nI = na'(\theta)\tau'(\theta) = n/\theta$.

Probability If $p = P(\mathscr{A})$ is the probability of an event $\mathscr{A}$ and $\mathbf{x}$ is the zero-one RV associated with $\mathscr{A}$, then $\mathbf{x}$ is an RV with point density

$$f(x, p) = p^x(1-p)^{1-x} = \exp\{x \ln p + (1-x)\ln(1-p)\} \qquad x = 0, 1$$

This is an exponential density with

$$a(p) = \ln \frac{p}{1-p} \qquad b(p) = \ln(1-p) \qquad q(x) = x \qquad \tau(p) = p$$

Hence, the ratio $\Sigma\, x_i/n = k/n$ is the most efficient estimator of p.

Sufficient Statistics and Completeness

We turn now to the problem of determining the best estimator $\hat{\boldsymbol{\theta}}$ of an arbitrary parameter θ. If the density $f(x, \theta)$ is of the exponential type, then $\hat{\boldsymbol{\theta}}$ exists, and it equals the most efficient estimator $g(\mathbf{X})$ in (9-92). Otherwise, $\hat{\boldsymbol{\theta}}$ might not exist, and if it does, there is no general method for determining it. We show next that for a certain class of densities, the search for the best estimator is simplified.

■ ***Definition.*** If the joint density $f(X, \theta)$ of the n samples $\mathbf{x}_i$ is of the form

$$f(X, \theta) = H(X)J[y(X), \theta] \tag{9-95}$$

where the functions $H(X)$ and $y(X)$ do not depend on θ, the function $y(\mathbf{X})$ is called a *sufficient statistic* of the parameter θ.

From the definition, it follows that if $y(\mathbf{X})$ is a sufficient statistic, then $ky(\mathbf{X})$ or, in fact, any function of $y(\mathbf{X})$ is a sufficient statistic. If $f(x, \theta)$ is a density of exponential type as in (9-91), then

$$f(X, \theta) = \prod h(x_i) \exp\{a(\theta) \sum q(x_i) - nb(\theta)\}$$

is of the form (9-95); hence, the sum $y(\mathbf{X}) = \Sigma\, q(\mathbf{x}_i)$ is a sufficient statistic of θ.

The importance of sufficient statistics in parameter estimation is based on the following theorem.

■ ***Sufficiency Theorem.*** If $\mathbf{z}$ is an arbitrary statistic and $\mathbf{y}$ is sufficient statistic of the parameter θ, then the regression line

$$\psi(y) = E\{\mathbf{z}|y\} = \int_{-\infty}^{\infty} z f_z(z|y)\, dz$$

is also a statistic; that is, it does not depend on θ.

▪ ***Proof.*** To prove this theorem, it suffices to show that the function $f_z(z|y)$ does not depend on θ. As we know,

$$f_z(z|y)\,dz = \frac{f_{yz}(y, z)\,dy\,dz}{f_y(y)\,dy} = \frac{P\{(\mathbf{y}, \mathbf{z}) \in D_{yz}\}}{P\{\mathbf{y} \in D_y\}} \tag{9-96}$$

where D_{yz} is a differential region and D_y a vertical strip in the yz plane (Fig. 9.19a). The transformation $y = y(X)$, $z = z(X)$ maps the region D_{yz} of the yz plane on the region Δ_{yz} of the sample space and the region D_y on the region Δ_y (Fig. 9.19b). The numerator and the denominator of (9-96) equal the integral of the density $f(X, \theta)$ in the regions Δ_{yz} and Δ_y, respectively. In these regions, $y(X) = y$ is constant; hence the term $J(y, \theta)$ can be taken outside the integral. This yields [see (9-95)],

$$P\{(\mathbf{y}, \mathbf{z}) \in D_{yz}\} = J(y, \theta) \int_{\Delta_{yz}} H(X)\,dX$$

$$P\{\mathbf{y} \in D_y\} = J(y, \theta) \int_{\Delta_y} H(X)\,dX$$

Inserting into (9-96) and canceling $J(y, \theta)$, we conclude that the function $f_z(z|y)$ does not depend on θ.

▪ ***Corollary 1.*** From the Rao-Blackwell theorem (6-58) it follows that if $\mathbf{z}$ is an unbiased estimator of θ, then the statistic $\hat{\boldsymbol{\theta}} = E\{\mathbf{z}|\mathbf{y}\}$ is also unbiased and its variance is smaller than σ_z^2. Thus if we know an unbiased estimator $\mathbf{z}$ of θ, using the sufficiency theorem, we can construct another unbiased estimator $E\{\mathbf{z}|\mathbf{y}\}$ with smaller variance.

▪ ***Corollary 2.*** The best estimator of θ is a function $\psi(\mathbf{y})$ of its sufficient statistic.

▪ ***Proof.*** Suppose that $\mathbf{z}$ is the best estimator of θ. If $\mathbf{z}$ does not equal $\psi(\mathbf{y})$, then the variance of the statistic $\hat{\boldsymbol{\theta}} = E\{\mathbf{z}|\mathbf{y}\}$ is smaller than σ_z^2. This, however, is impossible; hence, $\mathbf{z} = \psi(\mathbf{y})$.

Figure 9.19

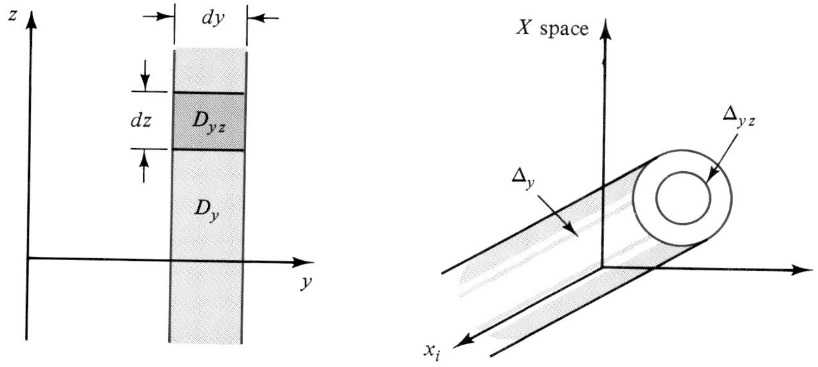

(a) (b)

It follows that to find the best estimator of θ, it suffices to consider only functions of its sufficient statistic $\mathbf{y}$. This simplifies the problem, but even with this simplification, there is no general solution. However, if the density $f_y(y, \theta)$ of the sufficient statistic $\mathbf{y}$ satisfies certain conditions related to the uniqueness problem in transform theory, finding the best estimator simply entails finding an unbiased estimator, as we show next.

COMPLETENESS Consider the integral

$$Q(\theta) = \int_{-\infty}^{\infty} q(y)k(y, \theta)\, dy \tag{9-97}$$

where $k(y, \theta)$ is a given function of y and θ is a parameter taking values in a region Θ. This integral assigns to any funciton $q(y)$ for which it converges a function $Q(\theta)$ defined for every θ in Θ. This function is called the *transform* of $q(y)$ generated by the kernel $k(y, \theta)$. A familiar example is the Laplace transform generated by the kernel $e^{-\theta y}$. We shall say that the kernel $k(y, \theta)$ is *complete* if the transform $Q(\theta)$ has a unique inverse transform. By this we mean that a specific $Q(\theta)$ is the transform of one and only one function $q(y)$. We show next that the notion of completeness leads to a simple determination of the best estimator of θ.

■ ***Definition.*** A sufficient statistic $\mathbf{y}$ is called complete if its density $f_y(y, \theta)$ is a complete kernel.

■ ***Theorem.*** If $\hat{\boldsymbol{\theta}}$ is a function $\hat{\boldsymbol{\theta}} = q(\mathbf{y})$ of the complete statistic $\mathbf{y}$ and its mean equals θ:

$$E\{\hat{\boldsymbol{\theta}}\} = \int_{-\infty}^{\infty} q(y)f_y(y, \theta)\, dy = \theta \tag{9-98}$$

then $\hat{\boldsymbol{\theta}}$ is the best unbiased estimator of θ.

■ ***Proof.*** Suppose that $\mathbf{z}$ is the best unbiased estimator of θ. As we have seen, $\mathbf{z} = \psi(\mathbf{y})$; hence,

$$E\{\mathbf{z}\} = \int_{-\infty}^{\infty} \psi(y)f_y(y, \theta)\, dy = \theta \tag{9-99}$$

The last two equations show that θ is the transform of the functions $q(y)$ and $\psi(y)$ generated by the kernel $f_y(y, \theta)$. This kernel is complete by assumption; hence, θ has a unique inverse. From this it follows that $q(y) = \psi(y)$; therefore, $\hat{\boldsymbol{\theta}} = \mathbf{z}$.

These results lead to the following conclusion: If $\mathbf{y}$ is a sufficient and complete statistic of θ, then to find the best unbiased estimator of θ, it suffices to find merely an unbiased estimator $\mathbf{w}$. Indeed, starting from $\mathbf{w}$, we form the RV $\mathbf{z} = E\{\mathbf{w}|\mathbf{y}\}$. This RV is an unbiased function of $\mathbf{y}$; hence, according to the sufficiency theorem, it is the best estimator of θ.

These conclusions are based on the completeness of $\mathbf{y}$. The problem of establishing completeness is not simple. For exponential type distributions, completeness follows from the uniqueness of the inversion of the Laplace transform. In other cases, special techniques must be used. Here is an illustration.

Example 9.20

We are given an RV $\mathbf{x}$ with uniform distribution in the interval $(0, \theta)$ (Fig. 9.20a), and we wish to find the best estimator $\hat{\boldsymbol{\theta}}$ of θ.

In this case,

$$f(X, \theta) = \frac{1}{\theta^n} \qquad 0 < x_1, \ldots, x_n < \theta$$

and zero otherwise. This density can be expressed in terms of the maximum $\mathbf{z} = \mathbf{x}_{\max}$ and the minimum $\mathbf{w} = \mathbf{x}_{\min}$ of $\mathbf{x}_i$. Indeed, $f(X, \theta) = 0$ iff $w < 0$ or $z > \theta$; hence,

$$f(X, \theta) = \frac{1}{\theta^n} U(w)U(\theta - z) \tag{9-100}$$

where U is the unit-step function. This density is of the form (9-95) with $y(X) = z$; hence, the function $\mathbf{y} = \mathbf{x}_{\max} = \mathbf{z}$ is a sufficient statistic of θ, and its density (see Example 7.6) equals

$$f_y(y, \theta) = \frac{n}{\theta^n} y^{n-1} \qquad 0 < y < \theta \tag{9-101}$$

as in Fig. 9.20b. Next we show that $\mathbf{y}$ is complete. It suffices to show that if

$$Q(\theta) = \frac{n}{\theta^n} \int_0^\theta q(y) y^{n-1}\, dy$$

then $Q(\theta)$ has a unique inverse $q(y)$. For this purpose, we multiply both sides by θ^n and differentiate with respect to θ. This yields

$$n\theta^{n-1}Q(\theta) + \theta^n Q'(\theta) = nq(\theta)\theta^{n-1}$$

Hence, $q(\theta) = Q(\theta) + (\theta/n)Q'(\theta)$ is the unique inverse of $Q(\theta)$.

To complete the determination of $\hat{\boldsymbol{\theta}}$, we must find an unbiased estimator of θ. From (9-101) it follows that $E\{\mathbf{y}\} = n\theta/(n + 1)$. This leads to the conclusion that the statistic

$$\hat{\boldsymbol{\theta}} = \frac{n+1}{n}\mathbf{y} = \frac{n+1}{n}\mathbf{x}_{\max}$$

is an unbiased estimator of θ. And since it is a function of the complete statistic $\mathbf{y}$, it is the best estimator. ■

Figure 9.20

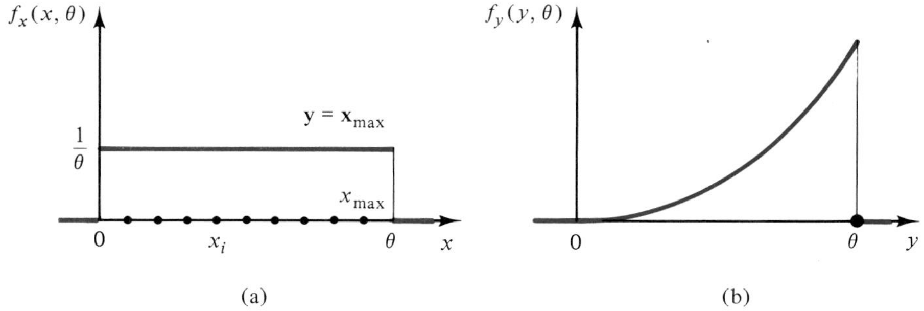

Let us turn, finally, to the use of completeness in establishing the independence between the sufficient statistic **y** and certain other statistics.

■ ***Theorem.*** If **z** is a statistic such that its density is a function $f_z(z)$ that does not depend on θ and **y** is a complete statistic of θ, then the RVs **z** and **y** are independent.

■ ***Proof.*** It suffices to show that $f_z(z|y) = f_z(z)$. The area of the density $f_y(y, \theta)$ of **y** equals 1. From this and the total probability theorem (6-31) it follows that

$$f_z(z) = \int_{-\infty}^{\infty} f_z(z|y) f_y(y, \theta)\, dy = \int_{-\infty}^{\infty} f_z(z) f_y(y, \theta)\, dy \tag{9-102}$$

The density $f_z(z)$ does not depend on θ by assumption, and the conditional density $f_z(z|y)$ does note depend on θ because **y** is a sufficient statistic of θ (sufficiency theorem). And since the kernel $f_y(y, \theta)$ is complete, we conclude from (9-102) that $f_z(z|y) = f_z(z)$.

Using the properties of quadratic forms, we have shown in Section 7-4 that the sample mean $\bar{\mathbf{x}}$ and the sample variance $\mathbf{s}^2$ of a normal RV are independent. This result follows directly from the theorem just proved. Indeed, $\bar{\mathbf{x}}$ is a complete statistic of the mean of **x** because its density is normal and a normal kernel is complete. Furthermore, the density of $\mathbf{s}^2$ does not depend on η. Hence, $\bar{\mathbf{x}}$ and $\mathbf{s}^2$ are independent.

Problems

9-1 The diameter of cylindrical rods coming out of a production line is a normal RV **x** with $\sigma = 0.1$ mm. We measure $n = 9$ units and find that the average of the measurements is $\bar{x} = 91$ mm. **(a)** Find c such that with a .95 confidence coefficient, the mean η of **x** is in the interval $\bar{x} \pm c$. **(b)** We claim that η is in the interval (90.95, 91.05). Find the confidence coefficient of our claim.

9-2 The length of a product is an RV **x** with $\sigma = 1$ mm and unknown mean. We measure four units and find that $\bar{x} = 203$ mm. **(a)** Assuming that **x** is a normal RV, find the .95 confidence interval of η. **(b)** The distribution of **x** is unknown. Using Tchebycheff's inequality, find c such that with confidence coefficient .95, η is in the interval $203 \pm c$.

9-3 We know from past records that the life length of type A tires is an RV **x** with $\sigma = 5{,}000$ miles. We test 64 samples and find that their average life length is $\bar{x} = 25{,}000$ miles. Find the .9 confidence interval of the mean of **x**.

9-4 We wish to determine the length a of an object. We use as an estimate of a the average $\bar{x}$ of n measurements. The measurement error is approximately normal with zero mean and standard deviation 0.1 mm. Find n such that with 95% confidence, $\bar{x}$ is within ± 0.2 mm of a.

9-5 An object of length a is measured by two persons using the same instrument. The instrument error is an $N(0, \sigma)$ RV where $\sigma = 1$ mm. The first person measures the object 25 times, and the average of the measurements is $\bar{x} =$ 12 mm. The second person measures the object 36 times, and the average of the measurements is $\bar{y} = 12.8$. We use as point estimate of a the weighed average $\hat{c} = a\bar{x} + b\bar{y}$. **(a)** Find a and b such that $\hat{c}$ is the minimum-variance unbiased estimate of c as in (7-37). **(b)** Find the 0.95 confidence interval of c.

9-6 In a statewide math test, 17 students obtained the following scores:

49 57 64 72 75 77 78 79 81 81 82 84 85 87 89 93 96

Assuming that the scores are approximately normal, find the .95 confidence interval of their mean **(a)** using (9-10); **(b)** using (9-12).

9-7 A grocer weighs 10 boxes of cereal, and the results yield $\bar{x} = 420$ g and $s = 12$ g for the sample mean and sample standard deviation respectively. He then claims with 95% confidence that the mean weight of all boxes exceeds c g. Assuming normality, find c.

9-8 The RV $\mathbf{x}$ is uniformly distributed in the interval $\theta - 2 < x < \theta + 2$. We observe 100 samples x_i and find that their average equals $\bar{x} = 30$. Find the .95 confidence interval of θ.

9-9 Consider an RV $\mathbf{x}$ with density $f(x) = xe^{-x}U(x)$. Predict with 95% confidence that the next value of x will be in the interval (a, b). Show that the length $b - a$ of this interval is minimum if a and b are such that

$$f(a) = f(b) \qquad P\{a < \mathbf{x} < b\} = .95$$

Find a and b.

9-10 (Estimation-prediction). The time to failure of electric bulbs of brand A is a normal RV with $\sigma = 10$ hours and unknown mean. We have used 20 such bulbs and have observed that the average $\bar{x}$ of their time to failure is 80 hours. We buy a new bulb of the same brand and wish to predict with 95% confidence that its time to failure will be in the interval $80 \pm c$. Find c.

9-11 The time to failure of an electric motor is an RV $\mathbf{x}$ with density $\beta e^{-\beta x}U(x)$. **(a)** Show that if $\bar{\mathbf{x}}$ is the sample mean of n samples of $\mathbf{x}$, then the RV $2n\bar{\mathbf{x}}/\beta$ has a $\chi^2(2n)$ distribution. **(b)** We test $n = 10$ motors and find that $\bar{x} = 300$ hours. Find the left .95 confidence interval of β. **(c)** The probability p that a motor will be good after 400 hours equals $p = P\{\mathbf{x} > 400\} = e^{-400\beta}$. Find the .95 confidence interval $p > p_0$ of p.

9-12 Suppose that the time between arrivals of patients in a dentist's office constitutes samples of an RV $\mathbf{x}$ with density $\theta e^{-\theta x}U(x)$. The 40th patient arrived 4 hours after the first. Find the .95 confidence interval of the mean arrival time $\eta = 1/\theta$.

9-13 The number of particles emitted from a radioactive substance in 1 second is a Poisson-distributed RV with mean λ. It was observed that in 200 seconds, 2,550 particles were emitted. Assuming that the numbers of particles in nonoverlapping intervals are independent, find the .95 confidence interval of λ.

9-14 Among 4,000 newborns, 2,080 are male. Find the .99 confidence interval of the probability $p = P\{\text{male}\}$.

9-15 In an exit poll, of 900 voters questioned, 360 responded that they favor a particular proposition. On this basis, it was reported that 40% of the voters favor the proposition. (a) Find the margin of error if the confidence coefficient of the results is .95. (b) Find the confidence coefficient if the margin of error is ±2%.

9-16 In a market survey, it was reported that 29% of respondents favor product A. The poll was conducted with confidence coefficient .95, and the margin of error was ±4%. Find the number of respondents.

9-17 We plan a poll for the purpose of estimating the probability p of Republicans in a community. We wish our estimate to be within ±.02 of p. How large should our sample be if the confidence coefficient of the estimate is .95?

9-18 A coin is tossed once, and heads shows. Assuming that the probability p of heads is the value of an RV $\mathbf{p}$ uniformly distributed in the interval (0.4, 0.6) find its Bayesian estimate (9-37).

9-19 The time to failure of a system is an RV $\mathbf{x}$ with density $f(x, \theta) = \theta e^{-\theta x}U(x)$. We wish to find the Bayesian estimate $\hat{\theta}$ of θ in terms of the sample mean $\bar{\mathbf{x}}$ of the n samples x_i of $\mathbf{x}$. We assume that θ is the value of an RV $\boldsymbol{\theta}$ with prior density $f_\theta(\theta) = ce^{-c\theta}U(\theta)$. Show that

$$\hat{\theta} = \frac{n}{c + n\bar{x}} \underset{n\to\infty}{\longrightarrow} \frac{1}{\bar{x}}$$

9-20 The RV $\mathbf{x}$ has a Poisson distribution with mean θ. We wish to find the Bayesian estimate $\hat{\theta}$ of θ under the assumption that θ is the value of an RV $\boldsymbol{\theta}$ with prior density $f_\theta(\theta) \sim \theta^b e^{-c\theta}U(\theta)$. Show that

$$\hat{\theta} = \frac{n\bar{x} + b + 1}{n + c}$$

9-21 Suppose that $\mathbf{x}$ is the yearly starting income of teachers with a bachelor's degree and $\mathbf{y}$ is the corresponding income of teachers with a master's degree. We wish to estimate the difference $\eta_x - \eta_y$ of their mean incomes. We question $n = 100$ teachers with a bachelor's degree and $m = 50$ teachers with a master's degree and find the following averages:

$$\bar{x} = 20\text{K} \qquad \bar{y} = 24\text{K} \qquad s_x = 3.1\text{K} \qquad s_y = 4\text{K}$$

Assuming that the RVs $\mathbf{x}$ and $\mathbf{y}$ are normal with the same variance, find the .95 confidence interval of $\eta_x - \eta_y$.

9-22 Suppose that the IQ scores of children in a certain grade are the samples of an $N(\eta, \sigma)$ RV $\mathbf{x}$. We test 10 children and obtain the following averages: $\bar{x} = 90$, $s = 5$. Find the .95 confidence interval of η and of σ.

9-23 The RVs $\mathbf{x}_i$ are i.i.d. and $N(0, \sigma)$. We observe that $x_1^2 + \cdots + x_{10}^2 = 4$. Find the .95 confidence interval of σ.

9-24 The readings of a voltmeter introduces an error $\boldsymbol{\nu}$ with mean 0. We wish to estimate its standard deviation σ. We measure a calibrated source $V = 3$ volts four times and obtain the values 2.90, 3.15, 3.05, and 2.96. Assuming that $\boldsymbol{\nu}$ is normal, find the .95 confidence interval of σ.

9-25 We wish to estimate the correlation between freshman grades and senior grades. We examine the records of 100 students and we find that their sample correlation coefficient is $\hat{r} = .45$. Using Fisher's auxiliary variable, find the 0.9 confidence interval of r.

9-26 Given the n paired samples $(\mathbf{x}_i, \mathbf{y}_i)$ of the RVs $\mathbf{x}$ and $\mathbf{y}$, we form their sample means $\bar{\mathbf{x}}$ and $\bar{\mathbf{y}}$ and the sample covariance

$$\hat{\boldsymbol{\mu}}_{11} = \frac{1}{n-1} \sum (\mathbf{x}_i - \bar{\mathbf{x}})(\mathbf{y}_i - \bar{\mathbf{y}})$$

Show that $E\{\hat{\boldsymbol{\mu}}_{11}\} = \mu_{11}$.

9-27 (a) (Cauchy-Schwarz inequality). Show that

$$\left(\sum_{i=1}^{n} a_i b_i\right)^2 \le \sum_{i=1}^{n} a_i^2 \sum_{i=1}^{n} b_i^2$$

(b) Show that if

$$\hat{r}^2 = \frac{\left[\sum (x_i - \bar{x})(y_i - \bar{y})\right]^2}{\sum (x_i - \bar{x})^2 \sum (y_i - \bar{y})^2} \qquad \text{then} \qquad |\hat{r}| \le 1$$

9-28 Given the 16 samples

$x_i = 61$	42	93	75	40	73	64	68
78	49	54	87	84	71	72	58

of the RV **x**, find the probability that its median is between 68 and 75 **(a)** exactly from (9-60); **(b)** approximately from (9-61).

9-29 The RVS $\mathbf{y}_1, \ldots, \mathbf{y}_5$ are the order statistics of the five samples $\mathbf{x}_1, \ldots, \mathbf{x}_5$ of **x**. Find the probability $P\{\mathbf{y}_1 < x_u < \mathbf{y}_5\}$ that the u-percentile x_u of **x** is between $\mathbf{y}_1$ and $\mathbf{y}_5$ for $u = .5, .4$, and $.3$.

9-30 We use as estimate of the median $x_{.5}$ of an RV **x** the interval (y_k, y_{k+1}) between the order statistics closest to the center of the range (y_1, y_n): $k \le n/2 \le k + 1$. Using (9-61), show that

$$P\{\mathbf{y}_{k-1} < x_{.5} < \mathbf{y}_{k+1}\} \simeq \sqrt{\frac{2}{\pi n}}$$

9-31 We use as the estimate of the distribution $F(x)$ of an RV **x** the empirical distribution $\hat{F}(x)$ in (9-67) obtained with n samples x_i. We wish to claim with 90% confidence that the unknown $F(x)$ is within $\pm.02$ from $F(x)$. Find n.

9-32 The RV **x** has the Erlang density $f(x) \sim c^4 x^3 e^{-cx} U(x)$. We observe the samples $x_i = 3.1, 3.4, 3.3$. Find the ML estimate $\hat{c}$ of c.

9-33 The RV **x** has the truncated exponential density $f(x) = ce^{-c(x-x_0)}U(x - x_0)$. Find the ML estimate $\hat{c}$ of c in terms of the n samples x_i of **x**.

9-34 The time to failure of a bulb is an RV **x** with density $ce^{-cx}U(x)$. We test 80 bulbs and find that 200 hours later, 62 of them are still good. Find the ML estimate of c.

9-35 The RV **x** has a Poisson distribution with mean θ. Show that the ML estimate of θ equals $\bar{x}$.

9-36 Show that if $L(x, \theta) = \ln f(x, \theta)$ is the likelihood function of an RV **x**, then

$$E\left\{\left|\frac{\partial L(\mathbf{x}, \theta}{\partial \theta}\right|^2\right\} = -E\left\{\frac{\partial^2 L(\mathbf{x}, \theta)}{\partial \theta^2}\right\}$$

9-37 The time to failure of a pump is an RV **x** with density $\theta e^{-\theta x}U(x)$. **(a)** Find the information I about θ contained in **x**. **(b)** We have shown in Example 9-24 that the ML estimator $\hat{\boldsymbol{\theta}}$ of θ equals $1/\bar{\mathbf{x}}$. Show that for $n > 2$

$$E\{\hat{\boldsymbol{\theta}}\} = \frac{n\theta}{n-1} \qquad \sigma_{\hat{\theta}}^2 = \frac{n^2\theta^2}{(n-1)^2(n-2)}$$

9-38 Show that if **y** is the best estimator of a parameter $\theta \ne 0$ and **z** is an arbitrary statistic with zero mean, then **y** and **z** are uncorrelated.

9-39 The RV **x** has the gamma density $\theta^2 x e^{-\theta x} U(x)$. Find the best estimator of the parameter $1/\theta$.

9-40 The RV $\mathbf{x}$ has Weibull distribution $x^{\theta-1}e^{-x^{\theta}/\theta}U(\theta)$. Show that the most efficient estimate of θ is the sum

$$\hat{\theta} = \frac{1}{n\theta} \sum x_i^{\theta}$$

9-41 Show that if the function $f(x, \theta)$ is of the exponential type as in (9-91), the ML estimator of the parameter $\tau = b'(\theta)/a'(\theta)$ is the best estimator of τ.

9-42 Show that if $\mathbf{x}_i$ are the samples of an *rv* $\mathbf{x}$ with density $e^{-(x-\theta)}U(x)$, the RV $\mathbf{w} = \min \mathbf{x}_i$ is a sufficient and complete statistic of θ.

9-43 Show that if the density $f(x, \theta)$ has a sufficient statistic $\mathbf{y}$ [see (9-95)] and the density $f_y(y, \theta)$ of $\mathbf{y}$ is known for $\theta = \theta_0$, then it is known for every θ.

9-44 If $f(x) = \theta e^{-\theta x}U(x)$ and $n = 2$, then the sum $\mathbf{y} = \mathbf{x}_1 + \mathbf{x}_2$ is a sufficient statistic of θ. Show that if $\mathbf{z} = \mathbf{x}_1$, then $E\{\mathbf{z}|\mathbf{y}\} = \mathbf{y}/2$, in agreement with the sufficiency theorem.

9-45 Suppose that $\mathbf{x}_i$ are the samples of an $N(\eta, 5)$ RV $\mathbf{x}$. **(a)** Show that the sum $\mathbf{y} = \mathbf{x}_1 + \cdots + \mathbf{x}_n$ is a sufficient statistic of η. **(b)** Show that if a_i are n constants such that $a_1 + \cdots + a_n = 0$ and $\mathbf{z} = a_1\mathbf{x}_1 + \cdots + a_n\mathbf{x}_n$, then the RVs $\mathbf{y}$ and $\mathbf{z}$ are independent.

10 Hypothesis Testing

Hypothesis testing is part of decision theory. It is based on statistical considerations and on other factors, often subjective, that are outside the scope of statistics. In this chapter, we deal only with the statistical part of the theory. In the first two sections, we develop the basic concepts using commonly used parameters, including tests of means, variances, probabilities, and distributions. In the next three sections, we present a variety of applications, including quality control, goodness-of-fit tests, and analysis of variance. The last section deals with optimality criteria, sequential testing, and likelihood ratios.

10-1 General Concepts

A hypothesis is an assumption. A statistical hypothesis is an assumption about the values of one or more parameters of a statistical model. Hypothesis testing is a process of establishing the validity of a hypothesis. In hypothesis testing, we are given an RV $\mathbf{x}$ modeling a physical quantity. The distribution of $\mathbf{x}$ is a function $F(x, \theta)$ depending on a parameter θ. We wish to test the hypothesis that θ equals a given number θ_0. This problem is fundamental in many areas of applied statistics. Here are several illustrations.

1. We know from past experience that under certain experimental conditions, the parameter θ equals θ_0. We modify various factors of the experiment, and we wish to establish whether these modifications have any effect on the value of θ. The modifications might be intentional (we try a new fertilizer), or they might be beyond our control (undesirable changes in a production process).
2. The hypothesis that $\theta = \theta_0$ might be the result of a theory to be verified.
3. The hypothesis might be a standard that we have established (expected productivity of a worker) or a desirable objective.

Terminology The assumption that $\theta = \theta_0$ will be denoted by H_0 and will be called the *null hypothesis*. The assumption that $\theta \neq \theta_0$ will be denoted by H_1 and will be called the *alternative hypothesis*. The set of values that θ might take under the alternative hypothesis will be denoted by Θ_1. If Θ_1 consists of a single point $\theta = \theta_1$, the hypothesis H_1 is called *simple;* otherwise, it is called *composite*. Typically, Θ_1 is one of the following three sets: $\theta \neq \theta_0$, $\theta > \theta_0$, or $\theta < \theta_0$. The null hypothesis is in most cases simple.

THE TEST The purpose of hypothesis testing is to establish whether experimental evidence supports rejecting the null hypothesis. The available evidence consists of the n samples $X = [x_1, \ldots, x_n]$ of the RV $\mathbf{x}$. Suppose that under the null hypothesis the joint density $f(X, \theta_0)$ of the samples $\mathbf{x}_i$ is negligible in a certain region D_c of the sample space, taking significant values only in the complement D_a of D_c. It is reasonable then to reject H_0 if X is in D_c and to accept it if X is in D_a. The set D_c is called the *critical region* of the test, and the set D_a is the *region of acceptance* of H_0. The terms "accept" and "reject" will be interpreted as follows: If $X \in D_c$, the evidence supports the rejection of H_0; if $X \notin D_c$, the evidence does not support the rejection of H_0. The test is thus specified in terms of the set D_c. The choice of this set depends on the nature of the decision errors. There are two types of errors, depending on the location of the observation vector X. We shall explain the nature of these errors and their role in the selection of the critical region of the test.

Suppose, first, that H_0 is true. If $X \in D_c$, we reject H_0 even though it is true—a *Type I error*. The Type I error probability is denoted by α and is called the *significance level* of the test. Thus

$$\alpha = P\{\mathbf{X} \in D_c | H_0\} \tag{10-1}$$

The difference $1 - \alpha$ equals the probability that we accept H_0 when true.

Suppose, next, that H_0 is false. If $X \notin D_c$, we accept H_0 even though it is false—a *Type II error*. The Type II error probability depends on the value of θ. It is thus a function $\beta(\theta)$ of θ called the *operating characteristic* (OC) of the test. Thus

$$\beta(\theta) = P\{\mathbf{X} \notin D_c | H_1\} \tag{10-2}$$

The difference

$$P(\theta) = 1 - \beta(\theta) = P\{\mathbf{X} \in D_c | H_1\} \tag{10-3}$$

equals the probability that we reject H_0 when it is false. The function $P(\theta)$ is called the *power of the test*. For brevity, we shall often identify the two types of errors by the expressions α error and $\beta(\theta)$ error, respectively.

To design an optimum test, we assign a value to α and select the critical region D_c so as to minimize the resulting β. If we succeed, the test is called *most powerful*. The critical region of such a test usually depends on θ. If it happens that a most powerful test is obtained with the same D_c for every $\theta \in \Theta_1$, the test is called *uniformly most powerful*.

Note Hypothesis testing belongs to decision theory. Statistical considerations lead merely to the following conclusions:

$$\begin{aligned} &\text{If } H_0 \text{ is true, then } P\{\mathbf{X} \in D_c\} = \alpha \\ &\text{If } H_0 \text{ is false, then } P\{\mathbf{X} \notin D_c\} = \beta(\theta) \end{aligned} \tag{10-4}$$

Guided by this, we reach a *decision:*

$$\text{Reject } H_0 \text{ iff } X \in D_c \tag{10-5}$$

This decision is not based only on (10-4). It takes into account our prior knowledge concerning the validity of H_0, the consequences of a wrong decision, and possibility other, often subjective factors.

Test Statistic

The critical region is a set D_c in the sample space. If it is properly chosen, the test is more powerful. This involves a search in the n-dimensional space. We shall use a simpler approach. Prior to any experimentation we select a function $g(X)$ and form the RV $\mathbf{q} = g(\mathbf{X})$. This RV will be called the *test statistic*. The function $g(X)$ may depend on θ_0 and on other known parameters, but it must be independent of θ. Only then is $G(X)$ a known number for a specific X. The test of a hypothesis involving a test statistic is simpler. The decision whether to reject H_0 is based not on the value of the vector X but on the value of the scalar $q = g(X)$.

To test the hypothesis H_0 using a test statistic, we find a region R_c on the real line, and we reject H_0 iff q is in R_c. The resulting error probabilities are

$$\alpha = P\{\mathbf{q} \in R_c | H_0\} = \int_{R_c} f_q(q, \theta_0)\, dq$$

$$\beta(\theta) = P\{\mathbf{q} \notin R_c | H_1\} = \int_{R_a} f_q(q, \theta)\, dz$$

where R_a is the region of acceptance of H_0. The density $f_q(q, \theta)$ of the test statistic can be expressed in terms of the function $g(X)$ and the joint density

$f(X, \theta)$ of the samples $\mathbf{x}_i$ [see (7-6)]. The critical region R_c is determined as before: We select α and search for a region R_c minimizing the resulting OC function $\beta(\theta)$.

In the next section, we design tests based on empirically chosen test statistics. In general, such tests are not most powerful no matter how R_c is chosen. In Section 10-6, we develop the conditions that a test statistic must satisfy in order that the resulting test be most powerful (Neyman-Pearson criterion) and show that many of the empirically chosen test statistics meet these conditions.

10-2 Basic Applications

In this section, we develop tests involving the commonly used parameters. The tests are based on the value of a test statistic $\mathbf{q} = g(\mathbf{X})$. The choice of the function $g(X)$ is more or less empirical. Optimality criteria are developed in Section 10-6. In the applications of this section, the density $f_q(q, \theta)$ of the test statistic has a single maximum at $q = q_{\max}$. To be concrete, we assume that $f_q(q, \theta)$ is concentrated on the right of $q_{\max}$ if $\theta > \theta_0$, and on its left if $\theta < \theta_0$ as in Fig. 10.1.

Our problem can be phrased as follows: We have an RV $\mathbf{x}$ with distribution $F(x, \theta)$. We wish to test the hypothesis $\theta = \theta_0$ against one of the alternative hypotheses $\theta \neq \theta_0$, $\theta > \theta_0$, or $\theta < \theta_0$. In all three cases, we shall use the same test statistic. To carry out the test, we select a function $g(X)$. We form the RV $\mathbf{q} = g(\mathbf{X})$ and determine its density $f_q(q, \theta)$. We next choose a value for the α error and determine the critical region R_c. We compute the resulting OC function $\beta(\theta)$. If $\beta(\theta)$ is unacceptably large, we increase the

Figure 10.1

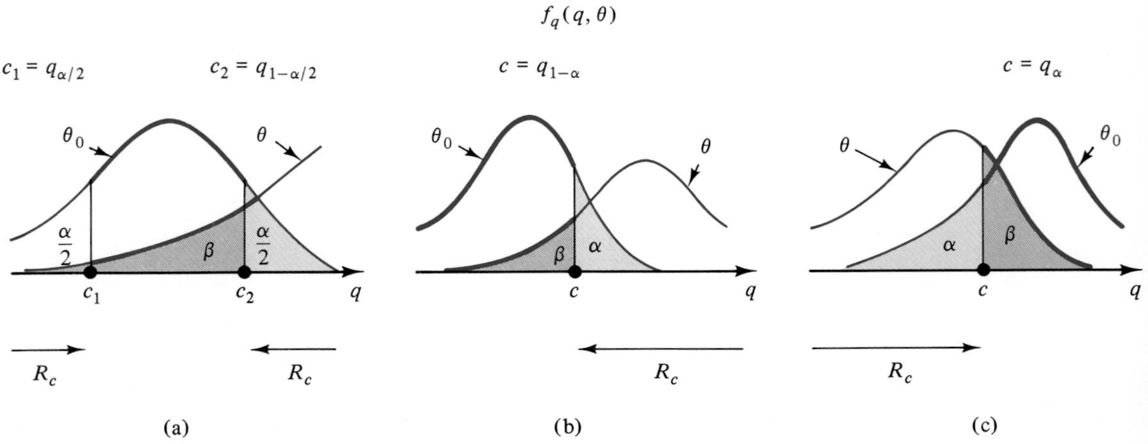

number n of the samples x_i. Finally, we observe the sample vector, compute the value $q = g(X)$ of the test statistic, and reject H_0 iff q is in the region R_c.

To complete the test, we must determine its critical region R_c. The result depends on the nature of the alternative hypothesis. We consider each of the three cases $\theta \neq \theta_0$, $\theta > 0$, $\theta < 0$, and we determine the corresponding OC function.

1. Suppose that H_1 is the hypothesis $\theta \neq \theta_0$. In this case, R_c consists of the two half-lines $q < c_1$ and $q > c_2$, as in Fig. 10.1a. The corresponding α error is the area under the two tails of the density $f_q(q, \theta_0)$:

$$\alpha = \int_{-\infty}^{c_1} f_q(q, \theta_0)\, dq + \int_{c_2}^{\infty} f_q(q, \theta_0)\, dq \qquad (10\text{-}6)$$

The constants c_1 and c_2 are so chosen as to minimize the length $c_2 - c_1$ of the interval (c_1, c_2). This leads to the condition $f_q(c_1, \theta_0) = f_q(c_2, \theta_0)$; however, the computations required to determine c_1 and c_2 are involved. To simplify the problem, we choose c_1 and c_2 such that the area of $f_q(q, \theta_0)$ under each of its tails equals $\alpha/2$. This yields

$$\int_{-\infty}^{c_1} f_q(q, \theta_0)\, dq = \frac{\alpha}{2} \qquad \int_{c_2}^{\infty} f_q(q, \theta_0)\, dq = \frac{\alpha}{2} \qquad (10\text{-}7a)$$

Denoting by q_u the u-percentile of the test statistic $\mathbf{q}$, we conclude that $c_1 = q_{\alpha/2}$, $c_2 = q_{1-\alpha/2}$. The resulting OC function $\beta(\theta)$ equals the area of $f_q(q, \theta)$ in the interval (c_1, c_2).

2. Suppose that H_1 is the hypothesis $\theta > \theta_0$. The critical region is now the half-line $q > c$ as in Fig. 10.1b, and the corresponding α error equals the area of the right tail of $f_q(q, \theta_0)$:

$$\alpha = \int_{c}^{\infty} f_q(q, \theta_0)\, dq \qquad c = q_{1-\alpha} \qquad (10\text{-}7b)$$

The OC function $\beta(\theta)$ equals the area of $f_q(q, \theta)$ in the region $q < c$.

3. Suppose that H_1 is the hypothesis $\theta < \theta_0$. The critical region is the half-line $q < c$ of Fig. 10.1c, and the α error equals

$$\alpha = \int_{-\infty}^{c} f_q(q, \theta_0)\, dq \qquad c = q_{\alpha} \qquad (10\text{-}7c)$$

The OC function is the area of $f_q(q, \theta)$ in the region $q > c$.

Summary To test the hypothesis $\theta = \theta_0$ against one of the alternative hypotheses $\theta \neq \theta_0$, $\theta > \theta_0$, or $\theta < \theta_0$, we proceed as follows:

1. Select the statistic $\mathbf{q} = g(\mathbf{X})$ and determine its density $f_q(q, \theta)$.
2. Assign a value to the significance level α and determine the critical region R_c for each case.
3. Observe the sample X and compute the function $q = g(X)$.

4. Reject H_0 iff q is in the region R_c.
5. Compute the OC function $\beta(\theta)$ for each case.

Here is the decision process and the resulting OC function $\beta(\theta)$ for each case. The numbers q_u are the u-percentiles of the test statistic $\mathbf{q}$ under hypothesis H_0.

$H_1: \theta \neq \theta_0$ Accept H_0 iff $c_i \le q \le c_2$ $c_1 = q_{\alpha/2}$ $c_2 = q_{1-\alpha/2}$

$$\beta(\theta) = \int_{c_1}^{c_2} f_q(q, \theta)\, dq \tag{10-8a}$$

$H_1: \theta > \theta_0$ Accept H_0 iff $q \le c$ $c = q_{1-\alpha}$

$$\beta(\theta) = \int_{-\infty}^{c} f_q(q, \theta)\, dq \tag{10-8b}$$

$H_1: \theta < \theta_0$ Accept H_0 iff $q \ge c$ $c = q_\alpha$

$$\beta(\theta) = \int_{c}^{\infty} f_q(q, \theta)\, dq \tag{10-8c}$$

Notes

1. The test of the simple hypothesis $\theta = \theta_0$ against the alternative simple hypothesis $\theta = \theta_1 > \theta_0$ is a special case of (10-8b); the test against $\theta = \theta_1 < \theta_0$ is a special case of (10-8c).
2. The test of the *composite* null hypothesis $\theta \le \theta_0$ against the alternative $\theta > \theta_0$ is identical to (10-8b); similarly, the test of the *composite* null hypothesis $\theta \ge \theta_0$ against $\theta < \theta_0$ is identical to (10-8c). For all three tests, the constant α is the Type I error probability only if $\theta = \theta_0$ when H_0 is true.
3. For the determination of the critical region of the test, knowledge of the density $f_q(q, \theta)$ for $\theta = \theta_0$ is necessary. For the determination of the OC function $\beta(\theta)$, knowledge of $f_q(q, \theta)$ for every $\theta \in \Theta_1$ is necessary.
4. If the statistic $\mathbf{q}$ generates a certain test, the same test can be generated by any function of $\mathbf{q}$.
5. We show in Section 10-6 that under the stated conditions, all tests in (10-8) are most powerful. Furthermore, the tests against $\theta > \theta_0$ or $\theta < \theta_0$ are uniformly most powerful. This is not true for the test against $\theta \neq \theta_0$. For a specific $\theta > \theta_0$, for example, the critical region $q > c$ yields a Type II error smaller than the integral in (10-8a).
6. The OC function $\beta(\theta)$ equals the Type II error probability. If it is too large, we increase α to its largest tolerable value. If $\beta(\theta)$ is still too large, we increase the number n of samples.
7. Tests based on (10-8) require knowledge of the *percentiles* q_u of the test statistic $\mathbf{q}$ for various values of u. However, all tests can be carried out in terms of the *distribution* $F_q(q, \theta)$ of $\mathbf{q}$. Indeed, suppose that we wish to test the hypothesis $\theta = \theta_0$ against $\theta \neq \theta_0$. From the monotonicity of distributions it follows that

$$q_{\alpha/2} \le q \le q_{1-\alpha/2} \text{ iff}$$

$$\frac{\alpha}{2} = F_q(q_{\alpha/2}, \theta_0) \le F_q(q, \theta_0) \le F_q(q_{1-\alpha/2}, \theta_0) = 1 - \frac{\alpha}{2}$$

Hence, the test (2-2a) is equivalent to the following test:

Determine the function $F_q(q, \theta_0)$ where q is the observed value of the test statistic

$$\text{Accept } H_0 \text{ iff } \frac{\alpha}{2} < F_q(q, \theta_0) < 1 - \frac{\alpha}{2}$$

This approach is used in tests based on computer simulation (see Section 8-3).

Mean

We have an RV $\mathbf{x}$ with mean η, and we wish to test the hypothesis

$$H_0: \eta = \eta_0 \qquad \text{against} \qquad H_1: \eta \neq \eta_0, \ \eta > \eta_0, \text{ or } \eta < \eta_0$$

Assuming that the variance of $\mathbf{x}$ is known, we use as the test statistic the RV

$$\mathbf{q} = \frac{\bar{\mathbf{x}} - \eta_0}{\sigma/\sqrt{n}} \tag{10-9}$$

where $\bar{\mathbf{x}}$ is the sample mean of $\mathbf{x}$. With the familiar assumptions, the RV $\bar{\mathbf{x}}$ is $N(\eta, \sigma/\sqrt{n})$. From this it follows that under hypothesis H_0, the test statistic $\mathbf{q}$ is $N(0, 1)$; hence, its percentile q_u equals the u-percentile z_u of the standard normal distribution. Setting $q_u = z_u$ in (10-8), we obtain the critical regions of Fig. 10.2. To find the corresponding OC functions, we must determine the density of $\mathbf{q}$ under hypothesis H_1. In this case, $\bar{\mathbf{x}}$ is $N(\eta, \sigma/\sqrt{n})$, and $\mathbf{q}$ is $N(\eta_q, 1)$ where

$$\eta_q = \frac{\eta - \eta_0}{\sigma/\sqrt{n}} \tag{10-10}$$

Since $z_{1-u} = -z_u$, (10-8) yields

$$H_1: \eta \neq \eta_0 \qquad \text{Accept } H_0 \text{ iff } z_{\alpha/2} \le q \le -z_{\alpha/2}$$

$$\beta(\eta) = G(-z_{\alpha/2} - \eta_q) - G(z_{\alpha/2} - \eta_q) \tag{10-11a}$$

$$H_1: \eta > \eta_0 \qquad \text{Accept } H_0 \text{ iff } q \le z_{1-\alpha}$$

$$\beta(\eta) = G(-z_\alpha - \eta_q) \tag{10-11b}$$

Figure 10.2

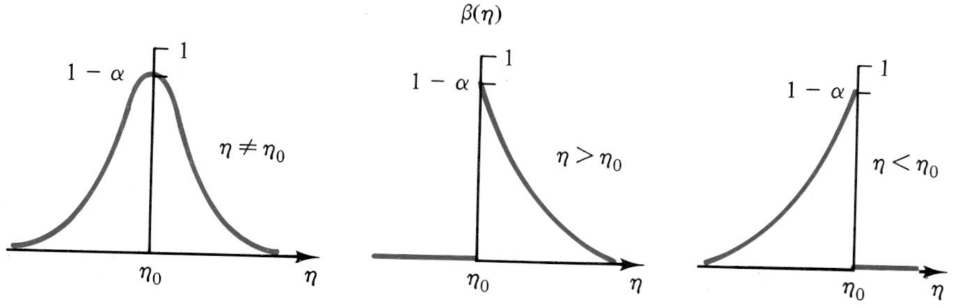

$$H_1: \eta < \eta_0 \qquad \text{Accept } H_0 \text{ iff } q \geq z_\alpha$$
$$\beta(\eta) = 1 - G(z_\alpha - \eta_q) \tag{10-11c}$$

The OC functions $\beta(\eta)$ are shown in Fig. 10.2 for each case.

Example 10.1

We receive a gold bar of nominal weight 8 oz., and we wish to test the hypothesis that its actual weight is indeed 8 oz. against the hypothesis that it is less than 8 oz. To do so, we measure the bar 10 times. The results of the measurements are the values

$$x_i = 7.86 \quad 7.90 \quad 7.93 \quad 7.95 \quad 7.96 \quad 7.97 \quad 7.98 \quad 8.01 \quad 8.02 \quad 8.04$$

of the RV $\mathbf{x} = \eta + \boldsymbol{\nu}$ where η is the actual weight of the bar and $\boldsymbol{\nu}$ is the measurement error, which we assume normal with zero mean and $\sigma = 0.1$. The test will be performed with confidence level $\alpha = .05$.

In this problem,

$$H_0: \eta = \eta_0 = 8 \qquad H_1: \eta < 8$$

$$\bar{x} = 7.96 \qquad \frac{\sigma}{\sqrt{n}} \simeq 0.032 \qquad q = \frac{\bar{x} - \eta_0}{\sigma/\sqrt{n}} = -1.25 \qquad z_\alpha = -1.645$$

Since -1.25 is not in the critical region $q < -1.645$, we accept the null hypothesis. The resulting OC function equals

$$\beta(\eta) = 1 - G\left(-1.65 - \frac{\eta - 8}{0.032}\right)$$ ■

If the variance of $\mathbf{x}$ is unknown, we use as test statistic the RV

$$\mathbf{q} = \frac{\bar{\mathbf{x}} - \eta_0}{\mathbf{s}/\sqrt{n}} \tag{10-12}$$

where $\mathbf{s}^2$ is the sample variance of $\mathbf{x}$. Under hypothesis H_0, this RV has a $t(n - 1)$ distribution. We can therefore use (10-8), provided that we set q_u equal to the $t_u(n - 1)$ percentile. To find $\beta(\eta)$, we must determine the distribution of $\mathbf{q}$ for $\eta \neq \eta_0$. This is a noncentral Student t distribution introduced in Chapter 7 [see (7A-9)].

Example 10.2

The mean η of past SAT scores in a school district equals 560. A class of 25 students is taught by a new instructor. The students take the test, and their scores x_i yield a sample mean $\bar{x} = 569$ and a sample standard deviation $s = 30$. Assuming normality, test the hypothesis $\eta = 560$ against the hypothesis $\eta \neq 560$ with significance level $\alpha = .05$. In this problem,

$$H_0: \eta_0 = 560 \qquad H_1: \eta \neq \eta_0$$

$$q = \frac{569 - 560}{30/\sqrt{25}} = 1.5 \qquad t_{1-\alpha/2}(24) = t_{.975}(24) = 2.06 = -t_{.025}$$

Thus q is in the interval $(-2.06, 2.06)$ of acceptance of H_0; hence, we accept the hypothesis that the average scores did not change.

EQUALITY OF TWO MEANS We have two normal RVs **x** and **y**, and we wish to test the hypothesis that their means are equal:

$$H_0: \eta_x = \eta_y \tag{10-13}$$

As in the problem of estimating the difference of two means, this test has two aspects.

Paired Samples Suppose, first, that both RVs can be observed at each trial. We then have n pairs of samples (x_i, y_i), as in (9-39). In this case, the RV

$$\mathbf{w} = \mathbf{x} - \mathbf{y}$$

is also observable, and its samples equal $w_i = x_i - y_i$. Under hypothesis H_0, the mean $\eta_w = \eta_x - \eta_y$ of **w** equals 0. Hence, (10-13) is equivalent to the hypothesis H_0: $\eta_w = 0$. We thus have a special case of the test of a mean considered earlier. Proceeding as in (10-9), we form the sample mean $\overline{\mathbf{w}} = \overline{\mathbf{x}} - \overline{\mathbf{y}}$ and use as test statistic the ratio

$$\frac{\overline{\mathbf{w}}}{\sigma_w/\sqrt{n}} = \frac{\overline{\mathbf{x}} - \overline{\mathbf{y}}}{\sigma_w/\sqrt{n}} \qquad \text{where } \sigma_w^2 = \sigma_x^2 + \sigma_y^2 - 2\mu_{xy} \tag{10-14}$$

Under hypotheses H_0, this RV is $N(0, 1)$; hence, the test of H_0 against one of the alternative hypothesis $\eta_w \neq 0$, $\eta_w > 0$, $\eta_w < 0$ is specified by (10-11) provided that we replace q by the ratio $\sqrt{n}(\bar{x} - \bar{y})/\sigma_w$ and η_q by $\sqrt{n}(\eta_x - \eta_y)/\sigma_w$.

If σ_w is unknown, we use as test statistic the ratio

$$\frac{\mathbf{w}}{\mathbf{s}_w/\sqrt{n}} \tag{10-15}$$

where $\mathbf{s}_w^2$ is the sample variance of the RV **w**, and proceed as in (10-12).

Example 10.3

The RVs **x** and **y** are defined on the same experiment $\mathcal{S}$. We know that $\sigma_x = 4$, $\sigma_y = 6$, and $\mu_{xy} = 13.5$, and we wish to test the hypothesis $\eta_x = \eta_y$ against the hypothesis $\eta_x \neq \eta_y$. Our decision will be based on the values of the 100 paired samples (x_i, y_i). We average these samples and find $\bar{x} = 90.7$, $\bar{y} = 89$.

In this problem, $n = 100$, $1 - \alpha/2 = .975$,

$$\sigma_w^2 = 36 + 16 - 27 = 25 \qquad \bar{w} = 1.1 \qquad \frac{\bar{w}}{\sigma_w/\sqrt{n}} = 3.4$$

Since $3.4 > z_{.975} \simeq 2$, we reject the hypothesis that $\eta_x = \eta_y$. ■

Sequential Samples We assume now that the RVs **x** and **y** cannot be sampled in pairs. This case arises in applications involving RVs that are observed under different experimental conditions. In this case, the available observations are $n + m$ independent samples, as in (9-41). Using these samples, we form their sample means $\bar{x}$, $\bar{y}$ and the RV

$$\overline{\mathbf{w}} = \overline{\mathbf{x}} - \overline{\mathbf{y}} \tag{10-16}$$

Known Variances Suppose, first, that the parameters σ_x and σ_y are known. To test the hypothesis $\eta_x = \eta_y$, we form the test statistic

$$\mathbf{q} = \frac{\overline{\mathbf{w}}}{\sigma_{\overline{w}}} \qquad \text{where } \sigma_{\overline{w}}^2 = \frac{\sigma_x^2}{n} + \frac{\sigma_y^2}{m} \tag{10-17}$$

If H_0 is true, then $\eta_{\overline{w}} = \eta_x - \eta_y = 0$; hence, the RV $\mathbf{q}$ is $N(0, 1)$. We can therefore use (10-11) with

$$\eta_q = \frac{\eta - \eta_y}{\sigma_{\overline{w}}}$$

Example 10.4

A radio transmitter transmits a signal of frequency η. We wish to establish whether the values η_x and η_y of η in two consecutive days are different. To do so, we measure η 20 times the first day and 10 times the second day. The average of the first day's readings $x_i = \eta_x + \nu_i$ equals 98.3 MHz, and the average of the second day's readings $y_j = \eta_y + \nu_j$ equals 98.32 MHz. The measurement errors ν_i and ν_j are the samples of a normal RV $\boldsymbol{\nu}$ with $\eta_\nu = 0$ and $\sigma_\nu = 0.04$ MHz. Based on these measurements, can we reject with significance level .05 the hypothesis that $\eta_x = \eta_y$?

In this problem, $n = 20$, $m = 10$, $z_{.975} \simeq 2$, $\bar{x} = 98.3$, $\bar{y} = 98.32$,

$$\sigma_{\overline{w}} = \sigma_\nu \sqrt{\frac{1}{n} + \frac{1}{m}} \simeq 0.015 \qquad \overline{w} = 0.02 \qquad \frac{\overline{w}}{\sigma_{\overline{w}}} = 1.3$$

Since 1.3 is not in the critical region (−2, 2), we accept the null hypothesis. ■

Unknown Variances Suppose that the parameters σ_x and σ_y are unknown. We shall carry out the test, under the assumption that $\sigma_x = \sigma_y = \sigma$. As we have noted in Section 9-2, only then is the distribution of the test statistic independent of σ. Guided by (9-47), we compute the sample variances s_x^2 and s_y^2 of $\mathbf{x}$ and $\mathbf{y}$, respectively, and form the test statistic

$$\mathbf{q} = \frac{\overline{\mathbf{w}}}{\hat{\boldsymbol{\sigma}}_{\overline{w}}} \qquad \text{where } \hat{\sigma}_{\overline{w}}^2 = \frac{(n-1)s_x^2 + (m-1)s_y^2}{n+m-2}\left(\frac{1}{n} + \frac{1}{m}\right) \tag{10-18}$$

is the estimate of the unknown variance $\sigma_{\overline{w}}^2$ of the RV $\overline{\mathbf{w}} = \overline{\mathbf{x}} - \overline{\mathbf{y}}$ [see (9-46)]. Under the null hypothesis, this statistic is identical to the estimator in (9-47) because if $\eta_x = \eta_y$, then $\eta_{\overline{w}} = 0$. Therefore, its density is $t(n + m - 2)$. This leads to the following test: Form the $n + m$ samples x_i and y_i; compute $\bar{x}$, $\bar{y}$, s_x^2, s_y^2; compute $\hat{\sigma}_{\hat{w}}$ from (10-18); use the test (10-11) with

$$q = \frac{\bar{x} - \bar{y}}{\hat{\sigma}_{\overline{w}}} \qquad \eta_q = \frac{\eta_x - \eta_y}{\hat{\sigma}_{\overline{w}}}$$

Replace the normal percentile z_u by $t_u(n + m - 2)$.

Example 10.5

The RVs $\mathbf{x}$ and $\mathbf{y}$ model the IQ scores of boys and girls in a school district. We wish to compare their means. To do so, we examine 40 boys and 30 girls and find the following averages:

Boys: $\bar{x} = 95$, $s_x = 6$
Girls: $\bar{y} = 93$, $s_y = 8$

Assuming that the RVs $\mathbf{x}$ and $\mathbf{y}$ are normal with equal variances, we shall test the hypothesis $H_0: \eta_x = \eta_y$ against $H_1: \eta_x \neq \eta_y$ with $\alpha = .05$.

In this problem, $n = 40$, $m = 30$, $t_{.975}(68) \simeq z_{.975} \simeq 2$,

$$\sigma_{\bar{w}}^2 = \frac{39s_x^2 + 29s_y^2}{68}\left(\frac{1}{40} + \frac{1}{30}\right) = 1.64^2 \qquad q = \frac{95 - 93}{1.64} = 1.22$$

Since $1.22 < 2$, we accept the hypothesis that the mean IQ scores of boys and girls are equal. ■

Mean-dependent Variances The preceding results of this section cannot be used if the distributions of $\mathbf{x}$ and $\mathbf{y}$ depend on a single parameter. In such problems, we must develop special techniques for each case. We shall consider next exponential distributions and Poisson distributions.

Exponential Distributions Suppose that

$$f_x(x) = \frac{1}{\theta_0} e^{-x/\theta_0} U(x) \qquad f_y(y) = \frac{1}{\theta_1} e^{-y/\theta_1} U(y)$$

In this case, $\eta_x = \theta_0$ and $\eta_y = \theta_1$; hence, our problem is to test the hypothesis $H_0: \theta_1 = \theta_0$ against $H_1: \theta_1 \neq \theta_0$. Clearly, the RVs $\mathbf{x}/\theta_0$ and $\mathbf{y}/\theta_1$ have a $\chi^2(2)$ distribution. From this it follows that the sum $n\bar{\mathbf{x}}/\theta_0$ of the n samples $\mathbf{x}_i/\theta_0$ has a $\chi^2(2n)$ distribution and the sum $m\bar{\mathbf{y}}/\theta_1$ of the m samples $\mathbf{y}_i/\theta_1$ has a $\chi^2(2m)$ distribution [see (7-87)]. Therefore, the ratio

$$\frac{n\bar{\mathbf{x}}/2n\theta_0}{m\bar{\mathbf{y}}/2m\theta_1} = \frac{\theta_1\bar{\mathbf{x}}}{\theta_0\bar{\mathbf{y}}}$$

has a Snedecor $F(n, m)$ distribution.

We shall use as test statistic the ratio $\mathbf{q} = \bar{\mathbf{x}}/\bar{\mathbf{y}}$. Under hypothesis H_0, $\mathbf{q}$ has an $F(n, m)$ distribution. This leads to the following test.

1. Form the sample means $\bar{x}$ and $\bar{y}$ and compute their ratio.
2. Select α and find the $\alpha/2$ and $1 - \alpha/2$ percentiles of the $F(n, m)$ distribution.
3. Accept H_0 iff $F_{\alpha/2}(n, m) < \bar{x}/\bar{y} < F_{1-\alpha/2}(n, m)$.

Example 10.6

We wish to examine whether two shipments of transistors have the same mean time to failure. We select 9 units from the first shipment and 16 units from the second and find that

$$x_1 + \cdots + x_9 = 324 \text{ hours} \qquad y_1 + \cdots + y_{16} = 380 \text{ hours}$$

Assuming exponential distributions, test the hypothesis $\theta_1 = \theta_0$ against $\theta_1 \neq \theta_0$ with $\alpha = .1$.

In this problem,

$$F_{.95}(9, 16) = 2.54 \qquad F_{.05}(9, 16) = 0.334 \qquad \bar{x} = 36 \qquad \bar{y} = 23.75$$

Since $\bar{x}/\bar{y} = 1.52$ is between 0.334 and 2.54, we accept the hypothesis that $\theta_1 = \theta_0$. ■

Probabilities

Given an event $\mathscr{A}$ with probability $p = P(\mathscr{A})$, we shall test the hypothesis $p = p_0$ against $p \neq p_0$, $p > p_0$, or $p < p_0$, using as test statistic the number $\mathbf{k}$ of successes of $\mathscr{A}$ in n trials. We start with the first case:

$$H_0: p = p_0 \qquad \text{against } H_1: p \neq p_0 \tag{10-19}$$

As we know, $\mathbf{k}$ is a binomially distributed RV with mean np and variance npq. This leads to the following test: Select the consistency level α. Compute the largest integer k_1 and the smallest integer k_2 such that

$$\sum_{k=0}^{k_1} \binom{n}{k} p_0^k q_0^{n-k} < \frac{\alpha}{2} \qquad \sum_{k=k_2}^{n} \binom{n}{k} p_0^k q_0^{n-k} < \frac{\alpha}{2} \tag{10-20}$$

Determine the number k of successes of $\mathscr{A}$ in n trials. Accept H_0 iff $k_1 \leq k \leq k_2$.

The resulting OC function equals

$$\beta(p) = P\{k_1 \leq \mathbf{k} \leq k_2 | H_1\} = \sum_{k=k_1}^{k_2} \binom{n}{k} p^k q^{n-k}$$

For large n, we can use the normal approximation. This yields

$$k_1 = np_0 + z_{\alpha/2}\sqrt{np_0q_0} \qquad k_2 = np_0 - z_{\alpha/2}\sqrt{np_0q_0}$$

$$\beta(p) = G\left(\frac{k_2 - np}{\sqrt{npq}}\right) - G\left(\frac{k_1 - np}{\sqrt{npq}}\right) \tag{10-21}$$

One-sided Tests We shall now test the hypothesis

$$H_0: p = p_0 \qquad \text{against} \qquad H_1: p > p_0 \tag{10-22}$$

The case against $p < p_0$ is treated similarly.

We determine the smallest integer k_2 such that

$$\sum_{k=k_2}^{n} \binom{n}{k} p_0^k q_0^{n-k} < \alpha$$

and we accept H_0 iff $k < k_2$. For large n,

$$k_2 = np_0 + z_{1-\alpha}\sqrt{np_0q_0} \qquad \beta(p) = G\left(\frac{k_2 - np}{\sqrt{npq}}\right) \tag{10-23}$$

Note that (10-22) is equivalent to the test of the composite hypothesis H_0': $p \leq p_0$ against H_1': $p > p_0$.

Example 10.7

We toss a given coin 100 times and observe that heads shows 64 times. Does this evidence support rejection of the fair-coin hypothesis with consistency level .05? In this problem, $np_0 = 50$, $\sqrt{np_0q_0} = 5$, $z_{.025} \simeq -2$; hence, the region of acceptance of H_0 is the interval

$$np_0 \pm 2\sqrt{np_0q_0} \simeq 50 \pm 10$$

Since 64 is outside this interval, we reject the fair-coin hypothesis. The resulting OC function equals

$$\beta(p) = G\left(\frac{60 - 100p}{10\sqrt{pq}}\right) - G\left(\frac{40 - 100p}{10\sqrt{pq}}\right)$$ ■

Rare Events Suppose now that $p_0 \ll 1$. If n is so large that $np_0 \gg 1$, we can use (10-21). We cannot do so if np_0 is of the order of 1. In this case, we use the Poisson approximation

$$\binom{n}{k} p^k q^{n-k} \simeq e^{-\lambda} \frac{\lambda^k}{k!} \qquad \lambda = np \tag{10-24}$$

Applying (10-24) to (10-19) and (10-20), we obtain the following test. Set $\lambda_0 = np_0$ and compute the largest integer k_1 and the smallest integer k_2 such that

$$e^{-\lambda_0} \sum_{k=0}^{k_1} \frac{\lambda_0^k}{k!} < \frac{\alpha}{2} \qquad e^{-\lambda_0} \sum_{k=k_2}^{\infty} \frac{\lambda_0^k}{k!} < \frac{\alpha}{2} \tag{10-25}$$

Accept H_0 iff $k_1 \le k \le k_2$. The resulting OC function equals

$$\beta(\lambda) = e^{-\lambda} \sum_{k=k_1}^{k_2} \frac{\lambda^k}{k!} \tag{10-26}$$

The one-sided alternatives $p > p_0$ and $p < p_0$ lead to similar results. Here is an example.

Example 10.8

A factory has been making radios using process A. Of the completed units, 0.6% are defective. A new process is introduced, and of the first 1,000 units, 5 are defective. Using $\alpha = 0.1$ as the significance level, test the hypothesis that the new process is better than the old one.

In this problem, $n = 1{,}000$, $\lambda_0 = 6$, and the objective is to test the hypothesis $H_0: p = p_0 = .006$ against $H_1: p < p_0$. To do so, we determine the largest integer k_1 such that

$$P\{\mathbf{k} \le k_1 | H_0\} = e^{-\lambda_0} \sum_{k=0}^{k_1} \frac{\lambda_0^k}{k!} < \alpha \tag{10-27}$$

and we accept H_0 iff $k > k_1$. In our case, $k = 5$, $\alpha = .1$, and (10-27) yields $k_1 = 2 < 5$; hence, we accept the null hypothesis. ■

EQUALITY OF TWO PROBABILITIES In the foregoing discussion, we assumed that the value p_0 of p under the null hypothesis was known. We now assume that p_0 is unknown. This leads to the following problem: Given two events $\mathscr{A}_0$ and $\mathscr{A}_1$ with probabilities $p_0 = P(\mathscr{A}_0)$ and $p_1 = P(\mathscr{A}_1)$, respectively, we wish to test the hypothesis $H_0: p_1 = p_0$ against $H_1: p_1 \ne p_0$. To do so, we perform the experiment $n_0 + n_1$ times, and we denote by $\mathbf{k}_0$ the number of successes of $\mathscr{A}_0$ in the first n_0 trials and by $\mathbf{k}_1$ the number of successes of $\mathscr{A}_1$ in the following n_1 trials. The observations are sequential; hence, the RVS $\mathbf{k}_0$

and $\mathbf{k}_1$ are independent. Sequential sampling is essential if $\mathscr{A}_0$ models a physical event under certain experimental conditions (a defective component in a manufacturing process, for example), and $\mathscr{A}_1$ models the same physical event under modified conditions.

We shall use as our test statistic the RV

$$\mathbf{q} = \frac{\mathbf{k}_0}{n_0} - \frac{\mathbf{k}_1}{n_1} \tag{10-28}$$

and, to simplify the analysis, we shall assume that the samples are large. With this assumption, the RV $\mathbf{q}$ is normal with

$$\eta_q = p_0 - p_1 \qquad \sigma_q^2 = \frac{p_0 q_0}{n_0} + \frac{p_1 q_1}{n_1} \tag{10-29}$$

Under the null hypothesis, $p_0 = p_1$; hence,

$$\eta_q = 0 \qquad \sigma_q^2 = p_0 q_0 \left(\frac{1}{n_0} + \frac{1}{n_1}\right) \tag{10-30}$$

This shows that we cannot use $\mathbf{q}$ to determine the critical region of the test because the numbers p_0 and q_0 are unknown. To avoid this difficulty, we replace in (10-30) the unknown parameters p_0 and q_0 by their empirical estimates $\hat{p}_0$ and $\hat{q}_0$ obtained under the null hypothesis.

To find $\hat{p}_0$ and q_0, we observe that if $p_0 = p_1$, then the events $\mathscr{A}_0$ and $\mathscr{A}_1$ have the same probability. We can therefore interpret the sum $k_0 + k_1$ as the number of successes of the same event $\mathscr{A}_0$ or $\mathscr{A}_1$ in $n_0 + n_1$ trials. This yields the empirical estimates

$$\hat{p}_0 = \frac{k_0 + k_1}{n_0 + n_1} = 1 - \hat{q}_0 \qquad \hat{\sigma}_q^2 = \hat{p}_0 \hat{q}_0 \left(\frac{1}{n_0} + \frac{1}{n_1}\right) \tag{10-31}$$

where $\hat{\sigma}_q$ is the corresponding estimate of σ_q. Thus under the null hypothesis, the RV $\mathbf{q}$ is $N(0, \hat{\sigma}_q)$; hence, its u-percentile equals $z_u \hat{\sigma}_q$. Applying (10-8a) to our case, we obtain the following test:

1. Determine the number k_0 and k_1 of successes of the events $\mathscr{A}_0$ and $\mathscr{A}_1$, respectively.
2. Compute the sample $q = k_0/n_0 - k_1/n_1$ of $\mathbf{q}$.
3. Compute $\hat{\sigma}_q$ from (10-31).
4. Accept H_0 iff $z_{\alpha/2}\hat{\sigma}_q \le q \le -z_{\alpha/2}\hat{\sigma}_q$.

Under hypothesis H_1, the RV $\mathbf{q}$ is normal with η_q and σ_q, as in (10-29). Assigning specific values to p_0 and p_1, we determine the OC function of the test from (10-8a).

Example 10.9

In a national election, we conducted an exit poll and found that of 200 men, 99 voted Republican and of 120 women, 45 voted Republican. Based on these results, we wish to test the hypothesis that the probability p_1 of female voters equals the probability p_0

of male voters with consistency level $\alpha = .05$. In this problem, $z_{.025} = -z_{.975} \simeq -2$,

$$q = \frac{99}{200} - \frac{45}{120} = .12 \qquad \hat{p}_0 = \frac{99 + 45}{200 + 120} \simeq .45 \qquad \hat{q}_0 = .55$$

$$\hat{\sigma}_q^2 = .45 \times .55 \left(\frac{1}{200} + \frac{1}{120}\right) \qquad \hat{\sigma}_q = 0.057$$

Since $0.12 > z_{.975}\hat{\sigma}_q \simeq 0.114$, the hypothesis that $p = p_0$ is rejected, but barely. ■

Poisson Distributions

The RV $\mathbf{x}$ is Poisson-distributed with mean λ. We wish to test the hypothesis $H_0: \lambda = \lambda_0$ against $H_1: \lambda \neq \lambda_0$. To do so, we form the n samples $\mathbf{x}_i$ and use as test statistic their sum:

$$\mathbf{q} = \mathbf{x}_1 + \cdots + \mathbf{x}_n$$

This sum is a Poisson-distributed RV with mean $n\lambda$. We next determine the largest integer k_1 and the smallest integer k_2 such that

$$e^{-n\lambda_0} \sum_{k=0}^{k_1} \frac{(n\lambda_0)^k}{k!} < \frac{\alpha}{2} \qquad e^{-n\lambda_0} \sum_{k=k_2}^{\infty} \frac{(n\lambda_0)^k}{k!} < \frac{\alpha}{2} \tag{10-32}$$

The left sides of these inequalities equal $P\{\mathbf{q} \leq k_1|H_0\}$ and $P\{\mathbf{q} \geq k_2|H_0\}$ respectively. This leads to the following test: Find the sum $q = x_1 + \cdots + x_n$ of the observations x_i; accept H_0 iff $k_1 \leq q \leq q_2$. The resulting OC function equals

$$\beta(\lambda) = P\{k_1 \leq \mathbf{k} \leq k_2|H_1\} = e^{-n\lambda} \sum_{k=k_1}^{k_2} \frac{(n\lambda)^k}{k!} \tag{10-33}$$

For large n, we can use the normal approximation. Under hypothesis H_0, $\eta_q = n\lambda_0$; $\sigma_q^2 = n\lambda_0$. Hence [see (10-11)],

$$k_1 \simeq n\lambda_0 + z_{\alpha/2}\sqrt{n\lambda_0} \qquad k_2 \simeq n\lambda_0 - z_{\alpha.2}\sqrt{n\lambda_0} \tag{10-34}$$

Example 10.10

The weekly highway accidents in a certain region are the values of a Poisson-distributed RV with mean $\lambda = \lambda_0 = 5$. We wish to examine whether a change in the speed limit from the present 55 mph to 65 mph will have any effect on λ. To do so, we monitor the number x_i of weekly accidents over a 10-week period and obtain

$$x_i = 5 \quad 2 \quad 9 \quad 6 \quad 3 \quad 7 \quad 9 \quad 1 \quad 6 \quad 8$$

Choosing $\alpha = .05$, we shall test the hypothesis $\lambda = 5$ against $\lambda \neq 5$.

In this problem, $z_{\alpha/2} \simeq -2$, $n\lambda_0 = 50$,

$$n\lambda_0 \pm 2\sqrt{n\lambda_0} \simeq 50 \pm 15 \qquad q = \sum x_i = 56$$

Since 56 is between 35 and 65, we accept the $\lambda = 5$ hypothesis. ■

EQUALITY OF TWO POISSON DISTRIBUTIONS We have two Poisson-distributed RVs **x** and **y** with means λ_0 and λ_1, respectively, and we wish to test the hypothesis H_0: $\lambda_1 = \lambda_0$ against H_1; $\lambda_1 \neq \lambda_0$ or $\lambda_1 > \lambda_0$ or $\lambda_1 < \lambda_0$ where λ_0 and λ_1 are two unknown constants. Our test will be based on the $n_0 + n_1$ samples x_i and y_i, obtained sequentially.

The exact solution of this problem is rather involved because the variances of **x** and **y** depend on their mean. To simplify the analysis, we shall consider only large samples. This leads to two simplifications: We can assume that the RVs **x** and **y** are normal, and we can replace the variances $\sigma_x^2 = \lambda_0$ and $\sigma_y^2 = \lambda_1$ by their empirical estimates.

Under hypothesis H_0, $\lambda_1 = \lambda_0$; hence, the RVs **x** and **y** have the same distribution. We can therefore interpret the $n_0 + n_1$ observations x_i and y_i as the samples of the same RV **x** or **y**. The resulting empirical estimate of λ_0 is

$$\hat{\lambda}_0 = \frac{1}{n_0 + n_1}\left(\sum_{i=1}^{n_0} x_i + \sum_{i=1}^{n_1} y_i\right)$$

Proceeding as in (10-17), we form the sample means $\bar{\mathbf{x}}$ and $\bar{\mathbf{y}}$ of the RVs **x** and **y** and use as test statistic the ratio

$$\mathbf{q} = \frac{\bar{\mathbf{x}} - \bar{\mathbf{y}}}{\hat{\sigma}_w} \qquad \text{where } \hat{\sigma}_w^2 = \hat{\lambda}_0\left(\frac{1}{n_0} + \frac{1}{n_1}\right) \tag{10-35}$$

is the empirical estimate of the variance of the difference $\mathbf{w} = \bar{\mathbf{x}} - \bar{\mathbf{y}}$. From this it follows that if H_0 is true, then $\mathbf{q}$ is approximately $N(0, 1)$. We can therefore use the test (10-11).

In particular, (10-11a) yields the following test of the hypothesis H_0: $\lambda_1 = \lambda_0$ against H_1: $\lambda_1 \neq \lambda_0$. Compute $\bar{x}$, $\bar{y}$, and $\hat{\sigma}_w$.

$$\text{Accept } H_0 \text{ iff } |\bar{x} - \bar{y}|/\hat{\sigma}_w < -z_{\alpha/2}.$$

Example 10.11

The number of absent boys and girls per week in a certain school district are the values x_i and y_k of two Poisson-distributed RVs **x** and **y** with means λ_0 and λ_1, respectively. We monitored the boys for 10 weeks and the girls for 8 weeks and obtained the following data:

x_i = 15	12	17	19	12	23	17	24	27	14
y_k = 13	10	16	23	19	10	13	8		

We shall test the hypothesis $\lambda_1 = \lambda_0$ against $\lambda_1 \neq \lambda_0$ with consistency level $\alpha = .05$.

In this problem, $-z_{\alpha/2} \simeq 2$,

$$\hat{\lambda}_0 = \frac{292}{18} = 16.2 \qquad \hat{\sigma}_{\bar{w}}^2 = 16.2\left(\frac{1}{10} + \frac{1}{8}\right) \qquad \sigma_{\bar{w}} = 1.91$$

$$\bar{x} - \bar{y} = \frac{180}{10} - \frac{112}{8} = 4 \qquad q = \frac{4}{1.91} = 2.09 > 2$$

Hence, we reject the null hypothesis, but barely. ■

Variance and Correlation

Given an $N(\eta, \sigma)$ RV $\mathbf{x}$, we wish to test the hypothesis

$$H_0: \sigma = \sigma_0 \qquad \text{against} \qquad H_1: \sigma \neq \sigma_0, \sigma > \sigma_0, \text{ or } \sigma < \sigma_0$$

Suppose, first, that the mean η of $\mathbf{x}$ is known. In this case, we use as test statistic the RV

$$\mathbf{q} = \sum_{i=1}^{n} \frac{(\mathbf{x}_i - \eta)^2}{\sigma_0^2} \tag{10-36}$$

Under hypothesis H_0, the RV $\mathbf{q}$ is $\chi^2(n)$. We can therefore use the tests (10-8) where q_u is now the $\chi_u^2(n)$ percentile. To find the corresponding OC function $\beta(\sigma)$, we must determine the distribution of $\mathbf{q}$ under hypothesis H_1. As we show in Problem 10-10, this is a scaled version of the $\chi^2(n)$ density.

Example 10.12

We wish to compare the accuracies of two measuring instruments. The error $\boldsymbol{\nu}_0$ of the first instrument is an $N(0, \sigma_0)$ RV where $\sigma_0 = 0.1$ mm, and the error $\boldsymbol{\nu}_1$ of the second is an $N(0, \sigma)$ RV with unknown σ. Our objective is to test the hypothesis $H_0: \sigma = \sigma_0 = 0.1$ mm against $H_1: \sigma \neq \sigma_0$ with $\alpha = .05$. To do so, we measure a standard object of length $\eta = 8$ mm 10 times and record the samples

$x_i = 8.15 \quad 7.93 \quad 8.22 \quad 8.04 \quad 7.85 \quad 7.95 \quad 8.06 \quad 8.12 \quad 7.86 \quad 7.92$

of the RV $\mathbf{x} = \eta + \boldsymbol{\nu}_1$. Inserting the results into (10-36), we obtain

$$q = \sum_{i=1}^{10} \frac{(x_i - 8)^2}{0.01} = 14.64$$

From Table 2 we find the percentiles

$$\chi^2_{.025}(10) = 3.25 \qquad \chi^2_{.975}(10) = 20.48$$

Since 14.64 is in the interval (3.25, 20.48) of acceptance of H_0, we accept the hypothesis that $\sigma = \sigma_0$. ■

If η is unknown, we use as test statistic the sum

$$\mathbf{q} = \sum_{i=1}^{n} \frac{(\mathbf{x}_i - \bar{\mathbf{x}})^2}{\sigma_0^2} \qquad \bar{\mathbf{x}} = \frac{1}{n}\sum_{i=1}^{n} \mathbf{x}_i \tag{10-37}$$

Under hypothesis H_0, the RV $\mathbf{q}$ is $\chi^2(n-1)$. We can therefore again use (10-8) where q_u is now the $\chi_u^2(n-1)$ percentile.

Example 10.13

Suppose now that the length η of the measured object in Example 10.12 is unknown. Inserting the 10 measurements x_i into (10-37), we obtain

$$\bar{x} = 8.01 \qquad q = \sum_{i=1}^{10} \frac{(x_i - \bar{x})^2}{0.01} = 14.54$$

From Table 2, we find that

$$\chi^2_{.025}(9) = 2.70 \qquad \chi^2_{.975}(9) = 19.02$$

Since 14.54 is in the interval (2.70, 19.02), we accept the hypothesis that $\sigma = \sigma_0$. ■

Equality of Two Variances The RV $\mathbf{x}$ is $N(\eta_x, \sigma_x)$, and the RV $\mathbf{y}$ is $N(\eta_y, \sigma_y)$. We shall test the hypothesis

$$H_0: \sigma_x = \sigma_y \qquad \text{against} \qquad H_1: \sigma_x \neq \sigma_y, \ \sigma_x > \sigma_y, \text{ or } \sigma_x < \sigma_y$$

Suppose, first, that the two means η_x and η_y are known. In this case, we use as test statistic the ratio

$$\mathbf{q} = \frac{\dfrac{1}{n}\sum_{i=1}^{n} (\mathbf{x}_i - \eta_x)^2}{\dfrac{1}{m}\sum_{i=1}^{m} (\mathbf{y}_i - \eta_y)^2} \tag{10-38}$$

where $\mathbf{x}_i$ are the n samples of $\mathbf{x}$ and $\mathbf{y}_i$ are the m samples of $\mathbf{y}$ obtained sequentially. From (7-106) it follows that under hypothesis H_0, the RV $\mathbf{q}$ has a Snedecor $F(n, m)$ distribution. We can therefore use the tests (10-8) where q_u is the $F_u(n, m)$ percentile.

If the means η_x and η_y are unknown, we use as test statistic the ratio

$$\mathbf{q} = \frac{\mathbf{s}_x^2}{\mathbf{s}_y^2} \tag{10-39}$$

where $\mathbf{s}_x^2$ is the sample variance of $\mathbf{x}$ obtained with the n samples $\mathbf{x}_i$ and $\mathbf{s}_y^2$ is the sample variance of $\mathbf{y}$ obtained with the m samples $\mathbf{y}_i$. From (7-106) it follows that if $\sigma_x = \sigma_y$, then the RV $\mathbf{q}$ is $F(n-1, m-1)$. We can therefore apply the tests (10-8) where q_u is the $F_u(n-1, m-1)$ percentile.

Example 10.14

We wish to compare the accuracies of two voltmeters. To do so, we measure an unknown voltage 10 times with the first instrument and 17 times with the second instrument. We compute the samples variances and find that

$$s_x = 8 \ \mu\text{V} \qquad s_y = 6.4 \ \mu\text{V}$$

Using these data, we shall test the hypothesis $H_0: \sigma_x = \sigma_y$ against $H_1: \sigma_x \neq \sigma_y$ with consistency level $\alpha = .1$. In this problem,

$$q = \frac{s_x^2}{s_y^2} = 1.56 \qquad n = 10 \qquad m = 17$$

From Table 4 we find

$$F_{.95}(9, 16) = 2.54 \qquad F_{.05}(9, 16) = \frac{1}{F_{.95}(16, 9)} = 0.334$$

Since 1.56 is in the interval (0.334, 2.54), we accept the hypothesis that $\sigma_x = \sigma_y$. ■

Correlation We wish to investigate whether two RVs $\mathbf{x}$ and $\mathbf{y}$ are uncorrelated. With r their correlation coefficient, our problem is to test the hypothesis $H_0: r = 0$ against $H_1: r \neq 0$. To solve this problem, we perform the underlying experiment n times and obtain the n *paired* samples (x_i, y_i). With these samples, we form the estimate $\hat{r}$ of r as in (9-55) and the corresponding value z of Fisher's auxiliary variable $\mathbf{z}$ as in (9-56):

$$\hat{r} = \frac{\Sigma(x_i - \bar{x})(y_i - \bar{y})}{\sqrt{\Sigma(x_i - \bar{x})^2 \Sigma(y_i - \bar{y})^2}} \qquad z = \frac{1}{2} \ln \frac{1 - \hat{r}}{1 + \hat{r}} \tag{10-40}$$

We shall use as test statistic the RV

$$\mathbf{q} = \mathbf{z}\sqrt{n - 3} \tag{10-41}$$

Under hypothesis H_0, the RV $\mathbf{z}$ is $N(0, 1/\sqrt{n - 3})$ [see (9-57)]. Hence, the test statistic $\mathbf{q}$ is $N(0, 1)$. We can therefore use the test (10-11) directly.

Example 10.15

We wish to examine whether the freshman grades $\mathbf{x}$ and the senior grades $\mathbf{y}$ are correlated. For this purpose, we examine the grades x_i and y_i of $n = 67$ students and compute $\hat{r}$, z, and q from (10-40) and (10-37). The results are

$$\hat{r} = 0.462 \qquad z = 0.5 \qquad q = 4$$

To test the hypothesis $r = 0$ against $r \neq 0$, we apply (10-8a). Since $q = 4 > |z_{\alpha/2}| \simeq 2$, we conclude that freshman and senior grades are correlated. ■

Distributions

We wish to examine whether the distribution function $F(x)$ of an RV $\mathbf{x}$ equals a given function $F_0(x)$. Later we present a method based on the χ^2 test. The following method is based on the empirical distribution (9-64).

KOLMOGOROFF-SMIRNOV TEST Our purpose is to test the hypothesis

$$H_0: F(x) = F_0(x) \qquad \text{against} \qquad H_1: F(x) \neq F_0(x) \tag{10-42}$$

For this purpose, we form the empirical estimate $\hat{F}(x)$ of $F(x)$ as in (9-64) and use as test statistic the maximum distance between $\hat{\mathbf{F}}(x)$ and $F_0(x)$:

$$\mathbf{q} = \max_x |\hat{\mathbf{F}}(x) - F_0(x)| \tag{10-43}$$

If H_0 is true, $E\{\hat{\mathbf{F}}(x)\} = F_0(x)$; hence $\mathbf{q}$ is small. It is therefore reasonable to reject H_0 iff the observed value q of $\mathbf{q}$ is larger than some constant c. To complete the test, it thus suffices to find c such that $P\{\mathbf{q} > c|H_0\} = \alpha$. Under hypothesis H_0, the statistic $\mathbf{q}$ equals the RV $\mathbf{w}$ in (9-66). Hence [see (9-68)],

$$\alpha = P\{\mathbf{q} > c|H_0\} \simeq 2e^{-2nc^2}$$

This yields the following test: Using the samples x_i of $\mathbf{x}$, form the empirical estimate $\hat{F}(x)$ of $F(x)$; plot the difference $\hat{F}(x) - F_0(x)$ (Fig. 10.3), and evaluate q from (10-43):

$$\text{Accept } H_0 \text{ iff } q < \sqrt{-\frac{1}{2n} \ln \frac{\alpha}{2}} = c \tag{10-44}$$

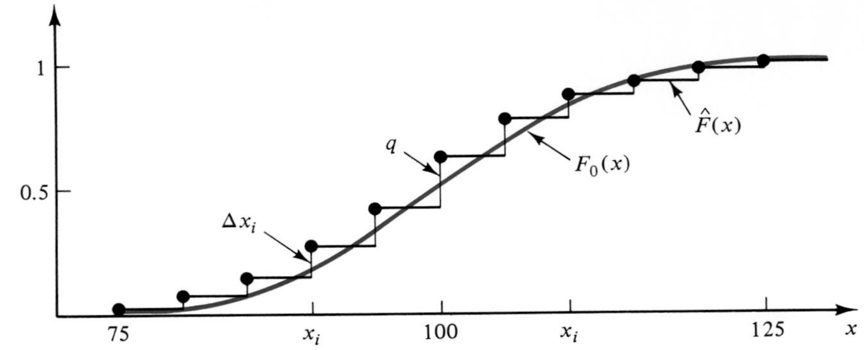

Figure 10.3

Example 10.16 We shall test the hypothesis that the tabulated IQ scores of Example 9.19 are the samples of a normal RV **x** with $\eta = 100$ and $\sigma = 10$. In the following table, we list the bunched samples x_i, the corresponding values of the empirical distribution $\hat{F}(x)$, the normal distribution $F_0(x)$, and their distance $\Delta(x) = |\hat{F}(x) - F_0(x)|$ (see also Fig. 10.3).

x_i	75	80	85	90	95	100	105	110	115	120	125
$\hat{F}(x_i)$	.025	0.75	1.50	.275	.425	.625	.775	.875	.925	.975	.100
$F_0(x_i)$	.006	.023	.067	.159	.308	.500	.691	.841	.933	.977	.994
$\Delta(x_i)$	.019	.052	.083	.116	.117	.125	.084	.034	.008	.002	.006

As we see from this table, $\Delta(x)$ is maximum for $x = 100$. Thus

$$c = \sqrt{-\frac{1}{80} \ln \frac{.05}{2}} = .217 \qquad q = \Delta(100) = .125$$

Since $.125 < .217$, we accept the hypothesis that the RV **x** is $N(100, 10)$. ■

EQUALITY OF TWO DISTRIBUTIONS We are given two independent RVs **x** and **y** with continuous distributions, and we wish to test the hypothesis

$$H_0: F_x(w) = F_y(w) \qquad \text{against} \qquad H_1: F_x(w) \neq F_y(w) \tag{10-45}$$

We shall give a simple test based on the special case $p_0 = .5$ of the test (10-19) of the equality of two probabilities.

Sign Test From the independence of the RVs **x** and **y** it follows that if H_0 is true, then

$$f_x(x)f_y(y) = f_x(y)f_y(x)$$

This shows that the joint density $f_{xy}(x, y)$ is symmetrical with respect to the line $x = y$; hence, the probability masses above and below this line are equal. In other words,

$$P\{\mathbf{x} > \mathbf{y}\} = P\{\mathbf{x} < \mathbf{y}\} = .5 \tag{10-46}$$

Thus under hypothesis H_0, the probability $p = P(\mathscr{A})$ of the event $\mathscr{A} = \{\mathbf{x} > \mathbf{y}\}$ equals .5. To test (10-45), we shall test first the hypothesis

$$H_0': p = .5 \qquad \text{against} \qquad H_1': p \neq .5 \tag{10-47}$$

To do so, we proceed as in (10-20) with $p_0 = .5$: Select a consistancy level α', and compute the largest integer k_1 and the smallest integer k_2 such that

$$\frac{1}{2^n}\sum_{k=0}^{k_1}\binom{n}{k} < \frac{\alpha'}{2} \qquad \frac{1}{2}\sum_{k=k_2}^{n}\binom{n}{k} < \frac{\alpha'}{2} \tag{10-48}$$

In this case, $k_2 = n - k_1$ because $\binom{n}{k} = \binom{n}{n-k}$. For large n, we can use the normal approximation (10-21). With $p_0 = .5$, this yields

$$k_1 = \frac{n}{2} + z_{\alpha'/2}\frac{\sqrt{n}}{2} \qquad k_2 = \frac{n}{2} - z_{\alpha'/2}\frac{\sqrt{n}}{2} \tag{10-49}$$

Form the n paired samples (x_i, y_i) ignoring all trials such that $x_i = y_i$. Determine the number k of trials such that $x_i > y_i$. This number equals the number k of times the event $\mathscr{A}$ occurs.

$$\text{Reject } H_0' \text{ iff } k < k_1 \qquad \text{or} \qquad k > k_2 \tag{10-50}$$

If H_0' is false, then H_0 is also false. This leads to the following test of H_0. Compute k, k_1, and k_2 as earlier.

$$\text{Reject } H_0 \text{ if } k < k_1 \qquad \text{or} \qquad k > k_2 \tag{10-51}$$

Note The tests of H_0 and H_0' are not equivalent because H_0' might be true even when H_0 is false. In fact, the corresponding error probabilities α, β and α', β' are different. Indeed, Type I error occurs if H_0' is rejected when true. In this case, H_0 might or might not be true; hence, $\alpha < \alpha'$. Type II error occurs if H_0' is not rejected when false. In this case, H_0 is false; however, it might be false even if H_0' is true; hence, $\beta > \beta'$.

Example 10.17

The RVs $\mathbf{x}$ and $\mathbf{y}$ model the salaries of equally qualified male and female employees in a certain industry. Using (10-50), we shall test the hypothesis that their distributions are equal. The consistency level α of the test is not to exceed the value $\alpha' = .05$.

To do so, we interrogate 84 pairs of employees and find that $x_i > y_i$ for 54 pairs, $x_i < y_i$ for 26 pairs, and $x_i = y_i$ for 4 pairs. Ignoring the last four cases, we have

$$n = 80 \qquad z_{\alpha'/2} \simeq -2 \qquad \frac{n}{2} \pm z_{\alpha'/2}\frac{\sqrt{n}}{2} = 40 \pm 8.94$$

Hence, $k_1 = 31$ and $k_2 = 49$. Since $k = 54 > 49$, we reject the null hypothesis.

We have assumed that the RVS x_i, y_i are i.i.d.; this is not, however, necessary. The sign test can be used even if the distributions of $\mathbf{x}_i$ and $\mathbf{y}_i$ depend on i. Here is an illustration.

Example 10.18 We wish to compare the effectiveness of two fertilizers. To do so, we select 50 farm plots, divide each plot into two parcels, and fertilize one half with the first fertilizer and the other half with the second. The resulting yields are the values (x_i, y_i) of the RVS $(\mathbf{x}_i, \mathbf{y}_i)$. The distributions of these RVS vary from plot to plot. We wish to test the hypothesis that for each plot, $\mathbf{x}_i$ and $\mathbf{y}_i$ have the same distribution. Our decision is based on the following data: $x_i > y_i$ for 12 plots and $x_i < y_i$ for 38 plots. We proceed as in (10-49) with $\alpha' = .05$. This yields the interval

$$\frac{n}{2} \pm 2\frac{\sqrt{n}}{2} = 25 \pm 7.07$$

Since $k = 12 < 18 = k_1$, we reject the null hypothesis. ■

10-3 Quality Control

In a manufacturing process, the quality of the plant output usually depends on the value of a measurable characteristic: the diameter of a shaft, the inductance of a coil, the life length of a system component. Due to a variety of "random" causes, the characteristic of interest varies from unit to unit. Its values are thus the samples of an RV $\mathbf{x}$. Under normal operating conditions, this RV satisfies certain specifications. For example, its mean equals a given number η_0. If these specifications are met during the production process, we say that the plant is *in control*. If they are not, the plant is *out of control*. A plant might go out of control for a variety of reasons: machine failure, faulty material, operator errors. The purpose of quality control is to detect whether, for whatever reasons, the plant goes out of control and, if it does, to find the errors and eliminate their causes. In many cases, this involves interruption of the production process.

Quality control is a fundamental discipline in industrial engineering that involves a variety of control methods depending on the nature of the product, the cost of inspection, the consequences of a wrong decision, the complexity of the analysis, and many other factors. We shall introduce the first principles of statistical quality control. Our analysis is a direct application of hypothesis testing.

The problem of statistical quality control can be phrased as follows: The distribution of the RV $\mathbf{x}$ is a known function $F(x, \theta)$ depending on a parameter θ. This parameter could, for example, be the mean of $\mathbf{x}$ or its variance. When the plant is in control, θ equals a specified number θ_0. The null hypothesis H_0 is the assumption that the plant is in control, that is, that $\theta = \theta_0$. The alternative hypothesis H_1 is the assumption that the plant is out of control, that is, that $\theta \neq \theta_0$. In quality control, we test the null hypothesis periodically until it is rejected. We then conclude with a certain significance level that the plant is out of control, stop the production process, and remove the cause of the error. The test proceeds as follows.

We form a sequence x_i of independent samples of $\mathbf{x}$. These samples are the measurements of production units selected in a variety of ways. Usually they form *groups* consisting of n units each. The units of each group are picked either sequentially or at random forming the samples

$$x_1, \ldots, x_n, \ldots, x_{m+1}, \ldots, x_{m+n}, \ldots$$

Using the n samples of each group, we form a test of the hypothesis H_0. We thus have a sequence of tests designed as in Section 10-2. The testing stops when the hypothesis H_0 is rejected.

Control Test Statistic

The decision whether to reject the hypothesis that the plant is in control is based on the values of a properly selected test statistic. Proceeding as before, we choose a significance level α and a function $\mathbf{q} = g(\mathbf{x}_1, \ldots, \mathbf{x}_n)$ of the n samples $\mathbf{x}_i$. We determine an interval (c_1, c_2) such that

$$P\{c_1 < \mathbf{q} < c_2 | H_0\} = 1 - \alpha \qquad (10\text{-}52)$$

We form the samples

$$q_m = g(x_{m+1}, \ldots, x_{m+n})$$

of $\mathbf{q}$. If q_m is between c_1 and c_2, we conclude that the plant is in control. If $q_m < c_1$ or $q_m > c_2$, we reject the null hypothesis; that is, we conclude that the plant is out of control. The constant c_1 is called the *upper control limit* (UCL); the constant c_2 is called the *lower control limit* (LCL). In Fig. 10.4 we show the values of q_m as the test progresses. The process terminates

Figure 10.4

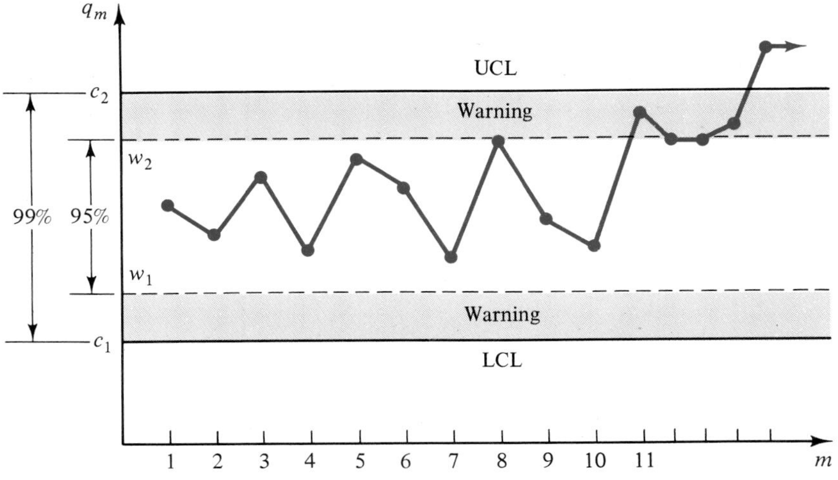

when q_m is outside the control interval (c_1, c_2). The graph so formed is called the *control chart*.

This test forms the basis of quality control. To apply it to a particular problem, it suffices to select the function $g(x_1, \ldots, x_n)$ appropriately. This we did in Section 10-2 for most parameters of interest. Special cases involving means, variances, and probabilities are summarized at the conclusion of this section.

Warning Limits If the time between groups is long, it is recommended that the testing be speeded up when q_m approaches the control limits. In such cases, we select a test interval (w_1, w_2) with $\alpha' > \alpha$:

$$P\{w_1 < \mathbf{q}_m < w_2 | H_0\} = 1 - \alpha'$$

and we speed up the process if q_m is in the shaded area of Fig. 10.4.

Control Error In quality control, two types of errors occur. The first is a false alarm: The plant is in control, but q_m crosses the control limits. The second is a faulty control: The plant goes out of control at $m = m_0$, but q_m remains between the control limits, crossing the limits at a later time.

False Alarm The control limits might be crossed even when the plant is in control. Suppose that they are crossed at the mth step (Fig. 10.5a). Clearly, m is a random number. It therefore defines an RV $\mathbf{z}$ taking the values 1, 2, . . . We maintain that

$$P\{\mathbf{z} = m | H_0\} = (1 - \alpha)^{m-1}\alpha \tag{10-53}$$

Indeed, $\mathbf{z} = m$ iff the statistic $\mathbf{q}$ is in the interval (c_1, c_2) in the first $m - 1$ tests and it crosses to the critical region at the mth step. Hence, (10-53) follows from (10-52). Thus $\mathbf{z}$ has a *geometric distribution,* and its mean [see (4-109)] equals

$$\eta_z = \frac{1}{\alpha} \tag{10-54}$$

This is the mean number of tests until a false alarm is given, and it suggests that if q_m crosses the control limits for $m > \eta_z$, the plant is probabily still in control. Such rare crossings might be interpreted merely as warnings.

Figure 10.5

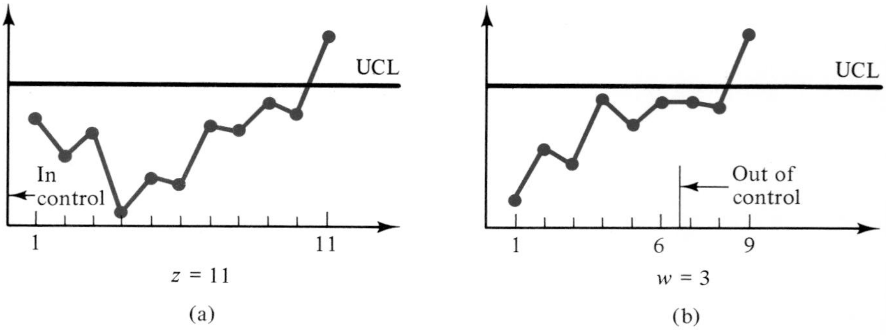

Faulty Control Suppose that the plant goes out of control at the mth test but q_m crosses the control limits not immediately but k tests later (Fig. 10.5b). The plant is then in operation for k test intervals while out of control. Again, k is a random number specifying the RV $\mathbf{w}$. This RV takes the values 1, 2, . . . with probability

$$P\{\mathbf{w} = k|H_1\} = \beta^{k-1}(\theta)[1 - \beta(\theta)] \tag{10-55}$$

where θ is the new value of the control parameter and

$$\beta(\theta) = P\{c_1 < \mathbf{q} < c_2|H_1\}$$

is the OC function of the test. Thus $\mathbf{w}$ has a geometric distribution with $p = 1 - \beta(\theta) = P(\theta)$, and its mean equals

$$\eta_w = \frac{1}{P(\theta)} \tag{10-56}$$

where $P(\theta)$ is the power of the test. If θ is distinctly different from θ_0, η_w is small. In fact, if $\beta(\theta) < .5$, then $\eta_w < 2$. As the difference $|\theta - \theta_0|$ decreases, η_w increases; for θ close to θ_0, η_w approaches the mean false alarm length η_z.

Control of the Mean

We shall construct a chart for controlling the mean η of $\mathbf{x}$. Suppose, first, that its standard deviation σ is known. We can then use as the test statistic the ratio $\mathbf{q}$ in (10-9):

$$\mathbf{q} = \frac{\bar{\mathbf{x}} - \eta_0}{\sigma/\sqrt{n}} \qquad q_m = \frac{\bar{x}_m - \eta_0}{\sigma/\sqrt{n}} \qquad \bar{x}_m = \frac{1}{n}\sum_{i=1}^{n} x_{m+i}$$

When the process is in control, $\eta = \eta_0$; hence, $\mathbf{q}$ is $N(0, 1)$, and the control limits are $\mp z_{\alpha/2}$. If we use as the test statistic the sample mean $\bar{x}_m$ of the samples x_{m+i} of the mth group, we obtain the chart of Fig. 10.6a with control

Figure 10.6

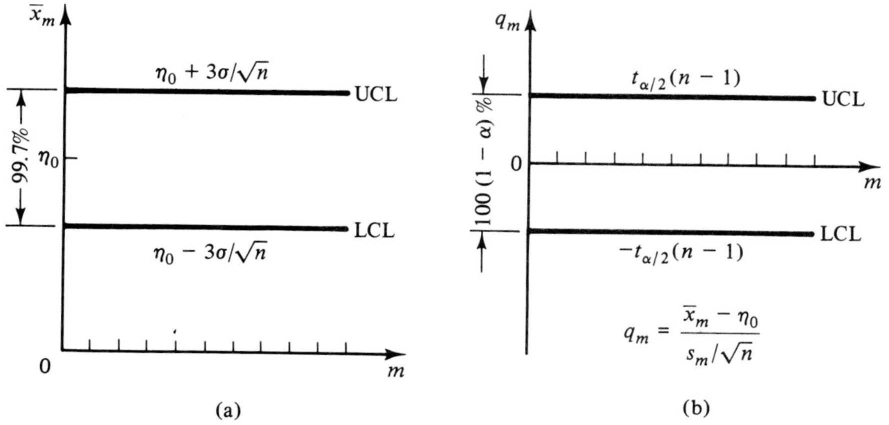

limits

$$\text{UCL} = \eta_0 - z_{\alpha/2}\sigma/\sqrt{n} \qquad \text{LCL} = \eta_0 + z_{\alpha/2}\sigma/\sqrt{n} \tag{10-57}$$

These limits depend on α. In the applications, it is more common to specify not the significance level α but the control interval directly. A common choice is the 3σ interval $\eta_0 \pm 3\sigma/\sqrt{n}$. Since

$$P\left\{\eta_0 - 3\frac{\sigma}{\sqrt{n}} < \bar{\mathbf{x}} < \eta_0 + 3\frac{\sigma}{\sqrt{n}} \,\middle|\, H_0\right\} = .997$$

the resulting consistency level α equals .003. The mean false alarm length of this test equals $1/\alpha = 333$.

Example 10.19

A factory manufactures cylindrical shafts. The diameter of the shafts is a normal RV with $\sigma = 0.2$ mm. When the manufacturing process is in control, $\eta = \eta_0 = 10$ mm. Design a chart for controlling η such that the mean false alarm length η_z equals 100. Use 25 samples for each test.

In this problem, $\alpha = 1/\eta_z = .01$, $z_{.995} = 2.576$, and $n = 25$. Inserting into (10-57), we obtain the interval

$$10 \pm 2.576\frac{0.2}{5} \simeq 10 \pm 0.1$$

If σ is unknown, we use as the test statistic the RV $\mathbf{q}$ in (10-12). Thus

$$\mathbf{q} = \frac{\bar{\mathbf{x}} - \eta_0}{\mathbf{s}/\sqrt{n}} \qquad q_m = \frac{\bar{x}_m - \eta_0}{s_m/\sqrt{n}} \qquad s_m^2 = \frac{1}{n-1}\sum_{i=1}^{n}(x_{m+i} - \bar{x}_m)^2$$

The corresponding chart is shown in Fig. 10.6b. Since the test statistic $\mathbf{q}$ has a $t(n-1)$ distribution, the control limits are

$$\text{LCL} = t_{\alpha/2}(n-1) \qquad \text{UCL} = -t_{\alpha/2}(n-1)$$

Example 10.20

Design a test for controlling η when σ is unknown. The design requirements are

$$\text{CL} = \pm 3 \qquad \text{False alarm length } \eta_z = 100$$

In this problem, $\alpha = 1/\eta_z = .01$; hence, the only unknown is the number n of samples. From Table 3 we see that $t_{.995}(n-1) = 3$ if $n - 1 = 13$; hence, $n = 14$.

In quality control, the tests are often one-sided. The corresponding charts have then one control limit, upper or lower. Let us look at two illustrations.

Control of the Standard Deviation

In a number of applications, it is desirable to keep the variability of the output product below a certain level. A measure of variability is the variance σ^2 of the control variable $\mathbf{x}$. The plant is in control if σ is below a specified level σ_0; it is out of control if $\sigma > \sigma_0$. This leads to the one-sided test H_0: $\sigma =$

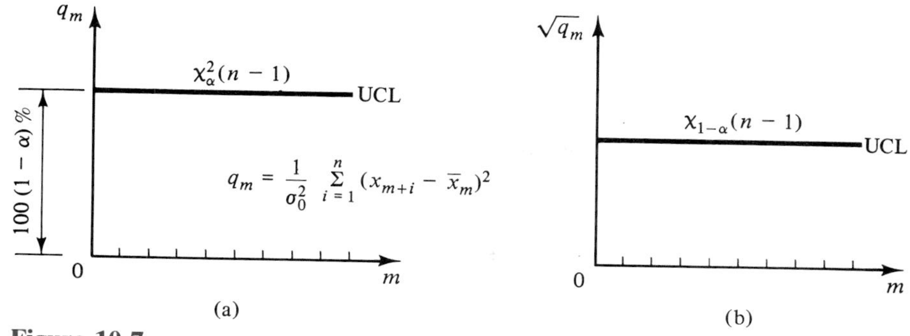

Figure 10.7

σ_0 against H_1: $\sigma > \sigma_0$ or, equivalently, H_0': $\sigma^2 = \sigma_0^2$ against H_1' $\sigma^2 > \sigma_0^2$. To carry out this test, we shall use as the test statistic the sum in (10-37):

$$\mathbf{q} = \sum_{i=1}^{n} \frac{(\mathbf{x}_i - \bar{\mathbf{x}})^2}{\sigma_0^2} \qquad q_m = \sum_{i=1}^{n} \frac{(x_{m+i} - \bar{x}_m)^2}{\sigma_0^2}$$

Under hypothesis H_0', the RV $\mathbf{q}$ has a $\chi^2(n-1)$ distribution. From this it follows that if $P\{\mathbf{q} > c | H_0'\} = \alpha$, then $c = \chi^2_{1-\alpha}(n-1)$; hence, the UCL equals $\chi^2_{1-\alpha}(n-1)$, and the chart of Fig. 10.7a results.

Note Suppose that $\mathbf{q}$ is a test statistic for the parameter θ and the corresponding control limits are c_1 and c_2. If $\tau = r(\theta)$ is a monotonic increasing function of θ, then $r(\mathbf{q})$ is a test statistic of the parameter τ, and the corresponding control limits are $r(c_1)$ and $r(c_2)$. This is a consequence of the fact that the events $\{c_1 < \mathbf{q} < c_2\}$ and $\{r(c_1) < r(\mathbf{q}) < r(c_2)\}$ are equal. Applying this to the parameter $\theta = \sigma^2$ with $r(\theta) = \sqrt{\theta}$, we obtain the chart of Fig. 10.7b for the control of the standard deviation σ of $\mathbf{x}$.

CONTROL OF DEFECTIVE UNITS In a number of applications, the purpose of quality control is to keep the proportion p of defective units coming out of a production line below a specified level p_0. A defective unit can be identified in a variety of ways, depending on the nature of the product. It might be a unit that does not perform an assigned function: a fuse, a switch, a broken item. It might also be a measurable characteristic, the value of which is not between specified tolerance limits; for example, a resistor with tolerance limits 1,000 ± 5% ohms is defective if $R < 950$ or $R > 1050$. In such problems, $p = \{\text{defective}\}$ is the probability that a unit is defective, and the objective is to test the hypothesis H_0: $p \le p_0$ against H_1: $p > p_0$ where p_0 is the probability of defective parts when the plant is in control. This test is identical to the test H_0': $p = p_0$ against H_1': $p > p_0$.

Proceeding as in (10-22), we use as the test statistic the number $\mathbf{x}$ of successes of the event $\mathscr{A} = \{\text{defective}\}$ in n trials: $\mathbf{q} = \mathbf{x}$ and $q_m = x_m$. Thus x_m is the number of defective units tested as the mth inspection. As we know, the RV $\mathbf{x}$ has a binomial distribution. This leads to the conclusion that the

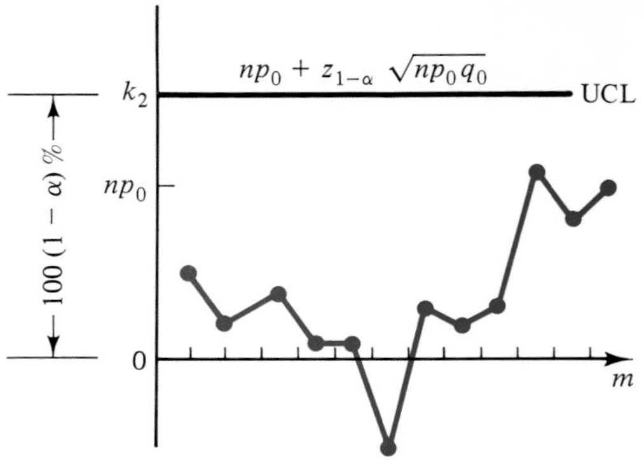

Figure 10.8

UCL is the smallest integer k_2 such that

$$P\{\mathbf{x} \geq k_2 | H_0'\} = \sum_{k=k_2}^{n} \binom{n}{k} p_0^k q_0^{n-k} < \alpha$$

For large n [see (10-23)],

$$\text{UCL} = k_2 \simeq np_0 + z_{1-\alpha}\sqrt{np_0q_0}$$

and the chart of Fig. 10.8 results. Thus if at the mth inspection, the number x_m of defective parts exceeds k_2, we decide with significance level α that the plant is out of control.

10-4 Goodness-of-Fit Testing

The objective of goodness-of-fit testing is twofold.

1. It seeks to establish whether data obtained from a real experiment fit a known theoretical model: Is a die fair? Is Mendel's theory of heredity valid? Does the life length of a transistor have a Weibull distribution?
2. It seeks to establish whether two or more sets of data obtained from the same experiment, performed under different experimental conditions, fit the same theoretical model: Is the proportion of Republicans among all voters the same as among male voters? Is the distribution of accidents with a 55-mph speed limit the same as with a 65-mph speed limit? Is the proportion of daily faulty units in a production line the same for all machines?

Goodness of fit is part of hypothesis testing based on the following seminal problem: We have a partition

$$A = [\mathscr{A}_1, \ldots, \mathscr{A}_m]$$

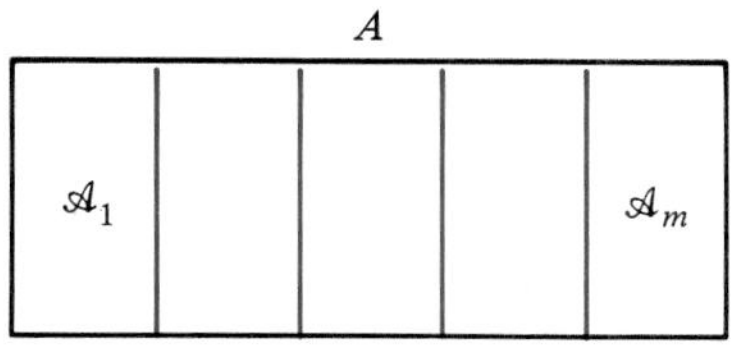

Figure 10.9

consisting of m events $\mathscr{A}_i$ (Fig. 10.9), and we wish to test the hypothesis that their probabilities $p_i = P(\mathscr{A}_i)$ have m given values p_{0i}:

$$H_0: p_i = p_{0i}, \text{ all } i \qquad \text{against} \qquad H_1: p_i \neq p_{0i}, \text{ some } i \tag{10-58}$$

This is a generalization of the test (10-19) involving a single event.

To carry out the test, we repeat the experiment n times and denote by $\mathbf{k}_i$ the number of successes of $\mathscr{A}_i$.

Pearson's Test Statistic We shall use as the test statistic the exponent of the normal approximation (7-71) of the generalized binomial distribution:

$$\mathbf{q} = \sum_{i=1}^{m} \frac{(\mathbf{k}_i - np_{0i})^2}{np_{0i}} \tag{10-59}$$

This choice is based on the following considerations: The RVs $\mathbf{k}_i$ have a binomial distribution with

$$E\{\mathbf{k}_i\} = np_i \qquad \sigma^2_{k_i} = np_iq_i \tag{10-60}$$

Hence, the ratio $\mathbf{k}_i/n \to p_i$ as $n \to \infty$. This leads to the conclusion that under hypothesis H_0, the difference $|k_i - np_{0i}|$ is small, and it increases as $|p_i - p_{0i}|$ increases.

The test proceeds as follows: Observe the numbers k_i and compute the sum in (10-59), select a significance level α, and find the percentile $q_{1-\alpha}$ of $\mathbf{q}$.

$$\text{Accept } H_0 \text{ iff } \sum_{i=1}^{m} \frac{(k_i - np_{0i})^2}{np_{0i}} < q_{1-\alpha} \tag{10-61}$$

Note that the computations in (10-59) are simplified if we expand the square and use the identities $\Sigma\, p_{0i} = 1$, $\Sigma\, k_i = n$. This yields

$$q = \sum_{i=1}^{m} \frac{k_i^2}{np_{0i}} - n \tag{10-62}$$

CHI-SQUARE TEST To carry out the test, we must determine the percentile q_u of Pearson's test statistic. This is, in general, a difficult problem involving the determination of a function of u depending on many parameters. In Sec. 8-3, we solve this problem numerically using computer simulation. In this section we use an approximation based on the assumption that n is large

■ ***Theorem.*** If n is large, then under hypothesis H_0, Pearson's test statistic

$$\mathbf{q} \quad \text{is} \quad \chi^2(m-1) \tag{10-63}$$

This is based on the following facts. For large n, the RVs $\mathbf{k}_i$ are normal. Hence [see (10-60)], if $p_i = p_{0i}$, the RVs $(\mathbf{k}_i - np_{0i})^2/np_{0i}q_{0i}$ are $\chi^2(1)$. However, they are not independent because $\Sigma\,(\mathbf{k}_i - np_{0i}) = 0$. Using these facts, we can show, proceeding as in the proof of (7-97), that $\mathbf{q}$ can be written as the sum of the squares of $m - 1$ independent $N(0, 1)$ RVs; the details, however, are rather involved [see (7A-6)].

We shall verify the theorem for $m = 2$. In this case [see (7-72)],

$$\mathbf{k}_1 + \mathbf{k}_2 = n \qquad p_{01} = 1 - p_{02} \qquad |\mathbf{k}_1 - np_{01}| = |\mathbf{k}_2 - np_{02}|$$

Hence,

$$\mathbf{q} = \frac{(\mathbf{k}_1 - np_{01})^2}{np_{01}} + \frac{(\mathbf{k}_2 - np_{02})^2}{np_{02}} = \frac{(\mathbf{k}_1 - np_{01})^2}{np_{01}p_{02}} \tag{10-64}$$

This shows that for $m = 2$, $\mathbf{q}$ is $\chi^2(1)$, in agreement with (10-63).

From (10-63) it follows that if n is large, Pearson's test takes the form

$$\text{Accept } H_0 \quad \text{iff} \quad q < \chi^2(m-1) \tag{10-65}$$

A decision based on (10-65) is called a *chi-square test*.

The Type II error probability depends on p_i and can be expressed in terms of the noncentral χ^2 distribution (see the Appendix to Chapter 7).

Example 10.21

We wish to test the hypothesis that a die is fair. To do so, we roll it 450 times and observe that the ith face shows k_i times where

$$k_i = 66 \quad 60 \quad 84 \quad 72 \quad 81 \quad 77$$

In this problem, $n = 450$, $p_{0i} = 1/6$, $np_{0i} = 75$, and

$$q = \sum_{i=1}^{6} \frac{(k_i - 75)^2}{75} = 7.68$$

Since $\chi^2_{.95}(5) = 11 > 7.68$, we accept the fair-die hypothesis with $\alpha = .05$. ■

The next example illustrates the use of the chi-square test in establishing the validity of a scientific theory, Mendel's theory of heredity.

Example 10.22

Peas are divided into the following four classes depending on their color and their shape: round-yellow, round-green, angular-yellow, angular green. According to the theory of heredity, the probabilities of the events of each class are

$$p_{0i} = \frac{9}{16}\ \frac{3}{16}\ \frac{3}{16}\ \frac{1}{16}$$

To test the validity of the theory, we examine 556 peas and find that the number of peas in each class equals

$$k_i = 315 \quad 108 \quad 101 \quad 32$$

Using these observations, we shall test the hypothesis that $p_i = p_{0i}$ with $\alpha = .05$.

In this problem, $n = 556$ and $np_{0i} = 312.75, 104.25, 104.25, 34.75$, $m = 4$, $\chi^2_{.95}(3) = 6.25$, and

$$q = \sum_{i=1}^{4} \frac{(k_i - np_{0i})^2}{np_{0i}} = 0.470$$

Since 6.25 is much larger than 0.470, the evidence strongly supports the null hypothesis. ■

Refined Data In general, the variations of a random experiment are due to causes that remain the same from trial to trial. In certain cases, we can isolate certain factors that might affect the data. This is done to refine the data and to improve the test. The next example illustrates this.

Example 10.23

In Mendel's theory, the ratio of yellow to green peas is 3:1. We wish to test the hypothesis that the theory is correct. We pick 478 peas from 10 plants and observe that 355 of them are yellow. What is our decision?

The problem is to test the hypothesis $p = p_0 = .75$ where p is the probability that a pea is yellow. To solve this problem, we use the chi-square test with $m = 2$ or, equivalently, the test (10-19). In our case, $k = 355$, $n = 478$, $np_0q_0 = 89.6$, and (10-14) yields

$$q = \frac{(k - np_0q_0)^2}{np_0q_0} = 0.136 < \chi^2_{.95}(1) = 3.84$$

Hence, we accept the 3:1 hypothesis with $\alpha = .05$.

Refinement There might be statistical differences among plants. This possibility leads to a refined chi-square test: We group the data according to plant and observe that the jth plant has n_j peas where

n_j = 36	39	19	97	37	26	45	53	64	62

of which

k_j = 25	32	14	70	24	20	32	44	50	44

are green. Using the data k_j and n_j of the jth plant, we form the 10 statistics

$$q_j = \frac{(k_j - n_jp_0q_0)^2}{n_jp_0q_0}$$

and we obtain

q_j = 0.59	1.03	0.02	0.34	2.03	0.05	0.36	1.82	0.33	0.54

To use the differences among plants, we shall use as the test statistic the sum

$$q = \sum_{j=1}^{10} q_j = \sum_{j=1}^{10} \frac{(k_j - n_jp_0q_0)^2}{n_jp_0q_0} = 7.19$$

The corresponding RVs $\mathbf{q}_j$ are independent, and each is $\chi^2(1)$; hence, their sum $\mathbf{q}$ is $\chi^2(10)$. Since $\chi^2_{.95}(10) = 18.3 > 7.19$, we accept the 3:1 hypothesis. ■

Incomplete Null Hypothesis

We have assumed that under the null hypothesis, the m probabilities $p_i = p_{0i}$ are known. In a number of applications, they are specified only partially. The following cases are typical.

Case 1 Only $m - r$ of the m numbers p_{0i} are known.

Case 2 The m numbers p_{0i} satisfy $m - r$ equations:

$$\phi_j(p_{01}, \ldots, p_{0m}) = 0 \qquad j = 1, \ldots, m - r \tag{10-66}$$

These equations might, for example, be the results of a scientific theory.

Case 3 The m numbers p_{0i} are functions of r unknown parameters θ_j:

$$p_{0i} = \psi_i(\theta_1, \ldots, \theta_r) \qquad i = 1, \ldots, m \tag{10-67}$$

In all cases, we have $r < m$ unknown parameters.

MODIFIED PEARSON'S STATISTIC To carry out the test, we find the *maximum likelihood* (ML) estimates of the r unknown parameters and use the given constraints to determine the ML estimates $\hat{p}_{0i}$ of all m probabilities p_{0i}. In case 1, we find the ML estimates of the r unknown probabilities p_{0i}. In case 2, we find the ML estimates of r of the m numbers p_{0i} and use the $m - r$ equations in (10-66) to find the ML estimates of the remaining $m - r$ probabilities. In case 3, we find the ML estimates θ_j of the r parameters θ_j and use the m equations in (10-66) to find the ML estimates $\hat{p}_{0i}$ of the m numbers p_{0i}.

Inserting the estimates so obtained into (10-59), we obtain the modified Pearson's statistic,

$$\hat{\mathbf{q}} = \sum_{i=1}^{m} \frac{(\mathbf{k}_i - n\hat{p}_{0i})^2}{n\hat{p}_{0i}} \tag{10-68}$$

where $\hat{p}_{0i} = p_{0i}$ for the values of i for which p_{0i} is known. To complete the test, we need the distribution of $\mathbf{q}$.

■ ***Theorem.*** If n is large, then under hypothesis H_0, the statistic

$$\mathbf{q} \quad \text{is} \quad \chi^2(m - r - 1) \tag{10-69}$$

The proof is based on the fact that the m parameter, $\hat{p}_{0i}$ satisfy r constraints and their sum equals 1.

This theorem leads to the following test: Find the ML estimates $\hat{p}_{0i}$ using the techniques of Section 9-5. Compute the sum $\hat{q}$ in (10-68):

$$\text{Accept } H_0 \quad \text{iff} \quad \hat{q} < \chi^2_{1-\alpha}(m - r - 1) \tag{10-70}$$

A decision based on (10-70) is called a *modified chi-square test.*

The foregoing is an approximation. An exact determination of the distribution of $\hat{\mathbf{q}}$ can be obtained by computer simulation.

Example 10.24

The probabilities p_i of the six faces $\{f_i\}$ of a die are unknown. We participate in a game betting on {even}. To examine our chances, we count the number k_i of times f_i

shows in 120 rolls. The results are

$$k_i = 18\ 16\ 15\ 24\ 22\ 25 \qquad \text{for } i = 1, \ldots, 6.$$

On the basis of these data, we shall test the hypothesis

$$H_0\text{: } P\{\text{even}\} = .5 \qquad \text{against} \qquad H_1\text{: } P\{\text{even}\} \neq .5$$

This is case 2 of a modified Pearson's test with $n = 120$, $m = 6$, $r = 1$. Under the null hypothesis, we have the constraints

$$P\{\text{even}\} = \hat{p}_{02} + \hat{p}_{04} + \hat{p}_{06} = 0.5 \qquad \sum_{i=1}^{6} \hat{p}_{0i} = 1 \tag{10-71}$$

We use as free parameters the probabilities p_{01}, p_{02}, p_{03}, and p_{04}. To find their ML estimates, we determine the maxima of the density

$$f(p_{01}, \ldots, p_{06}) = \gamma(p_{01})^{k_1} \cdots (p_{06})^{k_6}$$

subject to (10-71). With $L = \ln f$, (9-76) yields

$$\begin{aligned} \frac{\partial L}{\partial p_{01}} &= \frac{k_1}{p_{01}} - \frac{k_5}{p_{05}} = 0 \qquad & \frac{\partial L}{\partial p_{02}} &= \frac{k_2}{p_{02}} - \frac{k_6}{p_{06}} = 0 \\ \frac{\partial L}{\partial p_{03}} &= \frac{k_3}{p_{03}} - \frac{k_5}{p_{05}} = 0 \qquad & \frac{\partial L}{\partial p_{04}} &= \frac{k_4}{p_{04}} - \frac{k_6}{p_{06}} = 0 \end{aligned} \tag{10-72}$$

The solutions of the six equations in (10-71) and (10-72) yield the ML estimates of p_{0i}:

$$\hat{p}_{01} = .164 \qquad \hat{p}_{02} = .123 \qquad \hat{p}_{03} = .200 \qquad \hat{p}_{04} = .185 \qquad \hat{p}_{05} = .200 \qquad \hat{p}_{06} = .192$$

Inserting into (10-68), we obtain $\hat{q} = 0.834$. Since $\chi^2_{.95}(4) = 9.49 > 0.834$, we conclude that the evidence strongly supports the null hypothesis with $\alpha = .05$.

One might argue that we could have applied (10-19) to the event {even} with $k = k_2 + k_4 + k_6$. We could have; however, the resulting Type II error probability would be larger.

Test of Independence and Contingency Tables

We are given two events $\mathscr{B}$ and $\mathscr{C}$ with probabilities

$$b = P(\mathscr{B}) \qquad c = P(\mathscr{C})$$

and we wish to test the hypothesis that they are independent:

$$H_0\text{: } P(\mathscr{B} \cap \mathscr{C}) = bc \qquad \text{against} \qquad H_1\text{: } P(\mathscr{B} \cap \mathscr{C}) \neq bc \tag{10-73}$$

To do so, we form the four events $\mathscr{B} \cap \mathscr{C}$, $\mathscr{B} \cap \overline{\mathscr{C}}$, $\overline{\mathscr{B}} \cap \mathscr{C}$, and $\overline{\mathscr{B}} \cap \overline{\mathscr{C}}$ shown in Fig. 10.10a. These events form a partition of $\mathscr{S}$; hence, we can use the chi-square test, properly interpreted.

Figure 10.10

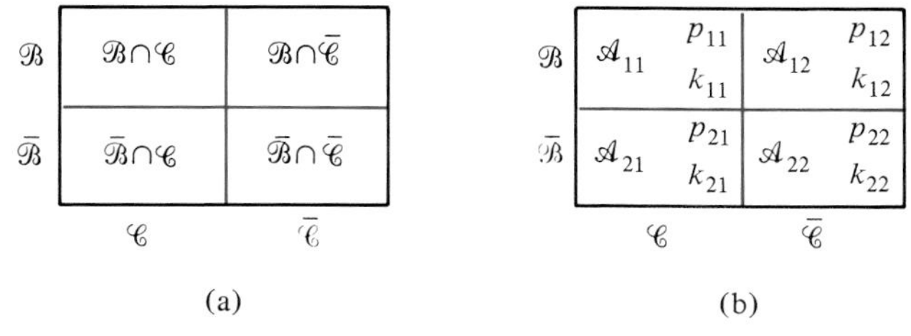

As preparation for the generalization of this important problem, we introduce the following notations. We identify the four events by $\mathscr{A}_{ij}$ where $i = 1, 2$ and $j = 1, 2$, their probabilities by p_{ij}, and their number of successes by k_{ij}. For example,

$$\mathscr{A}_{12} = \mathscr{B} \cap \overline{\mathscr{C}} \qquad p_{12} = P(\mathscr{A}_{12})$$

The number k_{12} equals the number of successes of the event $\mathscr{A}_{12}$, that is, the number of times $\mathscr{B}$ occurs but $\mathscr{C}$ does not occur. Figure 10.10b is a redrawing of Fig. 10.10a using these notations. This diagram is called a *contingency table*.

We know that $P(\overline{\mathscr{B}}) = 1 - b$ and $P(\overline{\mathscr{C}}) = 1 - c$. Furthermore, if the events $\mathscr{B}$ and $\mathscr{C}$ are independent, then the events $\mathscr{B}$ and $\overline{\mathscr{C}}$, $\overline{\mathscr{B}}$ and $\mathscr{C}$, and $\overline{\mathscr{B}}$ and $\overline{\mathscr{C}}$ are also independent. From this it follows that under hypothesis H_0,

$$p_{11} = bc \qquad p_{12} = b(1-c) \qquad p_{21} = (1-b)c \qquad p_{22} = (1-b)(1-c) \qquad (10\text{-}74)$$

Applying (10-70) to the four-element partition

$$A = [\mathscr{A}_{11}, \mathscr{A}_{12}, \mathscr{A}_{21}, \mathscr{A}_{22}]$$

we obtain the following test of (10-73):

$$\text{Accept } H_0 \text{ iff } \sum_{i=1}^{2}\sum_{j=1}^{2} \frac{(k_{ij} - np_{ij})^2}{np_{ij}} < \chi^2_{1-\alpha}(3) \qquad (10\text{-}75)$$

Example 10.25

It is known that in a certain district, 52% of all voters are male and 40% are Republican. We wish to test whether the proportion of Republicans among all voters equals the proportion of Republicans among males, that is, whether the events $\mathscr{B}$ = {Republicans} and $\mathscr{C}$ = {male} are independent. For this purpose, we poll 200 voters and find that

$$k_{11} = 35 \qquad k_{12} = 71 \qquad k_{21} = 43 \qquad k_{22} = 51$$

In this problem, $n = 200$, $b = .52$, $c = .4$. Hence,

$$p_{11} = .208 \qquad p_{12} = .312 \qquad p_{21} = .192 \qquad p_{22} = .288$$

Inserting into the sum in (10-75), we obtain $q = 3.54$. Since $\chi^2_{.95}(3) = 7.81 > 3.54$, we accept the hypothesis of independence with $\alpha = .05$. ■

Suppose now that the probabilities $b = P(\mathscr{B})$ and $c = P(\mathscr{C})$ are unknown. This is case 3 of the incomplete null hypothesis with $\theta_1 = b$ and $\theta_2 = c$. In our case, the constraints (10-67) are the four equations (10-74) expressing the probabilities p_{ij} in terms of the $r = 2$ unknown parameters b and c. The ML estimates of b and c are their empirical estimates $n_{\mathscr{B}}/n$ and $n_{\mathscr{C}}/n$, respectively, where $n_{\mathscr{B}} = k_{11} + k_{12}$ is the number of successes of the event $\mathscr{B} = \mathscr{A}_{11} \cup \mathscr{A}_{12}$ and $n_{\mathscr{C}} = k_{11} + k_{21}$ is the number of successes of the event $\mathscr{C} = \mathscr{A}_{21} \cup \mathscr{A}_{22}$. Thus

$$\hat{b} = \frac{k_{11} + k_{12}}{n} \qquad \hat{c} = \frac{k_{11} + k_{21}}{n} \qquad n = k_{11} + k_{12} + k_{21} + k_{22}$$

Replacing in (10-74) the probabilities b and c by their ML estimates $\hat{b}$ and $\hat{c}$, we obtain the ML estimates $\hat{p}_{ij}$ of p_{ij}. To complete the test, we form the modified Pearson's sum $\hat{\mathbf{q}}$ as in (10-68). In our case, $m - r - 1 = 1$, and the following test results:

$$\text{Accept } H_0 \text{ iff } \hat{q} = \sum_{i=1}^{2} \sum_{j=1}^{2} \frac{(k_{ij} - n\hat{p}_{ij})^2}{n\hat{p}_{ij}} < \chi^2_{1-\alpha}(1) \qquad (10\text{-}76)$$

Example 10.26

Engineers might have a bachelor's, master's, or doctor's degree. We wish to test whether the proportion of bachelor's degrees among all engineers is the same as the proportion among all civil engineers, that is, to test whether the events $\mathcal{B} = \{\text{bachelor}\}$ and $\mathcal{C} = \{\text{civil}\}$ are independent. For this purpose, we interrogate 200 engineers and obtain the contingency table of Fig. 10.11a.

In this problem, $n = 400$, $n_{\mathcal{B}} = 54 + 186 = 240$, $n_{\mathcal{C}} = 54 + 26 = 80$,

$$\hat{b} = .6 \qquad \hat{c} = .2 \qquad \hat{p}_{11} = .12 \qquad \hat{p}_{12} = .48 \qquad \hat{p}_{21} = .08 \qquad \hat{p}_{22} = .32$$

and (10-67) yields $\hat{q} = 2.34$. Since $\chi^2_{.95}(1) = 3.84 > 2.34$, we accept the assumption of independence with $\alpha = .05$.

GENERAL CONTINGENCY TABLES Refining the objective of Example 10-26, we shall test the hypothesis that each of the three events—engineers with bachelor's ($\mathcal{B}_1$), master's ($\mathcal{B}_2$), and doctor's ($\mathcal{B}_3$) degrees—is independent of each of the four events—civil ($\mathcal{C}_1$), electrical ($\mathcal{C}_2$), mechanical ($\mathcal{C}_3$), and chemical ($\mathcal{C}_4$) engineer. For this purpose, we interrogate the same 400 engineers and list the data in the table of Fig. 10.11b. This refinement is a special case of the following problem.

We are given two partitions

$$B = [\mathcal{B}_1, \ldots, \mathcal{B}_u] \qquad C = [\mathcal{C}_1, \ldots, \mathcal{C}_v]$$

consisting of the u events $\mathcal{B}_i$ and the v events $\mathcal{C}_j$ with respective probabilities

$$b_i = P(\mathcal{B}_i) \qquad 1 \le i \le u \qquad c_j = P(\mathcal{C}_j) \qquad 1 \le j \le v$$

We wish to test the hypothesis H_0 that each of the events $\mathcal{B}_i$ is independent of each of the events $\mathcal{C}_j$. For this purpose, we form a partition A consisting

Figure 10.11

	$\mathcal{C}$	$\overline{\mathcal{C}}$	
$\mathcal{B}$	54	186	240
$\overline{\mathcal{B}}$	26	134	160
	80	320	400

(a)

	$\mathcal{C}_1$	$\mathcal{C}_2$	$\mathcal{C}_3$	$\mathcal{C}_4$	
$\mathcal{B}_1$	54	101	50	35	240
$\mathcal{B}_2$	20	44	33	23	120
$\mathcal{B}_3$	6	15	12	7	40
	80	160	95	65	400

(b)

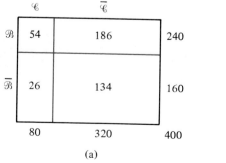

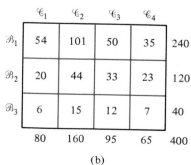

of the $m = uv$ events
$$\mathcal{A}_{ij} = \mathcal{B}_i \cap \mathcal{C}_j$$
(Fig. 10.12) and denote by p_{ij} their probabilities. Under the null hypothesis, $p_{ij} = b_i c_j$; hence, our problem is to test the hypothesis

$$H_0: p_{ij} = b_i c_j, \text{ all } i, j \quad \text{against} \quad H_1: p_{ij} \neq b_i c_j, \text{ some } i, j \quad (10\text{-}77)$$

We perform the experiment n times and observe that the event $\mathcal{A}_{ij}$ occurs k_{ij} times. This is the number of times both events $\mathcal{B}_i$ and $\mathcal{C}_j$ occur. For example, in the table of Fig. 10.11, k_{23} is the number of electrical engineers with a doctor's degree.

Suppose, first, that the probabilities b_i and c_j are known. In this case, the null hypothesis is completely specified because then $p_{ij} = b_i c_j$. Applying the chi-square test (10-63) to the partition A, we obtain the following test:

$$\text{Accept } H_0 \text{ iff } q = \sum_{i=1}^{u} \sum_{j=1}^{v} \frac{(k_{ij} - n b_i c_j)^2}{n b_i c_j} < \chi_{1-\alpha}^2(uv - 1) \quad (10\text{-}78)$$

If the probabilities b_i and c_j are unknown, we determine their ML estimates $\hat{b}_i$ and $\hat{c}_j$ and apply the modified chi-square test (10-70). The number $n_{\mathcal{B}_i}$ of occurrences of the event $\mathcal{B}_i$ is the sum of the entries k_{ij} of the ith row of the contingency table of Fig. 10.12, and the number $n_{\mathcal{C}_j}$ of occurrences of the event $\mathcal{C}_j$ equals the number of entries k_{ij} of the jth column. Hence,

$$\hat{b}_i = \frac{n_{\mathcal{B}_i}}{n} = \frac{1}{n} \sum_{j=1}^{v} k_{ij} \qquad \hat{c}_j = \frac{n_{\mathcal{C}_j}}{n} = \frac{1}{n} \sum_{i=1}^{u} k_{ij} \quad (10\text{-}79)$$

In (10-79), there are actually $r = u + v - 2$ unknown parameters because $\Sigma b_i = 1$ and $\Sigma c_j = 1$. Hence,

$$m - r - 1 = uv - (u + v - 2) - 1 = (u - 1)(v - 1)$$

Inserting into (10-70), we obtain the following test:

$$\text{Accept } H_0 \text{ if } \quad \hat{q} = \sum_{i=1}^{u} \sum_{j=1}^{v} \frac{k_{ij} - n\hat{b}_i\hat{c}_j}{n\hat{b}_i\hat{c}_j} < \chi_{1-\alpha}^2[(u-1)(v-1)] \quad (10\text{-}80)$$

Figure 10.12

B \ C	$\mathcal{C}_1$	$\cdots$	$\mathcal{C}_j$	$\mathcal{C}_v$
$\mathcal{B}_1$	$\mathcal{A}_{11}$			$\mathcal{A}_{1v}$
$\vdots$				
$\mathcal{B}_i$			$\mathcal{A}_{ij}$	
$\mathcal{B}_u$	$\mathcal{A}_{u1}$			$\mathcal{A}_{uv}$

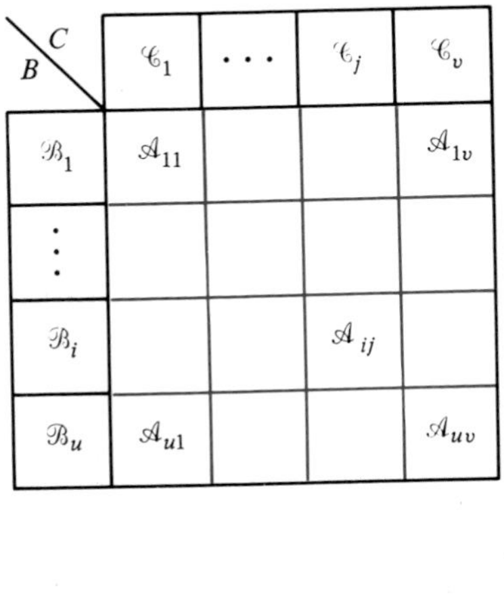

Example 10.27

We shall apply (10-80) to the data in Fig. 10.11b. In this problem, $n = 400$, and (10-79) yields

$$\hat{b}_i = .6 \quad .3 \quad .1 \qquad \hat{c}_j = .2 \quad .4 \quad .2375 \quad .1625$$

The resulting estimates $\hat{p}_{ij} = \hat{b}_i \hat{c}_j$ of p_{ij} equal

$$\hat{p}_{ij} = .12 \qquad .24 \qquad .1425 \qquad .0975$$
$$\hat{p}_{2j} = .06 \qquad .12 \qquad .0712 \qquad .0488$$
$$\hat{p}_{3j} = .02 \qquad .04 \qquad .0238 \qquad .0162$$

Hence, $\hat{q} = 5.94$. Since $u = 3$, $v = 4$, $(u - 1)(v - 1) = 6$, and $\chi^2_{.95}(6) = 12.59$, we conclude with consistency level $\alpha = .05$ that the proportion of engineers with bachelor's, master's, or doctor's degrees among the four groups considered is the same. ■

Goodness of Fit of Distributions

In Section 10-2, we presented a method for testing the hypothesis that the distribution $F(x)$ equals a given function $F_0(x)$ for all x:

$$H_0: F(x) = F_0(x), \text{ all } x \qquad \text{against} \qquad H_1: F(x) \neq F_0(x), \text{ some } x \quad (10\text{-}81)$$

In this section, we have a more modest objective. We wish to test the hypothesis that $F(x) = F_0(x)$ only at a set of $m - 1$ points a_i (Fig. 10.13):

$$H_0': F(a_i) = F_0(a_i), \text{ all } i \qquad \text{against} \qquad H_1': F(a_i) \neq F_0(a_i), \text{ some } i \quad (10\text{-}82)$$

Figure 10.13

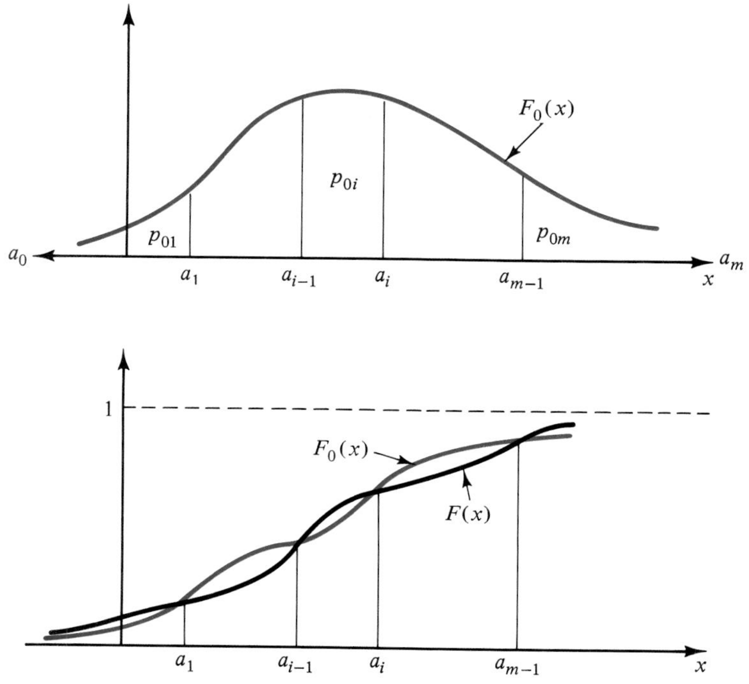

If $\mathbf{x}$ is of the discrete type, taking the values a_i, then the two tests are equivalent. If $\mathbf{x}$ is of the continuous type and H_0' is rejected, then H_0 is rejected; however, if H_0' is accepted, our conclusion is that the evidence does not support the rejection of H_0.

The simplified hypothesis H_0' is equivalent to (10-58); hence, we can apply the chi-square test. To do so, we form the m events

$$\mathcal{A}_i = \{a_{i-1} < \mathbf{x} \le a_i\} \qquad 1 \le i \le m \tag{10-83}$$

The probabilities $p_i = P(\mathcal{A}_i)$ of these events equal

$$P(\mathcal{A}_i) = F(a_i) - F(a_{i-1}) \qquad a_0 = -\infty \qquad a_m = +\infty \tag{10-84}$$

and under hypothesis H_0', $p_i = p_{0i} = F_0(a_i) - F_0(a_{i-1})$. Hence, (10-82) is equivalent to (10-58). This equivalence leads to the following test of H_0': Compute the m probabilities $p_{0i} = F_0(a_i) - F_0(a_{i-1})$. Determine the n samples x_i of $\mathbf{x}$, and count the number k_i of samples that are in the interval $(a_{i-1} < x < a_i)$. Form Pearson's statistic (10-59).

$$\text{Reject } H_0' \text{ iff } q > \chi^2_{1-\alpha}(m-1) \tag{10-85}$$

Example 10.28

We wish to test the hypothesis that the time to failure $\mathbf{x}$ of a system has an exponential distribution $F_0(x) = 1 - e^{-x/2}$, $x > 0$. Proceeding as in (10-83), we select the points 3, 6, 9 and form the events

$$\{\mathbf{x} \le 3\} \qquad \{3 < \mathbf{x} \le 6\} \qquad \{6 < \mathbf{x} \le 9\} \qquad \{\mathbf{x} > 9\}$$

In this problem,

$$p_{01} = F_0(3) = .221 \qquad\qquad p_{02} = F_0(6) - F_0(3) = .173$$
$$p_{03} = F_0(9) - F_0(6) = .134 \qquad p_{04} = 1 - F_0(9) = .472$$

We next determine the times to failure x_i of $n = 200$ systems. We observe that k_i are in the ith interval, where

$$k_i = 53 \qquad 42 \qquad 35 \qquad 70$$

Inserting into (10-59), we obtain

$$q = \sum_{i=1}^{4} \frac{(k_i - np_{0i})^2}{np_{0i}} = 12.22 > \chi^2_{.9}(3) = 11.34$$

Hence, we reject H_0', and therefore also H_0, with $\alpha = .1$. ■

Note The significance level α used in (10-85) is the Type I error probability of the test (10-82). The corresponding error α' of the test (10-81) is smaller. An increase of the number m of intervals brings α' closer to α decreasing, thus the resulting Type II error. This, however, results in a decrease of the number of samples k_i in each interval (a_{i-1}, a_i), thereby weakening the validity of the large n approximation (10-63). In most cases, the condition $k_i > 3$ is adequate. If this condition is not met in a particular test, we combine two or more intervals into one until it does.

INCOMPLETE NULL HYPOTHESIS We have assumed that the function $F_0(x)$ is known. In a number of cases, the distribution of $\mathbf{x}$ is a function $F_0(x, \theta_1, \ldots, \theta_r)$ of known form depending on r unknown parameters θ_j. In this case, the probabilities p_{0i} satisfy the m equations $p_{0i} = \psi_i$ as in (10-67) where

$$\psi_i(\theta_1, \ldots, \theta_r) = F_0(a_i, \theta_1, \ldots, \theta_r) - F_0(a_{i-1}, \theta_1, \ldots, \theta_r)$$

This is case 3 of the incomplete null hypothesis. To complete the test, we find the ML estimates $\hat{\theta}_j$ of θ_j and insert into (10-84). This yields the ML estimates of $\hat{p}_{0i}$. Using these estimates, we form the modified Pearson's statistic (10-68), and we apply (10-70) in the following form:

$$\text{Reject} \quad H_0 \quad \text{if} \quad \hat{q} > \chi^2_{1-\alpha}(m - r - 1) \qquad (10\text{-}86)$$

Example 10.29

We wish to test the hypothesis that the number of particles emitted from a radioactive substance in t_0 seconds is a Poisson-distributed RV with parameter $\theta = \lambda t_0$:

$$P\{\mathbf{x} = k\} = e^{-\theta}\frac{\theta^k}{k!} \qquad k = 0, 1, \ldots \qquad (10\text{-}87)$$

We shall carry out the analysis under the assumption that the numbers of particles in nonoverlapping intervals are independent. In Rutherford's experiment, 2,612 intervals were considered, and it was found that in n_k of these intervals, there were k particles where

$k =$	0	1	2	3	4	5	6	7	8	9	10	11	12	>12
$n_k =$	57	203	383	525	532	408	273	139	49	27	20	4	2	0

Thus in 57 intervals, there were no particles, in 203 intervals there was only one particle, and in no one interval there were more than 12 particles (Fig. 10.14). We next form the 12 events

$$\{\mathbf{x} = 0\}, \{\mathbf{x} = 1\}, \ldots, \{\mathbf{x} = 10\}, \{\mathbf{x} \geq 11\}$$

and determine the ML estimates $\hat{p}_{0i}$ of their probabilities p_{0i}. To do so, we find the ML estimate $\hat{\theta}$ of θ. As we know, $\hat{\theta}$ equals the empirical estimate $\bar{x}$ of $\mathbf{x}$ (see Problem

Figure 10.14

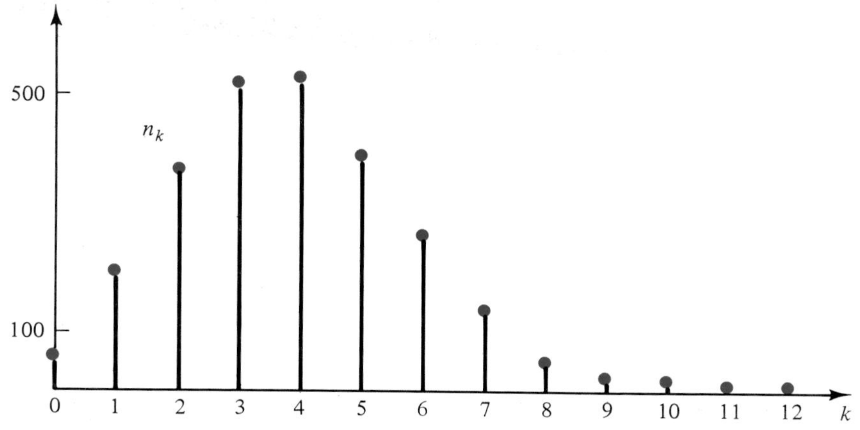

9-35). The RV $\mathbf{x}$ takes the values $k = 0, 1, 2, \ldots$ with multiplicity n_k; hence,

$$\hat{\theta} = \bar{x} = \frac{1}{n}\sum_{k=0}^{12} kn_k = 3.9 \qquad n = \sum_{k=0}^{12} n_k = 2{,}612$$

Replacing θ by $\hat{\theta}$ in (10-87), we obtain

$$\hat{p}_{0i} = e^{-\hat{\theta}}\frac{\hat{\theta}^k}{k!} \qquad i = k + 1 = 1, \ldots, 11 \qquad \hat{p}_{0i} = e^{-\hat{\theta}}\left(\frac{\hat{\theta}^{11}}{11!} + \frac{\hat{\theta}^{12}}{12!}\right) \qquad i = 12$$

and with $\hat{\theta} = 3.9$, this yields

$$\hat{p}_{0i} = .020\ .079\ .154\ .012\ .200\ .195\ .152\ .099\ .055\ .027\ .012\ .005\ .003$$

To find Pearson's test statistic $\hat{q}$, we observe that the numbers k_i in (10-68) are for this problem the numbers n_{i-1}. The resulting sum equals

$$\hat{q} = \sum_{i=1}^{m} \frac{(n_{i-1} - n\hat{p}_{0i})^2}{n_{i-1}\hat{p}_{0i}} = 10.67$$

We have $m = 12$ events and $r = 1$ unknown (namely, the parameter θ); hence, $\hat{\mathbf{q}}$ is $\chi^2(10)$. Since $\chi^2_{.95}(10) = 18.31 > \hat{q}$, we conclude that the evidence does not support the rejection of the Poisson distribution. ■

10-5 Analysis of Variance

The values of an RV $\mathbf{x}$ are usually random. In a number of cases, they also depend on certain observable causes. Such causes are called *factors*. Each factor is partitioned into several characteristics called *groups*. Analysis of variance (ANOVA) is the study of the possible effects of the groups of each factor on the mean η_x of $\mathbf{x}$. ANOVA is a topic in hypothesis testing. The null hypothesis is the assumption that the groups have no effect on the mean of $\mathbf{x}$; the alternative hypothesis is the assumption that they do. The test (10-13) of the equality of two means is a special case. In ANOVA, we study various tests involving *means;* the topic, however, is called analysis of *variance*. As we shall presently explain, the reason is that the tests are based on ratios of variances.

We shall illustrate with an example. Suppose that the RV $\mathbf{x}$ represents the income of engineers. We wish to study the possible effects of education, sex, and marital status on the mean of $\mathbf{x}$. In this problem, we have three factors. Factor 1 consists of the groups bachelor's degree, master's degree, and doctor's degree. Factor 2 consists of the groups male and female. Factor 3 consists of the groups married and unmarried.

In an ANOVA investigation, we have one-factor tests, two-factor tests, or many-factor tests. In a one-factor test, we study the effects of the groups of a single factor on the mean of $\mathbf{x}$. We refine the underlying experiment, forming a different RV for each group. We thus have m RVs $\mathbf{x}_j$ where m is the number of groups in the factor under consideration. In a two-factor test, we form mr RVS $\mathbf{x}_{jk}$ where m is the number of groups of factor 1 and r is the number of groups of factor 2.

In a one-factor problem where $\mathbf{x}$ is the income of engineers and the factor is education, the RVS $\mathbf{x}_1$, $\mathbf{x}_2$, $\mathbf{x}_3$ represent the income of engineers with a bachelor's, a master's, and a doctor's degree, respectively. If we wish to include the possible effect of sex, we have a two-factor tests. In this case, we have six RVS $\mathbf{x}_{jk}$ where $j = 1, 2, 3$ and $k = 1, 2$. For example, $\mathbf{x}_{31}$ represents the income of male engineers with a doctor's degree.

In the course of this investigation, we shall use samples of the RVS $\mathbf{x}_j$ and $\mathbf{x}_{jk}$. To avoid mixed subscripts, we shall identify the samples with superscripts, retaining the subscripts for the various groups. Thus $\mathbf{x}_j^i$ will mean the ith sample of the RV $\mathbf{x}_j$; similarly, x_{jk}^i will mean the ith sample of the RV $\mathbf{x}_{jk}$. The corresponding sample means will be identified by overbars. Thus

$$\bar{\mathbf{x}}_j = \frac{1}{n_j}\sum_{i=1}^{n_j} \mathbf{x}_j^i \qquad \bar{\mathbf{x}}_{jk} = \frac{1}{n_{jk}}\sum_{i=1}^{n_{jk}} \mathbf{x}_{jk}^i \tag{10-88}$$

where n_j and n_{jk} are the number of samples of the RVS $\mathbf{x}_j$ and $\mathbf{x}_{jk}$, respectively.

THE ANOVA PRINCIPLE We have shown in (7-95) that if the RVS $\mathbf{z}_i$ are uncorrelated with the same mean and variance and

$$\mathbf{Q}_0 = \sum_{j=1}^{m} (\mathbf{z}_j - \bar{\mathbf{z}})^2 \qquad \text{then} \qquad E\{\mathbf{Q}_0\} = (m - 1)\sigma^2 \tag{10-89}$$

Furthermore, if they are normal, then the ratio $\mathbf{Q}_0/\sigma^2$ has a $\chi^2(m - 1)$ distribution. The following result is a simple generalization.

Given m uncorrelated RVS $\mathbf{w}_j$ with the same variance σ^2 and means η_j, we form the sum

$$\mathbf{Q} = \sum_{j=1}^{m} (\mathbf{w}_j - \bar{\mathbf{w}})^2 \qquad \bar{\mathbf{w}} = \frac{1}{m}\sum_{j=1}^{m} \mathbf{w}_j \tag{10-90}$$

■ ***Theorem***

$$E\{\mathbf{Q}\} = (m - 1)\sigma^2 + e \tag{10-91}$$

where

$$e = \sum_{j=1}^{m} (\eta_j - \bar{\eta})^2 \qquad \bar{\eta} = \frac{1}{m}\sum_{j=1}^{m} \eta_j = E\{\bar{\mathbf{w}}\} \tag{10-92}$$

Furthermore, if the RVS $\mathbf{w}_j$ are normal and $e = 0$, then the ratio $\mathbf{Q}/\sigma^2$ has a $\chi^2(m - 1)$ distribution.

■ ***Proof.*** This result was established in (7A-14); however, because of its importance, we shall prove it directly. The RVS $\mathbf{z}_j = \mathbf{w}_j - \eta_j$ are uncorrelated, with zero mean and variance σ^2; furthermore, $\mathbf{w}_j - \bar{\mathbf{w}} = \mathbf{z}_j - \bar{\mathbf{z}} + \eta_j - \bar{\eta}$. Inserting into (10-90), we obtain

$$\mathbf{Q} = \sum_{j=1}^{m} [(\mathbf{z}_j - \bar{\mathbf{z}}) + (\eta_j - \bar{\eta})]^2$$

We next square both sides and take expected values. This yields

$$E\{\mathbf{Q}\} = E\left\{\sum_{j=1}^{m} (\mathbf{z}_j - \overline{\mathbf{z}})^2\right\} + \sum_{j=1}^{m} (\eta_j - \overline{\eta})^2$$

because $E\{(\mathbf{z}_j - \overline{\mathbf{z}})(\eta_j - \overline{\eta})\} = 0$, and (10-91) follows from (10-89). If $e = 0$, then $\mathbf{Q} = \mathbf{Q}_0$; hence, $\mathbf{Q}/\sigma^2$ is $\chi^2(m-1)$. If $e \neq 0$, then the distribution of $\mathbf{Q}/\sigma^2$ is noncentral $\chi^2(m-1, e)$.

This theorem is the essence of all ANOVA tests; based on the following consequence of (10-91). Suppose that $\hat{\boldsymbol{\sigma}}^2$ is an unbiased consistent estimator of σ^2. If n is sufficiently large, then $\mathbf{Q}$ is concentrated near its mean $(m-1)\sigma^2 + e$, and $\hat{\boldsymbol{\sigma}}^2$ is concentrated near its mean σ^2. Hence, if $e = 0$—that is, if the RVs $\mathbf{w}_j$ have the same mean—then the ratio $\mathbf{Q}/(m-1)\hat{\boldsymbol{\sigma}}^2$ is close to 1, and it increases as e increases.

One-Factor Tests

The m RVS

$$\mathbf{x}_1, \ldots, \mathbf{x}_j, \ldots, \mathbf{x}_m$$

represent the m groups of a factor. We assume that they are normal with the same variance. We shall test the hypothesis that their means $\eta_j = E\{\mathbf{x}_j\}$ are equal:

$$H_0: \eta_1 = \cdots = \eta_m \qquad \text{against} \qquad H_1: \eta_i \neq \eta_j, \text{ some } i, j \tag{10-93}$$

For this purpose, we sample the jth RV n_j times and obtain the samples x_j^i (Fig. 10.15). The total number of samples equals

$$N = n_1 + \cdots + n_m$$

We next form the sum

$$\mathbf{Q}_1 = \sum_{j=1}^{m} n_j(\overline{\mathbf{x}}_j - \overline{\mathbf{x}})^2 = \sum_{j=1}^{m} (\mathbf{w}_j - \overline{\mathbf{w}})^2 \tag{10-94}$$

Figure 10.15

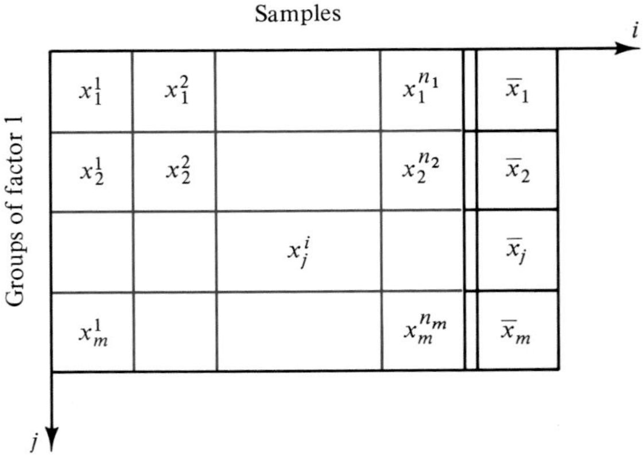

where $\bar{\mathbf{x}}_j$ is the average of the samples $\mathbf{x}_j^i$ of $\mathbf{x}_j$ as in (10-88), $\mathbf{w}_j = \sqrt{n_j}\bar{\mathbf{x}}_j$, and

$$\bar{\mathbf{x}} = \frac{1}{m}\sum_{j=1}^{m} \bar{\mathbf{x}}_j \qquad \bar{\mathbf{w}} = \frac{1}{m}\sum_{j=1}^{m} \mathbf{w}_j$$

The RVS $\bar{\mathbf{x}}_j$ are independent, with mean η_j and variance σ^2/n_j. From this it follows that the RVS $\mathbf{w}_j$ are independent, with mean $\eta_j\sqrt{n_j}$ and variance σ^2. Hence [see (10-91)],

$$E\{\mathbf{Q}_1\} = (m-1)\sigma^2 + e \qquad e = \sum_{j=1}^{m} n_j(\eta_j - \bar{\eta})^2 \tag{10-95}$$

The constant e equals the value of $\mathbf{Q}_1$ when the RVS $\bar{\mathbf{x}}_j$ and $\bar{\mathbf{x}}$ are replaced by their means η_j and $\bar{\eta}$.

To apply the ANOVA principle, we must find an unbiased estimator of σ^2. For this purpose, we form the sum

$$\mathbf{Q}_2 = \sum_{j=1}^{m}\sum_{i=1}^{n_j} (\mathbf{x}_j^i - \bar{\mathbf{x}}_j)^2 \tag{10-96}$$

For a specific j, the RVS $\mathbf{z}_i = \mathbf{x}_j^i$ are i.i.d., with variance σ^2 and sample mean $\bar{\mathbf{z}} = \bar{\mathbf{x}}_j$. Applying (10-89), we obtain

$$E\left\{\sum_{i=1}^{n_j} (\mathbf{x}_j^i - \bar{\mathbf{x}}_j)^2\right\} = (n_j - 1)\sigma^2$$

Hence,

$$E\{\mathbf{Q}_2\} = \sum_{j=1}^{m} (n_j - 1)\sigma^2 = (N - m)\sigma^2 \tag{10-97}$$

Reasoning as in the ANOVA principle, we conclude that the ratio

$$\mathbf{q} = \frac{\mathbf{Q}_1/(m-1)\sigma^2}{\mathbf{Q}_2/(N-m)\sigma^2} \tag{10-98}$$

is close to 1 if $e = 0$, and it increases as e increases. It is reasonable, therefore, to use $\mathbf{q}$ as our test statistic. To complete the test, we must find the distribution of $\mathbf{q}$.

■ ***Theorem.*** Under hypothesis H_0, the RV

$$\mathbf{q} = \frac{(N-m)\mathbf{Q}_1}{(m-1)\mathbf{Q}_2} \qquad \text{is} \qquad F(m-1, N-m) \tag{10-99}$$

■ ***Proof.*** If H_0 is true, the RVS $\mathbf{w}_j = \bar{\mathbf{x}}_j\sqrt{n_j}$ are i.i.d., with variance σ^2. Hence, as in (10-90), the ratio $\mathbf{Q}_1/\sigma^2$ is $\chi^2(m-1)$. Similarly, the ratio

$$\mathbf{y}_j = \frac{1}{\sigma^2}\sum_{i=1}^{n_j} (\mathbf{x}_j^i - \bar{\mathbf{x}}_j)^2 \qquad \text{is} \qquad \chi^2(n_j - 1)$$

From this and (7-87) it follows that the sum

$$\frac{\mathbf{Q}_2}{\sigma^2} = \sum_{j=1}^{m} \mathbf{y}_j \text{ is } \chi^2 \text{ with } \sum_{j=1}^{m} (n_j - 1) = N - m \tag{10-100}$$

degrees of freedom. Furthermore, $\mathbf{y}_j$ is independent of $\bar{\mathbf{x}}_j$ [see (7-97)], and $\mathbf{Q}_1$ is a function of $\bar{\mathbf{x}}_j$; hence, the RVs $\mathbf{Q}_1$ and $\mathbf{Q}_2$ are independent. Applying (7-106) to the RVs $\mathbf{Q}_1/\sigma^2$ and $\mathbf{Q}_2/\sigma^2$, we obtain (10-99).

A one-factor ANOVA test thus proceeds as follows: Observe the n_j samples x_j^i of each RV $\mathbf{x}_j$. Compute $\bar{x}_j$, $\bar{x}$, and the sums Q_1 and Q_2.

$$\text{Accept } H_0 \text{ iff } \frac{(N-m)Q_1}{(m-1)Q_2} < F_{1-\alpha}(m-1, N-m) \qquad \text{(10-101)}$$

where $F_u(m-1, N-m)$ is the u-percentile of the Snedecor distribution.

OC Function The Type II error probability is a function $\beta(e)$ depending only on e. To find it, we must determine the distribution of $\mathbf{q}$ under the alternative hypothesis. The distribution of $\mathbf{Q}_2/\sigma^2$ is $\chi^2(N-m)$ even if H_0 is not true because the value of $\mathbf{Q}_2$ does not change if we replace the RVs $\mathbf{x}_j^i$ by $\mathbf{x}_j^i - \eta_j$. If $e \neq 0$, the distribution of $\mathbf{Q}_1/\sigma^2$ is a noncentral $\chi^2(m-1, e)$ with eccentricity e (see the Appendix to Chapter 7). Hence, the ratio $\mathbf{q}$ has a noncentral $F(m-1, N-m, e)$ distribution.

Example 10.30

A factory acquires four machines producing electric motors. We wish to test the hypothesis that the mean times to failure of the motors are equal. We monitor $n_j = 7$, 5, 6, 4 motors from each machine and find the following times to failure x_j^i, in weeks.

Machine 1:	8.2	6.3	7.7	9.4	4.2	4.7	8.5	7.0	$\bar{x}_1 = 7.0$
Machine 2:	7.2	6.3	5.5	6.9	8.1				$\bar{x}_2 = 6.8$
Machine 3:	7.9	6.9	9.0	4.9	5.8	6.9			$\bar{x}_3 = 6.9$
Machine 4:	9.2	7.5	8.5	6.8					$\bar{x}_4 = 8.0$

In this problem, $m = 4$, $N = 22$, $\bar{x} = 7.175$,

$$Q_1 = 4.09 \qquad Q_2 = 5.68 \qquad q = \frac{18Q_1}{3Q_2} = 4.32$$

Since $F_{.95}(3, 18) = 3.16 < 4.32$, we reject the null hypothesis with $\alpha = .05$. ■

Example 10.31

We wish to test the hypothesis that parental education has no effect on the IQ of children. We have here three groups: grammar school (GS), high school (HS), and college (C). We select 10 children from each group and observe the following scores.

GS:	77	92	102	94	115	83	90	120	75	96	$\bar{x}_1 = 94.4$
HS:	88	99	80	105	92	112	92	108	79	123	$\bar{x}_2 = 97.8$
C:	102	90	95	118	76	108	80	122	86	110	$\bar{x}_3 = 99.5$

In this problem, $m = 3$, $n_1 = n_2 = n_3 = 10$, $N = 30$, $\bar{x} = 97.23$,

$$Q_1 = 134.9 \qquad Q_2 = 5{,}956 \qquad q = \frac{270Q_1}{2Q_2} = 3.06$$

Since $F_{.95}(2, 27) = 3.35 > 3.06$, we accept the null hypothesis with $\alpha = .05$. ■

Two-Factor Tests

In a two-factor test, we have mr RVS

$$\mathbf{x}_{jk} \qquad j = 1, \ldots, m \qquad k = 1, \ldots, r$$

The subscript j identifies the m groups of factor 1, and the subscript k identifies the r groups of factor 2. The purpose of the test is to determine the possible effects of each factor on the mean $\eta_{jk} = E\{\mathbf{x}_{jk}\}$ of $\mathbf{x}_{jk}$.

We introduce the averages

$$\mathbf{x}_{j.} = \frac{1}{r}\sum_{k=1}^{r}\mathbf{x}_{jk} \qquad \mathbf{x}_{.k} = \frac{1}{m}\sum_{j=1}^{m}\mathbf{x}_{jk} \qquad \mathbf{x}_{..} = \frac{1}{mr}\sum_{j=1}^{m}\sum_{k=1}^{r}\mathbf{x}_{jk} \tag{10-102}$$

Thus $\mathbf{x}_{j.}$ is the average of all RVS in the jth row of Fig. 10.16, and $\mathbf{x}_{.k}$ is the average of all RVS in the kth column. The average of all RVS (global average) is $\mathbf{x}_{..}$.

We shall study the effects of the groups on the means

$$\eta_{j.} = E\{\mathbf{x}_{j.}\} \qquad \eta_{.k} = E\{\mathbf{x}_{.k}\} \qquad \eta_{..} = E\{\mathbf{x}_{..}\} \tag{10-103}$$

of the RVS so formed. For this purpose, we introduce the constants (Fig. 10.17)

$$\alpha_j = \eta_{j.} - \eta_{..} \qquad \beta_k = \eta_{.k} - \eta_{..} \qquad \gamma_{jk} = \eta_{jk} - \eta_{..} - \alpha_j - \beta_k \tag{10-104}$$

Thus α_j is the deviation of the row mean $\eta_{j.}$ from the global mean $\eta_{..}$, β_k is the deviation of the column mean $\eta_{.k}$ from the global mean $\eta_{..}$, and γ_{jk} is the deviation of η_{jk} from the sum $\eta_{..} + \alpha_j + \beta_k$. Each of the sums of the m numbers α_j, the r numbers β_k, and the mr numbers γ_{jk} is 0.

In a two-factor test, the objective is to study the effects of the groups on the parameters α_j, β_k, and γ_{jk}. This problem has many aspects. We shall briefly discuss two cases related to the following concept.

Figure 10.16

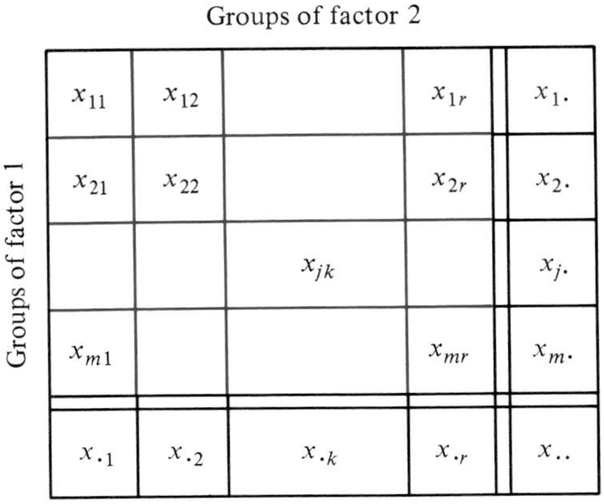

Groups of factor 2
Groups of factor 1
x_{11} x_{12} x_{1r} $x_{1.}$
x_{21} x_{22} x_{2r} $x_{2.}$
x_{jk} $x_{j.}$
x_{m1} x_{mr} $x_{m.}$
$x_{.1}$ $x_{.2}$ $x_{.k}$ $x_{.r}$ $x_{..}$

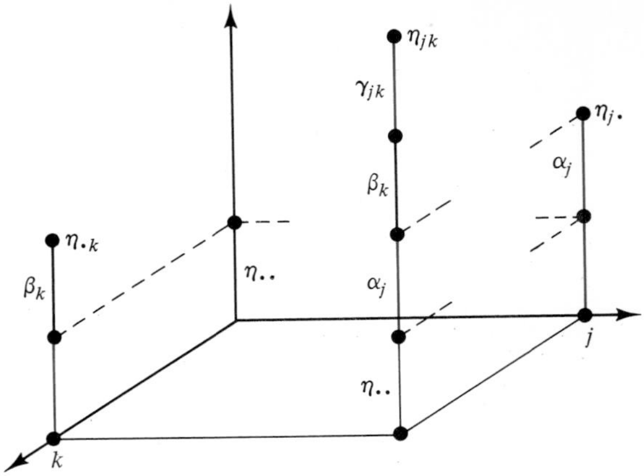

Figure 10.17

ADDITIVITY We shall say that two factors are *additive* if the corresponding parameter γ_{jk} is 0 or, equivalently, if

$$\eta_{jk} = \eta_{..} + \alpha_j + \beta_k \qquad (10\text{-}105)$$

In the engineer example, the condition $\gamma_{jk} = 0$ means that the variations in average income due to education are the same for male and female engineers and the variations due to sex are the same for engineers with bachelor's, master's, and doctor's degree.

As in the one-factor tests, the analysis of a two-factor test is based on the ANOVA principle. We start with the following problem: We assume that factors 1 and 2 of a two-factor problem are additive, and we wish to test the hypothesis that $\alpha_j = 0$, that is, that the mean η_{jk} of $\mathbf{x}_{jk}$ does not depend on j:

$$H_0: \gamma_{jk} = 0, \alpha_j = 0, \text{all } j \qquad \text{against} \qquad H_1: \gamma_{jk} = 0, \alpha_j \neq 0, \text{some } j \qquad (10\text{-}106)$$

One might argue that (10-106) could be treated as a one-factor problem: Ignore factor 2 and test the hypothesis that the groups of factor 1 have no effect on the mean of $\mathbf{x}_{jk}$. We could do so. The reason we choose a two-factor test is that as in Example 10.23, the possible variations within the groups of factor 2 lead to a more powerful test.

Proceeding as in (10-94), we form the sum

$$\mathbf{Q}_3 = r\sum_{j=1}^{m}(\mathbf{x}_{j.} - \mathbf{x}_{..})^2 = \sum_{j=1}^{m}(\mathbf{w}_j - \mathbf{w}_.)^2 \qquad \mathbf{w}_j = \mathbf{x}_{j.}\sqrt{r} \qquad (10\text{-}107)$$

This sum is essentially identical to $\mathbf{Q}_1$ if we replace sample means by column averages and n_j by r. This yields [see (10-95)]

$$E\{\mathbf{Q}_3\} = (m-1)\sigma^2 + e \qquad e = r\sum_{j=1}^{m}(\eta_{j.} - \eta_{..})^2 = r\sum_{j=1}^{m}\alpha_j^2 \qquad (10\text{-}108)$$

We next form the sum

$$\mathbf{Q}_4 = \sum_{j=1}^{m} \sum_{k=1}^{r} (\mathbf{x}_{jk} - \mathbf{x}_{j.} - \mathbf{x}_{.k} + \mathbf{x}_{..})^2 \tag{10-109}$$

Under the additivity assumption, $\gamma_{jk} = 0$; hence,

$$\eta_{jk} = \eta_{j.} + \eta_{.k} - \eta_{..}$$

for both hypotheses. From this it follows that if we subtract the mean from all RVs in (10-109), the sum $\mathbf{Q}_4$ remains the same; hence, it has the same distribution under both hypotheses.

Reasoning as in (7-97), we can show that the ratio

$$\frac{\mathbf{Q}_4}{\sigma^2} \text{ is } \chi^2[(m-1)(r-1)] \tag{10-110}$$

The details of the proof are rather involved and will be omitted. Since the mean of an RV with $\chi^2(n)$ distribution equals n, we conclude that

$$E\{\mathbf{Q}_4\} = (m-1)(r-1)\sigma^2 \tag{10-111}$$

Proceeding as in (10-98), we form the ratio

$$\mathbf{q} = \frac{\mathbf{Q}_3/(m-1)\sigma^2}{\mathbf{Q}_4/(m-1)(r-1)\sigma^2} \tag{10-112}$$

If H_0 is true, then $\alpha_j = 0$; hence, the eccentricity e in (10-108) is 0. From this it follows that under hypothesis H_0,

$$\frac{\mathbf{Q}_3}{\sigma^2} \text{ is } \chi^2(m-1) \tag{10-113}$$

Combining with (10-110), we conclude that if H_0 is true, then the ratio $\mathbf{q}$ in (10-112) is $F[m-1, (m-1)(r-1)]$. This yields the following test of (10-106):

$$\text{Accept } H_0 \text{ iff } \frac{(r-1)Q_3}{Q_4} < F_{1-\alpha}[m-1, (m-1)(r-1)] \tag{10-114}$$

Example 10.32

We shall examine the effects of sex and education on the yearly income x_{jk} of factory workers. In this problem, we have two factors. Factor 1, sex, consists of the two factors M and F. Factor 2, education, consists of the three factors GS, HS, and C. We observe one value of each RV $\mathbf{x}_{jk}$ and obtain the following list of incomes in thousands.

		GS	GC	C	
		x_{j1}	x_{j2}	x_{j3}	$x_{j.}$
M	x_{1k}	15	18	27	20
F	x_{2k}	14	16	24	18
	$x_{.k}$	14.5	17	25.5	$x_{..} = 19$

Assuming that there is no interaction between the groups (additivity), we shall test the hypothesis that $\alpha_j = 0$ and the hypothesis that $\beta_k = 0$.

(a) H_0: $\alpha_j = 0$. With $m = 2$ and $r = 3$, we obtain

$$Q_3 = 6 \qquad Q_4 = 1 \qquad q = \frac{2Q_3}{Q_4} = 12$$

Since $F_{.95}(1, 2) = 18.6 > q$, we conclude with $\alpha = .05$ that the evidence does not support the rejection of the hypothesis that sex has no effect on income.

(b) H_0: $\beta_k = 0$. We proceed similarly, interchanging j and k. We have the same Q_4, but $m = 3$ and $r = 2$. This yields

$$Q_3 = 133 \qquad Q_4 = 1 \qquad q = \frac{Q_3}{Q_4} = 133$$

Since $F_{.95}(2, 2) = 19 < q$, we reject the hypothesis that education has no effect on income.

Note that in both cases, the evidence is limited, and the resulting Type II error probability is large. ∎

TEST OF ADDITIVITY In (10-106), we assumed that the two factors are additive, and we tested the hypothesis that the first factor has no effect on η_{jk}. Now we test the hypothesis that the two factors are additive:

$$H_0\colon \gamma_{jk} = 0, \text{ all } j, k \qquad \text{against} \qquad H_1\colon \gamma_{jk} \neq 0, \text{ some } j, k \tag{10-115}$$

Unlike (10-106), the test of (10-115) cannot be carried out in terms of a single sample. We shall use n samples for each RV $\mathbf{x}_{jk}$. This yields the mrn samples (Fig. 10.18)

$$\mathbf{x}_{jk}^i \qquad i = 1, \ldots, n \qquad j = 1, \ldots, m \qquad k = 1, \ldots, r$$

Figure 10.18

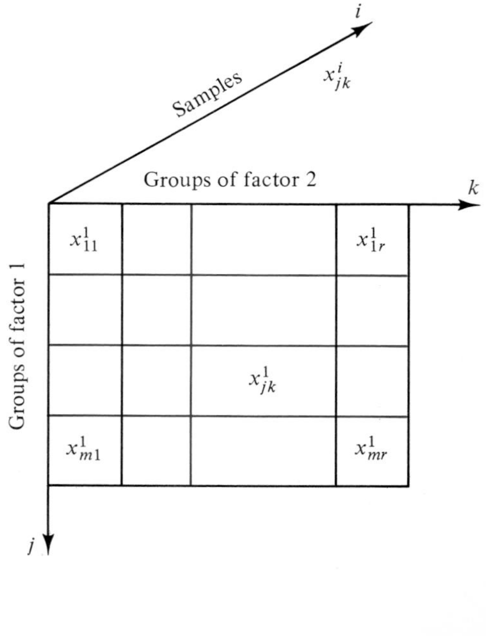

The test is based again in the ANOVA principle. We form the sum

$$\mathbf{Q}_5 = n \sum_{j=1}^{m} \sum_{k=1}^{r} (\bar{\mathbf{x}}_{jk} - \bar{\mathbf{x}}_{j.} - \bar{\mathbf{x}}_{.k} + \bar{\mathbf{x}}_{..})^2 \qquad (10\text{-}116)$$

where overbars indicate sample averages and dots row and column averages as in (10-102). The sum $\mathbf{Q}_5$ is identical to the sum $\mathbf{Q}_4$ in (10-109) provided that we replace the RVS $\mathbf{x}_{jk}$ by $\bar{\mathbf{x}}_{jk}\sqrt{n}$. From this and (10-110) it follows that if H_0 is true, then

$$\frac{\mathbf{Q}_5}{\sigma^2} \text{ is } \chi^2[(m-1)(r-1)] \qquad (10\text{-}117)$$

If H_0 is not true, that is, if $\gamma_{jk} \neq 0$, then the ratio $\mathbf{Q}_5/\sigma^2$ has a noncentral χ^2 distribution, and

$$E\{\mathbf{Q}_5\} = (m-1)(r-1)\sigma^2 + e \qquad (10\text{-}118)$$

where again, e is the value of $\mathbf{Q}_5$ if all RVS are replaced by their means:

$$e = n \sum_{j=1}^{m} \sum_{k=1}^{r} (\eta_{jk} - \eta_{j.} - \eta_{.k} + \eta_{..})^2 = \sum_{j=1}^{m} \sum_{k=1}^{r} \gamma_{jk}^2$$

We next form the sum

$$\mathbf{Q}_6 = \sum_{j=1}^{m} \sum_{k=1}^{r} \sum_{i=1}^{n} (\mathbf{x}_{jk}^i - \bar{\mathbf{x}}_{jk})^2 \qquad (10\text{-}119)$$

For specific j and k, the RVS $\mathbf{x}_{jk}^i$ are i.i.d., with sample mean $\bar{\mathbf{x}}_{jk}$; hence [see (7-97)],

$$\frac{1}{\sigma^2} \sum_{i=1}^{n} (\mathbf{x}_{jk}^i - \bar{\mathbf{x}}_{jk})^2 \text{ is } \chi^2(n-1)$$

The quadratic form $\mathbf{Q}_6/\sigma^2$ is the sum of mr such terms; therefore [see (7-87)],

$$\frac{\mathbf{Q}_6}{\sigma^2} \text{ is } \chi^2[mr(n-1)] \qquad (10\text{-}120)$$

Combining with (10-117), we conclude that the ratio

$$\mathbf{q} = \frac{\mathbf{Q}_5/(m-1)(r-1)}{\mathbf{Q}_4/mr(n-1)} \text{ is } F[(m-1)(r-1),\ mr(n-1)] \qquad (10\text{-}121)$$

This leads to the following test of (10-115):

$$\text{Accept } H_0 \text{ iff } q < F_{1-\alpha}[(m-1)(r-1),\ mr(n-1)] \qquad (10\text{-}122)$$

10-6 Neyman-Pearson, Sequential, and Likelihood Ratio Tests

In the design of a hypothesis test, we assign a value α to the Type I error probability and search for a region D_c of the sample space minimizing the resulting Type II error probability. Such a test is called most powerful. A simpler approach is the determination of a test based on a test statistic: We select a function $\mathbf{q} = g(\mathbf{X})$ of the sample vector $\mathbf{X}$ and search for a region R_c of the real line minimizing β. We shall say that the statistic $\mathbf{q}$ is *most powerful* if a test so designed is most powerful. In our earlier discussion, we selected the function $g(X)$ empirically. In general, such a choice of $g(X)$ does not lead to a most powerful test. In the following, we determine the function $g(X)$ such that a test based on the test statistic $g(\mathbf{X})$ is most powerful, and we determine the properties of the corresponding critical region. The analysis will involve the simple hypothesis $\theta = \theta_0$ against the simple hypothesis $\theta = \theta_1$.

We denote by $f(X, \theta) = f(x_1, \theta) \cdots f(x_n, \theta)$ the joint density of the n samples $\mathbf{x}_i$ of $\mathbf{x}$, and we form the ratio

$$\mathbf{r} = r(\mathbf{X}) = \frac{f(\mathbf{X}, \theta_0)}{f(\mathbf{X}, \theta_1)} \tag{10-123}$$

We shall show that $\mathbf{r}$ is a most powerful test statistic.

NEYMAN-PEARSON CRITERION Suppose that D_c is the critical region of the test of the hypothesis

$$H_0: \theta = \theta_0 \qquad \text{against} \qquad H_1: \theta = \theta_1 \tag{10-124}$$

We maintain that the test is most powerful iff the region D_c is such that

$$r(X) \le c \quad \text{for } X \in D_c \quad \text{and} \quad r(X) > c \quad \text{for } X \notin D_c \tag{10-125}$$

Thus $r = c$ on the boundary of D_c, $r < c$ in the interior of D_c, and $r > c$ outside D_c.

The constant c is specified in terms of the Type I error probability

$$\alpha = P\{\mathbf{X} \in D_c | H_0\} = P\{\mathbf{r} \le c | H_0\} \tag{10-126}$$

The resulting Type II error probability equals

$$\beta = P\{\mathbf{X} \notin D_c | H_1\} = P\{\mathbf{r} > c | H_1\} \tag{10-127}$$

▪ ***Proof.*** Suppose that D'_c is the critical region of another test and the corresponding error probabilities equal α' and β'. It suffices to show that

$$\text{if} \qquad \alpha' = \alpha \qquad \text{then} \qquad \beta' > \beta$$

In Fig. 10.19, we show the sets $D_c = A \cup B$ and $D'_c = A \cup B'$ where $A = D_c \cap D'_c$. Under hypothesis H_0, the probability masses in the regions B and B' are equal; hence, we can assign to each differential element ΔV_i of B centered at X a corresponding element $\Delta V'_i$ of B' centered at X', with the same mass

$$f(X, \theta_0)\Delta V_i = f(X', \theta_0)\Delta V'_i$$

Under hypothesis H_1, the masses in the regions ΔV_i and $\Delta V'_i$ equal $f(X, \theta_1)\Delta V_i$ and $f(X', \theta_1)\Delta V'_i$, respectively. Hence, as we replace ΔV_i by $\Delta V'_i$, the resulting increase in β equals

$$\Delta\beta = f(X, \theta_1)\Delta V_i - f(X', \theta_1)\Delta V'_i$$

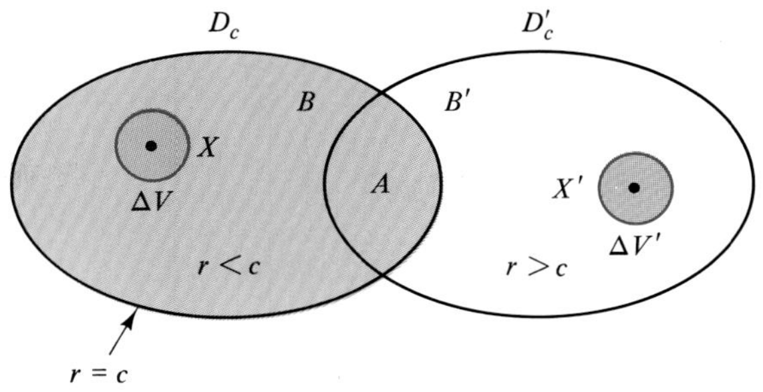

Figure 10.19

But

$$f(X, \theta_0) < cf(X, \theta_1) \qquad f(X', \theta_0) > cf(X', \theta_1)$$

Therefore,

$$\Delta\beta_i > \frac{1}{c}\,[f(X, \theta_0)\Delta V - f(x', \theta_0)] = 0$$

And since $\beta' \simeq \beta + \Sigma\Delta\beta_i$, we conclude that $\beta' > \beta$.

The ratio **r** will be called the NP test statistic. It is a statistic because the constants θ_0 and θ_1 are known. From the Neyman-Pearson criterion it follows that **r** is a most powerful test statistic. The corresponding critical region is the interval $r < c$ where $c = r_{1-\alpha}$ is the $1 - \alpha$ percentile of **r** [see (10-126)]. The most powerful test of (10-124) thus proceeds as follows: Observe X and compute the ratio r.

$$\text{Reject} \quad H_0 \quad \text{iff} \quad \frac{f(X, \theta_0)}{f(X, \theta_1)} \le r_{1-\alpha} \tag{10-128}$$

Note There is a basic difference between the NP test statistic and the test statistics considered earlier. The NP statistic **r** is specified in terms of the density of **x**, and it is in all cases most powerful. In general this is not true for the empirically chosen test statistics. However, as the following illustrations suggest, for most cases of interest empirical test statistics generate most powerful tests.

To carry out the test in (10-128) we must determine the distribution of **r**. This we can do, in principle, with the techniques of Section 7-1; however, the computations are not always simple. The problem is simplified if **r** is of the form

$$\mathbf{r} = \psi(\mathbf{q}) \tag{10-129}$$

where **q** is a statistic with known distribution. We can then replace the test (10-128) with a test based on **q**. Such a test is most powerful, and its critical region is determined as follows: Suppose, first, that the function $\psi(q)$ is monotonically increasing as in Fig. 10.20a. In this case, $r \le c$ iff $q \le c_a$;

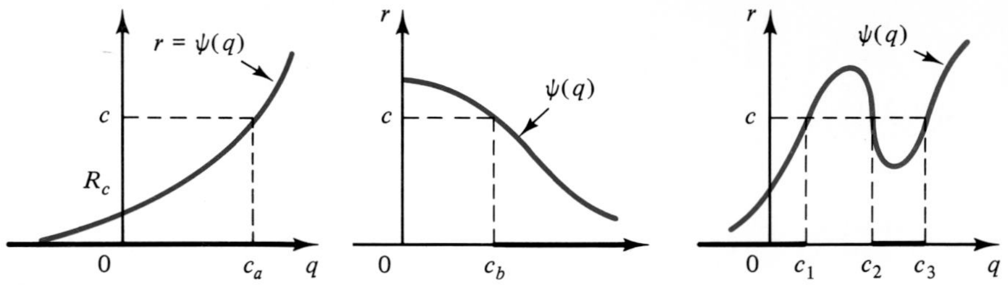

Figure 10.20

hence,

$$\alpha = P\{\mathbf{r} \le c|H_0\} = P\{\mathbf{q} \le c_a|H_0\} \qquad \text{(10-130a)}$$

Denoting by q_u the u-percentile of $\mathbf{q}$, we conclude that $c_a = q_\alpha$. Thus H_0 is rejected iff $q < q_\alpha$. Suppose, next, that $\psi(q)$ is monotonically decreasing (Fig. 10.20b). We now have

$$\alpha = P\{\mathbf{r} \le c|H_0\} = P\{\mathbf{q} \ge c_b|H_0\} \qquad c_b = q_{1-\alpha} \qquad \text{(10-130b)}$$

H_0 is rejected iff $q \ge q_{1-\alpha}$.

The general case is not as simple because the critical region R_c might consist of several segments. For the curve of Fig. 10.20c, R_c consists of the half-line $q \le c_1$ and the interval $c_2 \le q \le c_3$; hence,

$$\alpha = P\{\mathbf{q} \le c_1|H_0\} + P\{c_2 \le \mathbf{q} \le c_3|H_0\}$$

To find R_c, we assign values to c from 0 to 1, determine the corresponding points c_i, and compute the sum until its value equals α.

Example 10.33 The RV $\mathbf{x}$ is $N(\theta, \sigma)$ where σ is a known constant. In this case, the NP ratio r equals

$$r = \exp\left\{\frac{1}{2\sigma^2}\left[\sum (x_i - \theta_1)^2 - \sum (x_i - \theta_0)^2\right]\right\} = \exp\left\{\frac{n}{2\sigma^2}\left[(\theta_1^2 - \theta_0^2) - 2(\theta_1 - \theta_0)\bar{x}\right]\right\} \qquad \text{(10-131)}$$

This shows that the NP test statistic is a function $\mathbf{r} = \psi(\bar{\mathbf{x}})$ of the sample mean $\bar{\mathbf{x}}$ of $\mathbf{x}$. From this it follows that $\bar{\mathbf{x}}$ is a most powerful test statistic of the mean θ of $\mathbf{x}$. To find the corresponding critical region, we use (10-130): If $\theta_1 < \theta_0$, then $\chi(\bar{x})$ is a monotonically increasing function; hence, we reject H_0 iff $\bar{x} < q_\alpha$ where q_u is the u-percentile of $\bar{\mathbf{x}}$ under hypothesis H_0. If $\theta_1 > \theta_0$, then $\psi(\bar{x})$ is monotonically decreasing; hence, we reject H_0 iff $\bar{x} > q_{1-\alpha}$.

Note that the test against the simple hypothesis $\theta = \theta_1 < \theta_0$ is the same as the test against the composite hypothesis $\theta < \theta_0$ because the critical region $\bar{x} < q_\alpha$ does not depend on θ_1. And since the test is most powerful for every $\theta_1 < \theta_0$, we conclude that $\bar{x} < q_\alpha$ is the uniformly most powerful test of the hypothesis $\theta = \theta_0$ against $\theta < \theta_0$. Similarly, $\bar{x} > q_{1-\alpha}$ is the uniformly most powerful test of the hypothesis $\theta = \theta_0$ against $\theta > \theta_0$. ■

Exponential Type Distributions The results of Example 10.33 can be readily generalized. Suppose that the RV **x** has an exponential type distribution

$$f(x, \theta) = h(x) \exp \{a(\theta)\phi(x) - b(\theta)\}$$

as in (9-91). In this case, the NP ratio equals

$$r = \exp \{[a(\theta_0) - a(\theta_1)]G(X) - [b(\theta_0) - b(\theta_1)]\} \tag{10-132}$$

where $G(X) = \Sigma\phi(x_i)$. Thus **r** is a function $\psi(\mathbf{q})$ of the test statistic $\mathbf{q} = G(\mathbf{X})$; hence, **q** is a most powerful test statistic of the hypothesis $\theta = \theta_0$ against $\theta = \theta_1$. We shall now examine tests against the composite hypothesis $\theta < \theta_0$ or $\theta > \theta_0$.

Suppose that $a(\theta)$ is a monotonically increasing function of θ. If $\theta_1 < \theta_0$ [see (10-132)], the function $\psi(q)$ is also monotonically increasing. From this it follows that the test statistic **q** generates the uniformly most powerful test of the hypothesis $\theta = \theta_0$ against $\theta < \theta_1$, and its critical region is $G(X) \le q_\alpha$ where q_u is the u-percentile of $G(\mathbf{X})$ under hypothesis H_0. If $\theta_1 > \theta_0$, then $\psi(q)$ is monotonically decreasing, and the test against $\theta > \theta_0$ is uniformly most powerful with critical region $G(X) \ge q_{1-\alpha}$. If $a(\theta)$ is monotonically decreasing, $G(\mathbf{X})$ is again the uniformly most powerful test statistic against the hypothesis $\theta < \theta_0$ or $\theta > \theta_0$, and the corresponding critical regions are $G(X) \ge q_{1-\alpha}$ and $G(X) \le q_\alpha$, respectively.

Example 10.34 The RV **x** is $N(\eta\ , \theta)$ where η is a known constant. In this case,

$$f(x, \theta) = \frac{1}{\sqrt{2\pi}} \exp \left\{-\frac{1}{2\theta^2}(x-\eta)^2 - \ln \theta\right\}$$

This is an exponential density with

$$a(\theta) = -\frac{1}{2\theta^2} \qquad \phi(x) = (x-\eta)^2 \qquad \mathbf{q} = \sum(\mathbf{x}_i - \eta)^2$$

Thus $a(\theta)$ is a monotonically increasing function of θ; hence, the statistic **q** generates the uniformly most powerful tests of the hypothesis $\theta = \theta_0$ against $\theta < \theta_0$ or $\theta > \theta_1$. The corresponding critical regions are $q \le q_\alpha$ and $q \ge q_{1-\alpha}$, respectively. To complete the test, we must find the distribution of **q** under hypothesis H_0. As we know, if $\theta = \theta_0$, then the RV

$$\frac{\mathbf{q}}{\theta_0^2} = \frac{1}{\theta_0^2}\sum (\mathbf{x}_i - \eta)^2 \qquad \text{is} \qquad \chi^2(n)$$

Hence, $q_u = \theta_0^2\chi_u^2(n)$. This leads to the following tests:

$$H_1\colon \theta < \theta_0 \qquad \text{Reject } H_0 \text{ iff } \Sigma(x_i - \eta)^2 < \theta_0^2\chi_\alpha^2(n)$$

$$H_1\colon \theta > \theta_0 \qquad \text{Reject } H_0 \text{ iff } \Sigma(x_i - \eta)^2 > \theta_0^2\chi_{1-\alpha}^2(n)$$

We have thus reestablished earlier results [see (10-36)]. However, using the NP criterion, we have shown that the tests are uniformly most powerful. ■

Sufficient Statistics Suppose that the parameter θ has a sufficient statistic $y(X)$; that is, the function $f(X, \theta)$ is of the form

$$f(X, \theta) = H(X)J[y(X), \theta]$$

as in (9-95). In this case, the NP ratio equals

$$r = \frac{f(X, \theta_0)}{f(X, \theta_1)} = \frac{J[y(X), \theta_0)]}{J[y(X), \theta_1)]}$$

Thus $\mathbf{r}$ is a function of $y(X)$; hence, the sufficient statistic $\mathbf{q} = y(\mathbf{X})$ of θ is a most powerful test statistic of the hypothesis $\theta = \theta_0$ against $\theta = \theta_1$.

Example 10.35

We have shown in Example 9.29 that if $\mathbf{x}$ is uniform in the interval $(0, \theta)$, then the maximum $\mathbf{q} = \mathbf{x}_{\max}$ of its samples $\mathbf{x}_i$ is a sufficient statistic of θ and its distribution [see (9-100)] equals

$$F_q(q, \theta) = \frac{1}{\theta^n} q^n U(\theta - q)$$

From this it follows that $\mathbf{q}$ is the most powerful test statistic of θ, and

$$u = P\{\mathbf{q} \le q_u | H_0\} = F_q(q_u, \theta_0) = \frac{1}{\theta_0^n} q_u^n$$

Hence, $q_u = \theta_0^n \sqrt{u}$. Furthermore [see (10-142)],

$$r = \left(\frac{\theta_1}{\theta_0}\right)^n U(\theta - q)$$

Since r is a function of q, we shall determine the critical region R_c of the test directly in terms of $\mathbf{q}$. If $\theta_1 < \theta_0$, then R_c is the interval $q < c$ where

$$\alpha = P\{\mathbf{q} \le c | H_0\} = \left(\frac{c}{\theta_0}\right)^n \qquad c = \theta_0 \sqrt[n]{\alpha}$$

The corresponding β error equals

$$\beta = P\{\mathbf{q} \ge c | H_1\} = 1 - \left(\frac{c}{\theta_1}\right)^n = 1 - \alpha \left(\frac{\theta_0}{\theta_1}\right)^n$$

If $\theta_1 > \theta_0$, then R_c is the interval $q > c$, and

$$\alpha = P\{\mathbf{q} > c | H_0\} = 1 - \left(\frac{c}{\theta_0}\right)^n \qquad c = \theta_0 \sqrt[n]{1 - \alpha}$$

$$\beta = P\{\mathbf{q} \le c | H_1\} = \left(\frac{c}{\theta_1}\right)^n = (1 - \alpha) \left(\frac{\theta_0}{\theta_1}\right)^n$$

■

Sequential Hypothesis Testing

In all tests considered so far, the number n of samples used was specified in advance, either directly in terms of the complexity of the test or indirectly in terms of the error probabilities α and β. Now we consider a different approach. We continue the test until the data indicate whether we should accept one or the other of the two hypotheses. Suppose that we wish to test whether a coin is fair. If at the 20th toss, heads has shown 9 times, we decide that the coin is fair; if heads has shown 16 times, we decide that it is loaded; if heads has shown 13 times, we consider the evidence inconclusive and continue the test. In this approach, the length of the test—that is, the number n of trials required until a decision is reached—is random. It might be arbitrarily large. However, for the same error probabilities, it is on the average smaller than the number required for a fixed-length test. Such a test is called *sequential*.

We shall describe a sequential test of two simple hypotheses:

$$H_0: \theta = \theta_0 \quad \text{against} \quad H_1: \theta = \theta_1$$

The method is based on the NP statistic (10-123).

We form the ratio

$$\mathbf{r}_m = \frac{f(\mathbf{X}_m, \theta_0)}{f(\mathbf{X}_m, \theta_1)} \qquad \text{where } f(X_m, \theta) = f(x_1, \theta) \cdots f(x_m, \theta)$$

is the joint density of the first m samples $\mathbf{x}_1, \ldots, \mathbf{x}_m$ of $\mathbf{x}$.

We select *two* constants c_0 and c_1 such that $c_0 < c_1$.

At the mth trial, we reject H_0 if $r_m < c_0$; we accept H_0 if $r_m > c_1$; we continue the test if $c_0 < r_m < c_1$. Thus the test is completed at the nth step (Fig. 10.21) iff

$$c_0 < r_m < c_1 \qquad \text{for every } m < n \text{ and } r_n < c_0 \text{ or } r_n > c_1 \tag{10-133}$$

The ratio r_m can be determined recursively:

$$r_m = r_{m-1}\frac{f(X_m, \theta_0)}{f(X_m, \theta_1)}$$

To carry out the test, we must express the constants c_0 and c_1 in terms of the error probabilities α and β. This is a difficult task. We shall give only an approximation.

■ ***Theorem.*** In a sequential test carried out with the constants c_0 and c_1, the Type I and Type II error probabilities α and β satisfy the inequalities

$$\frac{\alpha}{1-\beta} \le c_0 \qquad \frac{\beta}{1-\alpha} \le \frac{1}{c_1} \tag{10-134}$$

Figure 10.21

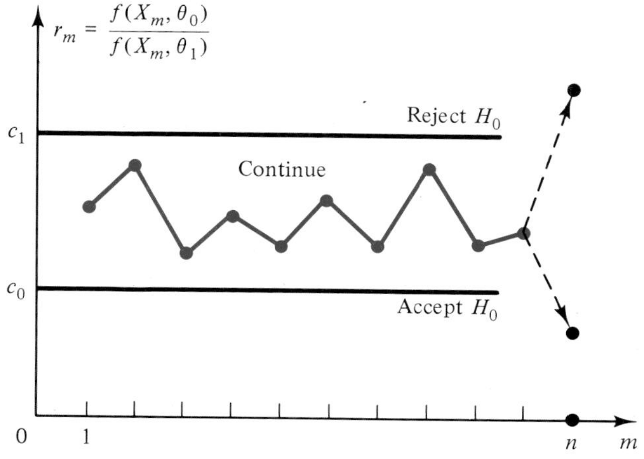

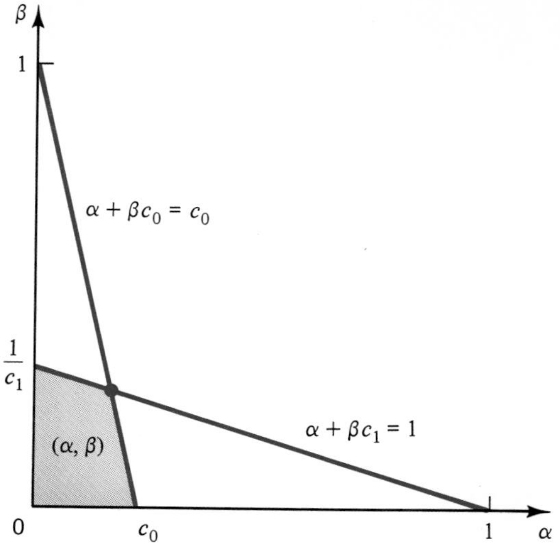

Figure 10.22

The proof is outlined in Problem 10-30. These inequalities show that the point (α, β) is in the shaded region of Fig. 10.22 bounded by the lines $\alpha + \beta_{c_0} = c_0$ and $\alpha + \beta c_1 = 1$. Guided by this, we shall use for c_0 and c_1 the constants

$$c_0 = \frac{\alpha}{1-\beta} \qquad c_1 = \frac{1-\alpha}{\beta} \tag{10-135}$$

In the construction of Fig. 10.22, (α, β) is the intersection of the two lines. From (10-134) it follows that with c_0 and c_1 so chosen, the resulting error probabilities are in the shaded region; hence, in most cases, the choice is conservative.

Example 10.36

The RV $\mathbf{x}$ is $N(\eta, 2)$. We shall test sequentially the hypothesis

$$h_0\text{: } \eta = 20 \qquad \text{against} \qquad H_1\text{: } \eta = 24$$

with $\alpha = .05$, $\beta = .1$.

Denoting by $\bar{x}_m$ the average of the first m samples x_i of $\mathbf{x}$, we conclude from (10-131) with $\theta_0 = 20$, $\theta_1 = 24$, and $\sigma = 2$ that

$$r_m = \exp\left\{\frac{m}{8}\,[(24^2 - 20^2) - 2(24 - 20)\bar{x}_m]\right\} = \exp\{-m(\bar{x}_m - 22)\}$$

From (10-135) it follows that

$$c_0 = \frac{.05}{.9} \qquad c_1 = \frac{.95}{.1} \qquad lnc_0 = -2.89 \qquad lnc_1 = 2.25$$

And since $lnr_m = -m(\bar{x}_m - 22)$, we conclude that $c_0 < r_m < c_1$ iff $-2.89 <$

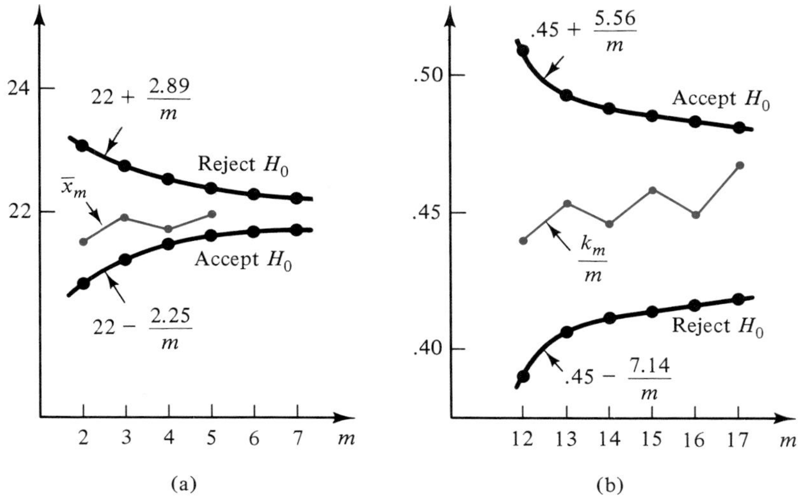

Figure 10.23

$-m(\bar{x}_m - 22) < 2.25$, that is, iff

$$22 - \frac{2.25}{m} < \bar{x}_m < 22 + \frac{2.89}{m}$$

Thus the test terminates when $\bar{x}_m$ crosses the boundary lines $22 + 2.25/m$ and $22 - 2.89/m$ of the uncertainty interval (Fig. 10.23a). We accept H_0 *if* $\bar{x}_m$ crosses the lower line; we reject H_0 if it crosses the upper line. ■

Example 10.37

Suppose that $\mathscr{A}$ is an event with probability p. We shall test the hypothesis

$$H_0\colon p = .5 \qquad \text{against} \qquad H_1\colon p = .4$$

with $\alpha = .05$, $\beta = .1$.

We form the zero-one RV **x** associated with the event $\mathscr{A}$. This RV is of the discrete type with point density $p^x(1-p)^{1-x}$ where $x = 0$ or 1. Hence, $f(k_m, p) \sim p^{k_m}(1-p)^{m-k_m}$, and

$$r_m = \left(\frac{p_0}{p_1}\right)^{k_m}\left(\frac{1-p_0}{1-p_1}\right)^{m-k_m} = \left(\frac{5}{4}\right)^{k_m}\left(\frac{5}{6}\right)^{m-k_m}$$

where k_m is the number of successes of $\mathscr{A}$ in the first m trials. In this case, r_m is a monotonically increasing function of the sample mean $\bar{x}_m = k_m/m$; hence, $-2.89 < \ln r_m < 2.25$ iff

$$0.45 - \frac{7.14}{m} < \bar{x}_m < 0.45 + \frac{5.56}{m}$$

This establishes the boundaries of the uncertainty region of the test as in Fig. 10.23b. ■

Likelihood Ratio Tests

So far, we have considered tests of a simple null hypothesis H_0 against an alternative hypothesis H_1, simple or composite, and we have assumed that the parameter θ was scalar. Now we shall develop a general method for testing any hypothesis H_0, simple or composite, involving the vector $\theta = [\theta_1, \ldots, \theta_k]$.

Consider an RV $\mathbf{x}$ with density $f(x, \theta)$. We shall test the hypothesis H_0 that θ is in a region Θ_0 of the k-dimensional parameter space against the hypothesis H_1 that θ is in a region Θ_1 of that space:

$$H_0: \theta \in \Theta_0 \qquad \text{against} \qquad H_1: \theta \in \Theta_1 \tag{10-136}$$

The union $\Theta = \Theta_0 \cup \Theta_1$ of the sets Θ_0 and Θ_1 is the parameter space.

For a given sample vector X, the joint density $f(X, \theta)$ is a function of the parameter θ. The maximum of this function as θ ranges over the parameter space Θ will be denoted by θ_m. In the language of Section 9-5, $f(X, \theta)$ is the likelihood function and θ_m the maximium likelihood estimate of θ. The maximum of $f(X, \theta)$ as θ ranges over the set Θ_0 will be denoted by θ_{m0}. If H_0 is the simple hypothesis $\theta = \theta_0$, then the region Θ_0 consists of the single point θ_0 and $\theta_{m0} = \theta_0$.

With θ_m and θ_{m0} so defined, we form the ratio

$$\lambda = \frac{f(X, \theta_{m0})}{f(X, \theta_m)} \tag{10-137}$$

A test of (10-136) based on the corresponding statistic $\boldsymbol{\lambda}$ is called the likelihood ratio (LR) test. We shall determine its critical region. Clearly, $f(X, \theta_{m0}) \le f(X, \theta_m)$; hence, for every X,

$$0 \le \lambda \le 1 \tag{10-138}$$

From this it follows that the density of $\boldsymbol{\lambda}$ is 0 outside the interval (0, 1). We maintain that it is concentrated near 1 if $\theta \in \Theta_0$ and to the left of 1 if $\theta \in \Theta_1$, as in Fig. 10.24. The data X are the observed samples of an RV $\mathbf{x}$ with distribution $f(x, \bar{\theta})$ where $\bar{\theta}$ is a specific unknown number. The point θ_m is the ML estimate of $\bar{\theta}$; therefore (see Section 9-5), it is close to $\bar{\theta}$ if n is large.

Figure 10.24

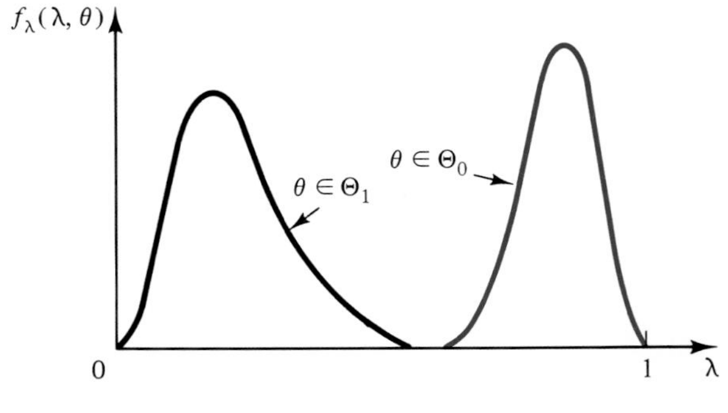

Under hypothesis H_0, $\bar{\theta}$ is in the set Θ_0; hence, with probability close to 1, θ_m equals θ_{m0} and $\lambda = 1$. If $\bar{\theta}$ is in Θ_1, then θ_m is different from θ_{m0} and λ is less than 1. These observations form the basis of the LR test. The test proceeds as follows: Observe X and find the extremes θ_m and θ_{m0} of the likelihood function $f(X, \theta)$.

$$\text{Reject } H_0 \quad \text{iff} \quad \frac{f(X, \theta_{m0})}{f(X, \theta_m)} < c \tag{10-139}$$

To complete the test, we must find c. If H_0 is the simple hypothesis $\theta = \theta_0$, then c equals the corresponding percentile λ_α of $\boldsymbol{\lambda}$. If, however, H_0 is composite, the Type I error probability $\alpha(\theta) = P\{\boldsymbol{\lambda} \le c|H_0\}$ is no longer a constant. In this case, we select a constant α_0 as before and determine c such that $\alpha(\theta)$ does not exceed α_0. Thus c is the smallest number such that

$$P\{\boldsymbol{\lambda} \le c\} \le \alpha_0 \qquad \text{for every } \theta \in \Theta_0 \tag{10-140}$$

In many cases, $P\{\boldsymbol{\lambda} \le c|H_0\}$ is maximum for $\theta = \theta_0$. In such cases, c is determined as in the simple hypothesis $\theta = \theta_0$; that is, the constant c is such that $P\{\boldsymbol{\lambda} \le c|\theta = \theta_0\} = \alpha_0$.

Example 10.38

The RV $\boldsymbol{\theta}$ has an exponential density $f(x, \theta) = \theta e^{-\theta x} U(x)$. We shall test the hypothesis

$$H_0: \theta \le \theta_0 \qquad \text{against} \qquad H_1: \theta > \theta_0$$

using the LR test. In this case (Fig. 10.25a),

$$f(X, \theta) = \theta^n e^{-n\bar{x}\theta} \qquad \theta > 0$$

As we see,

$$\theta_m = \frac{1}{\bar{x}} \qquad \theta_{m0} = \begin{cases} 1/\bar{x} & \bar{x}\theta_0 > 1 \\ \theta_0 & \bar{x}\theta_0 < 1 \end{cases}$$

$$\lambda = \frac{\theta_{m0}^n e^{-n\bar{x}\theta_{m0}}}{\theta_m^n e^{-n\bar{x}\theta_m}} = \begin{cases} 1 & \bar{\mathbf{x}}\theta_0 > 1 \\ (\bar{x}\theta_0)^n e^{-n(\theta_0\bar{x}-1)} & \bar{x}\theta_0 < 1 \end{cases}$$

Figure 10.25

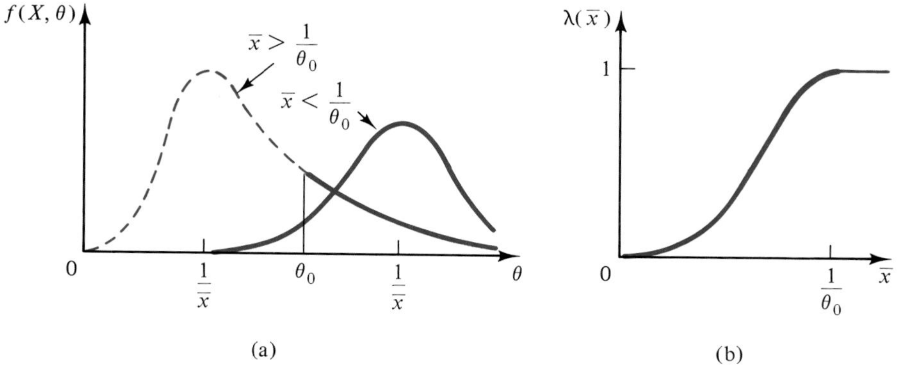

Thus λ is an increasing function of $\bar{x}$ (Fig. 10.25b). Hence, in this case, the LR test is equivalent to a test with test statistic $\mathbf{q} = \bar{\mathbf{x}}$. Since $\lambda \le c$ iff $\bar{x} \le c_1$, the corresponding critical region is $\bar{x} \le c_1$ where c_1 is such that $P\{\bar{\mathbf{x}} \le c_1 | \theta_0\} = \alpha_0$. ■

ASYMPTOTIC PROPERTIES OF LIKELIHOOD RATIOS In an LR test, we must find the distribution of the LR $\boldsymbol{\lambda}$; in general, this is not a simple problem. The problem is simplified if we can find a function $\mathbf{q} = \psi(\boldsymbol{\lambda})$ with known distribution. If $\psi(\lambda)$ is monotonically increasing or decreasing, then the LR test $\lambda \le c$ is equivalent to the test $q \le c_1$ or $q \ge c_1$. Example 10-38 illustrated this. Of particular interest in the transformation $\mathbf{w} = -2 \ln \boldsymbol{\lambda}$. As the next theorem shows, for large n, the RV $\mathbf{w}$ has a χ^2 distribution regardless of the form of $f(x, \theta)$. We shall need the following concept.

Free Parameters Suppose that the distribution of $\mathbf{x}$ depends on k parameters $\theta_1, \ldots, \theta_k$. We shall say that θ_i is a free parameter if its possible values in the parameter space Θ are noncountable. The number of free parameters of Θ will be denoted by N. We define similarly the number N_0 of free parameters of Θ_0. Suppose that $\mathbf{x}$ is a normal RV with mean η and variance σ^2 where η is any number and $\sigma > 0$. In this case, Θ is the half-plane $-\infty < \eta < \infty$, $\sigma > 0$; hence, $N = 2$. If under the null hypothesis, $\eta = \eta_0$ and $\sigma > 0$, then only σ is a free parameter, and $N_0 = 1$. If $\eta \ge \eta_0$ and $\sigma > 0$, then both parameters are free, and $N_0 = 2$. Finally, if $\eta = \eta_0$ and $\sigma = \sigma_0$, then $N_0 = 0$.

■ ***Theorem.*** If $N > N_0$ and n is large, then under hypothesis H_0, the statistic

$$\mathbf{w} = -2 \ln \boldsymbol{\lambda} \quad \text{is} \quad \chi^2(N - N_0) \tag{10-141}$$

The proof of this rather difficult theorem will not be given. We shall demonstrate its validity with two examples.

Example 10.39

The RV $\mathbf{x}$ is normal with mean η and known variance v. We shall test the hypothesis

$$H_0\colon \eta = \eta_0 \quad \text{against} \quad H_1\colon \eta \ne \eta_0 \tag{10-142}$$

using the LR test, and we shall show that the result agrees with (10-141). In this problem, we have one unknown parameter and

$$f(X, \eta) \sim \exp\left\{-\frac{1}{2v} \sum (x_i - \eta)^2\right\}$$

In this problem, $\eta_{m0} = \eta_0$ because H_0 is the simple hypothesis $\eta = \eta_0$. To find η_m, we observe that

$$\sum (x_i - \eta)^2 = \sum (x_i - \bar{x})^2 + n(\bar{x} - \eta)^2 \tag{10-143}$$

Thus $f(X, \eta)$ is maximum if the term $(\bar{x} - \eta)^2$ is minimum, that is, if $\eta = \eta_m = \bar{x}$. From this it follows that

$$\lambda = \frac{\exp\left\{-\frac{1}{2v} \sum (x_i - \eta_0)^2\right\}}{\exp\left\{-\frac{1}{2v} \sum (x_i - \bar{x})^2\right\}} = \exp\left\{-\frac{n}{2v} (\bar{x} - \eta_0)^2\right\} \tag{10-144}$$

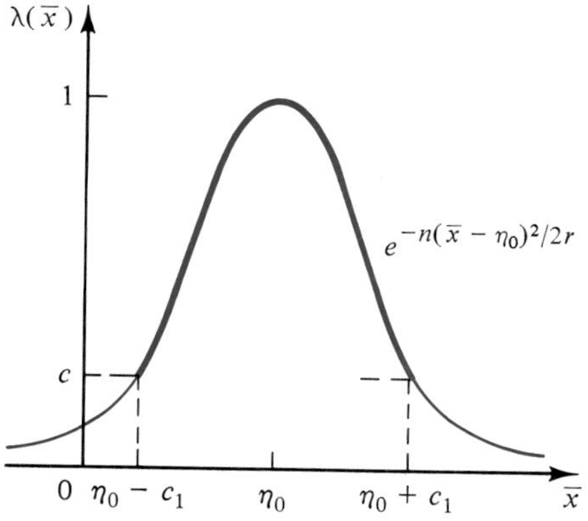

Figure 10.26

In Fig. 10.26, we plot λ as a function of $\bar{x}$. As we see, $\lambda \le c$ iff $|\bar{x} - \eta_0| \ge c_1$. This shows that the LR test is equivalent to the test based on the test statistic $\mathbf{q}$ of (10-9).

To verify (10-141) we observe that,

$$\mathbf{w} = -2 \ln \boldsymbol{\lambda} = \frac{(\bar{\mathbf{x}} - \eta_0)^2}{v/n}$$

Under hypothesis H_0, the RV $\bar{\mathbf{x}}$ is normal with mean η_0 and variance v/n; hence, $\mathbf{w}$ is $\chi^2(1)$. This agrees with (10-141) because $N = 1$ and $N_0 = 0$. ∎

Example 10.40

We shall again test (10-142), but now we assume that both parameters η are v are unknown. Reasoning as in Example (9-27), we find

$$\eta_{m0} = \eta_0 \qquad v_{m0} = \frac{1}{n}\sum (x_i - \eta_0)^2 \qquad \eta_m = \bar{x} \qquad v_m = \frac{1}{n}\sum (x_i - \bar{x})^2 \tag{10-145}$$

The resulting LR equals

$$\lambda = \frac{v_{m0}^{-n/2} \exp\left\{-\dfrac{1}{2v_{m0}} \sum (x_i - \eta_0)^2\right\}}{v_m^{-n/2} \exp\left\{-\dfrac{1}{2v_m} \sum (x_i - \bar{x})^2\right\}} = \left(\frac{v_m}{v_{m0}}\right)^{n/2}$$

This yields [see (10-143)]

$$\lambda = (1 + y^2)^{-n/2} \qquad y^2 = \frac{(\bar{x} - \eta_0)^2}{v_m}$$

Thus λ is a decreasing function of $|y|$; hence, $\lambda \le c$ iff $|y| \ge c_1$.

Large n In this example, $N = 2$ and $N_0 = 1$, to verify (10-141) we must show that the RV $-2 \ln \boldsymbol{\lambda}$ is $\chi^2(1)$.

Suppose that H_0 is true. We form the RVS

$$\mathbf{v}_m = \frac{1}{n}\sum (\mathbf{x}_i - \bar{\mathbf{x}})^2 \qquad \mathbf{y}^2 = \frac{(\bar{\mathbf{x}} - \eta_0)^2}{\mathbf{v}_m}$$

As n increases, the variance of $\mathbf{v}_m$ approaches 0 [see (7-99)]. Hence, for large n

$$\mathbf{v}_m \simeq E\{\mathbf{v}_m\} = \frac{n-1}{n} v \simeq v$$

Furthermore, $\bar{\mathbf{x}} - \eta_0 \to 0$ as $n \to \infty$. From this it follows that

$$\mathbf{y}^2 \simeq \frac{(\bar{x} - \eta_0)^2}{v} \ll 1 \qquad -2 \ln \boldsymbol{\lambda} = n \ln (1 + \mathbf{y}^2) \simeq n\mathbf{y}^2 \simeq \frac{(\bar{\mathbf{x}} - \eta_0)^2}{v/n}$$

with probability close to 1. This agrees with (10-141) because the RV $\bar{\mathbf{x}} - \eta_0$ is normal with 0 mean and variance v/n. ■

Problems

10-1 We are given an RV $\mathbf{x}$ with mean η and standard deviation $\sigma = 2$, and we wish to test the hypothesis $\eta = 8$ against $\eta = 8.7$ with $\alpha = .01$ using as test statistic the sample mean $\bar{x}$ of n samples. **(a)** Find the critical region R_c of the test and the resulting β if $n = 64$. **(b)** Find n and R_c if $\beta = .05$.

10-2 A new car is introduced with the claim that its average mileage in highway driving is at least 28 miles per gallon. Seventeen cars are tested, and the following mileage is obtained:

19 20 24 25 26 26.8 27.2 27.5
28 28.2 28.4 29 30 31 32 33.3 35

Can we conclude with significance level at most .05 that the claim is true?

10-3 The weights of cereal boxes are the values of an RV $\mathbf{x}$ with mean η. We measure 64 boxes and find that $\bar{x} = 7.7$ oz. and $s = 1.5$ oz. Test the hypothesis H_0: $\eta = 8$ oz. against H_1: $\eta \neq 8$ oz. with $\alpha = .1$ and $\alpha = .01$.

10-4 We wish to examine the effects of a new drug on blood pressure. We select 250 patients with about the same blood pressure and separate them into two groups. The first group of 150 patients receives the drug in tablets, and the second group of 100 patients is given identical tablets without the drug. At the end of the test period, the blood pressure of each group, modeled by the RVs $\mathbf{x}$ and $\mathbf{y}$, respectively, is measured, and it is found that $\bar{x} = 130$, $s_x = 5$ and $\bar{y} = 135$, $s_y = 8$. Assuming normality, test the hypothesis $\eta_x = \eta_y$ against $\eta_x \neq \eta_y$ with $\alpha = .05$.

10-5 Brand A batteries cost more than brand B batteries. Their life lengths are two RVs $\mathbf{x}$ and $\mathbf{y}$. We test 16 batteries of brand A and 26 batteries of brand B and find these values, in hours:

$$\bar{x} = 4.6 \qquad s_x = 1.1 \qquad \bar{y} = 4.2 \qquad s_y = 0.9$$

Test the hypothesis $\eta_x = \eta_y$ against $\eta_x > \eta_y$ with $\alpha = .05$.

10-6 Given r RVs $\mathbf{x}_k$ with the same variance σ^2 and with means $E\{\mathbf{x}_k\} = \eta_k$, we wish to test the hypothesis

$$H_0\text{: } \sum_{k=1}^{r} c_k \eta_k = 0 \qquad \text{against} \qquad H_1\text{: } \sum_{k=1}^{r} c_k \eta_k \neq 0$$

where c_k are r given constants. To do so, we observe n_k samples x_{k_i} of the kth

RV $\mathbf{x}_k$ and form their respective sample means $\bar{x}_k$. Carry out the test using as the test statistic the sum $\mathbf{y} = \Sigma\, c_k \bar{\mathbf{x}}_k$.

10-7 A coin is tossed 64 times, and heads shows 22 times. Test the hypothesis that the coin is fair with significance level .05.

10-8 We toss a coin 16 times, and heads shows k times. If k is such that $k_1 \le k \le k_2$, we accept the hypothesis that the coin is fair with significance level $\alpha = .05$. Find k_1 and k_2 and the resulting β error: **(a)** using (10-20); **(b)** using the normal approximation (10-21).

10-9 In a production process, the number of defective units per hour is a Poisson-distributed RV $\mathbf{x}$ with parameter $\lambda = 5$. A new process is introduced, and it is observed that the hourly defectives in a 22-hour period are

$x_i = 3 \quad 0 \quad 5 \quad 4 \quad 2 \quad 6 \quad 4 \quad 1 \quad 5 \quad 3 \quad 7 \quad 4 \quad 0 \quad 8 \quad 3 \quad 2 \quad 4 \quad 3 \quad 6 \quad 5 \quad 6 \quad 9$

Test the hypothesis $\lambda = 5$ against $\lambda < 5$ with $\alpha = .05$.

10-10 Given an $N(\eta, \sigma)$ RV $\mathbf{x}$ with known η, we test the hypothesis $\sigma = \sigma_0$ against $\sigma > \sigma_0$ using as the test statistic the sum $\mathbf{q} = \dfrac{1}{\sigma_0^2} \sum (\mathbf{x}_i - \eta)^2$. Show that the resulting OC function $\beta(\sigma)$ equals the area of the $\chi^2(n)$ density from 0 to $\chi^2_{1-\alpha}(n)\sigma_0^2/\sigma^2$ (Fig. P10.10).

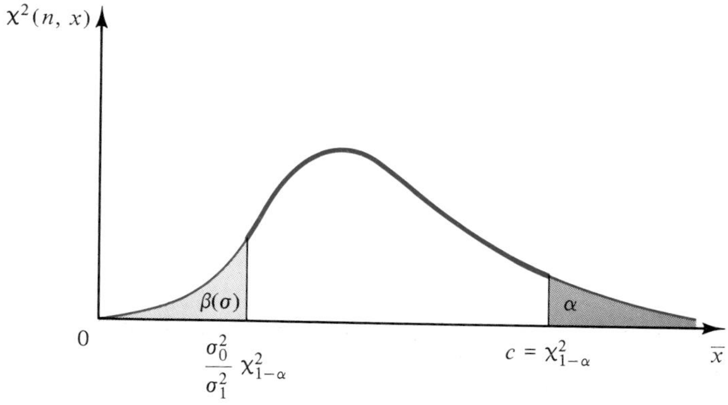

Figure P10.10

10-11 The RVs $\mathbf{x}$ and $\mathbf{y}$ model the grades of students and their parental income in thousands of dollars. We observe the grades x_i of 17 students and the income y_i of their parents and obtain

x_i	50	55	59	63	66	68	69	70	70	72	72	75	79	84	89	93	96
y_i	65	17	70	20	45	15	55	30	25	28	42	28	18	28	32	75	32

Compute the empirical correlation coefficient $\hat{r}$ of $\mathbf{x}$ and $\mathbf{y}$ [see (10-40)], and test the hypothesis $r = 0$ against $r \neq 0$.

10-12 It is claimed that the time to failure of a unit is an RV **x** with density $3x^2e^{-3x^3}U(x)$ (Weibull). To test the claim, we determine the failure times x_i of 80 units and form the empirical distribution $\hat{F}(x)$ as in (10-43), and we find that the maximum of the distance between $\hat{F}(x)$ and the Weibull distribution $F_0(x)$ equals 0.1. Using the Kolmogoroff-Smirnov test, decide whether the claim is justified with $\alpha = .1$.

10-13 We wish to compare the accuracies of two measuring instruments. We measure an object 80 times with each instrument and obtain the samples (x_i, y_i) of the RVs $\mathbf{x} = c + \boldsymbol{\nu}_a$ and $\mathbf{y} = c + \boldsymbol{\nu}_b$. We then compare x_i and y_i and find that $x_i > y_i$ 36 times, $x_i < y_i$ 42 times, and $x_i = y_i$ 2 times. Test the hypothesis that the distributions $F_a(\nu)$ and $F_b(\nu)$ of the errors are equal with $\alpha = .1$.

10-14 The length of a product is an RV **x** with $\sigma = 0.2$. When the plant is in control, $\eta = 30$. We select the values $\alpha = .05$, $\beta\ (30.1) = .1$ for the two error probabilities and use as the test statistic the sample mean $\bar{\mathbf{x}}$ of n samples of **x**. **(a)** Find n and design the control chart. **(b)** Find the probability that when the plant goes out of control and $\eta = 30.1$ the chart will cross the control limits at the next test.

10-15 A factory produces resistors. Their resistance is an RV **x** with standard deviation σ. When the plant is in control, $\sigma > 20\Omega$; when it is out of control, $\sigma > 20\Omega$. **(a)** Design a control chart using 10 samples at each test and $\alpha = .01$. Use as test statistic the RV $\mathbf{y} = \sqrt{\Sigma(\mathbf{x}_i - \bar{\mathbf{x}})^2}$. **(b)** Find the probability p that when the plant is in control, the chart will not cross the control limits before the 25th test.

10-16 A die is tossed 102 times, and the ith face shows $k_i = 18, 15, 19, 17, 13$, and 20 times. Test the hypothesis that the die is fair with $\alpha = .05$ using the chi-square test.

10-17 A utility proposes the construction of an atomic plant. The county objects claiming that 55% of the residents oppose the plant, 35% approve, and 10% express no opinion. To test this claim, the utility questions 400 residents and finds that 205 oppose the plant, 150 approve, and 45 express no opinion. Do the results support the county's claim at the 5% significance level?

10-18 A computer prints out 1,000 numbers consisting of the 10 integers $j = 0, 1, \ldots, 9$. The number n_j of times j appears equals

$n_j = 85 \quad 110 \quad 118 \quad 91 \quad 78 \quad 105 \quad 122 \quad 94 \quad 101 \quad 96$

Test the hypothesis that the numbers j are uniformly distributed between 0 and 9, with $\alpha = .05$.

10-19 The RVs **x** and **y** take the values 0 and 1 with $P\{\mathbf{x} = 0\} = p_x$, $P\{\mathbf{y} = 0\} = p_y$. A computer generates the following paired samples (x_i, y_i) of these RVs:

x_i	0	1	1	0	1	0	0	1	1	0	1
	0	1	1	1	0	0	1	0	1	0	0
	1	1	0	1	0	1	1	0	0	1	1
y_i	0	0	1	0	1	1	0	0	1	0	1
	1	0	0	1	0	1	0	0	1	0	1
	1	0	1	0	1	1	0	0	0	0	1

Using $\alpha = .05$, test the following hypotheses: **(a)** $p_x = .5$ against $p_x \neq .5$; **(b)** $p_y = .5$ against $p_y \neq .5$; **(c)** the RVs $\mathbf{x}$ and $\mathbf{y}$ are independent.

10-20 Suppose that under the null hypothesis, the probabilities p_i satisfy the m equations (10-67). Show that for large n the sum

$$\hat{q} = \sum_{i=1}^{m} \frac{(k_i - np_i)^2}{np_i} \text{ is minimum if } \sum_{i=1}^{m} \frac{k_i}{p_i} \frac{\partial p_i}{\partial \theta_j} = 0$$

This shows that the ML estimates of the parameters θ_j minimize the modified Pearson's sum $\hat{q}$.

10-21 The duration of telephone calls is an RV $\mathbf{x}$ with distribution $F(x)$. We monitor 100 calls x_i and observe that $x_i < 7$ minutes for every i and the number n_k of calls of duration between $k - 1$ and k equals 24, 20, 16, 15, 11, 8, 6 for $k = 1, 2, \ldots, 7$, respectively. Test the hypothesis that $F(x) = (1 - e^{-\theta x})U(x)$ with $\alpha = .1$. **(a)** Assume that $\theta = 0.25$. **(b)** Assume that θ is an unknown parameter, as in (10-67).

10-22 The RVs $\mathbf{x}_j^i$ are i.i.d. and $N(\eta, \sigma)$. Show that if

$$\mathbf{Q} = \sum_{i=1}^{m} \sum_{j=1}^{n_j} (\mathbf{x}_j^i - \bar{\mathbf{x}})^2 \qquad \text{then} \qquad \mathbf{Q} = \mathbf{Q}_1 + \mathbf{Q}_2$$

where $\mathbf{Q}_1$ and $\mathbf{Q}_2$ are the sums in (10-94) and (10-96).

10-23 Three sections of a math class, taught by different teachers, take the test and their grades are as follows:

Section 1:	38	45	51	54	58	62	65	66	69	71	73	74	76	79	83	86	90	96
Section 2:	42	46	53	57	61	66	68	72	74	77	80	82	87	91	96			
Section 3:	41	50	56	60	65	67	70	73	75	80	81	84	92					

(a) Assuming normality and equal variance, test the hypothesis that there is no significant difference in the average grades of the three sections with $\alpha = .05$ [see (10-101)]. **(b)** Using (10-97) with $N = 46$ and $m = 3$, estimate the common variance of the section grades.

10-24 We devide the pupils in a certain grade into 12 age groups and 3 weight groups, and we denote by $\mathbf{x}_{jk}$ the grades of pupils in the jth weight group and the kth age group. Selecting one pupil from each of the 36 sets so formed, we observe the following grades.

k	1	2	3	4	5	6	7	8	9	10	11	12
x_{1k}	60	66	70	74	77	80	80	82	85	89	92	96
x_{2k}	58	65	69	73	75	77	78	80	83	87	91	93
x_{3k}	58	66	69	72	75	77	79	79	81	84	88	92

Assuming normality and additivity, test the hypothesis that neither weight nor age has any effect on grades.

10-25 An RV $\mathbf{x}$ has the Erlang distribution

$$f(x) = \frac{x^m}{m!} e^{-x} U(x)$$

Using the NP criterion, test the hypothesis $m = 0$ against $m = 1$ (Fig. P10.25) with $\alpha = .25$ in terms of a single sample x of $\mathbf{x}$. Find the resulting Type II error β.

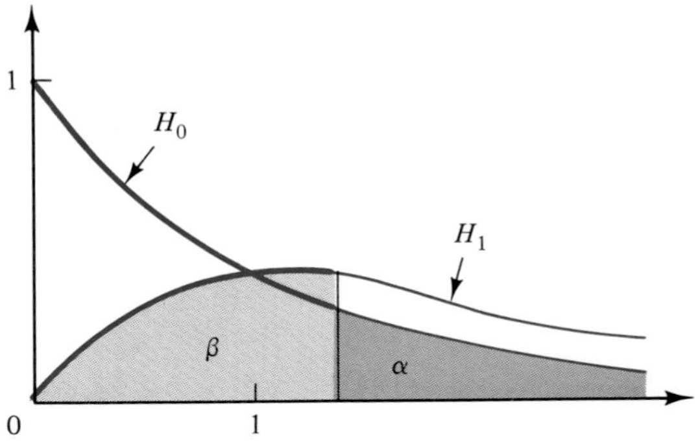

Figure P10.25

10-26 Suppose that $f_r(r, \theta)$ is the density of the NP ratio $\mathbf{r} = f(\mathbf{X}, \theta_0)/f(\mathbf{X}, \theta_1)$. **(a)** Show that $f_r(r, \theta_0) = rf_r(r, \theta_1)$. **(b)** Find $f_r(r, \theta)$ for $n = 1$ if $f(x, \theta) = \theta e^{-\theta x} U(x)$ and verify (a).

10-27 Given an RV $\mathbf{x}$ with $f(x, \theta) = \theta^2 x e^{-\theta x} U(x)$, we wish to test the hypothesis $\theta = \theta_0$ against $\theta < \theta_0$ using as test statistic the sums $\mathbf{q} = \mathbf{x}_1 + \cdots + \mathbf{x}_n$. **(a)** Show that the test

$$\text{Reject } H_0 \text{ iff } q < c = \frac{\chi^2_\alpha(4n)}{2\theta_0}$$

is most powerful. **(b)** Find the resulting OC function $\beta(\theta)$.

10-28 Given an RV $\mathbf{x}$ with density $\theta x^{\theta-1} U(x - 1)$ and samples x_i, we wish to test the hypothesis $\theta = 2$ against $\theta = 1$. **(a)** Show that the NP criterion leads to the critical region $x_1 \cdots x_n < c$. **(b)** Assuming that n is large, express the constant c and the Type II error probability β in terms of α.

10-29 We wish to test the hypothesis H_0 that the distribution of an RV $\mathbf{x}$ is uniform in the interval (0, 1) against the hypothesis H_1 that it is $N(1.25, 0.25)$, as in Fig. P10.29, using as our data a single observation x of $\mathbf{x}$. Assuming $\alpha = .1$, determine the critical region of the test satisfying the NP criterion, and find β.

10-30 We denote by A_m, R_m, and U_m the regions of acceptance, rejection, and uncertainty, respectively, of H_0 at the mth step in sequential testing. **(a)** Show that the sets A_m and R_m, $m = 1, 2, \ldots$ are disjoint. **(b)** Show that

$$\alpha = \sum_{m=1}^{\infty} \int_{R_m} f(X_m, \theta_0)\, dX_m \le c_0 \sum_{m=1}^{\infty} \int_{R_m} f(X_m, \theta_1) dX_m = c_0(1 - \beta)$$

$$1 - \alpha = \sum_{m=1}^{\infty} \int_{A_m} f(X_m, \theta_0)\, dX_m \ge c_1 \sum_{m=1}^{\infty} \int_{A_m} f(X_m, \theta_1)\, dX_m = c_1\beta$$

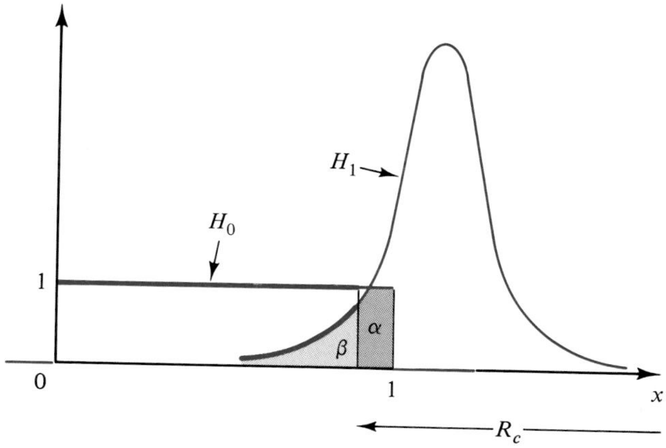

Figure P10.29

10-31 Given an event $\mathscr{A}$ with $p = P(\mathscr{A})$, we wish to test the hypothesis $p = p_0$ against $p \neq p_0$ in terms of number k of successes of $\mathscr{A}$ in n trials, using the LR test. Show that

$$\lambda = n^n \frac{p_0^k(1 - p_0)^{n-k}}{k^k(n - k)^{n-k}}$$

10-32 The number $\mathbf{x}$ of particles emitted from a radioactive substance in 1 second is a Poisson RV with mean θ. In 50 seconds, 1,058 particles are emitted. Test the hypothese $\theta_0 = 20$ against $\theta \neq 20$ with $\alpha = .05$ using the asymptotic approximation (10-141).

10-33 Using (10-146) and Example 10-40, show that the ANOVA test (10-101) is a special case of the LR test (10-139).

11 The Method of Least Squares

A common problem in many applications is the determination of a function $\phi(x)$ fitting in some sense a set of n points (x_i, y_i). The function $\phi(x)$ depends on $m < n$ parameters λ_i, and the problem is to find λ_i or, equivalently, to solve the overdetermined system $y_i = \phi(x_i)$, $i = 1, \ldots, n$. This problem can be given three interpretations. The first is deterministic: The coordinates x_i and y_i of the given points are n known numbers. The second is statistical: The abscissa x_i is a known number (controlled variable), but the ordinate y_i is the value of an RV $\mathbf{y}_i$ with mean $\phi(x_i)$. The third is a prediction: The numbers x_i and y_i are the samples of two RVs $\mathbf{x}$ and $\mathbf{y}$, and the objective is to find the best predictor $\hat{\mathbf{y}} = \phi(\mathbf{x})$ of $\mathbf{y}$ in terms of $\mathbf{x}$. We investigate all three interpretations and show that the results are closely related.

11-1 Introduction

We are given n points

$$(x_1, y_1), \ldots, (x_n, y_n)$$

on a plane (Fig. 11.1) with coordinates the $2n$ arbitrary numbers x_i and y_i. These numbers need not be different. We might have several points on a

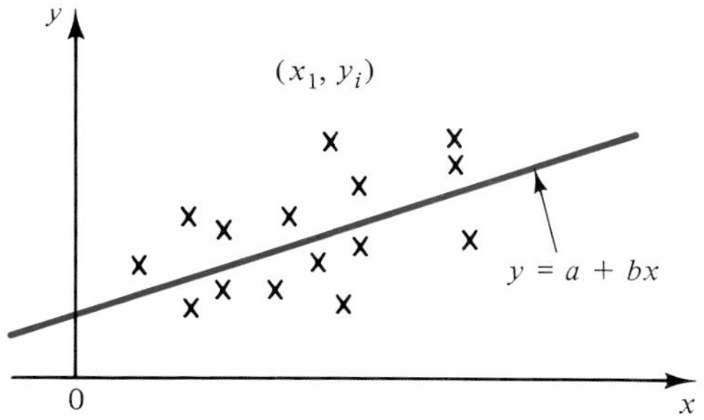

Figure 11.1

horizontal or vertical line; in fact, some of the points might be identical. Our first objective is to find a straight line $y = a + bx$ that fits "y on x" in the sense that the differences (errors)

$$y_i - (a + bx_i) = \nu_i \tag{11-1}$$

are as small as possible. The unknowns are the two constants a and b; their determination depends on the error criterion.

This is a special case of the problem of fitting a set of points with a general curve $\phi(x)$ depending on a number of unknown parameters. This problem is fundamental in the theory of measurements, and it arises in all areas of applied sciences. The curve $\phi(x)$ could be the statement of a physical law or an empirical function used for interpolation or extrapolation. Suppose that y is the temperature inside the earth at a distance x from the surface and that theoretical considerations lead to the conclusion that $y = a + bx$ where a and b are unknown parameters. To determine a and b, we measure the temperature at n depths x_i. The numbers x_i are *controlled variables* in the sense that their values are known *exactly*. The temperature measurements y_i, however, involve errors, as in (11-1). We shall develop techniques for estimating the parameters a and b. These estimates can then be used to determine the temperature $y_0 = a + bx_0$ at a new depth x_0. An example of empirical curve fitting is the stock market. We denote by y_i the price of a stock at time x_i, and we fit a straight line $y = a + bx$, or a higher-order curve, through the n observations (x_i, y_i). The line is then used to predict the price y_0 of the stock at some future time x_0.

THREE INTERPRETATIONS The curve-fitting problem can be given the following interpretations, depending on our assumptions about the points (x_i, y_i).

Deterministic In the first interpretation, x_i and y_i are viewed as pairs of known numbers. These numbers might be the results of measurements involving random errors; however, this is not used in the curve-fitting process, and the goodness of fit is not interpreted in a statistical sense.

Statistical In the second interpretation, the abscissas x_i are known numbers (controlled variables), but the ordinates y_i are the observed values of n RVS $\mathbf{y}_i$ with expected values $a + bx_i$.

Prediction In the third interpretation, x_i and y_i are the samples of two RVS $\mathbf{x}$ and $\mathbf{y}$. The sum $a + b\mathbf{x}$ is the linear predictor of $\mathbf{y}$ in terms of $\mathbf{x}$, and the function $\phi(\mathbf{x})$ is its nonlinear predictor. The constants a and b and the function $\phi(x)$ are determined in terms of the statistical properties of the RVS $\mathbf{x}$ and $\mathbf{y}$.

We shall develop all three interpretations. The results are closely related. The deterministic interpretation is not, strictly, a topic in statistics. It is covered here, however, because in most cases of interest, the data x_i and y_i are the results of observations involving random errors.

REGRESSION LINE This term is used to characterize the straight line $a + bx$, or the curve $\phi(x)$, that fits y on x in the sense intended. The underlying analysis is called *regression theory*.

Note The errors ν_i are the deviations of y_i from $a + bx_i$ (Fig. 11.2a). In this case, $y = a + bx$ is a line fitting y on x. We could similarly search for a line $x = \alpha + \beta y$ that fits x on y. In this case, the errors $\mu_i = x_i - (\alpha + \beta y_i)$ are the deviations of x_i from $\alpha + \beta y_i$ (Fig. 11.2b). The errors can, of course, be defined in other ways. For example, we can consider as errors the distances d_i of the points (x_i, y_i) from the line $Ax + By + C = 0$ (Fig. 11.2c). We shall discuss only the first case.

Overdetermined Systems The curve-fitting problem is a problem of solving somehow a system of n equations involving $m < n$ unknowns. For $m = 2$, this can be phrased as follows: Consider the system

$$a + bx_i = y_i \qquad i = 1, \ldots, n$$

where x_i and y_i are given numbers and a and b are two unknowns. Clearly, if the points x_i and y_i are not on a straight line, this system does not have a

Figure 11.2

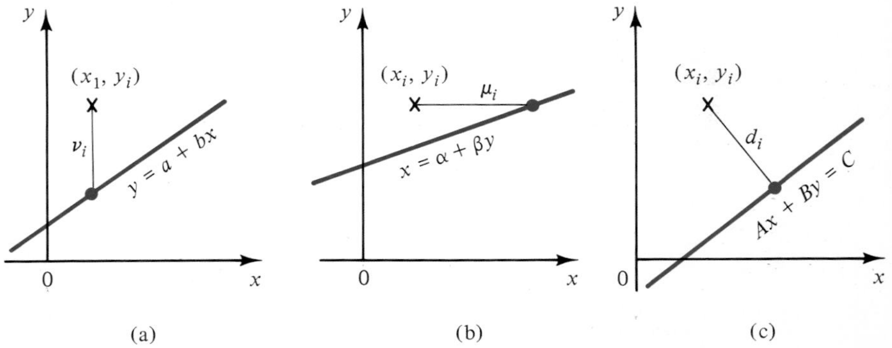

solution. To find a and b, we form the system

$$y_i - (a + bx_i) = \nu_i \qquad i = 1, \ldots, n$$

and we determine a and b so as to minimize in some sense the "errors" ν_i.

11-2 Deterministic Interpretation

We are given n points (x_i, y_i), and we wish to find a straight line $y = a + bx$ fitting these points in the sense of minimizing the *least square* (LS) error

$$Q = \Sigma \nu_i^2 \qquad (11\text{-}2)^*$$

where $\nu_i = y_i - (a + bx_i)$. This error criterion is used primarily because of its computational simplicity. We start with two special cases.

Horizontal Line Find a constant a_0 such that the line (Fig. 11.3a) $y = a_0$ is the LS fit of the n points (x_i, y_i) in the sense of minimizing the sum

$$Q_0 = \Sigma(y_i - a_0)^2$$

This case arises in problems in which y_i are the measured values of a constant a_0 and $\nu_i = y_i - a_0$ are the measurement errors. The abscissas x_i might be the times of measurement; their values, however, are not relevant in the determination of a_0.

Clearly, Q_0 is minimum if

$$\frac{\partial Q_0}{\partial a_0} = -2\Sigma\,(y_i - a_0) = 0$$

This yields

$$a_0 = \frac{1}{n}\sum y_i = \bar{y} \qquad Q_0 = \sum y_i^2 - n(\bar{y})^2 \qquad (11\text{-}3)$$

Thus a_0 is the average of the n numbers y_i.

* The notation Σ will mean $\sum_{i=1}^{n}$.

Figure 11.3

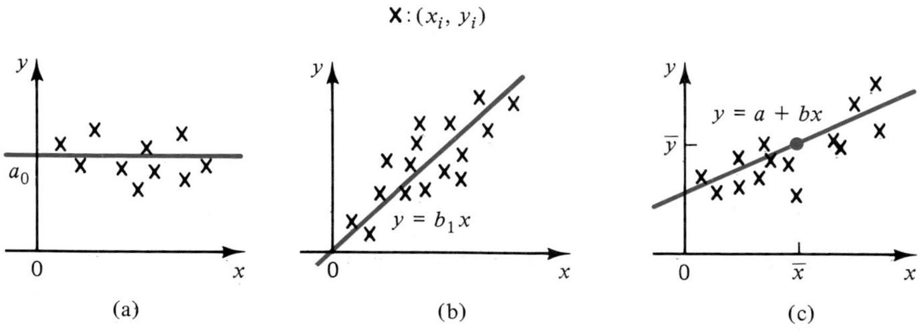

Homogeneous Line Find a constant b_1 such that the line $y = b_1 x$ (Fig. 11.3b) is the best LS fit of the n points (x_i, y_i).

In this case,

$$Q_1 = \Sigma(y_i - b_1 x_i)^2 \tag{11-4}$$

$$\frac{\partial Q_1}{\partial b_i} = -2\Sigma\,(y_i - b_1 x_i)x_i = 0 \tag{11-5}$$

Hence, Q_1 is minimum if

$$b_1 = \frac{\Sigma x_i y_i}{\Sigma x_i^2} \tag{11-6}$$

From (11-5) it follows that the LS error equals

$$Q_1 = \Sigma(y_i - b_i x_i)y_i = \Sigma \nu_i y_i$$

Geometric Interpretation The foregoing can be given a simple geometric interpretation. We introduce the vectors

$$X = [x_1, \ldots, x_n] \qquad Y = [y_1, \ldots, y_n] \qquad N = [\nu_1, \ldots, \nu_n]$$

and we denote by (X, Y) the inner product of X and Y and by $|X|$ the magnitude of X. Thus

$$(X, Y) = \Sigma x_i y_i \qquad (X, X) = \Sigma x_i^2 = |X|^2 \tag{11-7}$$

Clearly,

$$Q_1 = |N|^2 \qquad \text{where } N = Y - b_1 X$$

is the error vector (Fig. 11.4). The length $|N|$ of N depends on b_1, and it is minimum if N is orthogonal to X, that is, if

$$(N, X) = \Sigma(y_i - b_1 x_i)x_i = 0 \tag{11-8}$$

in agreement with (11-5). The linear LS fit can thus be phrased as follows: Q_1 is minimum if the error vector N is orthogonal to the data vector X (orthogonality principle) or, equivalently, if $b_1 X$ is the projection of Y on X (projection theorem).

GENERAL LINE Find the LS fit of the points (x_i, y_i) by the line (Fig. 11.3c)

$$y = a + bx$$

In this case, the square error

$$Q = \Sigma[y_i - (a + bx_i)]^2 \tag{11-9}$$

Figure 11.4

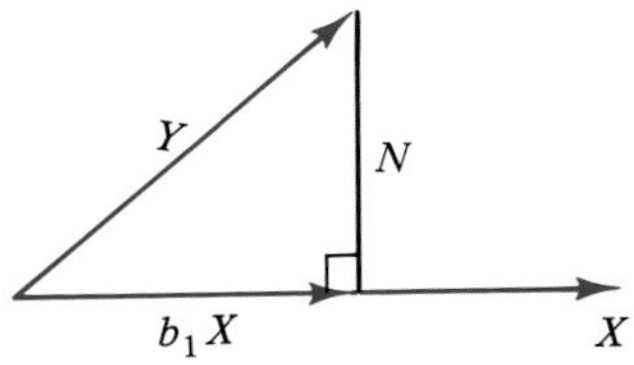

is a function of the two constants a and b, and it is minimum if

$$\frac{\partial Q}{\partial a} = -2\Sigma[y_i - (a + bx_i)] = 0 \qquad (11\text{-}10a)$$

$$\frac{\partial Q}{\partial b} = -2\Sigma[y_i - (a + bx_i)]x_i = 0 \qquad (11\text{-}10b)$$

This yields the system

$$na + b\Sigma x_i = \Sigma y_i \qquad (11\text{-}11a)$$

$$a\Sigma x_i + b\Sigma x_i^2 = \Sigma x_i y_i \qquad (11\text{-}11b)$$

Denoting by $\bar{x}$ and $\bar{y}$ the averages of x_i and y_i, respectively, we conclude from (11-11) that

$$a = \bar{y} - b\bar{x} \qquad (11\text{-}12a)$$

$$b = \frac{n\Sigma x_i y_i - \Sigma x_i \Sigma y_i}{n\Sigma x_i^2 - (\Sigma x_i)^2} = \frac{\Sigma(x_i - \bar{x})(y_i - \bar{y})}{\Sigma(x_i - \bar{x})^2} = \frac{\Sigma(x_i - \bar{x})y_i}{\Sigma x_i^2 - n(\bar{x})^2} \qquad (11\text{-}12b)$$

This yields

$$y - \bar{y} = b(x - \bar{x}) \qquad (11\text{-}13)$$

Hence, the regression line $y = a + bx$ passes through the point $(\bar{x}, \bar{y})$, and its slope b equals the slope b_1 of the homogeneous line fitting the *centered* data $(y_i - \bar{y}, x_i - \bar{x})$.

Note [see (11-10)] that

$$\Sigma[y_i - (a + bx_i)](a + bx_i) = 0$$

From this it follows that

$$Q = \Sigma[y_i - (a + bx_i)]y_i \qquad (11\text{-}14)$$

The LS error Q does not depend on the location of the origin because [see (11-12)]

$$Q = \Sigma[(y_i - \bar{y}) - b(x_i - \bar{x})]^2 = \Sigma(y_i - \bar{y})^2 - b^2\Sigma(x_i - \bar{x})^2 \qquad (11\text{-}15)$$

The ratio

$$\frac{Q}{\Sigma(y_i - \bar{y})^2} = 1 - r^2 \qquad r^2 = \frac{[\Sigma(x_i - \bar{x})(y_i - \bar{y})]^2}{\Sigma(x_i - \bar{x})^2\Sigma(y_i - \bar{y})^2}$$

is a normalized measure of the deviation of the points (x_i, y_i) from a straight line. Clearly, (see Problem 9-27)

$$0 \le |r| \le 1 \qquad (11\text{-}16)$$

and $|r| = 1$ iff $Q = 0$, that is, iff $y_i = a + bx_i$ for every i.

Example 11.1

We wish to fit a line to the points

x_i	2.3	3	3.4	4.2	4.2	5.1	6
	6	7.2	8	9	9.8	11	12
y_i	56	52	57	57	61	67	73
	73	70	82	89	86	99	105

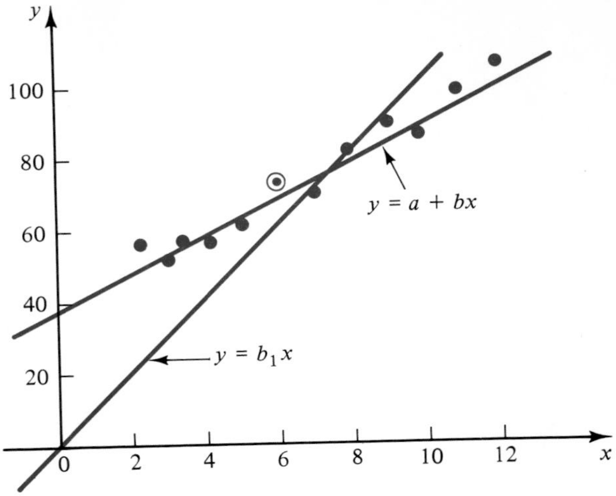

Figure 11.5

***Case 1:* $y = b_1 x$** From the data we obtain

$$\Sigma x_i^2 = 717 \qquad \Sigma x_i y_i = 7{,}345 \qquad \Sigma y_i^2 = 78{,}973$$

Hence (Fig. 11.5),

$$b_1 = 10.24 \qquad Q_1 = \Sigma y_i^2 - b_1 \Sigma x_i y_i = 3{,}760$$

***Case 2:* $y = a + bx$** The data yield $\bar{x} = 6.514$, $\bar{y} = 73.357$. Inserting into (11-12) and (11-14), we obtain

$$a = 38.67 \qquad b = 5.32 \qquad Q = 150$$ ■

Multiple Linear Regression

The straight line fit is a special case of the problem of multiple linear regression involving higher-order polynomials or other controlled variables. In this problem, the objective is to find a linear function

$$y = c_1 w_1 + \cdots + c_m w_m \tag{11-17}$$

of the m controlled variables w_k fitting the points

$$w_{1i}, \ldots, w_{mi}, y_i \qquad i = 1, \ldots, n$$

This is the homogeneous linear regression. The nonhomogeneous linear regression is the sum

$$c_o + c_1 w_1 \cdots + c_m w_m$$

This, however, is equivalent to (11-17) if we replace the term c_o by the product $c_o w_o$ where $w_o = 1$ [see also (11-72)].

As the following special cases show, the variables w_k might be functionally related: If $w_k = x^k$, we have the problem of fitting the points (x_i, y_i)

by a polynomial. If $w_k = \cos \omega_k x$, a fit by a trigonometric sum results. We discuss first these cases.

Parabola We wish to fit the n points (x_i, y_i) by the parabola

$$y = A + Bx + Cx^2 \tag{11-18}$$

This is a special case of (11-17) with

$$m = 3 \qquad w_1 = 1 \qquad w_2 = x \qquad w_3 = x^2$$

Our objective is to find the constants $A = c_1$, $B = c_2$, $C = c_3$ so as to minimize the sum

$$Q = \Sigma[y_i - (A + Bx_i + Cx_i^2)]^2 \tag{11-19}$$

To do so, we set

$$\begin{aligned} \frac{\partial Q}{\partial A} &= -2\Sigma[y_i - (A + Bx_i + Cx_i^2)] = 0 \\ \frac{\partial Q}{\partial B} &= -2\Sigma[y_i - (A + Bx_i + Cx_i^2)]x_i = 0 \\ \frac{\partial Q}{\partial C} &= -2\Sigma[y_i - (A + Bx_i + Cx_i^2)]x_i^2 = 0 \end{aligned} \tag{11-20}$$

This yields the system

$$\begin{aligned} nA + B\Sigma x_i + C\Sigma x_i^2 &= \Sigma y_i \\ A\Sigma x_i + B\Sigma x_i^2 + C\Sigma x_i^3 &= \Sigma x_i y_i \\ A\Sigma x_i^2 + B\Sigma x_i^3 + C\Sigma x_i^4 &= \Sigma x_i^2 y_i \end{aligned} \tag{11-21}$$

Solving, we obtain A, B, and C.

From (11-20) it follows that the LS error equals

$$Q = \Sigma[y_i - (A + Bx_i + Cx_i^2)]y_i \tag{11-22}$$

Example 11.2

Fit a parabola to the points (Fig. 11.6)

x_i	0.1	0.2	0.3	0.4	0.5	0.6	0.7	0.8	0.9	1
y_i	1.31	1.15	0.98	1.27	1.40	1.41	1.60	1.92	1.75	2.04

In this case,

$$\Sigma x_i = 55 \qquad \Sigma x_i^2 = 3.85 \qquad \Sigma x_i^3 = 3.02 \qquad \Sigma x_i^4 = 2.53$$

$$\Sigma y_i = 14.83 \qquad \Sigma x_i y_i = 8.98 \qquad \Sigma x_i^2 y_i = 6.68 \qquad \Sigma y_i^2 = 23.03$$

and (11-21) yields

$$A = 1.196 \qquad B = 0.314 \qquad C = 1.193 \qquad Q = 0.14$$

■

GENERAL CASE Given the n points $(w_{1i}, \ldots, w_{mi}, y_i)$, we wish to find the m constants c_k such that the sum

$$Q = \Sigma[y_i - (c_1 w_{1i} + \cdots + c_m w_{mi})]^2 \tag{11-23}$$

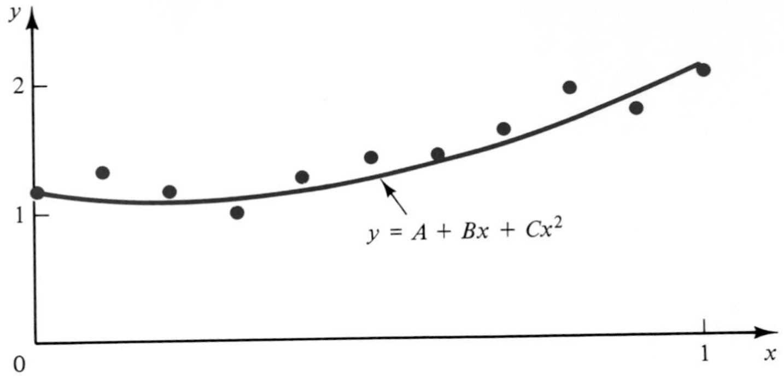

Figure 11.6

is minimum. This is the problem of fitting the given points by the function in (11-17). In the context of linear equations, the problem can be phrased as follows: We wish to solve the system

$$c_1 w_{1i} + \cdots + c_m w_{mi} = y_i \qquad 1 \le i \le n$$

of n equations involving $m < n$ unknowns $c_1, \ldots, c_m$. To do so, we introduce the n errors

$$y_i - (c_1 w_{1i} + \cdots + c_m w_{mi}) = \nu_i \qquad 1 \le i \le n \tag{11-24}$$

and we determine c_k such as to minimize the sum in (11-23). Clearly, Q is minimum if

$$\frac{\partial Q}{\partial c_k} = -2\Sigma[y_i - (c_1 w_{1i} + \cdots + c_m w_{mi})]w_{ki} = 0 \qquad 1 \le k \le m \tag{11-25}$$

This yields the system

$$\begin{aligned} c_1 \Sigma w_{1i}^2 + \cdots + c_m \Sigma w_{mi} w_{1i} &= \Sigma w_{1i} y_i \\ \cdots \qquad\qquad & \cdots \\ c_1 \Sigma w_{1i} w_{mi} + \cdots + c_m \Sigma w_{mi}^2 &= \Sigma w_{mi} y_i \end{aligned} \tag{11-26}$$

Solving, we obtain c_k. The resulting LS error equals

$$Q = \Sigma \nu_i^2 = \Sigma \nu_i y_i = \Sigma y_i^2 - c_1 \Sigma w_{1i} y_i - \cdots - c_m \Sigma w_{mi} y_i \tag{11-27}$$

Note that (11-25) is equivalent to the system

$$\Sigma \nu_i w_{ki} = 0 \qquad k = 1, \ldots, m \tag{11-28}$$

This is the orthogonality principle for the general linear regression.

Nonlinear Regression

We wish to find a function

$$y = \phi(x) \tag{11-29}$$

fitting the n points (x_i, y_i). Such a function can be used to describe economically a set of empirical measurement, to extrapolate beyond the observed data, or to determine a number of parameters of a physical law in terms of noisy observations. The problem has meaning only if we impose certain restrictions on the function $\phi(x)$. Otherwise, we can always find a curve fitting the given points exactly. We might require, for example, that $\phi(x)$ be smooth in some sense or that it depend on a small number of unknown parameters.

Suppose, first, that the unknown function $\phi(x)$ is of the form

$$\phi(x) = c_1 q_1(x) + \cdots + c_m q_m(x) \tag{11-30}$$

where $q_k(x)$ are m known functions. Clearly, $\phi(x)$ is a nonlinear function of x. However, it is linear in the unknown parameters c_k. This is thus a *linear* regression problem and can be reduced to (11-17) with the transformation $w_k = q_k(x)$. A special case is the fit of the points (x_i, y_i) by the polynomial

$$y = c_1 + c_2 x + \cdots + c_m x^{m-1}$$

This is an extension of the parabolic fit considered earlier. Let us look at another case.

Trigonometric Sums We wish to fit the n points (x_i, y_i) by the sum

$$y = c_1 \cos \omega_1 x + \cdots + c_m \cos \omega_m x \tag{11-31}$$

where the frequencies ω_k are known. This is a special case of (11-17) with

$$w_k = \cos \omega_k x \qquad w_{ki} = \cos \omega_k x_i$$

Inserting into (11-26), we conclude that the coefficients c_k of the LS fit are the solutions of the system

$$\begin{aligned} c_1 \Sigma \cos^2 \omega_1 x_i + \cdots + c_m \Sigma \cos \omega_m x_i \cos \omega_1 x_i &= \Sigma y_i \cos \omega_1 x_i \\ \cdots \qquad\qquad & \cdots \\ c_1 \Sigma \cos \omega_m x_i \cos \omega_1 x_i + \cdots + c_m \Sigma \cos^2 \omega_m x_i &= \Sigma y_i \cos \omega_m x_i \end{aligned} \tag{11-32}$$

Sums involving sines and cosines lead to similar results.

Example 11.3

We wish to fit the curve $y = c_1 + c_2 \cos 2.5x$ to the points (Fig. 11.7)

x_i	0.1	0.2	0.3	0.5	0.5	0.6	0.7	0.8	0.9	1.0
y_i	9	10.1	9.1	8.0	7.4	6.9	6	5.0	4.6	4.2

This is a special case of (11-31) with $\omega_1 = 0$ and $\omega_2 = 2.5$. Inserting into (11-32), we obtain the system

$$\begin{aligned} nc_1 + c_2 \Sigma \cos 2.5x_i &= \Sigma y_i \\ c_1 \Sigma \cos 2.5x_i + c_2 \Sigma \cos^2 2.5x_i &= \Sigma y_i \cos 2.5x_i \end{aligned}$$

This yields

$$\begin{aligned} 10c_1 + 1.48c_2 &= 70.3 \qquad & c_1 &= 6.562 \\ 1.48c_1 + 3.88c_2 &= 22.0 \qquad & c_2 &= 3.159 \end{aligned}$$

■

Next we examine two examples of nonlinear problems that can be linearized with a log transformation.

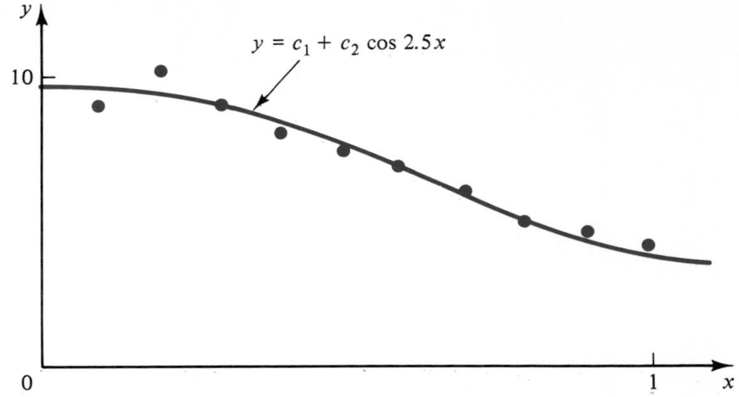

Figure 11.7

Example 11.4 We wish to fit the curve $y = \gamma e^{-\lambda x}$ to the points (Fig. 11.8)

x_i	1	2	3	4	5	6	7	8	9	10
y_i	81	55	42	29	20	15	11	7	5	3

so as to minimize the sum

$$Q = \Sigma(y_i - \gamma e^{-\lambda x_i})^2 \tag{11-33}$$

Figure 11.8

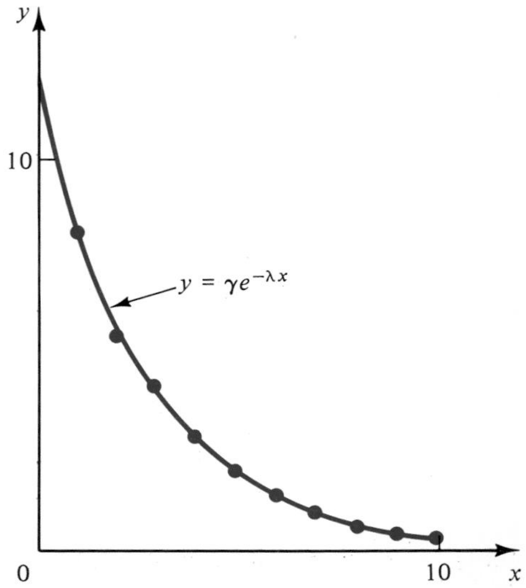

On a log scale, the curve is a straight line $\ln y = \ln \gamma - \lambda x$. This line is of the form $z = a + bx$ where

$$z = \ln y \qquad a = \ln \gamma \qquad b = -\lambda$$

Hence, the LS fit is given by (11-12). This yields

$$\lambda = -\frac{\Sigma(x_i - \bar{x}) \ln y_i}{\Sigma(x_i - \bar{x})^2} \qquad \ln \gamma = \frac{1}{n}\sum \ln y_i + \lambda\bar{x} \tag{11-34}$$

Inserting the given data into (11-34), we obtain

$$\lambda = 0.355 \qquad \ln \gamma = 4.78 \qquad y = 119e^{-0.355x}$$

Note that the constants γ and λ so obtained do not minimize (11-33); they minimize the sum

$$\Sigma[z_i - (a + bx_i)]^2 = \Sigma[\ln y_i - (\ln \gamma - \lambda x_i)]^2$$ ■

Example 11.5

We wish to fit the curve $y = \gamma x^\beta$ to the points (Fig. 11.9)

x_i	1	2	3	4	5	6	7	8	9	10
y_i	2.0	3.0	3.8	5.0	6.5	7.0	7.6	8.3	9.1	9.8

This is again a nonlinear problem; however, on a log-log scale, it is a straight line

$$\ln y = \ln \gamma + \beta \ln x$$

of the form $z = a + bw$ where

$$z = \ln y \qquad a = \ln \gamma \qquad b = \beta \qquad w = \ln x$$

Figure 11.9

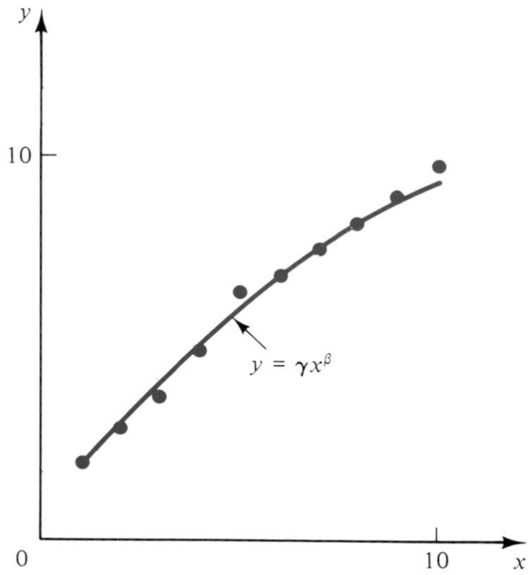

With $\overline{\ln x_i}$ the average of $\ln x_i$, (11-12) yields

$$\beta = \frac{\Sigma(\ln x_i - \overline{\ln x_i}) \ln y_i}{\Sigma(\ln x_i - \overline{\ln x_i})^2} = 0.715$$

$$\ln \gamma = \frac{1}{n}\sum \ln y_i - \beta\, \overline{\ln x_i} = 0.64 \qquad y = 1.9x^{0.715}$$

The constants γ and λ so obtained minimize the sum

$$\Sigma[\ln y_i - (\ln \gamma + \beta \ln x_i)]^2.$$ ■

PERTURBATIONS In the general curve-fitting problem, the regression curve is a nonlinear function $y = \phi(x, \lambda, \mu, \ldots)$ of known form depending on a number of unknown parameters. In most cases, this problem has no closed-form solution. We shall give an approximate solution based on the assumption that the unknown parameters $\lambda, \mu, \ldots$ are close to the known constants $\lambda_0, \mu_0, \ldots$

Suppose, first, that we have a single unknown parameter λ. In this case, $y = \phi(x, \lambda)$ and our problem is to find λ such as to minimize the sum

$$\Sigma[y_i - \phi(x_i, \lambda)]^2 \tag{11-35}$$

If the unknown λ is close to a known constant λ_0, we can linearize the problem (Fig. 11.10) using the approximation (truncated Taylor series)

$$\phi(x, \lambda) \simeq \phi(x, \lambda_0) + (\lambda - \lambda_0)\phi_\lambda(x, \lambda_0) \qquad \phi_\lambda = \frac{\partial \phi}{\partial \lambda} \tag{11-36}$$

Indeed, with

$$z = y - \phi(x, \lambda_0) \qquad w = \phi_\lambda(x, \lambda_0)$$

the nonlinear equation $y = \varphi(x, \lambda)$ is equivalent to the homogeneous linear equation $z = (\lambda - \lambda_0)w$. Our problem, therefore, is to find the slope $\lambda - \lambda_0$ of

Figure 11.10

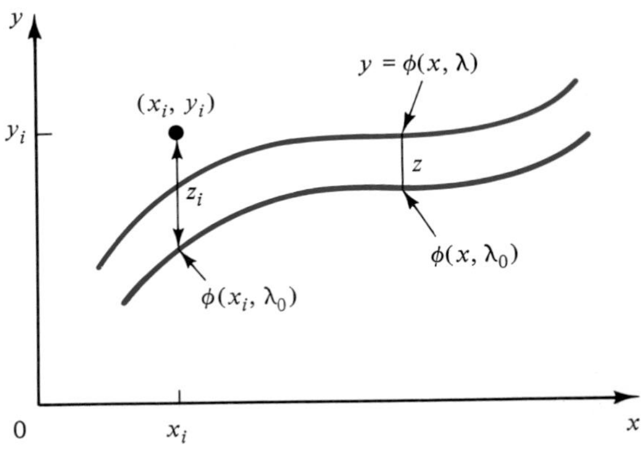

this equation so as to fit the points

$$z_i = y_i - \phi(x_i, \lambda_0) \qquad w_i = \phi_\lambda(x_i, \lambda_0)$$

in the sense of minimizing the LS error

$$\Sigma[y_i - \phi(x_i, \lambda_0) - (\lambda - \lambda_0)\phi_\lambda(x_i, \lambda_0)]^2 = \Sigma[z_i - (\lambda - \lambda_0)w_i]^2$$

Reasoning as in (11-6), we conclude with $b_1 = \lambda - \lambda_0$ that

$$\lambda - \lambda_0 = \frac{\Sigma z_i w_i}{\Sigma w_i^2} = \frac{\Sigma[y_i - \phi(x_i, \lambda_0)]\phi_\lambda(x_i, \lambda_0)}{\Sigma \phi_\lambda^2(x_i, \lambda_0)} \qquad (11\text{-}37)$$

Suppose next that the regression curve is a function $\phi(x, \lambda, \mu)$ depending on two parameters. The problem now is to find the LMS solution of the overdetermined nonlinear system

$$\phi(x_i, \lambda, \mu) = y_i \qquad i = 1, \ldots, n \qquad (11\text{-}38)$$

that is, to find λ and μ such as to minimize the sum

$$\Sigma[y_i - \phi(x_i, \lambda, \mu)]^2 \qquad (11\text{-}39)$$

We assume again that the optimum values of λ and μ are near the known constants λ_0 and μ_0. This assumption leads to the Taylor approximation

$$\phi(x, \lambda, \mu) \simeq \phi(x, \lambda_0, \mu_0) + (\lambda - \lambda_0)\phi_\lambda(x, \lambda_0, \mu_0) + (\mu - \mu_0)\phi_\mu(x, \lambda_0, \mu_0) \qquad (11\text{-}40)$$

where $\varphi_\lambda = \partial\varphi/\partial\lambda$ and $\varphi_\mu = \partial\varphi/\partial\mu$. Inserting into (11-38) and using the transformations

$$z = y - \phi(x, \lambda_0, \mu_0) \qquad w_1 = \phi_\lambda(x, \lambda_0, \mu_0) \qquad w_2 = \phi_\mu(x, \lambda_0, \mu_0)$$

we obtain the overdetermined system

$$z_i = c_i w_{1i} + c_2 w_{2i} \qquad c_1 = \lambda - \lambda_0 \qquad c_2 = \mu - \mu_0 \qquad (11\text{-}41)$$

This system is of the form (11-17); hence, the LS values of c_1 and c_2 are determined from (11-26). The solution can be improved by iteration.

Example 11.6

(a) We wish to fit the curve $y = 5 \sin \lambda x$ to the 13 points

x_i	0	0.5	1	1.5	2	2.5	3	3.5	4	4.5	5	5.5	6
y_i	4	6	18	30	46	42	52	40	44	34	20	36	10

of Fig. 11.11. As we see from the figure, a sine wave with period 12 is a reasonable fit. We can therefore use as our initial guess for the unknown λ the value $\lambda_0 = 2\pi/12$.

In this problem, $\phi(x, \lambda) = 5 \sin \lambda x$, $\phi_\lambda(x, \lambda) = 5x \cos \lambda x$,

$$z = y - 5 \sin \lambda_0 x \qquad w = 5x \cos \lambda_0 x$$

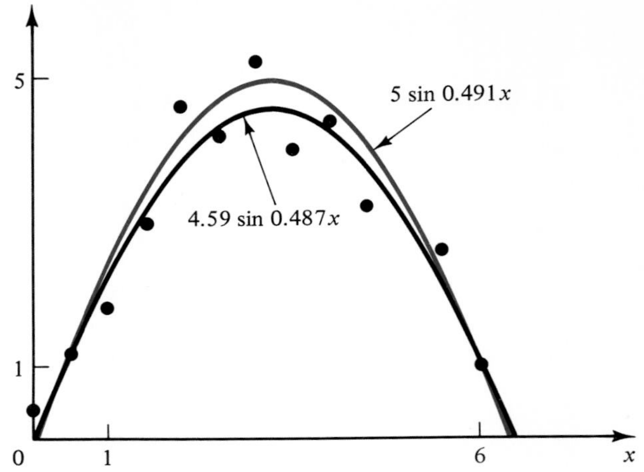

Figure 11.11

Hence [see (11-37)],

$$\lambda = \lambda_0 + \frac{5\,\Sigma x_i \cos \lambda_0 x_i (y_i - 5 \sin \lambda_0 x_i)}{25\,\Sigma x_i^2 \cos^2 \lambda_0 x_i} = 0.491$$

and the curve $y = 5 \sin 0.491x$ results.

(b) We now wish to fit the curve $y = \mu \sin \lambda x$ to the same points, using as initial guesses for the unknown parameters λ and μ the values

$$\lambda_0 = \frac{\pi}{6} \qquad \mu_0 = 5$$

With

$$\phi(x, \lambda, \mu) = \mu \sin \lambda x \qquad \phi_\lambda = \mu x \cos \lambda x \qquad \phi_\mu = \sin \lambda x$$

(11-40) yields

$$y = \mu \sin \lambda x \simeq \mu_0 \sin \lambda_0 x + (\mu - \mu_0) \sin \lambda_0 x + (\lambda - \lambda_0)\mu_0 x \cos \lambda_0 x$$

and the system

$$y = c_1 w_1 + c_2 w_2 \qquad w_1 = \sin \lambda_0 x \qquad w_2 = \mu_0 x \cos \lambda_0 x$$

results where $c_1 = \mu$, $c_2 = \lambda - \lambda_0$. Inserting into (11-26), we obtain

$$c_1\,\Sigma \sin^2 \lambda_0 x_i + c_2\,\Sigma \mu_0 x_i \sin \lambda_0 x_1 \cos \lambda_0 x_i = \Sigma y_i \sin \lambda_0 x_i$$

$$c_1\,\Sigma \mu_0 x_i \sin \lambda_0 x_i \cos \lambda_0 x_i + c_2\,\Sigma \mu_0^2 x_i^2 \cos^2 \lambda_0 x_i = \Sigma y_i \mu_0 x_i \cos \lambda_0 x_i$$

Hence, $c_1 = 4.592$, $c_2 = -0.0368$,

$$\mu = 4.592 \qquad \lambda = 0.487 \qquad y = 4.592 \sin 0.487x$$ ■

11-3 Statistical Interpretation

In Section 11-2, we interpreted the data (x_i, y_i) as deterministic numbers. In many applications, however, x_i and y_i are the values of two variables x and y related by a physical law $y = \phi(x)$. It is often the case that the values x_i of the controlled variable x are known exactly but the values y_i of y involve random errors. For example, x_i could be the precisely measured water temperature and y_i the imperfect measurement of the solubility η_i of a chemical substance. In this section, we use the random character of the errors to improve the estimate of the regression line $\phi(x)$. For simplicity, we shall assume that $\phi(x)$ is the straight line $a + bx$ involving only the parameters a and b. The case of multiple linear regression can be treated similarly.

We shall use the following probabilistic model. We are given n independent RVS $\mathbf{y}_1, \ldots, \mathbf{y}_n$ with the same variance σ^2 and with mean

$$E\{\mathbf{y}_i\} = \eta_i = a + bx_i \tag{11-42}$$

where x_i are known constants (controlled variables) and a and b are two unknown parameters. We can write $\mathbf{y}_i$ as a sum

$$\mathbf{y}_i = a + bx_i + \boldsymbol{\nu}_i \qquad E\{\boldsymbol{\nu}_i\} = 0 \tag{11-43}$$

where $\boldsymbol{\nu}_i$ are n independent RVS with variance σ^2. In the context of measurements, η_i is a quantity to be measured and $\boldsymbol{\nu}_i$ is the measurement error. Our objective is to estimate the parameters a and b in terms of the observed values y_i of the RVS $\mathbf{y}_i$. This is thus a parameter estimation problem differing from earlier treatments in that $E\{\mathbf{y}_i\}$ is not a constant but depends on i. We shall estimate a and b using the maximum likelihood (ML) method for normal RVS and the minimum variance method for arbitrary RVS and shall show that the estimates are identical. Furthermore, they agree with the LS solution (11-12) of the deterministic curve-fitting problem.

MAXIMUM LIKELIHOOD Suppose that the RVS $\mathbf{y}_i$ are normal. In this case, their joint density equals

$$\frac{1}{(\sigma\sqrt{2\pi})^n} \exp\left\{-\frac{1}{2\sigma^2}\sum [y_i - (a + bx_i)]^2\right\} \tag{11-44}$$

This density is a function of the parameters a and b. We shall find the ML estimators $\hat{\mathbf{a}}$ and $\hat{\mathbf{b}}$ of these parameters.

The right side of (11-44) is maximum if the sum

$$Q = \Sigma[y_i - (a + bx_i)]^2$$

is minimum. Clearly, Q equals the LS error in (11-9) hence, it is minimum if the parameters a and b are the solutions of the system (11-11). Replacing in (11-12) the average $\bar{y}$ of the numbers y_i by the average $\bar{\mathbf{y}}$ of the n RVS $\mathbf{y}_i$ we obtain the estimators

$$\hat{\mathbf{a}} = \bar{\mathbf{y}} - \hat{\mathbf{b}}\bar{x} \qquad \hat{\mathbf{b}} = \frac{\Sigma(x_i - \bar{x})\mathbf{y}_i}{\Sigma(x_i - \bar{x})^2} \tag{11-45}$$

As we shall see, these estimators are unbiased.

MINIMUM VARIANCE We assume again that the RVS $\mathbf{y}_i$ are independent with the same variance σ^2, but we impose no restrictions on their distribution. We shall determine the unbiased linear minimum variance estimators (best estimators)

$$\hat{\mathbf{a}} = \Sigma\alpha_i\mathbf{y}_i \qquad \hat{\mathbf{b}} = \Sigma\beta_i\mathbf{y}_i \tag{11-46}$$

of the regression coefficients a and b. Our objective is thus to find the $2n$ constants α_i and β_i satisfying the following requirements:

The expected values of $\hat{\mathbf{a}}$ and $\hat{\mathbf{b}}$ equal a and b respectively:

$$E\{\hat{\mathbf{a}}\} = \Sigma\alpha_i\eta_i = a \qquad E\{\hat{\mathbf{b}}\} = \Sigma\beta_i\eta_i = b \tag{11-47}$$

and their variances

$$\sigma_{\hat{a}}^2 = \sigma^2\Sigma\alpha_i^2 \qquad \sigma_{\hat{b}}^2 = \sigma^2\Sigma\beta_i^2 \tag{11-48}$$

are minimum.

■ ***Gauss-Markoff Theorem.*** The best estimators of a and b are given by (11-45) or, equivalently, by (11-46), where

$$\alpha_i = \frac{1}{n} - \beta_i\bar{x} \qquad \beta_i = \frac{x_i - \bar{x}}{\Sigma(x_i - \bar{x})^2} \tag{11-49}$$

■ ***Proof.*** Since $E\{\mathbf{y}_i\} = a + bx_i$, (11-47) yields

$$\Sigma\alpha_i(a + bx_i) = a \qquad \Sigma\beta_i(a + bx_i) = b$$

Rearranging terms, we obtain

$$(\Sigma\alpha_i - 1)a + (\Sigma\alpha_i x_i)b = 0 \qquad (\Sigma\beta_i)a + (\Sigma\beta_i x_i - 1)b = 0$$

This must be true for any a and b; hence,

$$\Sigma\alpha_i = 1 \qquad \Sigma\alpha_i x_i = 0 \tag{11-50a}$$

$$\Sigma\beta_i = 0 \qquad \Sigma\beta_i x_i = 1 \tag{11-50b}$$

Thus our problem is to minimize the sums $\Sigma\alpha_i^2$ and $\Sigma\beta_i^2$ subject to the constraints (11-50a) and (11-50b), respectively. The first two constraints contain only α_i, and the last two only β_i. We can therefore minimize each sum separately. Proceeding as in (7-37) we obtain (see Problem 11-8)

$$\alpha_i = \frac{nx_i\bar{x} - \Sigma x_i^2}{n(n\bar{x}^2 - \Sigma x_i^2)} \qquad \beta_i = \frac{x_i - \bar{x}}{\Sigma x_i^2 - n(\bar{x})^2}$$

and (11-49) follows.

Note that

$$\sum \beta_i^2 = \frac{\sum (x_i - \bar{x})^2}{\left[\sum (x_i - \bar{x})^2\right]^2} = \frac{1}{\sum(x_i - \bar{x})^2}$$

$$\sum \alpha_i\beta_i = \frac{1}{n}\sum \beta_i - \bar{x}\sum \beta_i^2 = \frac{-\bar{x}}{\sum(x_i - \bar{x})^2} \tag{11-51}$$

$$\sum \alpha_i^2 = \sum \left(\frac{1}{n} - \beta_i\bar{x}\right)^2 = \frac{1}{n} + \frac{(\bar{x})^2}{\sum(x_i - \bar{x})^2}$$

Variance and Interval Estimates From (11-51) and (11-48) it follows that

$$\sigma_{\hat{b}}^2 = \frac{\sigma^2}{\Sigma(x_i - \bar{x})^2} \qquad \sigma_{\hat{a}}^2 = \frac{\sigma^2}{n} + \frac{\sigma^2(\bar{x})^2}{\Sigma(x_i - \bar{x})^2} \tag{11-52}$$

$$\text{Cov}(\hat{\mathbf{a}}, \hat{\mathbf{b}}) = \sigma^2 \Sigma\alpha_i\beta_i = \frac{-\sigma^2\bar{x}}{\Sigma(x_i - \bar{x})^2} \tag{11-53}$$

Furthermore, the sum $\hat{\boldsymbol{\eta}}_k = \hat{\mathbf{a}} + \hat{\mathbf{b}}x_k$ is an unbiased estimator of the sum $a + bx_k$, and its variance equals

$$\sigma_{\hat{\eta}_k}^2 = \frac{\sigma^2}{n} + \frac{\sigma^2(x_k - \bar{x})^2}{\Sigma(x_i - \bar{x})^2} \tag{11-54}$$

We shall determine the confidence intervals of the parameters a, b, and η_k under the assumption that n is large. Since the RVS $\hat{\mathbf{a}}$, $\hat{\mathbf{b}}$, and $\hat{\boldsymbol{\eta}}$ are sums of the independent RVS $\mathbf{y}_i$, we conclude that they are nearly normal; hence, the $\gamma = 2u - 1$ confidence intervals of the parameters a, b, and η_k equal

$$\hat{a} \pm z_u\sigma_{\hat{a}} \qquad \hat{b} \pm z_u\sigma_{\hat{b}} \qquad \hat{\eta}_k \pm z_u\sigma_{\hat{\eta}_k} \tag{11-55}$$

respectively. These estimates can be used if σ^2 is known. If σ^2 is unknown, we replace it by its estimate $\hat{\sigma}^2$. To find $\hat{\sigma}^2$, we replace the parameters η_i in the sum $\Sigma(y_i - \eta_i)^2$ by their estimates. This yields the sum

$$\Sigma\epsilon_i^2 \qquad \epsilon_i = y_i - \hat{\eta}_i = y_i - (\hat{a} + \hat{b}x_i) \tag{11-56}$$

Reasoning as in the proof of (11-49) we conclude that the sum $\hat{\boldsymbol{\eta}}_x = \hat{\mathbf{a}} + \hat{\mathbf{b}}x$ is an unbiased estimator of η_x and its variance equals (see Problem 11-12)

$$\hat{\sigma}^2 = \frac{1}{n-2}\sum (y_i - \hat{\eta}_i)^2 \tag{11-57}$$

is an unbiased estimate of σ^2. Replacing the unknown σ^2 in (11-52) and (11-54) by its estimate $\hat{\sigma}^2$, we obtain the large sample estimates of $\alpha_{\hat{a}}$, $\sigma_{\hat{b}}$, and $\hat{\sigma}_{\eta_k}$. The corresponding confidence intervals are given by (11-55). Note, finally, that tests of various hypotheses about a, b, and η_k are obtained as in Section 10-2 (see Problem 11-9).

Regression Line Estimate Consider the sums $\eta_x = a + bx$ and $\mathbf{y}_x = a + bx + \boldsymbol{\nu}_x$ where a and b are the constants in (11-42), x is an arbitrary constant, and $\boldsymbol{\nu}_x$ is an RV with zero mean (Fig. 11.12). Thus

$$\mathbf{y}_x = \eta_x + \boldsymbol{\nu}_x \qquad \eta_x = a + bx \tag{11-58}$$

We shall estimate the ordinate η_x of the straight line $a + bx$ for $x \neq x_i$ in terms of the n data points (x_i, y_i). This problem has two interpretations depending on the nature of the RV $\mathbf{y}_x$.

First Interpretation The RV $\mathbf{y}_x$ is the result of the measurement of the sum $\eta_x = a + bx$, and $\boldsymbol{\nu}_x$ is the measurement error. This sum might represent a physical quantity, for example, the temperature of the earth at depth x. In this case, the quantity of interest is the estimate $\hat{\eta}_x$ of η_x. We shall use for $\hat{\eta}_x$ the sum

$$\hat{\eta}_x = \hat{a} + \hat{b}x \tag{11-59}$$

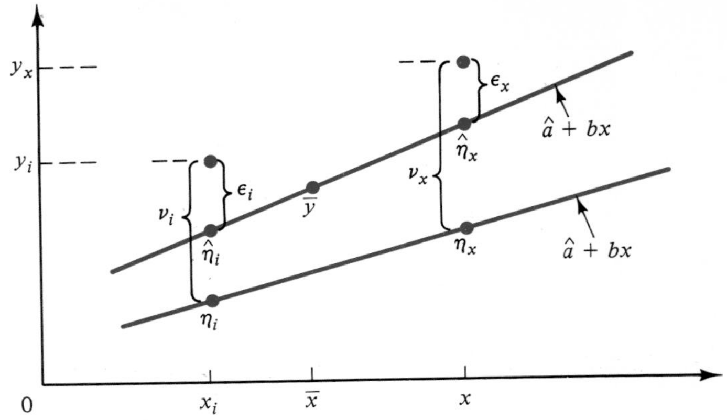

Figure 11.12

where $\hat{a}$ and $\hat{b}$ are the estimates of a and b given by (11-46). Reasoning as before, we conclude that $\hat{\boldsymbol{\eta}}_x$ is an unbiased estimator of η_x, and its variance equals

$$\sigma^2_{\hat{\eta}_x} = \frac{\sigma^2}{n} + \frac{\sigma^2(x - \bar{x})^2}{\Sigma(x_i - \bar{x})^2} \tag{11-60}$$

Second Interpretation The RV $\mathbf{y}_x$ represents a physical quantity of interest, the value of a stock at time x, for example, and η_x is its mean. The sum $\eta_x + \boldsymbol{\nu}_x$ relates $\mathbf{y}_x$ to the controlled variable x, and the quantity of interest is the value y_x of $\mathbf{y}_x$. We thus have a prediction problem: We wish to predict the value y_x of $\mathbf{y}_x$ in terms of the values y_i of the n RVs $\mathbf{y}_i$ in (11-43). As the predictor of the RV $\mathbf{y}_x = \eta_x + \boldsymbol{\nu}_x$ we shall use the sum

$$\hat{\mathbf{y}}_x = \hat{\boldsymbol{\eta}}_x = \hat{\mathbf{a}} + \hat{\mathbf{b}}x \tag{11-61}$$

Assuming that $\boldsymbol{\nu}_x$ is independent of the n RVs $\boldsymbol{\nu}_i$ in (11-43), we conclude from (11-60) that the MS value of the prediction error $\mathbf{y}_x - \hat{\mathbf{y}}_x = \boldsymbol{\nu}_x - (\hat{\boldsymbol{\eta}}_x - \eta_x)$ equals

$$E\{(\mathbf{y}_x - \hat{\mathbf{y}}_x)^2\} = E\{\boldsymbol{\nu}_x^2\} + \sigma^2_{\hat{\eta}_x} = \sigma^2_{\nu_x} + \frac{\sigma^2}{n} + \frac{\sigma^2(x - \bar{x})^2}{\Sigma(x_i - \bar{x})^2} \tag{11-62}$$

The two interpretations of the regression line estimate $\hat{a} + \hat{b}x$ can thus be phrased as follows: This line is the estimate of the line $\eta_x = a + bx$; the variance of this estimation is given by (11-60). It is also the predicted value of the RV $\mathbf{y}_x = a + bx + \boldsymbol{\nu}_x$; the MS prediction error is given by (11-62).

Example 11.7 We are given the following 11 points:

x_i	1	2	3	4	5	6	7	8	9	10	11	$\bar{x} = 6$
y_i	7	5	11	10	15	12	16	19	22	20	25	$\bar{y} = 14.73$

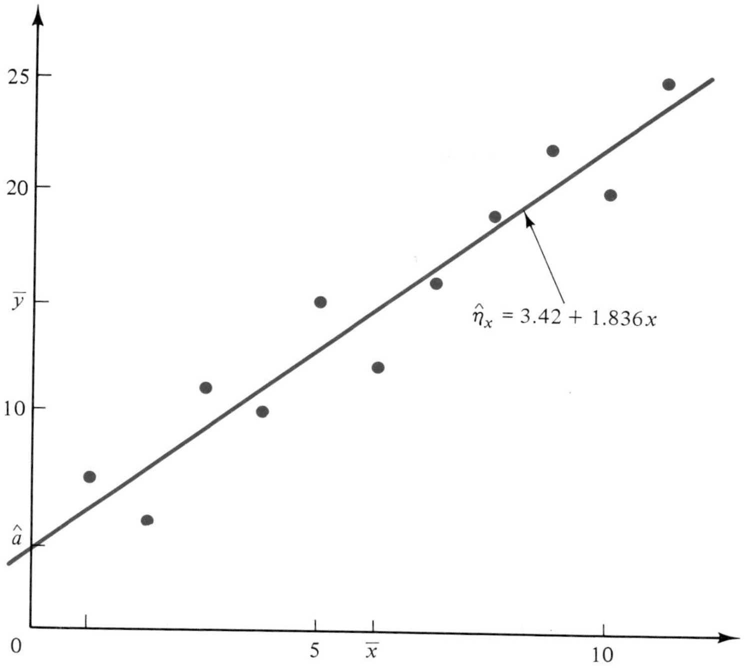

Figure 11.13

Find the point estimates of a, b, and σ^2.

Using the formulas

$$\beta_i = \frac{x_i - \bar{x}}{\Sigma(x_i - \bar{x})^2} \qquad \hat{b} = \Sigma\beta_i y_i \qquad \hat{a} = \bar{y} - \hat{b}\bar{x}$$

we find $\hat{b} = 1.836$, $\hat{a} = 3.714$,

$$\hat{\sigma}^2 = \frac{1}{9}\sum_{i=1}^{11}(y_i - 3.714 - 1.836x_i)^2 = 3.42 \qquad \hat{\sigma} = 1.85$$

In Fig. 11.13 we show the points (x_i, y_i) and the estimate $\hat{\eta}_x = 3.714 + 1.836x$ of $a + bx$. ∎

11-4 Prediction

In the third interpretation of regression, the points (x_i, y_i) are the samples of two RVS $\mathbf{x}$ and $\mathbf{y}$, and the problem is to estimate $\mathbf{y}$ in terms of $\mathbf{x}$. Multiple regression is the estimation of $\mathbf{y}$ in terms of several RVS $\mathbf{w}_k$. This topic is an extension of the nonlinear and linear regression introduced in Sections 5-2 and 6-3.

Linear Prediction

We start with a review of the results of Section 5-2. We wish to find two constants a and b such that the sum $\hat{\mathbf{y}} = a + b\mathbf{x}$ is the *least mean square* (LMS) predictor of the RV $\mathbf{y}$ in the sense of minimizing the MS value

$$Q = E\{(\mathbf{y} - \hat{\mathbf{y}})^2\} = \int_{-\infty}^{\infty}\int_{-\infty}^{\infty} [y - (a + bx)]^2 f(x, y)\, dx\, dy \qquad (11\text{-}63)$$

of the prediction error $\boldsymbol{\nu} = \mathbf{y} - \hat{\mathbf{y}}$. Clearly, Q is minimum if

$$\frac{\partial Q}{\partial a} = -2E\{\mathbf{y} - (a + b\mathbf{x})\} = 0 \qquad (11\text{-}64)$$
$$\frac{\partial Q}{\partial b} = -2E\{[\mathbf{y} - (a + b\mathbf{x})]\mathbf{x}\} = 0$$

This yields the system

$$a + b\eta_x = \eta_y \qquad a\eta_x + bE\{\mathbf{x}^2\} = E\{\mathbf{xy}\} \qquad (11\text{-}65)$$

Solving, we obtain

$$a = \eta_y - b\eta_x \qquad (11\text{-}66a)$$
$$b = \frac{E\{\mathbf{xy}\} - \eta_x\eta_y}{E\{\mathbf{x}^2\} - \eta_x^2} = \frac{\mu_{11}}{\sigma_x^2} = r\frac{\sigma_y}{\sigma_x} \qquad (11\text{-}66b)$$

Note that

$$\boldsymbol{\nu} = \mathbf{y} - (a + b\mathbf{x}) = (\mathbf{y} - \eta_y) - b(\mathbf{x} - \eta_x) \qquad E\{\boldsymbol{\nu}(a + b\mathbf{x})\} = 0$$

Hence, the LMS error equals

$$Q = E\{\boldsymbol{\nu}^2\} = E\{\boldsymbol{\nu}\mathbf{y}\} = E\{[(\mathbf{y} - \eta_y) - b(\mathbf{x} - \eta_x)]^2\}$$
$$= \sigma_y^2 - b\sigma_x^2 = \sigma_y^2(1 - r^2)$$

Thus the ratio Q/σ_y^2 is a normalized measure of the LMS prediction error Q, and it equals 0 iff $|r| = 1$.

If the RVs $\mathbf{x}$ and $\mathbf{y}$ are uncorrelated, then

$$r = 0 \qquad Q = \sigma_y^2 \qquad b = 0 \qquad \hat{\mathbf{y}} = a = \eta_y$$

In this case, the predicted value of $\mathbf{y}$ equals its mean; that is, the observed $\mathbf{x}$ does not improve the prediction.

The solution of the prediction problem is thus based on knowledge of the parameters η_x, η_y, σ_x, σ_y, and r. If these parameters are unknown, they can be estimated in terms of the data (x_i, y_i). The resulting estimates are $\eta_x = \bar{x}$, $\eta_y = \bar{y}$,

$$E\{\mathbf{x}^2\} \simeq \frac{1}{n}\sum x_i^2 \qquad E\{\mathbf{y}^2\} \simeq \frac{1}{n}\sum y_i^2 \qquad r^2 \simeq \frac{\left[\sum (x_i - \bar{x})(y_i - \bar{y})\right]^2}{\sum (x_i - \bar{x})^2 \sum (y_i - \bar{y})^2}$$

If these approximations are inserted into the two equations in (11-66), the two equations in (11-12) result. This shows the connection between the prediction problem and the deterministic curve-fitting problem considered in Section 11-2.

GENERALIZATION We are given $m + 1$ RVS

$$\mathbf{w}_1, \ldots, \mathbf{w}_m, \mathbf{y} \tag{11-67}$$

We perform the underlying experiment once and observe the values w_k of the m RVS $\mathbf{w}_k$. Using this information, we wish to predict the value y of the RV $\mathbf{y}$. In the linear prediction problem, the predicted value of $\mathbf{y}$ is the sum

$$\hat{y} = c_1 w_1 + \cdots + c_m w_m \tag{11-68}$$

and our problem is to find the m constants c_k so as to minimize the MS value

$$Q = E\{[\mathbf{y} - (c_1\mathbf{w}_1 + \cdots + c_m\mathbf{w}_m)]^2\} \tag{11-69}$$

of the prediction error $\boldsymbol{\nu} = \mathbf{y} - (c_1\mathbf{w}_1 + \cdots + c_m\mathbf{w}_m)$.

Clearly, Q is minimum if

$$\frac{Q}{\partial c_k} = -2E\{[\mathbf{y} - (c_1\mathbf{w}_1 + \cdots + c_m\mathbf{w}_m)]\mathbf{w}_k\} = 0 \tag{11-70}$$

This yields the system

$$\begin{aligned} &c_1E\{\mathbf{w}_1^2\} + \cdots + c_mE\{\mathbf{w}_m\mathbf{w}_1\} = E\{\mathbf{w}_1\mathbf{y}\} \\ &\cdots \\ &c_1E\{\mathbf{w}_1\mathbf{w}_m\} + \cdots + c_mE\{\mathbf{w}_m^2\} = E\{\mathbf{w}_m\mathbf{y}\} \end{aligned} \tag{11-71}$$

Solving, we obtain c_k.

The nonhomogeneous linear predictor of $\mathbf{y}$ is the sum

$$\gamma_0 + \gamma_1\mathbf{w}_1 + \cdots + \gamma_m\mathbf{w}_m$$

This can be considered as a special case of (11-68) if we replace the constant γ_0 by the product $\gamma_0\mathbf{w}_0$ where $\mathbf{w}_0 = 1$. Proceeding as in (11-71), we conclude that the constants γ_k are the solutions of the system

$$\begin{aligned} &\gamma_0 + \gamma_1E\{\mathbf{w}_1\} + \cdots + \gamma_mE\{\mathbf{w}_m\} = E\{\mathbf{y}\} \\ &\gamma_0E\{\mathbf{w}_1\} + \gamma_1E\{\mathbf{w}_1^2\} + \cdots + \gamma_mE\{\mathbf{w}_m\mathbf{w}_1\} = E\{\mathbf{w}_1\mathbf{y}\} \\ &\cdots \\ &\gamma_0E\{\mathbf{w}_m\} + \gamma_1E\{\mathbf{w}_1\mathbf{w}_m\} + \cdots + \gamma_mE\{\mathbf{w}_m^2\} = E\{\mathbf{w}_m\mathbf{y}\} \end{aligned} \tag{11-72}$$

From this it follows that if $E\{\mathbf{w}_k\} = E\{\mathbf{y}\} = 0$, then $\gamma_0 = 0$ and $\gamma_k = c_k$ for $k \neq 0$.

Orthogonality Principle Two RVS $\mathbf{x}$ and $\mathbf{y}$ are called orthogonal if $E\{\mathbf{xy}\} = 0$. From (11-70) it follows that

$$E\{\boldsymbol{\nu}\mathbf{w}_k\} = 0 \qquad k = 1, \ldots, m \tag{11-73}$$

Multiplying the kth equation by an arbitrary constant d_k and adding, we obtain

$$E\{\boldsymbol{\nu}(d_1\mathbf{w}_1 + \cdots + d_m w_m)\} = 0 \tag{11-74}$$

Thus in the linear prediction problem, the prediction error $\boldsymbol{\nu} = \mathbf{y} - \hat{\mathbf{y}}$ is orthogonal to the "data" $\mathbf{w}_k$ and to any *linear* function of the data (orthogonality principle). This result can be used to obtain the system in (11-71) directly, thereby avoiding the need for minimizing Q.

From (11-74) it follows that

$$E\{\boldsymbol{\nu}\hat{\mathbf{y}}\} = E\{(\mathbf{y} - \hat{\mathbf{y}})\hat{\mathbf{y}}\} = 0$$

Hence, the LMS prediction error Q equals

$$Q = E\{(\mathbf{y} - \hat{\mathbf{y}})^2\} = E\{(\mathbf{y} - \hat{\mathbf{y}})\mathbf{y}\} = E\{\mathbf{y}^2\} - E\{\hat{\mathbf{y}}^2\} \qquad (11\text{-}75)$$

As we see from (11-71), the determination of the linear predictor $\hat{\mathbf{y}}$ of $\mathbf{y}$ is based on knowledge of the joint moments of the $m + 1$ RVs in (11-67). If these moments are unknown, they can be approximated by their empirical estimates

$$E\{\mathbf{w}_j\mathbf{w}_k\} \simeq \frac{1}{n}\sum w_{ji}w_{ki} \qquad E\{\mathbf{w}_j\mathbf{y}\} \simeq \frac{1}{n}\sum w_{ji}y_i$$

where w_{ji} and y_i are the samples of the RVs $\mathbf{w}_j$ and $\mathbf{y}$, respectively. Inserting the approximations into (11-71), we obtain the system (11-26).

Nonlinear Prediction

The nonlinear predictor of an RV $\mathbf{y}$ in terms of the m RVs $\mathbf{x}_1, \ldots, \mathbf{x}_m$ is a function

$$\phi(x_1, \ldots, x_m) = \phi(X) \qquad X = [x_1, \ldots, x_n]$$

minimizing the MS prediction error

$$e = E\{[\mathbf{y} - \phi(\mathbf{X})]^2\} \qquad (11\text{-}76)$$

As we have shown in Section 6-3, if $m = 1$, then $\phi(x) = E\{\mathbf{y}|x\}$. For an arbitrary m, the function $\phi(X)$ is given by the conditional mean of $\mathbf{y}$ assuming X:

$$\boxed{\phi(X) = E\{\mathbf{y}|X\} = \int_{-\infty}^{\infty} yf(y|X)\,dy \qquad (11\text{-}77)}$$

To prove this, we shall use the identity [see (7-21)]

$$E\{\mathbf{z}\} = E\{E\{\mathbf{z}|\mathbf{X}\}\} \qquad (11\text{-}78)$$

With $\mathbf{z} = [\mathbf{y} - \phi(\mathbf{X})]^2$, this yields

$$e = E\{E\{[\mathbf{y} - \phi(\mathbf{X})]^2|\mathbf{X}\}\} \qquad (11\text{-}79)$$

The right side is a multiple integral involving only positive quantities; hence, it is minimum if the conditional mean

$$E\{[\mathbf{y} - \phi(X)]^2|X\} = \int_{-\infty}^{\infty} [y - \phi(X)]^2 f(y|X)\,dy \qquad (11\text{-}80)$$

is minimum. In this integral, the function $\varphi(X)$ is a constant (independent of the variable of integration y). Reasoning as in (6-52), we conclude that the integral is minimum if $\phi(X)$ equals the conditional mean of $\mathbf{y}$.

Note For normal RVS, nonlinear and linear predictors are identical because the conditional mean $E\{\mathbf{y}|X\}$ is a linear function of the components x_i of X (see Problem 11-16). This is a simple extension of (6-41).

Orthogonality Principle We have shown that in linear prediction, the error is orthogonal to the data and to any *linear* function of the data. We show next that in nonlinear prediction, the error $\mathbf{y} - \phi(\mathbf{X})$ is orthogonal to any function $q(\mathbf{X})$, *linear* or *nonlinear*, of the data $\mathbf{X}$. Indeed, from (11-77) it follows, as in (6-49), that

$$E\{[\mathbf{y} - \phi(\mathbf{X})]q(\mathbf{X})\} = E\{q(\mathbf{X})E\{\mathbf{y} - \phi(\mathbf{X})|\mathbf{X}\}\}$$

From (11-77) and the linearity of expected values it follows that

$$E\{\mathbf{y} - \phi(\mathbf{X})|X\} = E\mathbf{y}|X\} - E\{\phi(\mathbf{X})|X\} = E\{\mathbf{y}|X\} - \phi(X)$$

Hence [see (11-78)],

$$E\{[\mathbf{y} - \phi(\mathbf{X})]q(\mathbf{X})\} = 0 \qquad (11\text{-}81)$$

for any $q(\mathbf{X})$. This shows that the error $\mathbf{y} - \phi(\mathbf{X})$ is orthogonal to $q(\mathbf{X})$.

Problems

11-1 Fit the lines $y = b_1x$, $y = a + bx$, and $y = \gamma x^\beta$ to the points

x_i	0	1	2	3	4	5	6	7	8	9	10	11
y_i	1	3	3	5	5	8	7	8	9	11	13	15

and sketch the results. Find and compare the corresponding errors

$$\Sigma(y_i - b_1x_i)^2 \qquad \Sigma[y_i - (a + bx_i)]^2 \qquad \Sigma(y_i - \gamma x_i^\beta)^2$$

11-2 Fit the parabola $y = A + Bx + Cx^2$ to the following points

x_i	0	1	2	3	4	5	6	7	8	9	10
y_i	0	1	5	10	17	25	40	50	65	80	98

and sketch the results.

11-3 Here are the average grades x_i, y_i, and z_i of 15 students in their freshman, sophomore, and senior year, respectively:

x_i	2.8	2.2	3.0	3.5	3.6	3.4	1.9	3.4
	1.5	2.7	3.9	4.0	3.4	2.6	2.2	
y_i	2.6	2.9	3.4	2.5	3.4	2.9	2.5	3.6
	2.5	2.6	3.6	3.7	3.3	3.2	2.7	
z_i	3.1	3.4	3.8	2.4	4.0	3.4	2.9	3.8
	2.6	3.1	2.9	3.9	3.8	3.7	2.4	

(a) Find the LS fit of the plane $z = c_1 + c_2x + c_3y$ to the points (x_i, y_i, z_i). **(b)** Use (a) to predict the senior grade z of a student if his grades at the freshman and sophmore year are $x = 3.9$ and $y = 3.8$.

11-4 Fit the curve $y = a + b \sin \frac{\pi}{10} x$ to the following points

x_i	0	1	2	3	4	5	6	7	8	9	10
y_i	4	7.5	10	12.2	13.5	14	13.6	12	9.9	7	3.9

and sketch the results.

11-5 Fit the curve $y = a \sin \omega x$ to the points

x_i	0	1	2	3	4	5	6	7	8	9	10	11	12
y_i	2	7	11	12	6	1	−5	−9	−10	−6	−1	2	4

using perturbation with initial guess $a_0 = 10$, $\omega_0 = \pi/5$. Sketch the results.

11-6 We measure the angles α, β, γ of a triangle and obtain the n triplets (x_i, y_i, z_i) where $x_i - \alpha$, $y_i - \beta$, $z_i - \gamma$ are the measurement errors. **(a)** Find the LS estimates $\hat{\alpha}, \hat{\beta}, \hat{\gamma}$ of α, β, γ deterministically. **(b)** Assuming that the errors are the samples of three independent $N(0, \sigma_x)$, $N(0, \sigma_y)$, $N(0, \sigma_z)$ RVs, find the ML estimates of α, β, γ.

11-7 The RVs $\mathbf{y}_i$ are $N(a + bx_i, \sigma)$. Show that if

$$\hat{\mathbf{b}} = \frac{\Sigma(x_i - \bar{x})\mathbf{y}_i}{\Sigma(x_i - \bar{x})^2} \qquad \hat{\mathbf{a}} = \bar{\mathbf{y}} - \hat{\mathbf{b}}\bar{x} \qquad \text{then} \qquad E\{\hat{\mathbf{b}}\} = b, \qquad E\{\hat{\mathbf{a}}\} = a$$

11-8 **(a)** Find n numbers α_i such that the sum $I = \Sigma\alpha_i^2$ is minimum subject to the constraints $\Sigma\alpha_i = 1$, $\Sigma\alpha_i x_i = 0$, where x_i are n given constants. **(b)** Find β_i such that the sum $J = \Sigma\beta_i^2$ is minimum subject to the constraints $\Sigma\beta_i = 0$, $\Sigma\beta_i x_i = 1$.

11-9 The n RVs $\mathbf{y}_i$ are independent and $N(a + bx_i, \sigma)$. Test the hypothesis $b = b_0$ against $b \neq b_0$ in terms of the data y_i. Use as the test statistic the sum $\hat{b} = \Sigma\beta_i y_i$ in (11-46). **(a)** Assume that σ is known. **(b)** Assume that σ is unknown.

11-10 The RVs $\mathbf{y}_1, \ldots, \mathbf{y}_n$ are independent with the same variance and with mean

$$E\{\mathbf{y}_i\} = A + Bx_i + Cx_i^2$$

where x_i are n known constants. **(a)** Find the best linear estimates

$$\hat{A} = \Sigma\alpha_i y_i \qquad \hat{B} = \Sigma\beta_i y_i \qquad \hat{C} = \Sigma\gamma_i y_i$$

of the parameters A, B, and C. **(b)** Show that if the RVs $\mathbf{y}_i$ are normal, then the ML estimates of A, B, and C satisfy the system (11-21).

11-11 Suppose that the temperature of the earth distance x from the surface equals $\theta_x = a + bx$. We measure θ_x at 10 points x_i and obtain the values $y_i = a + bx_i + \nu_i$ where ν_i are the measurement errors, which we assume i.i.d. and $N(0, \sigma)$. The results in meters for x and degrees C for θ_x are

x_i	10	52	110	153	200	245	310	350	450	600
y_i	26.2	27.1	28.6	29.9	31.4	32.6	34.1	35.1	37.5	40.2

(a) Find the best unbiased estimates of a and b and test the hypothesis $a = 0$ against $a \neq 0$ if $\sigma = 1$. **(b)** If σ is unknown, estimate it and find the 0.95 confidence interval of a and b. **(c)** Estimate θ_x at $x = 800$ m and find its 0.95 confidence interval if $\sigma = 1$.

11-12 **(a)** Show that the RVs $\bar{\mathbf{y}}$ and $\hat{\mathbf{b}}$ in (11-45) are independent. **(b)** Show that the RV $\frac{1}{\sigma^2}\Sigma(\hat{\boldsymbol{\eta}}_i - \eta_i)^2$ is $\chi^2(2)$. **(c)** With $\boldsymbol{\epsilon}_i = \mathbf{y}_i - \hat{\boldsymbol{\eta}}_i = \boldsymbol{\nu}_i - (\hat{\boldsymbol{\eta}}_i - \eta_i)$, as in (11-56), show that the RVs $\boldsymbol{\epsilon}_i$ and $\hat{\boldsymbol{\eta}}_i = \hat{\mathbf{a}} + \hat{\mathbf{b}}x_i$ are uncorrelated. **(d)** Show that the RV $\frac{1}{\sigma^2}\Sigma\boldsymbol{\epsilon}_i^2$ is $\chi^2(n-2)$.

11-13 (Weighted least squares) The RVs $\mathbf{y}_i$ are independent and normal, with mean $a + bx_i$ and variance $\sigma_i^2 = \sigma^2/w_i$. Show that the ML estimates $\hat{a}$ and $\hat{b}$ of a and b are the solutions of the system

$$a\Sigma w_i + b\Sigma w_i x_i = \Sigma w_i y_i \qquad a\Sigma w_i x_i + b\Sigma w_i x_i^2 = \Sigma w_i x_i y_i$$

11-14 The RVs $\mathbf{x}$ and $\mathbf{y}$ are such that

$$\eta_x = 3 \qquad \eta_y = 4 \qquad \sigma_x = 2 \qquad \sigma_y = 8 \qquad r_{xy} = 0.5$$

(a) Find the homogeneous predictor $\hat{\mathbf{y}} = a\mathbf{x}$ of $\mathbf{y}$ and the MS prediction error $Q = E\{(\mathbf{y} - a\mathbf{x})^2\}$. **(b)** Find the nonhomogeneous predictor $\hat{\mathbf{y}}_0 = \gamma_0 + \gamma_1\mathbf{x}$ and the error $Q = E\{[\mathbf{y} - (\gamma_0 + \gamma_1\mathbf{x})]^2\}$.

11-15 The RVs $\mathbf{x}$ and $\mathbf{y}$ are jointly normal with zero mean. Suppose that $\hat{\mathbf{y}} = a\mathbf{x}$ is the LMS predictor of $\mathbf{y}$ in terms of $\mathbf{x}$ and $Q = E\{(\mathbf{y} - a\mathbf{x})^2\}$ is the MS error. **(a)** Show that the RVs $\mathbf{y} - a\mathbf{x}$ and $\mathbf{x}$ are independent. (b) Show that the conditional density $f(y|x)$ is a normal curve with mean ax and variance Q.

11-16 The RVs $\mathbf{y}, \mathbf{x}_1, \ldots, \mathbf{x}_n$ are jointly normal with zero mean. Show that if $\hat{\mathbf{y}} = a_1\mathbf{x}_1 + \cdots + a_n\mathbf{x}_n$ is the linear MS predictor of $\mathbf{y}$ in terms of $\mathbf{x}_i$, then

$$E\{\mathbf{y}|\mathbf{x}_1, \ldots, \mathbf{x}_n\} = a_1\mathbf{x}_1 + \cdots + a_n\mathbf{x}_n$$

This shows that for normal RVs, nonlinear and linear predictors are identical. The proof is based on the fact that for normal RVs with zero mean, uncorrelatedness is equivalent to independence.

12

Entropy

Entropy is rarely treated in books on statistics. It is viewed as an arcane subject related somehow to uncertainty and information and associated with thermodynamics, statistical mechanics, or coding. In this chapter, we argue that entropy is a basic concept precisely defined within a probabilistic model and that all its properties follow from the axioms of probability. We show that like probability, the empirical interpretation of entropy is based on the properties of long sequences generated by the repetition of a random experiment. This leads to the notion of typical sequences and offers, in our view, the best justification of the principle of maximum entropy and of the use of entropy in statistics.

12-1

Entropy of Partitions and Random Variables

Entropy, as a scientific concept, was introduced first in thermodynamics (Clausius, 1850). Several years later, it was given a probabilistic interpretation in the context of statistical mechanics (Boltzmann, 1877). In 1948, Shannon established the connection between entropy and typical sequences. This led to the solution of a number of basic problems in coding and data trans-

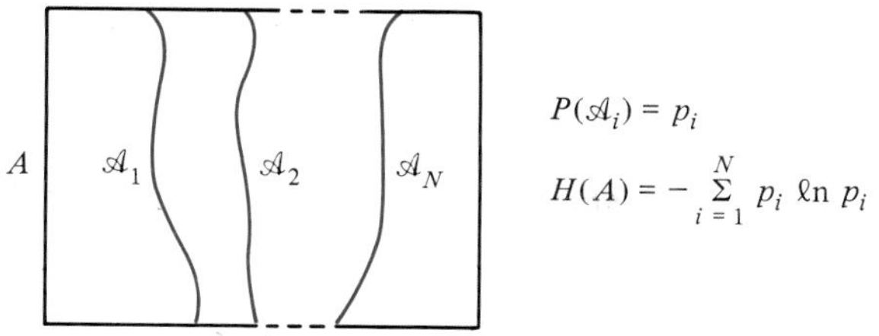

Figure 12.1

mission. Jaynes (1957) used the method of maximum entropy to solve a number of problems in physics and, more recently, in a variety of other areas involving the solution of ill-posed problems. In this chapter, we examine the relationship between entropy and statistics, and we use the principle of maximum entropy to estimate unknown distributions in terms of known parameters (see also Section 8-2).

▪ ***Definition.*** Given a probabilistic model $\mathscr{S}$ and a partition $A = [\mathscr{A}_1, \ldots, \mathscr{A}_N]$ of $\mathscr{S}$ consisting of the N events $\mathscr{A}_i$ (Fig. 12.1), we form the sum

$$H(A) = -\sum_{i=1}^{N} p_i \ln p_i \qquad p_i = P(\mathscr{A}_i) \tag{12-1}$$

This sum is called the *entropy* of the partition A.

Example 12.1

Consider a coin with $P\{h\} = p$ and $P\{t\} = q$. The events $\mathscr{A}_1 = \{h\}$ and $\mathscr{A}_2 = \{t\}$ form a partition $A = [\mathscr{A}_1, \mathscr{A}_2]$ with entropy

$$H(A) = -(p \ln p + q \ln q)$$

If $p = q = .5$, then $H(A) = \ln 2 = 0.693$; if $p = .25$ and $q = .75$, then $H(A) = 0.562$. ■

Example 12.2

In the fair-die experiment, the elementary events $\{f_i\}$ form a partition A with entropy

$$H(A) = -\left(\frac{1}{6}\ln\frac{1}{6} + \cdots + \frac{1}{6}\ln\frac{1}{6}\right) = \ln 6 = 1.79$$

In the same experiment, the events $\mathscr{B}_1 = \{\text{even}\}$ and $\mathscr{B}_2 = \{\text{odd}\}$ form a partition B with entropy

$$H(B) = -\left(\frac{1}{2}\ln\frac{1}{2} + \frac{1}{2}\ln\frac{1}{2}\right) = \ln 2 = 0.693$$ ■

In an arbitrary experiment $\mathscr{S}$, an event $\mathscr{A}$ and its complement $\overline{\mathscr{A}}$ form a partition $A = [\mathscr{A}, \overline{\mathscr{A}}]$ consisting of the two events $\mathscr{A}_1 = \mathscr{A}$ and $\mathscr{A}_2 = \overline{\mathscr{A}}$. The

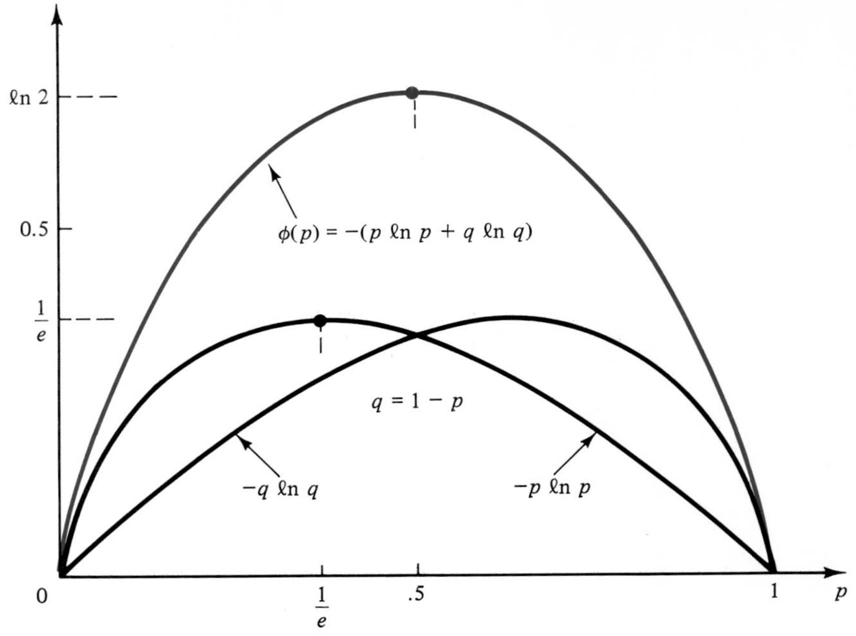

Figure 12.2

entropy of this partition equals

$$H(A) = -(p \ln p + q \ln q) \qquad \text{where } p = P(\mathscr{A}) \qquad q = P(\overline{\mathscr{A}})$$

In Fig. 12.2, we plot the functions $-p \ln p$, $-q \ln q$, and their sum

$$\phi(p) = -(p \ln p + q \ln q) \qquad q = 1 - p \tag{12-2}$$

The function $\phi(p)$ tends to 0 as p approaches 0 or 1 because

$$p \ln p \to 0 \qquad \text{for} \qquad p \to 0 \qquad \text{and for} \qquad p \to 1$$

Furthermore, $\phi(p)$ is symmetrical about the point $p = .5$, and it is maximum for $p = .5$. Thus the entropy of the partition $[\mathscr{A}, \overline{\mathscr{A}}]$ is maximum if the events $\mathscr{A}$ and $\overline{\mathscr{A}}$ have the same probability. We show next that this is true in general. The proof will be based on the following.

A Basic Inequality If c_i is a set of N numbers such that

$$c_1 + \cdots + c_N = 1 \qquad c_i \geq 0$$

then

$$-\sum_{i=1}^{N} p_i \ln p_i \leq -\sum_{i=1}^{N} p_i \ln c_i \tag{12-3}$$

■ ***Proof.*** From the convexity of the function $\ln z$ it follows (Fig. 12.3) that

$$\ln z \leq z - 1 \tag{12-4}$$

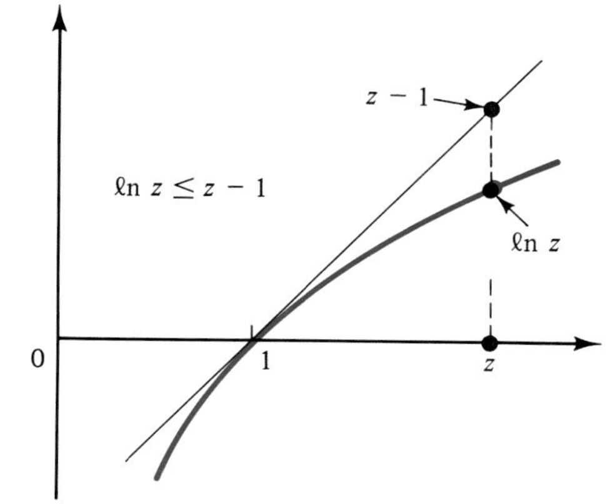

Figure 12.3

With $z = c_i/p_i$, this yields

$$\sum_{i=1}^{N} p_i \ln \frac{c_i}{p_i} \le \sum_{i=1}^{N} p_i \left(\frac{c_i}{p_i} - 1\right) = \sum_{i=1}^{N} c_i - \sum_{i=1}^{N} p_i = 0$$

Hence,

$$0 \ge \sum_{i=1}^{N} p_i \ln \frac{c_i}{p_i} = \sum_{i=1}^{N} p_i \ln c_i - \sum_{i=1}^{N} p_i \ln p_i$$

and (12-3) results.

■ ***Theorem.*** The entropy of a partition consisting of N events is maximum if all events have the same probabilities, that is, if $p_i = 1/N$:

$$H(A) = -\sum_{i=1}^{N} p_i \ln p_i \le -\sum_{i=1}^{N} \frac{1}{N} \ln \frac{1}{N} = \ln N \tag{12-5}$$

■ ***Proof.*** Setting $c_i = 1/N$ in (12-3), we obtain

$$-\sum_{i=1}^{N} p_i \ln p_i \le -\sum_{i=1}^{N} p_i \ln \frac{1}{N} = \ln N$$

From the theorem it follows that

$$0 \le H(A) \le \ln N \tag{12-6}$$

Furthermore, $H(A) = \ln N$ iff $p_1 = \cdots = p_N$ and

$$H(A) = 0 \qquad \text{iff } p_i = \begin{cases} 1 & i = k \\ 0 & i \ne k \end{cases} \tag{12-7}$$

The following two properties of entropy are consequences of the convexity of the function $-p \ln p$ of Fig. 12.2.

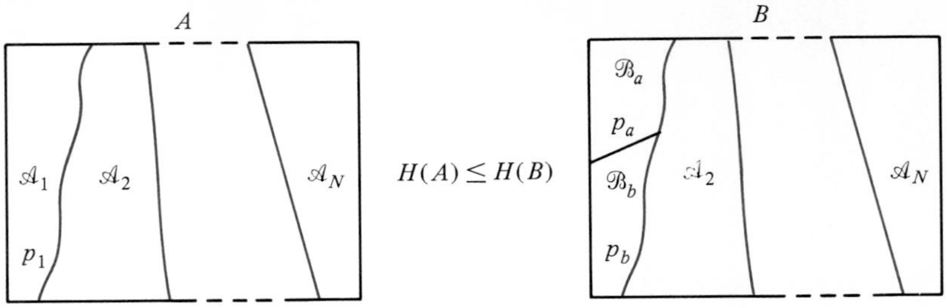

Figure 12.4

Property 1 The partitions A and B of Fig. 12.4 consist of N and $N + 1$ events, respectively. The $N - 1$ events $\mathcal{A}_2, \ldots, \mathcal{A}_N$ are the same in both partitions. Furthermore, the events $\mathcal{B}_a$ and $\mathcal{B}_b$ are disjoint and $\mathcal{A}_1 = \mathcal{B}_a \cup \mathcal{B}_b$; hence,

$$p_1 = P(\mathcal{A}_1) = P(\mathcal{B}_a) + P(\mathcal{B}_b) = p_a + p_b$$

We shall show that

$$H(A) < H(B) \tag{12-8}$$

■ ***Proof.*** From the convexity of the function $w(p) = -p \ln p$ it follows that

$$w(p_a + p_b) < w(p_a) + w(p_b) \tag{12-9}$$

Hence,

$$-p_1 \ln p_1 - \sum_{i=2}^{N} p_i \ln p_i \leq -(p_a \ln p_a + p_b \ln p_b) - \sum_{i=2}^{N} p_i \ln p_i$$

and (12-8) results because the left side equals $H(A)$ and the right side equals $H(B)$.

Example 12.3

The partition A consists of three events with probabilities

$$p_1 = .55 \qquad p_2 = .30 \qquad p_3 = .15$$

and its entropy equals $H(A) = 0.915$. Replacing the event $\mathcal{A}_1$ by the events $\mathcal{B}_a$ and $\mathcal{B}_b$, we obtain the partition B where $p_a = P(\mathcal{B}_a) = .38$, $p_b = P(\mathcal{B}_b) = .17$. The entropy of the partition so formed equals

$$H(B) = 1.3148 > H(A)$$

in agreement with (12-8). ■

Property 2 The partitions A and C of Fig. 12.5 consist of N events each. The $N - 2$ events $\mathcal{A}_2, \ldots, \mathcal{A}_N$ are the same in both partitions. Furthermore, $\mathcal{A}_1 \cup \mathcal{A}_2 = \mathcal{C}_a \cup \mathcal{C}_b$ and

$$p_1 = P(\mathcal{A}_1) \qquad p_2 = P(\mathcal{A}_2) \qquad p_a = P(\mathcal{C}_a) \qquad p_b = P(\mathcal{C}_b)$$

We shall show that if $p_1 < p_a < p_b < p_2$, then

$$H(A) < H(C) \tag{12-10}$$

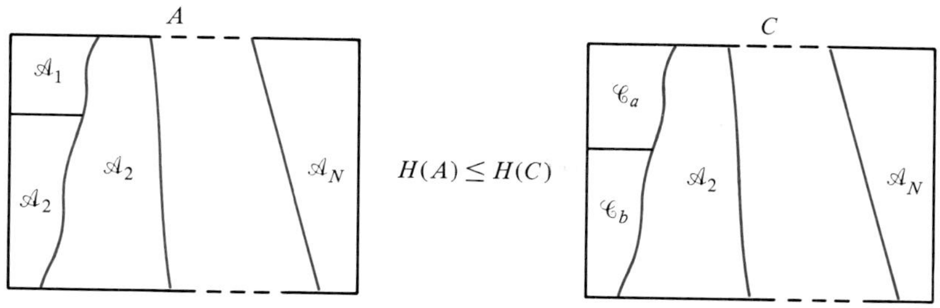

Figure 12.5

■ ***Proof.*** Clearly, $p_1 + p_2 = p_a + p_b$; hence, $p_a = p_1 + \varepsilon < p_2 - \varepsilon = p_b$. From the convexity of $w(p)$ it follows that if $\varepsilon > 0$, then

$$w(p_1) + w(p_2) < w(p_1 + \varepsilon) + w(p_2 - \varepsilon)$$

Hence,

$$-p_1 \ln p_1 - p_2 \ln p_2 - \sum_{i=3}^{N} p_i \ln p_i \leq -p_a \ln p_a - p_b \ln p_b - \sum_{i=3}^{N} p_i \ln p_i$$

and (12-10) results because the left side equals $H(A)$ and the right side equals $H(B)$.

Example 12.4

Suppose that A is the partition in Example 12-3 and C is such that

$$p_a = P(\mathscr{C}_a) = .52 \qquad p_b = P(\mathscr{C}_b) = .33$$

In this case, $H(C) = 0.990 > H(A) = 0.915$, in agreement with (12-10). ■

We can use property 2 to give another proof of the theorem (12-5). Indeed, if the events $\mathscr{A}_i$ of A do not have the same probability, we can construct another partition C as in (12-10), with larger entropy. From this it follows that if $H(A)$ is maximum, all the events of A must have the same probability.

Note The concept of entropy is usually introduced as a measure of uncertainty about the occurrence of the events $\mathscr{A}_i$ of a partition A, and the sum (12-1), used to quantify this measure, is *derived* from a number of postulates that are based on the heuristic notion of the properties of uncertainty. We follow a different approach. We view (12-1) as the *definition* of entropy, and we derive the relationship between $H(A)$ and the number of typical sequences. As we show in Section 12-3, this relationship forms the conceptual justification of the method of maximum entropy, and it shows the connection between entropy and relative frequency.

Random Variables

Consider a *discrete type* RV $\mathbf{x}$ taking the values x_i with probability $p_i = P\{\mathbf{x} = x_i\}$. The events $\{\mathbf{x} = x_i\}$ are mutually exclusive, and their union equals $\mathcal{S}$. Hence, they form the partition

$$A_x = [\mathcal{A}_1, \ldots, \mathcal{A}_N] \qquad \mathcal{A}_i = \{\mathbf{x} = x_i\}$$

The entropy of this partition is by definition the entropy $H(\mathbf{x})$ of the RV $\mathbf{x}$:

$$H(\mathbf{x}) = H(A_x) = -\sum_{i=1}^{N} p_i \ln p_i \qquad p_i = P\{\mathbf{x} = x_i\} \tag{12-11}$$

Thus $H(\mathbf{x})$ does not depend on the values x_i of $\mathbf{x}$; it depends only on the probabilities p_i.

Conversely, given a partition A, we can construct an RV $\mathbf{x}_A$ such that

$$\mathbf{x}_A(\zeta) = x_i \qquad \text{for} \qquad \zeta \in \mathcal{A}_i \tag{12-12}$$

where x_i are distinct numbers but otherwise arbitrary. Clearly, $\{\mathbf{x}_A = x_i\} = \mathcal{A}_i$; hence, $H(\mathbf{x}_A) = H(A)$.

Entropy as Expected Value We denote by $f(x)$ the point density of the discrete type RV $\mathbf{x}$. Thus $f(x)$ is different from 0 at the points x_i and $f(x_i) = P\{\mathbf{x} = x_i\} = p_i$. Using the function $f(x)$, we construct the RV $\ln f(\mathbf{x})$. This RV is a function of the RV $\mathbf{x}$, and its mean [see (4-94)] equals

$$E\{\ln f(\mathbf{x})\} = \sum_{i=1}^{N} p_i \ln f(x_i)$$

Comparing with (12-11), we conclude that the entropy of the RV $\mathbf{x}$ equals

$$H(\mathbf{x}) = -E\{\ln f(\mathbf{x})\} \tag{12-13}$$

Example 12.5

In the die experiment, the elementary events $\{f_i\}$ form a partition A. We construct the RV $\mathbf{x}_A$ such that $\mathbf{x}(f_i) = i$ as in (12-12). Clearly, $f(x)$ is different from 0 at the points $x = 1, \ldots, 6$, and $f(x_i) = p_i$; hence,

$$H(\mathbf{x}_A) = -E\{\ln f(\mathbf{x}_A)\} = -\sum_{i=1}^{6} p_i \ln p_i = H(A)$$

■

CONTINUOUS TYPE RVS The entropy $H(\mathbf{x})$ of a continuous type RV cannot be defined directly as the entropy of a partition because the events $\{\mathbf{x} = x_i\}$ are noncountable, and $P\{\mathbf{x} = x_i\} = 0$ for every x_i. To avoid this difficulty, we shall define $H(\mathbf{x})$ as expected value, extending (12-13) to continuous type RVS. Note that unlike (12-11), the resulting expression (12-14) is not consistent with the interpretation of entropy as a measure of uncertainty about the values of $\mathbf{x}$; however, as we show in Section 12-2, it leads to useful applications.

■ ***Definition.*** Denoting by $f(x)$ the density of $\mathbf{x}$, we form the RV $-\ln f(\mathbf{x})$. The expected value of this RV is by definition the entropy $H(\mathbf{x})$ of $\mathbf{x}$:

$$H(\mathbf{x}) = -E\{\ln f(\mathbf{x})\} = -\int_{-\infty}^{\infty} f(x) \ln f(x)\, dx \qquad (12\text{-}14)$$

Note that $f(x) \ln f(x) \to 0$ as $f(x) \to 0$; hence, the integral is 0 in any region of the x axis where $f(x) = 0$.

Example 12.6

The RV $\mathbf{x}$ has an exponential distribution:

$$f(x) = \alpha e^{-\alpha x} U(x)$$

In this case, $E\{\mathbf{x}\} = 1/\alpha$ and $\ln f(x) = \ln \alpha - \alpha x$ for $x > 0$; hence,

$$H(\mathbf{x}) = -E\{\ln \alpha - \alpha \mathbf{x}\} = -\ln \alpha + 1 = \ln \frac{e}{\alpha}$$ ■

Example 12.7

If $f(\mathbf{x})$ is $N(\eta, \sigma)$, then

$$f(x) = \frac{1}{\sigma \sqrt{2\pi}} e^{-(x-\eta)^2/2\sigma^2} \qquad \ln f(x) = -\ln \sigma \sqrt{2\pi} - \frac{(x-\eta)^2}{2\sigma^2}$$

Hence,

$$H(\mathbf{x}) = -E\{\ln f(\mathbf{x})\} = E\{\ln \sigma \sqrt{2\pi}\} + E\left\{\frac{(\mathbf{x}-\eta)^2}{2\sigma^2}\right\}$$

And since $E\{(\mathbf{x}-\eta)^2\} = \sigma^2$, we obtain

$$H(\mathbf{x}) = \ln \sigma \sqrt{2\pi} + \frac{1}{2} = \ln \sigma \sqrt{2\pi e}$$ ■

Example 12.8

The RV $\mathbf{x}$ is uniform in the interval $(0, c)$. Thus $f(x) = 1/c$ for $0 < x < c$ and 0 elsewhere; hence,

$$H(\mathbf{x}) = -E\left\{\ln \frac{1}{c}\right\} = -\frac{1}{c}\int_0^c \ln \frac{1}{c}\, dx = \ln c$$ ■

Joint Entropy Extending (12-13), we define the joint entropy $H(\mathbf{x}, \mathbf{y})$ of two RVs $\mathbf{x}$ and $\mathbf{y}$ as the expected value of the RV $-\ln f(\mathbf{x}, \mathbf{y})$ where $f(x, y)$ is the joint density of $\mathbf{x}$ and $\mathbf{y}$. Suppose, first, that the RVs $\mathbf{x}$ and $\mathbf{y}$ are of the discrete type, taking the values x_i and y_j with probabilities

$$P\{\mathbf{x} = x_i, \mathbf{y} = y_j\} = p_{ij} \qquad i = 1, \ldots, M \qquad j = 1, \ldots, N$$

In this case, the function $f(x, y)$ is different from 0 at the points (x_i, y_j), and $f(x_i, y_j) = p_{ij}$. Hence,

$$H(\mathbf{x}, \mathbf{y}) = -E\{\ln f(\mathbf{x}, \mathbf{y})\} = -\sum_{i,j} f(x_i, y_j) \ln f(x_i, y_j) \qquad (12\text{-}15)$$

If the RVs $\mathbf{x}$ and $\mathbf{y}$ are of the continuous type, then

$$H(\mathbf{x}, \mathbf{y}) = -E\{\ln f(\mathbf{x}, \mathbf{y})\} = -\int_{-\infty}^{\infty}\int_{-\infty}^{\infty} f(x, y) \ln f(x, y)\, dx\, dy \qquad (12\text{-}16)$$

Example 12.9 If the RVS $\mathbf{x}$ and $\mathbf{y}$ are jointly normal as in (5-100) (see Problem 12-15), their joint density equals

$$H(\mathbf{x}, \mathbf{y}) = \ln (2\pi e \sigma_1 \sigma_2 \sqrt{1 - r^2})$$ ■

Note that if the RVS $\mathbf{x}$ and $\mathbf{y}$ are independent, their joint entropy equals the sum of their "marginal entropies" $H(\mathbf{x})$ and $H(\mathbf{y})$:

$$H(\mathbf{x}, \mathbf{y}) = H(\mathbf{x}) + H(\mathbf{y}) \tag{12-17}$$

Indeed, in this case,

$$f(x, y) = f_x(x)f_y(y) \qquad \ln f(x, y) = \ln f_x(x) + \ln f_y(y)$$

and (12-17) results.

12-2 Maximum Entropy and Statistics

We shall use the concept of entropy to determine the distribution $F(x)$ of an RV $\mathbf{x}$ or of other unknown quantities of a probabilistic model. The known information, if it exists, is in the form of parameters providing only partial specification of the model. It is assumed that no observations are available.

Suppose that $\mathbf{x}$ is a continuous type RV with unknown density $f(x)$. We know the second moment $E\{\mathbf{x}^2\} = \theta$ of $\mathbf{x}$, and we wish to determine $f(x)$. Thus our problem is to find a positive function $f(x)$ of unit area such that

$$\int_{-\infty}^{\infty} x^2 f(x)\, dx = \theta \tag{12-18}$$

Clearly, this problem does not have a unique solution because there are many densities satisfying (12-18). Nevertheless, invoking the following principle, we shall find a solution.

Principle of Maximum Entropy The unknown density must be such as to maximize the entropy

$$H(\mathbf{x}) = -\int_{-\infty}^{\infty} f(x) \ln f(x)\, dx \tag{12-19}$$

of the RV $\mathbf{x}$.

As we shall see, this condition leads to the unique function

$$f(x) = \gamma e^{-x^2/2\theta}$$

Thus the principle of maximum entropy (ME) leads to the conclusion that the RV $\mathbf{x}$ must be *normal*. This is a remarkable conclusion! However, it is based on a principle the validity of which we have not established.

The justification of the ME principle is usually based on the relationship between entropy and uncertainty. Entropy is a measure of uncertainty; if $f(x)$ is unknown, then our uncertainty about the values of $\mathbf{x}$ is maximum; hence, $f(x)$ must be such as to maximize $H(\mathbf{x})$. This reasoning is heuristic because uncertainty is not a precise concept. We shall give another justification based on the relationship between entropy and typical sequences. This justification is also heuristic; however, it shows the connection between entropy and relative frequency, a concept that is central in the applications of statistics. The ME principle has been used in a variety of physical problems, and in many cases, the results are in close agreement with the observations. In the last analysis, this is the best justification of the principle.

Method of Maximum Entropy

All ill-posed problems dealing with the specification of a probabilistic model can be solved with the method of maximum entropy. However, the usefulness of the solution varies greatly from problem to problem and is greatest in applications dealing with averages of very large samples (statistical mechanics, for example). We shall consider the problem of determining the distribution of one or more RVs under the assumption that certain statistical averages are known. As we shall see, this problem has a simple analytic solution. In other applications the ME method might involve very complex computations. Our development is based on the following version of (12-3).

A Basic Inequality If $c(x)$ is a function such that

$$\int_{-\infty}^{\infty} c(x)\,dx = 1 \qquad c(x) \ge 0$$

and $f(x)$ is the density of an RV $\mathbf{x}$, then

$$-\int_{-\infty}^{\infty} f(x)\ln f(x)\,dx \le -\int_{-\infty}^{\infty} f(x)\ln c(x)\,dx \tag{12-20}$$

Equality holds iff $f(x) = c(x)$.

■ ***Proof.*** From the inequality $\ln z \le z - 1$ it follows with $z = c(x)/f(x)$ that

$$-\int_{-\infty}^{\infty} f(x)\ln\frac{c(x)}{f(x)}\,dx \le \int_{-\infty}^{\infty} f(x)\left[\frac{c(x)}{f(x)} - 1\right]dx = 0$$

Hence,

$$0 \ge \int_{-\infty}^{\infty} f(x)\ln c(x)\,dx - \int_{-\infty}^{\infty} f(x)\ln f(x)\,dx$$

and (12-20) results.

We shall use this inequality to determine the ME solution of various ill-posed problems. The density of $\mathbf{x}$ so obtained will be denoted by $f_0(x)$ and will be called the ME density. Thus $f_0(x)$ maximizes the integral

$$H(\mathbf{x}) = -\int_{-\infty}^{\infty} f(x)\ln f(x)\,dx$$

The corresponding value of $H(\mathbf{x})$ will be denoted by $H_0(\mathbf{x})$.

■ ***Fundamental Theorem.*** (a) If the mean $E\{g(\mathbf{x})\} = \theta$ of a function $g(\mathbf{x})$ of the RV $\mathbf{x}$ is known, its ME density $f_0(x)$ is an exponential

$$f_0(x) = \gamma e^{-\lambda g(x)} \tag{12-21}$$

where γ and λ are two constants such that

$$\gamma \int_{-\infty}^{\infty} e^{-\lambda g(x)}\, dx = 1 \qquad \gamma \int_{-\infty}^{\infty} g(x) e^{-\lambda g(x)}\, dx = \theta \tag{12-22}$$

■ ***Proof.*** If $f_0(x)$ is given by (12-21), then

$$\ln f_0(x) = \ln \gamma - \lambda g(x)$$

The corresponding entropy $H_0(\mathbf{x})$ equals

$$-\int_{-\infty}^{\infty} f_0(x) \ln f_0(x)\, dx = -\int_{-\infty}^{\infty} f_0(x)[\ln \gamma - \lambda g(x)]\, dx$$

Hence,

$$H_0(\mathbf{x}) = -\ln \gamma + \lambda\theta \tag{12-23}$$

To show that $f_0(x)$ is given by (12-21), it therefore suffices to show that if $f(x)$ is any other function such that

$$\int_{-\infty}^{\infty} f(x)\, dx = 1 \qquad \int_{-\infty}^{\infty} g(x) f(x)\, dx = \theta$$

the corresponding entropy is less than $-\ln \gamma + \lambda\theta$. To do so, we set $c(x) = f_0(x)$ in (12-20). This yields

$$-\int_{-\infty}^{\infty} f(x) \ln f(x)\, dx \le -\int_{-\infty}^{\infty} f(x) \ln f_0(x)\, dx = -\int_{-\infty}^{\infty} f(x)[\ln \gamma - \lambda g(x)]\, dx$$

Thus $H(\mathbf{x}) \le \ln \gamma + \lambda\theta$, and the proof is complete.

(b) Suppose now that in addition to the information that $E\{g(\mathbf{x})\} = \theta$, we require that $f(x) = 0$ for $x \notin R$ where R is a specified region of the x-axis. Replacing in (a) the entire axis by the region R, we conclude that all results hold. Thus

$$f_0(x) = \begin{cases} \gamma e^{-\lambda g(x)} & x \in R \\ 0 & x \in R \end{cases} \tag{12-24}$$

and $H_0(\mathbf{x}) = -\ln \gamma + \lambda\theta$. The constant λ and γ are again determined from (12-22) provided that the region of integration is the set R.

(c) If no information about $f(x)$ is known, the ME density is a constant. This follows from (12-24) with $g(x) = 0$. In this case, the problem has a solution only if the region R in which $f(x) \ne 0$ has finite length. Thus if no information about $f(x)$ is given, $f_0(x)$ does not exist; that is, we can find an $f(x)$ with arbitrary large entropy. Suppose, however, that we require that $f(x) \ne 0$ only in a region R consisting of a number of intervals of total length c. In this case, (12-24) yields

$$f_0(x) = \begin{cases} 1/c & x \in R \\ 0 & x \in R \end{cases} \tag{12-25}$$

Note that the theorem does not establish the existence of a ME density. It states only that if $f_0(x)$ exists, it is the function in (12-21).

Discrete Type RVS Suppose, finally, that the RV $\mathbf{x}$ takes the values x_i with probability p_i. We shall determine the ME values p_{0i} of p_i under the assumption that the mean $E\{g(\mathbf{x})\} = \theta$ of the function $g(\mathbf{x})$ is known. Thus our problem is to find p_i such as to maximize the entropy

$$H(\mathbf{x}) = -E\{\ln f(\mathbf{x})\} = -\sum p_i \ln p_i \tag{12-26}$$

subject to the constraints

$$\sum p_i = 1 \qquad \sum p_i g(x_i) = \theta \tag{12-27}$$

where $g(x_i)$ are known numbers.

We maintain that

$$p_{0i} = \gamma e^{-\lambda g(x_i)} \tag{12-28}$$

where γ and λ are two constants such that

$$\gamma \sum e^{-\lambda g(x_i)} = 1 \qquad \gamma \sum e^{-\lambda g(x_i)} g(x_i) = \theta \tag{12-29}$$

This follows from (12-24) if we use for R the set of points x_i. However, we shall give a direct proof based on (12-3).

■ ***Proof.*** If p_{0i} is given by (12-28), then $\ln p_{0i} = \ln \gamma - \lambda g(x_i)$. Hence, the corresponding entropy $H_0(\mathbf{x})$ equals

$$-\sum p_{0i} \ln p_{0i} = -\ln \gamma \sum p_{0i} + \lambda \sum p_{0i} g(x_i) = -\ln \gamma + \lambda\theta$$

It therefore suffices to show that if p_i is another set of probabilities satisfying (12-27), the corresponding entropy is less than $H_0(\mathbf{x})$. To do so, we set $c_i = p_{0i}$ in (12-3). This yields

$$-\sum p_i \ln p_i \le -\sum p_i \ln p_{0i} = -\sum p_i[\ln \gamma - \lambda g(x_i)] = H_0(\mathbf{x})$$

and the proof is complete.

ILLUSTRATIONS Suppose, first that $E\{\mathbf{x}^2\} = \theta$. In this case, (12-21) yields

$$f_0(x) = \gamma e^{-\lambda x^2} \tag{12-30}$$

Thus if the second moment θ of an RV $\mathbf{x}$ is known, $\mathbf{x}$ is $N(0, \sqrt{\theta})$. Hence,

$$\gamma = \frac{1}{\sqrt{2\pi\theta}} \qquad \lambda = \frac{1}{2\theta} \qquad H_0(\mathbf{x}) = \ln \sqrt{2\pi e\theta}$$

If the variance σ^2 of $\mathbf{x}$ is specified, then $\mathbf{x}$ is $N(\eta, \sigma)$ where η is an arbitrary constant. This follows from (12-21) with $g(x) = (x - \eta)^2$.

Example 12.10

Consider a collection of particles moving randomly in a certain region. If they are in statistical equilibrium, the x component v_x of their velocity can be considered as the sample of an RV $\mathbf{v}_x$ with distribution $f(v_x)$. We maintain that if the average kinetic energy

$$N_x = E\left\{\frac{1}{2} m\mathbf{v}_x^2\right\}$$

of the particles is specified, the ME density of $\mathbf{v}_x$ equals

$$f_0(v_x) = \sqrt{\frac{m}{4\pi N_x}} \exp\left\{-\frac{mv_x^2}{4N_x}\right\}$$

Indeed, this follows from (12-30) with $\theta = E\{\mathbf{v}_x^2\} = 2N_x/m$. Thus $\mathbf{v}_x$ is $N(0, \sigma)$ where $\sigma^2 = 2N_x/m$ equals the average kinetic energy per unit mass. The same holds for v_y and v_z. ■

Suppose that $\mathbf{x}$ is an RV with mean $E\{\mathbf{x}\} = \theta$ and such that $f(x) = 0$ for $x < 0$. In this case, $f_0(x)$ is given by (12-24) where R is the region $x > 0$ and $g(x) = x$. This yields

$$f_0(x) = \gamma e^{-\lambda x} U(x) \qquad \gamma = \lambda = \frac{1}{\theta} \tag{12-32}$$

Thus the ME density of *a* positive RV with specified mean is exponential.

Example 12.11 Using the ME method, we shall determine the atmospheric pressure $P(z)$ as a function of the distance z from the ground knowing only the ratio N/m of the energy N over the mass m of a column of air. Assuming statistical equilibrium, we can interpret N as the energy and m as the mass of all particles in a vertical cylinder C of unit cross section (Fig. 12.6). The location z of each particle can be considered as the sample of an RV $\mathbf{z}$ with density $f(z)$. We shall show that

$$f_0(z) = \frac{mg}{N} e^{-mgz/N} U(z) \tag{12-33}$$

where g is the acceleration of gravity.

The probability that a particle is in a cylindrical segment Δ between z and dz equals $f(z)\,dz$; hence, the average mass in Δ equals $mf(z)\,dz$. Since the energy of a unit mass, distance z from the ground, equals gz, we conclude that the energy of the mass in the region Δ equals $gzmf(z)\,dz$, and the total energy N equals

$$N = \int_0^\infty mgzf(z)\,dz = E\{mg\mathbf{z}]$$

With $\theta = E\{\mathbf{z}\} = N/mg$, (12-33) follows from (12-32). The atmospheric pressure $P(z)$ equals the weight of the air in the cylinder C above z:

$$P(z) = \int_z^\infty mgf_0(z)\,dz = mge^{-mgz/N}$$ ■

Figure 12.6

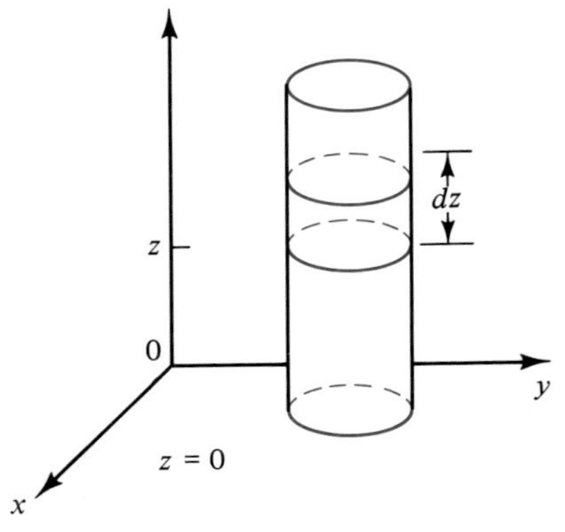

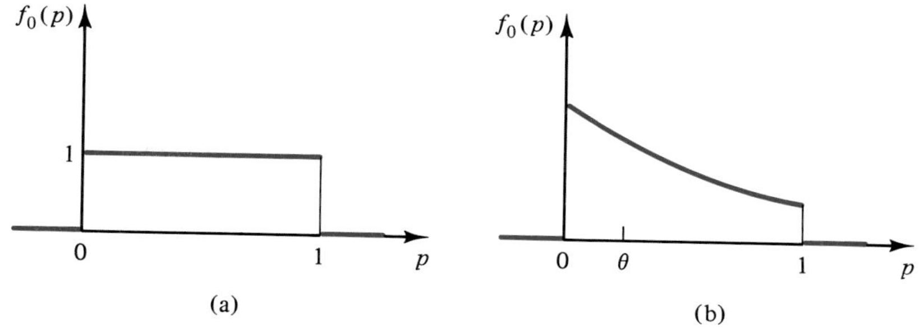

Figure 12.7

Example 12.12

In a coin experiment, the probability $p = P\{h\}$ that heads will show is the value of an RV $\mathbf{p}$ with unknown density $f(p)$. We wish to find its ME form $f_0(p)$.

(a) Clearly, $f(p) = 0$ outside the interval (0, 1); hence, if nothing else is known, then [see (12-25)] $f_0(p) = 1$, as in Fig. 12.7a.

(b) We assume that $E\{\mathbf{p}\} = \theta = 0.236$. In this case, (12-24) yields

$$f_0(p) = \gamma e^{-\lambda p} \qquad 0 < p < 1 \tag{12-34}$$

where γ and λ are such that

$$\gamma \int_0^1 e^{-\lambda p}\, dp = 1 \qquad \gamma \int_0^1 p e^{-\lambda p}\, dp = 0.236$$

Solving for γ and λ, we find $\gamma = 1.7$, $\lambda = 1.2$ (Fig. 12.7b). ■

Example 12.13

(Brandeis Die†). In the die experiment, we are given the information that the average number of faces up equals 4.5. Using this information, we shall determine the ME values p_{0i} of the probabilities $p_i = P\{f_i\}$. To do so, we introduce the RV $\mathbf{x}$ such that $\mathbf{x}(f_i) = i$. Clearly,

$$E\{\mathbf{x}\} = \sum_{i=1}^{6} i p_i = 4.5$$

With $g(x) = x$ and $x_i = i$, it follows from (12-28) that

$$p_{0i} = \gamma e^{-\lambda i} \qquad i = 1, \ldots, 6 \tag{12-35}$$

where the constants γ and λ are such that

$$\gamma \sum_{i=1}^{6} e^{-\lambda i} = 1 \qquad \gamma \sum_{i=1}^{6} i e^{-\lambda i} = 4.5$$

To solve this system, we plot the ratio

$$\eta(w) = \frac{w^{-1} + 2w^{-2} + \cdots + 6w^{-6}}{w^{-1} + w^{-2} + \cdots + w^{-6}} \qquad w = e^{\lambda}$$

and we find the value of w such that $\eta(w) = 4.5$. This yields $w \simeq 0.68$ (see Fig. 12.8). Hence, $\gamma \simeq 0.036$,

$p_{01} \simeq .054$ $\quad p_{02} \simeq .079$ $\quad p_{03} \simeq .114$ $\quad p_{04} \simeq .165$ $\quad p_{05} \simeq .240$ $\quad p_{06} \simeq .348$ ■

† E. T. Jaymes, Brandeis lectures, 1962.

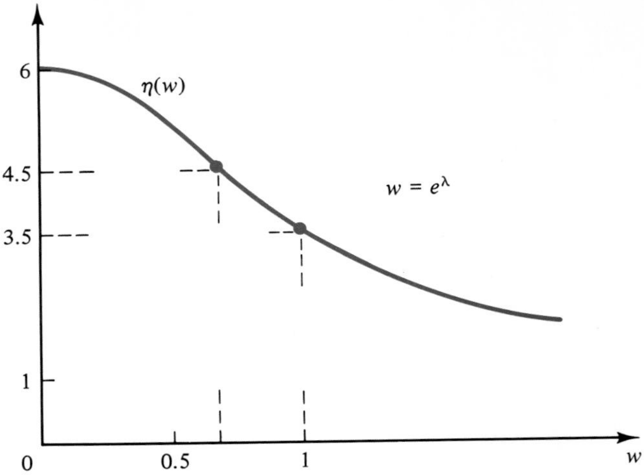

Figure 12.8

GENERALIZATION Now let us consider the problem of determining the density of an RV $\mathbf{x}$ under the assumption that the expected values

$$\theta_k = E\{g_k(\mathbf{x})\} = \int_{-\infty}^{\infty} g_k(x) f(x)\, dx \qquad k = 1, \ldots, n \tag{12-36}$$

of n functions $g_k(\mathbf{x})$ of $\mathbf{x}$ are known.

Reasoning as in (12-21), we can show that the ME density of $\mathbf{x}$ is the function

$$f_0(x) = \gamma \exp\{-[\lambda_1 g_1(x) + \cdots + \lambda_n g_n(x)]\} \tag{12-37}$$

The $n + 1$ constants γ and λ_i are determined from the area condition

$$\gamma \int_{\infty}^{\infty} \exp\left\{-\sum_{k=1}^{n} \lambda_k g_k(x)\right\} dx = 1 \tag{12-38}$$

and the n equations [see (12-36)]

$$\gamma \int_{-\infty}^{\infty} g_k(x) \exp\left\{-\sum_{k=1}^{n} \lambda_k g_k(x)\right\} dx = \theta_k \tag{12-39}$$

The proof is identical to the proof of (12-21) if we replace the terms $\lambda g(x)$ and $\lambda\theta$ by the sums $\sum_{k=1}^{n} \lambda_k g_k(x)$ and $\sum_{k=1}^{n} \lambda_k \theta_k$, respectively. The resulting entropy equals

$$H_0(\mathbf{x}) = -\ln \gamma + \sum_{k=1}^{n} \lambda_k \theta_k \tag{12-40}$$

Example 12.14 Given the first two moments

$$\theta_1 = E\{\mathbf{x}\} \qquad \theta_2 = E\{\mathbf{x}^2\}$$

of the RV $\mathbf{x}$, we wish to find $f_0(x)$.

In this problem, $g_1(x) = x$, $g_2(x) = x^2$, and (12-37) yields

$$f_0(x) = \gamma e^{-\lambda_1 x - \lambda_2 x^2}$$

This shows that $f_0(x)$ is a normal density with mean $\eta = \theta_1$ and variance $\sigma^2 = \theta_2 - \theta_1^2$. ■

Partition Function The partition function is by definition the integral

$$Z(\lambda_1, \ldots, \lambda_n) = \int_{-\infty}^{\infty} \exp\left\{-\sum_{k=1}^{n} \lambda_k g_k(x)\right\} dx \tag{12-41}$$

As we see from (12-38), $Z = 1/\gamma$. Furthermore,

$$\frac{\partial Z}{\partial \lambda_i} = -\int_{-\infty}^{\infty} g_i(x) \exp\left\{-\sum_{k=1}^{n} \lambda_k g_k(x)\right\} dx \tag{12-42}$$

Comparing with (12-39), we conclude that

$$-\frac{1}{Z}\frac{\partial Z}{\partial \lambda_k} = \theta_k \qquad k = 1, \ldots, n \tag{12-43}$$

This is a system of n equations equivalent to (12-39); however, it involves only the n parameters λ_k.

Consider, for example, the coin experiment where

$$E\{\mathbf{p}\} = \int_0^1 pf(p)\,dp = \theta$$

is a given number. In this case [see (12-34)],

$$Z = \frac{1}{\gamma} = \int_0^1 e^{-\lambda p}\,dp = \frac{1 - e^{-\lambda}}{\lambda}$$

and with $n = 1$, (12-43) yields

$$-\frac{1}{Z}\frac{\partial Z}{\partial \lambda} = \frac{1 - e^{-\lambda} - \lambda e^{-\lambda}}{\lambda(1 - e^{-\lambda})} = \theta$$

To find $f_0(p)$ for a given θ, it suffices to solve this equation for λ.

Discrete Type RVs Consider, finally, a discrete type RV $\mathbf{x}$ taking the values x_i with probability p_i. We shall determine the ME values p_{0i} of p_i under the assumption that we know the n constants

$$\theta_k = E\{g_k(\mathbf{x})\} = \sum_i p_i g_k(x_i) \qquad k = 1, \ldots, n \tag{12-44}$$

Reasoning as in (12-28), we obtain

$$p_{0i} = \frac{1}{Z} \exp\{-\lambda_1 g_1(x_i) - \cdots - \lambda_n g_n(x_i)\}$$

where

$$Z = \sum_i \exp\{-\lambda_1 g_1(x_i) - \cdots - \lambda_n g_n(x_i)\} \tag{12-45}$$

is the discrete form of the partition function. From this and (12-44), it follows that Z satisfies the n equations

$$-\frac{1}{Z}\frac{\partial Z}{\partial \lambda} = \theta_k \qquad k = 1, \ldots, n \tag{12-46}$$

Thus to determine p_{0i}, it suffices to form the sum in (12-45) and solve the system (12-46) for the n parameters λ_k.

In Example 12.13, we assumed that $E\{\mathbf{x}\} = \eta$. In this case, $n = 1$,

$$Z = \sum_{i=1}^{6} e^{-\lambda i} \qquad -\frac{1}{Z}\frac{\partial Z}{\partial \lambda} = \frac{\sum_{i=1}^{6} ie^{-\lambda i}}{\sum_{i=1}^{6} e^{-\lambda i}} = \eta$$

To determine the probabilities p_{0i}, it suffices to solve the last equation for λ.

12-3 Typical Sequences and Relative Frequency

We commented earlier on the subjective interpretation of entropy as a measure of uncertainty about the occurrence of the events $\mathscr{A}_i$ of a partition A at a single trial. Next we give a different interpretation based on the relationship between the entropy $H(A)$ of A and the number n_t of typical sequences, that is, sequences that are most likely to occur in a large number of trials. This interpretation shows the equivalence between entropy and relative frequency, and it establishes the connection between the model concept $H(A)$ and the real world.

Typical sequences were introduced in Section 8-2 in the context of a partition consisting of an event $\mathscr{A}$ and its complement $\overline{\mathscr{A}}$. Here we generalize to arbitrary partitions. As preparation, let us review the analysis of the two-event partition $A = [\mathscr{A}, \overline{\mathscr{A}}]$.

In the experiment $\mathscr{S}_n$ of repeated trials, the sequence

$$s_j = \mathscr{A}\overline{\mathscr{A}}\overline{\mathscr{A}} \cdots \mathscr{A} \qquad j = 1, \ldots, 2^n \tag{12-47}$$

is an event with probability

$$P(s_j) = p^k q^{n-k} \qquad p = P(\mathscr{A}) = 1 - q \tag{12-48}$$

where k is the number of successes of $\mathscr{A}$. The number n_s of such sequences equals 2^n. We know from the empirical interpretation of probability that if n is large, we expect with near certainty that $k \simeq np$. We can thus divide the 2^n sequences of the form (12-47) into two groups. The first group consists of all sequences such that $k \simeq np$. These sequences will be called *typical* and will be identified by the letter t. The second group, called *atypical,* consists of all sequences that are not typical.

NUMBER OF TYPICAL SEQUENCES We show next that the number n_t of typical sequences can be expressed in terms of the entropy

$$H(A) = -(p \ln p + q \ln q)$$

of the partition A. This number is empirical because it is based on the empirical formula $k \simeq np$. As we shall show, it can be given a precise interpretation based on the law of large numbers.

To determine n_t, we shall first find the probability of a typical sequence t_j. Clearly, $P(t_j)$ is given by (12-48) where now

$$k \simeq np \qquad n - k \simeq n - np = nq$$

Inserting into (12-48), we obtain

$$P(t_j) \simeq p^{np}q^{nq} = e^{np \ln p + nq \ln q} = e^{-nH(A)} \tag{12-49}$$

The union $\mathcal{T}$ of all typical sequences is an event in the space $\mathcal{S}_n$. This event will almost certainly occur at a single trial of $\mathcal{S}_n$ because it consists of all sequences with $k \simeq np$. Hence, $P(\mathcal{T}) \simeq 1$. And since all typical sequences have the same probability, we conclude that $1 \simeq P(\mathcal{T}) \simeq n_t P(t_j)$. This yields

$$n_t \simeq e^{nH(A)} \tag{12-50}$$

We shall now compare n_t to the number 2^n of all sequences of the form (12-47). If $p = .5$, then $H(A) = \ln 2$; hence, $n_t \simeq e^{n \ln 2} = 2^n$. If $p \neq .5$, then $H(A) < \ln 2$, and for large n,

$$n_t \simeq e^{nH(A)} \ll e^{n \ln 2} = 2^n \tag{12-51}$$

This shows (Fig. 12.9) that if $p \neq .5$, then the number n_t of typical sequences is much smaller than the number 2^n of all sequences even though $P(\mathcal{T}) \simeq 1$. Thus if the experiment $\mathcal{S}_n$ is repeated a large number of times, most sequences that will occur are typical (Fig. 12.10).

Figure 12.9

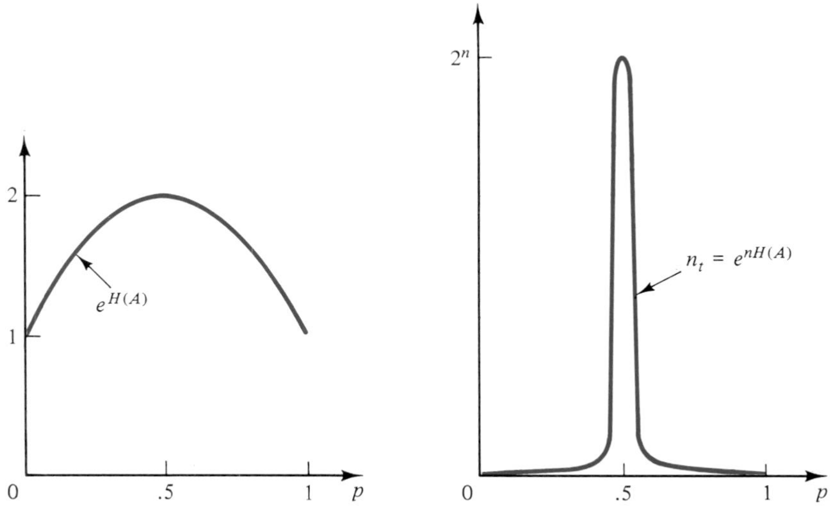

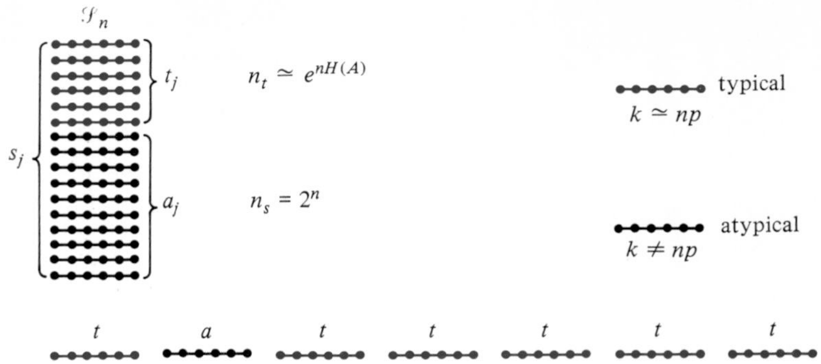

Figure 12.10

Note Since $P(\mathcal{T}) \simeq 1$, most—but not all—atypical sequences have a small probability of occurrence. If $p < .5$, then $p/q < 1$, hence, the sequence $p^k q^{n-k}$ is decreasing as k increases. This shows that all atypical sequences with $k < np$ are more likely than the typical sequences. However, the number of such sequences is small compared to n_t (see also Fig. 12.11).

Typical Sequences and Bernoulli Trials We shall reexamine the preceding concepts in the context of Bernoulli trials using the following definitions. We shall say that a sequence of the form (12-47) is typical if the number k of successes of $\mathcal{A}$ is in the interval (Fig. 12.11)

$$k_a < k < k_b \qquad k_a = np - 3\sqrt{npq} \qquad k_b = np + 3\sqrt{npq} \tag{12-52}$$

We shall use the De Moivre–Laplace approximation to determine the probability of the set $\mathcal{T}$ consisting of all typical sequences so defined. Since k is in the $\pm 3\sqrt{npq}$ interval centered at np, it follows from (3-30) that

$$P(\mathcal{T}) = \sum_{k=k_a}^{k_b} \binom{n}{k} p^k q^{n-k} \simeq 2G(3) - 1 = .997 \tag{12-53}$$

Figure 12.11

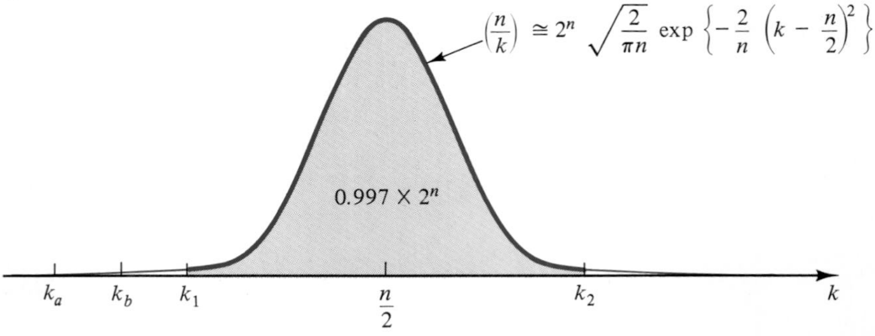

This shows that if the experiment $\mathcal{S}_n$ is repeated a large number of times, in 99.7% of the cases, we will observe only typical sequences. The number of such sequences is

$$n_t = \sum_{k=k_a}^{k_b} \binom{n}{k} \tag{12-54}$$

To find this sum, we set $p = q = 1/2$ in (12-53). This yields the approximation

$$\frac{1}{2^n} \sum_{k=k_1}^{k_2} \binom{n}{k} \simeq .997 \qquad k_1 = \frac{n}{2} - 3 \times \frac{\sqrt{n}}{2} \qquad k_2 = \frac{n}{2} + 3 \times \frac{\sqrt{n}}{2} \tag{12-55}$$

and it shows that $.997 \times 2^n$ of the 2^n sequences of the form (12-47) are in the interval (k_1, k_2) of Fig. 12.11. If $p \neq .5$, then the interval (k_a, k_b) in (12-52) is outside the interval (k_1, k_2); hence n_t is less than $.003 \times 2^n$.

The preceding analysis can be used to give a precise interpretation of (12-52) in the form of a limit. With k_a and k_b as in (12-52), it can be shown that the ratio

$$\frac{n_t}{n} = \frac{1}{n} \sum_{k=k_a}^{k_b} \binom{n}{k} \tag{12-56}$$

tends to $H(A)$ as $n \to \infty$. The proof, however, is not simple.

ARBITRARY PARTITIONS Consider an arbitrary partition $A = [\mathcal{A}_1, \ldots, \mathcal{A}_N]$ consisting of the N events $\mathcal{A}_i$. In the experiment $\mathcal{S}_n$ of repeated trials, we observe sequences of the form

$$s_j = \mathcal{B}_1 \cdots \mathcal{B}_k \cdots \mathcal{B}_n \qquad j = 1, \ldots, N^n \tag{12-57}$$

where $\mathcal{B}_k$ is any one of the events $\mathcal{A}_i$ of A. The sequence s_j is an event in the space $\mathcal{S}_n$, and its probability equals

$$P(s_j) = p_1^{k_1} \cdots p_N^{k_N} \qquad p_i = P(\mathcal{A}_i) \tag{12-58}$$

where k_i is the number of successes of $\mathcal{A}_i$. If $k_i \simeq np_i$, for every i then s_j is a typical sequence t_j, and its probability equals

$$P(t_j) = p_1^{np_1} \cdots p_N^{np_N} = e^{np_1 \ln p_1 + \cdots + np_N \ln p_N} = e^{-nH(A)} \tag{12-59}$$

where $H(A) = -(p_1 \ln p_1 + \cdots + p_N \ln p_N)$ is the entropy of the partition A. The union $\mathcal{T}$ of all typical sequences is an event in the space $\mathcal{S}_n$, and for large n, its probability equals almost 1 because almost certainly $k_i \simeq np_i$. From this and (12-59) it follows that the number n_t of typical sequences equals

$$n_t \simeq e^{nH(A)} \tag{12-60}$$

as in (12-50). If the events $\mathcal{A}_i$ are not equally likely, $H(A) < \ln N$, and (12-60) yields

$$n_t \simeq e^{nH(A)} \ll e^{n \ln N} = N^n \tag{12-61}$$

This shows that the number of typical sequences is much smaller than the number N^n of all sequences of the form (12-57).

Maximum Entropy and Typical Sequences We shall reexamine the concept of maximum entropy in the context of typical sequences, limiting the discussion to the determination of the probabilities p_i of a partition. As we see from (12-60), the entropy $H(A)$ of the partition A is maximum iff the number of typical sequences generated by the events of A is maximum. Thus the ME principle can be stated as the principle of maximizing the number of typical sequences. Since typical sequences are observable quantities, this equivalence gives a physical interpretation to the concept of entropy.

We comment finally on the relationship between the ME principle and the principle of insufficient reason. Suppose that nothing is known about the probabilities p_i. In this case, $H(A)$ is maximum iff [see (12-5)]

$$p_1 = \cdot \cdot \cdot = p_N = \frac{1}{N}$$

as in (1-8). The resulting number of typical sequences is N^n. If the unknown numbers p_i satisfy various constraints of the form $\Sigma p_i g_k(x_i) = \theta_k$ as in (12-44), no solution can be obtained with the classical method. The ME principle leads to a solution that maximizes the number of typical sequences subject to the given constraints.

Concluding Remarks

In the beginning of our development, we stated that the principal applications of probability involve averages of mass phenomena. This is based on the empirical interpretation $p \simeq k/n$ of the theoretical concept $p = P(\mathcal{A})$. We added, almost in passing, that this interpretation leads to useful results only if the following condition is satisfied: The ratio k/n must approach a constant as n increases, and this constant must be the same for any subsequence of trials. The notion of typical sequences shows that this apparently mild condition imposes severe restrictions on the class of phenomena for which it holds. It shows that of the N^n sequences, that we can form with the N elements of a partition A, only $2^{nH(A)}$ are likely to occur; most of the remaining sequences are nearly impossible.

Four Interpretations of Entropy We conclude with a summary of the similarities between the various interpretations of probability and entropy.

Probability In Chapter 1, we introduced the following interpretations of probability.
Axiomatic: $P(\mathcal{A})$ is a number p assigned to an event $\mathcal{A}$ of an experiment $\mathcal{S}$.
Empirical: In n repetitions of the experiment $\mathcal{S}$,

$$p \simeq \frac{k}{n} \tag{12-62}$$

Subjective: $P(\mathcal{A})$ is a measure of our uncertainty about the occurrence of $\mathcal{A}$ in a single performance of $\mathcal{S}$.
Principle of insufficient reason: If $\mathcal{S}$ is the union of N events $\mathcal{A}_i$ of a partition A and nothing is known about the probabilities $p_i = P(\mathcal{A}_i)$, then $p_i = 1/N$.

Entropy The results of this chapter lead to the following interpretations of $H(A)$:

Axiomatic: $H(A)$ is a number $H(A) = -\Sigma p_i \ln p_i$ assigned to a partition A of $\mathcal{S}$.

Empirical: This interpretation involves the repeated performance not of the experiment $\mathcal{S}$ but of the experiment $\mathcal{S}_n$. In this experiment, a specific typical sequence t_j is an event with probability $e^{-nH(A)}$. Applying (12-62) to this event, we conclude that if in m repetitions of $\mathcal{S}_n$ the event t_j occurs m_j times and m is large, then $P(t_j) \simeq e^{-nH(A)} \simeq m_j/m$; hence,

$$H(A) \simeq -\frac{1}{n} \ln \frac{m_j}{m} \tag{12-63}$$

This approximation relates the model concept $H(A)$ to the observation m_j and can be used in principle to determine $H(A)$ experimentally. It is, however, impractical.

Subjective: The number $H(A)$ equals our uncertainty about the occurrence of the events $\mathcal{A}_i$ of A in a single performance of $\mathcal{S}$.

Principle of maximum entropy: The unknown probabilities p_i must be such as to maximize $H(A)$, or equivalently, to maximize the number n_t of typical sequences. This yields $p_i = 1/N$ and $n_t = N^n$ if nothing is known about the probabilities p_i.

Problems

12-1 In the die experiment, $P\{\text{even}\} = .4$. Find the ME values of the probabilities $p_i = P\{f_i\}$.

12-2 In the die experiment, the average number of faces up equals 2.21. Find the ME values of the probabilities $p_i = P\{f_i\}$.

12-3 Find the ME density of an RV $\mathbf{x}$ if $\mathbf{x} = 0$ for $|x| > 1$ and $E\{\mathbf{x}\} = 0.31$.

12-4 It is observed that the duration of the telephone calls is a number $\mathbf{x}$ between 1 and 5 minutes and its mean is 3 min 37 sec. Find its ME density.

12-5 It is known that the range of an RV $\mathbf{x}$ is the interval (8, 10). Find its ME density if $\eta_x = 9$ and $\sigma_x = 1$.

12-6 The density $f(x)$ of an RV $\mathbf{x}$ is such that

$$\int_{-\pi}^{\pi} f(x)dx = 1 \qquad \int_{-\pi}^{\pi} f(x) \cos x dx = 0.5$$

Find the ME form of $f(x)$.

12-7 The number $\mathbf{x}$ of daily car accidents in a city does not exceed 30, and its mean equals 3. Find the ME values of the probabilities $P\{\mathbf{x} = k\} = p_k$.

12-8 We are given a die with $P\{\text{even}\} = .5$ and are told that the mean of the number $\mathbf{x}$ of faces up equals 4. Find the ME values of $p_i = P\{\mathbf{x} = i\}$.

12-9 Suppose that $\mathbf{x}$ is an RV with entropy $H(\mathbf{x})$ and $\mathbf{y} = 3\mathbf{x}$. Express the entropy $H(\mathbf{y})$ of $\mathbf{y}$ in terms of $H(\mathbf{x})$. **(a)** If $\mathbf{x}$ is of discrete type, **(b)** if $\mathbf{x}$ is of continuous type.

12-10 Show that if $c(x, y)$ is a positive function of unit volume and $\mathbf{x}, \mathbf{y}$ are two RVS with joint density $f(x, y)$, then

$$E\{\ln f(\mathbf{x}, \mathbf{y})\} \le -E\{\ln c(\mathbf{x}, \mathbf{y})\}$$

12-11 Show that if the expected values $\theta_k = E\{g_k(\mathbf{x}, \mathbf{y})\}$ of the m functions $g_k(\mathbf{x}, \mathbf{y})$ of the RVS $\mathbf{x}$ and $\mathbf{y}$ are known, then their ME density equals

$$f(x, y) = \gamma \exp\{-\lambda_1 g_1(x, y) - \cdots - \lambda_n g_n(x, y)\}$$

12-12 Find the ME density of the RVS $\mathbf{x}$ and $\mathbf{y}$ if $E\{\mathbf{x}^2\} = 4$, $E\{\mathbf{y}^2\} = 4$, and $E\{\mathbf{xy}\} = 3$.

12-13 Show that if the RVS $\mathbf{z}$ and $\mathbf{w}$ are jointly normal as in (5-100), then $H(\mathbf{z}, \mathbf{w}) = \ln (2\pi e \sigma_z \sigma_w \sqrt{1 - r^2})$.

12-14 **(a)** The RVS $\mathbf{x}$ and $\mathbf{y}$ are $N(0, 2)$ and $N(0, 3)$, respectively. Find the maximum of their joint entropy $H(\mathbf{x}, \mathbf{y})$. **(b)** The joint entropy of the RVS $\mathbf{x}$ and $\mathbf{y}$ is maximum subject to the constraints $E\{\mathbf{x}^2\} = 4$ and $E\{\mathbf{y}^2\} = 9$. Show that these RVS are normal and independent.

12-15 Suppose that $\mathbf{x}_1 = \mathbf{x} + \mathbf{y}$, $\mathbf{y}_1 = \mathbf{x} - \mathbf{y}$. Show that if the RVS $\mathbf{x}$ and $\mathbf{y}$ are of the discrete type, then $H(\mathbf{x}_1, \mathbf{y}_1) = H(\mathbf{x}, \mathbf{y})$, and if they are of the continuous type, then $H(\mathbf{x}_1, \mathbf{y}_1) = H(\mathbf{x}, \mathbf{y}) + \ln 2$.

12-16 The joint entropy of n RVS $\mathbf{x}_i$ is by definition $H(\mathbf{x}_1, \ldots, \mathbf{x}_n) = -E\{\ln f(\mathbf{x}_1, \ldots, \mathbf{x}_n)\}$. Show that if the RVS $\mathbf{x}_i$ are the samples of $\mathbf{x}$, then $H(\mathbf{x}_1, \ldots, \mathbf{x}_n) = nH(\mathbf{x})$.

12-17 In the experiment of two fair dice, A is a partition consisting of the events $\mathcal{A}_1 = \{\text{seven}\}$, $\mathcal{A}_2 = \{\text{eleven}\}$, and $\mathcal{A}_3 = \overline{\mathcal{A}_1 \cup \mathcal{A}_2}$. **(a)** Find its entropy. **(b)** The dice were rolled 100 times. Find the number of typical and atypical sequences formed with the events $\mathcal{A}_1$, $\mathcal{A}_2$, and $\mathcal{A}_3$.

12-18 (Coding and entropy). We wish to transmit pictures consisting of rectangular arrays of two-level spots through a binary channel. If we identify the black spots with 0 and the white spots with 1, the required time is T seconds. We are told that 83% of the spots are black and 17% are white. Show that by proper coding of the spots, the time of transmission can be reduced to $0.65T$ seconds.

Tables

In the following tables, we list the standard normal distribution $G(z)$ and the u-percentiles

$$z_u \qquad \chi_u^2(n) \qquad t_u(n) \qquad F_u(m, n)$$

of the standard normal, the chi-square, the Student t, and the Snedecor F distributions. The u-percentile x_u of a distribution $F(x)$ is the value x_u of x such that (Fig. T.1)

$$u = F(x_u) = \int_{-\infty}^{x_u} f(x)\,dx$$

Thus x_u is the inverse of the function $u = F(x)$. If $f(x)$ is even, $F(-x) = 1 - F(x)$ and $x_{1-u} = -x_u$. It suffices, therefore, to list $F(x)$ for $x \geq 0$ only and x_u for $u \geq .5$ only.

In Table 1a, we list the normal distribution $G(z)$ for $0 \leq z \leq 3$. For $z > 3$, we can use the approximation

$$G(z) \simeq 1 - \frac{1}{z\sqrt{2\pi}} e^{-z^2/2}$$

In Table 1b, we list the z_u-percentiles of the $N(0, 1)$ distribution $G(z)$. The x_u-percentile of the $N(\eta, \sigma)$ distribution $G\left(\frac{x - \eta}{\sigma}\right)$ is $x_u = \eta + z_u\sigma$.

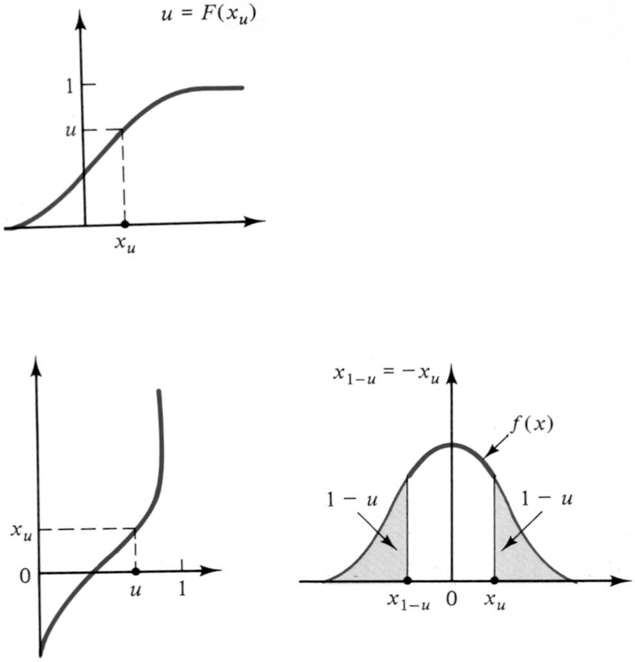

Figure T.1

In Table 2, we list the $\chi^2_u(n)$ percentiles. This is a number depending on u and on the parameter n. For large n, the $\chi^2(n)$ distribution $F_\chi(x)$ approaches a normal distribution with mean n and variance $2n$. Hence,

$$u = F_\chi(\chi^2_u) \simeq G\left(\frac{\chi^2_u - n}{\sqrt{2n}}\right) \qquad \chi^2_u(n) \simeq n + z_u\sqrt{2n}$$

The following is a better approximation

$$\chi^2_u(n) \simeq \frac{1}{2}\,(z_u + \sqrt{2n - 1})^2$$

In Table 3, we list the $t_u(n)$-percentiles. For large n, the $t(n)$ distribution $F_t(x)$ approaches a normal distribution with mean zero and variance $n/(n - 2)$. Hence,

$$u = F_t(t_u) \simeq G\left(\frac{t_u}{\sqrt{n/(n-2)}}\right) \qquad t_u(n) \simeq z_u\sqrt{\frac{n}{n-2}}$$

The $F_u(m, n)$ percentiles depend on the two parameters m and n and are determined in terms of their values for $u \geq .5$ because $F_u(m, n) = 1/F_{1-u}(n, m)$. They are listed in Table 4 for $u = .95$ and $u = .99$. Note that

$$F_{2u-1}(1, n) = t^2_u(n) \qquad \text{and} \qquad F_u(m, n) \simeq \frac{1}{m}\,\chi^2_u(m) \qquad \text{for } m \gg 1$$

Table 1a
$G(x)$

$$G(x) = \frac{1}{\sqrt{2\pi}} \int_{-\infty}^{x} e^{-y^2/2}\, dy$$

x	$G(x)$	x	$G(x)$	x	$G(x)$	x	$G(x)$
0.05	.51944	0.80	.78814	1.55	.93943	2.30	.98928
0.10	.53983	0.85	.80234	1.60	.94520	2.35	.99061
0.15	.55962	0.90	.81594	1.65	.95053	2.40	.99180
0.20	.57926	0.95	.82894	1.70	.95543	2.45	.99286
0.25	.59871	1.00	.84134	1.75	.95994	2.50	.99379
0.30	.61791	1.05	.85314	1.80	.96407	2.55	.99461
0.35	.63683	1.10	.86433	1.85	.96784	2.60	.99534
0.40	.65542	1.15	.87493	1.90	.97128	2.65	.99597
0.45	.67364	1.20	.88493	1.95	.97441	2.70	.99653
0.50	.69146	1.25	.89435	2.00	.97726	2.75	.99702
0.55	.70884	1.30	.90320	2.05	.97982	2.80	.99744
0.60	.72575	1.35	.91149	2.10	.98214	2.85	.99781
0.65	.74215	1.40	.91924	2.15	.98422	2.90	.99813
0.70	.75804	1.45	.92647	2.20	.98610	2.95	.99841
0.75	.77337	1.50	.93319	2.25	.98778	3.00	.99865

Figure T.2

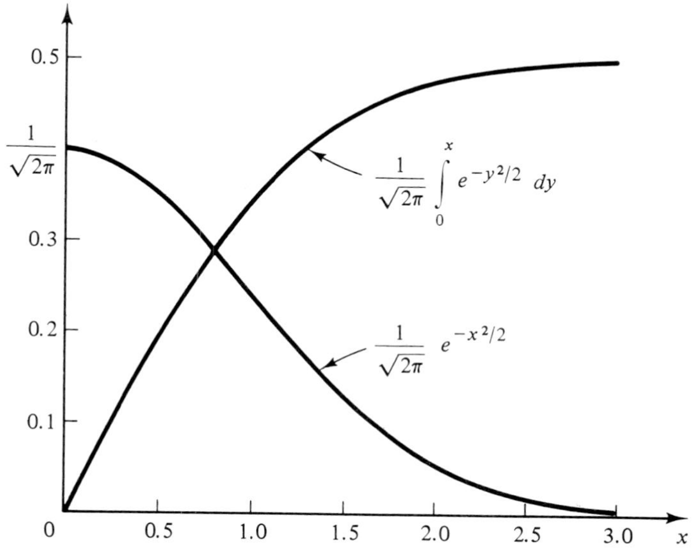

Table 1b
z_u

$$u = \frac{1}{\sqrt{2\pi}} \int_{-\infty}^{z_u} e^{-y^2/2}\, dy$$

u	.90	.925	.95	.975	.99	.995	.999	.9995
z_u	1.282	1.440	1.645	1.967	2.326	2.576	3.090	3.291

Table 2

$\chi^2_u(n)$ $\qquad$ $\chi^2_4(.95) = 9.49$

n \ u	.005	.01	.025	.05	.1		.9	.95	.975	.99	.995
1	0.00	0.00	0.00	0.00	0.02		2.71	3.84	5.02	6.63	7.88
2	0.01	0.02	0.05	0.10	0.21		4.61	5.99	7.38	9.21	10.60
3	0.07	0.11	0.22	0.35	0.58		6.25	7.81	9.35	11.34	12.84
4	0.21	0.30	0.48	0.71	1.06		7.78	9.49	11.14	13.28	14.86
5	0.41	0.55	0.83	1.15	1.61		9.24	11.07	12.83	15.09	16.75
6	0.68	0.87	1.24	1.64	2.20		10.64	12.59	14.45	16.81	18.55
7	0.99	1.24	1.69	2.17	2.83		12.02	14.07	16.01	18.48	20.28
8	1.34	1.65	2.18	2.73	3.49		13.36	15.51	17.53	20.09	21.96
9	1.73	2.09	2.70	3.33	4.17		14.68	16.92	19.02	21.67	23.59
10	2.16	2.56	3.25	3.94	4.87		15.99	18.31	20.48	23.21	25.19
11	2.60	3.05	3.82	4.57	5.58		17.28	19.68	21.92	24.73	26.76
12	3.07	3.57	4.40	5.23	6.30		18.55	21.03	23.34	26.22	28.30
13	3.57	4.11	5.01	5.89	7.04		19.81	22.36	24.74	27.69	29.82
14	4.07	4.66	5.63	6.57	7.79		21.06	23.68	26.12	29.14	31.32
15	4.60	5.23	6.26	7.26	8.55		22.31	25.00	27.49	30.58	32.80
16	5.14	5.81	6.91	7.96	9.31		23.54	26.30	28.85	32.00	34.27
17	5.70	6.41	7.56	8.67	10.09		24.77	27.59	30.19	33.41	35.72
18	6.26	7.01	8.23	9.39	10.86		25.99	28.87	31.53	34.81	37.16
19	6.84	7.63	8.91	10.12	11.65		27.20	30.14	32.85	36.19	38.58
20	7.43	8.26	9.59	10.85	12.44		28.41	31.41	34.17	37.57	40.00
22	8.6	9.5	11.0	12.3	14.0		30.8	33.9	36.8	40.3	42.8
24	9.9	10.9	12.4	13.8	15.7		33.2	36.4	39.4	43.0	45.6
26	11.2	12.2	13.8	15.4	17.3		35.6	38.9	41.9	45.6	48.3
28	12.5	13.6	15.3	16.9	18.9		37.9	41.3	44.5	48.3	51.0
30	13.8	15.0	16.8	18.5	20.6		40.3	43.8	47.0	50.9	53.7
40	20.7	22.2	24.4	26.5	29.1		51.8	55.8	59.3	63.7	66.8
50	28.0	29.7	32.4	34.8	37.7		63.2	67.5	71.4	76.2	79.5

For $n \geq 50$: $\qquad \chi^2_u(n) \simeq \frac{1}{2}(z_u + \sqrt{2n-1})^2$

Table 3

$t_u(n)$ $\qquad$ $t_{.99}(11) = 2.72$

$n \backslash u$	.9	.95	.975	.99	.995
1	3.08	6.31	12.7	31.8	63.7
2	1.89	2.92	4.30	6.97	9.93
3	1.64	2.35	3.18	4.54	5.84
4	1.53	2.13	2.78	3.75	4.60
5	1.48	2.02	2.57	3.37	4.03
6	1.44	1.94	2.45	3.14	3.71
7	1.42	1.90	2.37	3.00	3.50
8	1.40	1.86	2.31	2.90	3.36
9	1.38	1.83	2.26	2.82	3.25
10	1.37	1.81	2.23	2.76	3.17
11	1.36	1.80	2.20	2.72	3.11
12	1.36	1.78	2.18	2.68	3.06
13	1.35	1.77	2.16	2.65	3.01
14	1.35	1.76	2.15	2.62	2.98
15	1.34	1.75	2.13	2.60	2.95
16	1.34	1.75	2.12	2.58	2.92
17	1.33	1.74	2.11	2.57	2.90
18	1.33	1.73	2.10	2.55	2.88
19	1.33	1.73	2.09	2.54	2.86
20	1.33	1.73	2.09	2.53	2.85
22	1.32	1.72	2.07	2.51	2.82
24	1.32	1.71	2.06	2.49	2.80
26	1.32	1.71	2.06	2.48	2.78
28	1.31	1.70	2.05	2.47	2.76
30	1.31	1.70	2.05	2.46	2.75

For $n \geq 30$: $\qquad t_u(n) \simeq z_u \sqrt{\dfrac{n}{n-2}}$

Table 4a

$F_{.95}(m, n)$ $\qquad$ $F_{.95}(5, 8) = 3.69$

n \ m	1	2	3	4	5	6	8	10	20	30	40
1	161	200	216	225	230	234	239	242	248	250	251
2	18.5	29.0	19.2	19.2	19.3	19.3	19.4	19.4	19.4	19.5	19.5
3	10.1	9.55	9.28	9.12	9.01	8.94	8.85	8.79	8.66	8.62	8.59
4	7.71	6.94	6.59	6.39	6.26	6.16	6.04	5.96	5.80	5.75	5.72
5	6.61	5.79	5.41	5.19	5.05	4.95	4.82	4.74	4.56	4.50	4.46
6	5.99	5.14	4.76	4.53	4.39	4.28	4.15	4.06	3.87	3.81	3.77
7	5.59	4.74	4.35	4.12	3.97	3.87	3.73	3.64	3.44	3.38	3.34
8	5.32	4.46	4.07	3.84	3.69	3.58	3.44	3.35	3.15	3.08	3.04
9	5.12	4.26	3.86	3.63	3.48	3.37	3.23	3.14	2.94	2.86	2.83
10	4.96	4.10	3.71	3.48	3.33	3.22	3.07	2.98	2.77	2.70	2.66
12	4.75	3.89	3.49	3.26	3.11	3.00	2.85	2.75	2.54	2.47	2.43
14	4.60	3.74	3.34	3.11	2.96	2.85	2.70	2.60	2.39	2.31	2.27
16	4.49	3.63	3.24	3.01	2.85	2.74	2.59	2.49	2.28	2.19	2.15
18	4.41	3.55	3.16	2.93	2.77	2.66	2.51	2.41	2.19	2.11	2.06
20	4.35	3.49	3.10	2.87	2.71	2.60	2.45	2.35	2.12	2.04	1.99
30	4.17	3.32	2.92	2.69	2.53	2.42	2.27	2.16	1.93	1.84	1.79
40	4.08	3.23	2.84	2.61	2.45	2.34	2.18	2.08	1.84	1.74	1.69
50	4.03	3.18	2.79	2.56	2.40	2.29	2.13	2.03	1.78	1.69	1.63
60	4.00	3.15	2.76	2.53	2.37	2.25	2.10	1.99	1.75	1.65	1.59
70	3.98	3.13	2.74	2.50	2.35	2.23	2.07	1.97	1.72	1.62	1.57

Table 4b

$F_{.99}(m, n)$ $\qquad$ $F_{.99}(5, 8) = 6.63$

1	4,052	4,999	5,403	5,625	5,764	5,859	5,982	6,056	6,209	6,261	6,287
2	98.5	99.0	99.2	99.3	99.3	99.3	99.4	99.4	99.4	99.5	99.5
3	34.1	30.8	29.5	28.7	28.2	27.9	27.5	27.2	26.7	26.5	26.4
4	21.2	18.0	16.7	16.0	15.5	15.2	14.8	14.5	14.0	13.8	18.7
5	16.3	13.3	12.1	11.4	11.0	10.7	10.3	10.1	9.55	9.38	9.29
6	13.7	10.9	9.78	9.15	8.75	8.47	8.10	7.87	7.40	7.23	7.14
7	12.2	9.55	8.45	7.85	7.46	7.19	6.84	6.62	6.16	5.99	5.91
8	11.3	8.65	7.59	7.01	6.63	6.37	6.03	5.81	5.36	5.20	5.12
9	10.6	8.02	6.99	6.42	6.06	5.80	5.47	5.26	4.81	4.65	4.57
10	10.0	7.56	6.55	5.99	5.64	5.39	5.06	4.85	4.41	4.25	4.17
12	9.33	6.93	5.95	5.41	5.06	4.82	4.50	4.30	3.86	3.70	3.62
14	8.86	6.51	5.56	5.04	4.70	4.46	4.14	3.94	3.51	3.35	3.27
16	8.53	6.23	5.29	4.77	4.44	4.20	3.89	3.69	3.26	3.10	3.02
18	8.29	6.01	5.09	4.58	4.25	4.01	3.71	3.51	3.08	2.92	2.84
20	8.10	5.85	4.94	4.43	4.10	3.87	3.56	3.37	2.94	2.78	2.69
30	7.56	5.39	4.51	4.02	3.70	3.47	3.17	2.98	2.55	2.39	2.30
40	7.31	5.18	4.31	3.83	3.51	3.29	2.99	2.80	2.37	2.20	2.11
50	7.17	5.06	4.20	3.72	3.41	3.19	2.89	2.70	2.27	2.10	2.01
60	7.08	4.98	4.13	3.65	3.34	3.12	2.82	2.63	2.20	2.03	1.94
70	7.01	4.92	4.08	3.60	3.29	3.07	2.78	2.59	2.15	1.98	1.89

Answers and Hints for Selected Problems

CHAPTER 2

2-2 $\{4 \leq t \leq 10\}, \{7 \leq t \leq 8\}, \{t < 4\}, \{t < 4\} \cup \{8 < t\}$

2-3 $\binom{10}{3} = 120$

2-5 **(a)** $\overline{\mathscr{A} \cup \mathscr{B} \cup \mathscr{C}}$;

(f) $(\mathscr{A} \cap \mathscr{B}) \cap \overline{\mathscr{C}}$

2-6 .6, .8, .4, .3

2-9 $\binom{36}{13} \Big/ \binom{52}{13} = .0036$

2-11 $\binom{90}{18}\binom{10}{2} \Big/ \binom{100}{20} = .318$

2-12 $P(\mathscr{B}) = .4$

2-16 $p_1 = 20/30$; $p_2 = 8/11$

2-18 $P(\mathscr{A}) = 1$

2-20 $p_1 = 2.994 \times 10^{-3}$; $p_2 = 2 \times 10^{-6}$

2-22 $2^n - (n + 1)$

2-24 $p = .607$

CHAPTER 3

3-1 $p_1 = .031$, $p_2 = .016$

3-2 $p = \dfrac{7}{128}$

3-3 $p_1 = .5$, $p_2 = .322$

3-5 46, 656, 7,776; $P(\mathscr{A}) = .167$, $P(\mathscr{B}) = .842$

3-6 $p_1 = .818$, $p_2 = .086$, $p_3 = .86$

3-7 $p_1 = .201$

3-10 $p = .164$

3-12 $n = 9604$

3-13 .924

3-14 **(a)** .919; **(b)** .931

3-15 $p = .06$

3-16 **(a)** $p_1 = .29757$, $p_2 = .26932$; **(b)** $p_1 = .29754$, $p_2 = .26938$

3-17 **(a)** $p = .011869$; **(b)** $p \simeq .011857$

3-19 $p = .744$

CHAPTER 4

4-2 .003, .838, .683, .683

4-3 54.3%

4-5 $x_{.95} = 5.624$

4-8 **(a)** 6.7%; **(b)** 2.2 months

4-9 Differentiate the identity $c \int_0^x e^{-cy}\, dy = 1 - e^{-cx}$ with respect to c.

4-10 **(a)** .04; **(b)** 1.5×10^6 kw

4-11 **(a)** 4.6%; **(b)** 20%

4-12 **(a)** geometric;

(b) $\sum_{k=30}^{\infty} 0.5 \times .95^k = .95^{30-k}$

4-15 **(a)** .05; **(b)** .186

4-17 Uniform in the interval $-9 < x < 3$

4-19 **(a)** $\dfrac{1}{3\sqrt[3]{y^2}} f_x(\sqrt[3]{y})$

(b) $\dfrac{1}{4\sqrt[4]{y^3}} [f_y(\sqrt[4]{y}) + f_y(-\sqrt[4]{y})]$

(c) $[f_x(y) + f_x(-y)]U(y)$;

(d) $f_x(y)U(y)$; $P\{\mathbf{y} = 0\} = F_x(0)$

4-23 Uniform in the interval (3, 5)

4-24 $a = \pm 0.5, \qquad b = \mp 2.5$

4-26 $E\{\mathbf{x}\} = 6$

4-29 $E\{\mathbf{P}\} = 12.1 \times 1.004$ watts

4-30 **(a)** $f_y(y) = \begin{cases} \dfrac{1}{3\sqrt[3]{y^2}} & 729 < y < 1{,}331 \\ 0 & \text{otherwise} \end{cases}$;

(b) $\eta_x = 1{,}000$, $\sigma_x^2 = 1/3$;

(c) $E\{\mathbf{y}\} = 1{,}010$

4-31 Use the approximation $g(x) \simeq g(\eta) + (x - \eta)g'(\eta)$.

CHAPTER 5

5-1 **(a)** $N(0, 1/2), N(0, 1/2)$;

(b) $\gamma = 4/\pi$; **(c)** .163

5-4 Use the inequalities $F_{xy}(x, y) \le F_x(x)$, $F_{xy}(x, y) \le F_y(y)$.

5-5 **(a)** Use the identities $f(x, y) = f_x(x)f_y(y) = f(y, x)$;

(b) Note that $E\{\mathbf{y}^3\} = E\{\mathbf{x}^3/\mathbf{z}^3\}$.

5-6 Show that $E\{(\mathbf{x} - \mathbf{y})^2\} = 0$.

5-8 $\eta_y = \sigma^2; \qquad \sigma_y^2 = 2\sigma^4$

5-9 $\frac{1}{n}\sum (x_i + y_i) = \frac{1}{n}\sum x_i + \frac{1}{n}\sum y_i$;

$\frac{1}{n}\sum x_i y_i \ne \left(\frac{1}{n}\sum x_i\right)\left(\frac{1}{n}\sum y_i\right)$

5-13 Show that $a_1 = r\sigma_y/\sigma_x$, $a_2 = r\sigma_x/\sigma_y$.

5-14 $\eta_y = 0$, $\sigma_y = 8\sqrt{15}$; $a = 0$, $b = 12$

5-16 $\eta_w = 220$ watts; $\sigma_w = 11.7$ watts

5-18 $\Phi(s) = c^2/(c^2 - s^2)$, $\eta_x = 0$, $\sigma_x = \sqrt{2}/c$

5-19 Apply (5-73) to the moment function (5-103).

5-22 $f(x)$: isosceles triangle; base $0 \le z \le 2$; $f_s(s)$: isosceles triangle; base $-1 \le w \le 1$; $f_s(s)$: right triangle; base $0 \le s \le 1$.

5-23 **(a)** isosceles triangle with base $1{,}800 \le R \le 2{,}200$;

(b) $p = .125$

5-24 **(a)** isosceles triangle with base $-20 \le z \le 20$;

(b) $p_2 = P\{-5 \le \mathbf{z} \le 4\}$

5-28 **(a)** Use (5-92);

(b) $f_z(z) = \int_0^1 f_z(z - y) f_y(y)\, dy$

5-29 $f_z(z) = \begin{cases} c^2 \int_z^{\infty} e^{-c\alpha} e^{c(z-\alpha)}\, d\alpha & z > 0 \\ c^2 \int_0^{\infty} e^{-c\alpha} e^{c(z-\alpha)}\, d\alpha & z < 0 \end{cases}$

5-30 Set $\mathbf{z} = \mathbf{x}/\mathbf{y}$; show that $m_2 + m_4 = F_z(0)$; use (5-107).

CHAPTER 6

6-1 **(a)** $p_1 = .55$; **(b)** $p_2 = .82$

6-2 **(a)** 32.6%; **(b)** 32.8 years.

6-3 $f_z(z|\mathbf{w} = 5) : N(2.5, 3\sqrt{2})$; $E\{\mathbf{z}|\mathbf{w} = 5\} = 2.5$

6-4 $p_1 = .2863, \qquad p_2 = .1055$

6-5 $\varphi(x) = x - x^2/2$

6-6 **(a)** $\varphi(x) = x$; **(b)** $E\{[\mathbf{y} - \varphi(\mathbf{x})]\mathbf{x}\} = 0$

6-7 Show that $E\{\mathbf{xy}\} = E\{\mathbf{xz}\}$.

6-9 It follows from (6-21) and (6-54); $\gamma = 5/3$

6-10 **(a)** $P\{\mathbf{x} > t\} = .25e^{-4t} + .75e^{-6t}$;

(b) .355

6-12 $R(x) = (1 + cx)^{1/c} e^{-x}$

6-14 $p = \dfrac{R_x(t)F_y(t)}{F_x(t) + F_y(t)R_x(t)} \qquad p_1 = R_x(t)$

6-15 Use (6-66) and (6-82).

CHAPTER 7

7-1 Note that $E\{(c_1\mathbf{x}_1 + \cdots + c_n\mathbf{x}_n)^2\} \ge 0$ for any c_i.

7-5 Note that
$$\mathbf{y} = \mathbf{x}_i - \bar{\mathbf{x}} = \mathbf{x}_i\left(1 - \frac{1}{n}\right) + \frac{1}{n}\sum_{i \ne k=1}^{n} \mathbf{x}_k.$$

7-7 $E\{e^{s\mathbf{z}}|n\} = E\{e^{s(\mathbf{x}_1+\cdots+\mathbf{x}_n)}\} = \Phi_x^n(s)$
$$E\{e^{s\mathbf{z}}\} = E\{E\{e^{s\mathbf{z}}|\mathbf{n}\}\} = E\{\Phi_x^{\mathbf{n}}(s)\}$$
$$= \sum_{k=1}^{\infty} p_k\Phi_x^k(s)$$

7-8 **(a)** Use (5-63) and Problem 7-7 with $\Phi_k(s) = c/(s + c)$; **(b)** $E\{\mathbf{z}\} = E\{E\{\mathbf{z}|\mathbf{n}\}\} = E\{\mathbf{n}/c\} = Np/c = 200$

7-10 **(a)** Use Problem 5-8; **(b)** Show that $E\{(\mathbf{x}_i - \mathbf{x}_{i-1})^2\} = 2\sigma^2$.

7-13 Note that $\mathbf{y}_k < \mathbf{x}_{.5}$ iff $\mathbf{x}_i \le x_{.5}$ for $i \ge k$, that is, iff the event $\{\mathbf{x} \le x_{.5}\}$ occurs at least k times.

7-14 With $n = 5$ and $k = 3$, (7-47) yields $f_t(t) = 3 \times 10^{-3}(t - 55)^2(65 - t)^2$

7-15 $p \simeq 2G(\sqrt{3}) - 1 = .916$

7-16 Note that $E\{\mathbf{x}_i^2\} = \sigma^2, \qquad E\{\mathbf{x}_i^4\} = 3\sigma^4$.

7-17 **(a)** Note that $\mathbf{x}_n = mc$ iff heads shows k times; **(b)** Use (3-27); **(c)** $P\{\mathbf{x}_{50} > 6c\} \simeq G(6/7) \simeq .82$

7-19 Use the CLT and Problem 4-32.

7-20 Apply the CLT to the RVs $\ln \mathbf{x}_i$ and use Problem 4-32.

7-22 Note that $f_y(y) = 2y\chi^2(n, y^2)$.

7-23 Use (7-97) and Problem 7-22.

7-24 $\Phi_{\bar{x}}(s) = \dfrac{c^n}{(c - s/n)^n}$
$$f_{\bar{x}}(x) = \frac{c^n n^n}{(n-1)!} x^{n-1} e^{-ncx} U(x)$$

7-25 $f_z(z) = \gamma\sqrt{z}\, e^{-z/2\sigma}U(z)$

7-26 Use (7-89) and Problem 7-25.

7-28 Note that the RV $2\mathbf{x}$ is $F(4, 6)$.

7-30 **(a)** Use (7-38) and (7-97); **(b)** Show that the RV $\bar{\mathbf{w}}$ is normal with $\eta_{\bar{w}} = \eta_x - \eta_y$, $\sigma_{\bar{w}}^2 = \sigma^2(1/n + 1/m)$ and the sum
$$\frac{n-1}{\sigma^2}\mathbf{s}_x^2 + \frac{m-1}{\sigma^2}\mathbf{s}_y^2 \text{ is } \chi^2(n + m - 2)$$

7-31 If $A^2 = A$, then [see (7A-3)] $D^2 = TAT^*TAT^* = TA^2T^* = TAT^* = D$; hence, $\lambda_i^2 = \lambda_i$

7-32 Show that
$$\Phi_z(s) = \frac{1}{(1 - 2s)^{m+n}} \exp\left\{\frac{s(e_1 + e_2)}{1 - 2s}\right\}$$

CHAPTER 9

9-1 **(a)** $c = 0.115$ cm; **(b)** $n = 16$

9-2 **(a)** $202 < \eta < 204$; **(b)** 2.235 cm

9-3 $21{,}400 < \eta < 28{,}600$

9-4 $n = 1$

9-5 **(a)** $a = 25/61$, $b = 36/61$; **(b)** 12.47 ± 0.26

9-7 $c = 413$ g

9-8 $29.77 < \theta < 30.23$

9-9 $a = 0.8$, $b = 4$

9-10 If $\mathbf{w} = \mathbf{x} - \bar{\mathbf{x}}$, then $\sigma_w^2 = (1 + 1/20)\sigma^2$; $c \simeq 20.5$

9-12 $0.076 < \eta < 0.146$

9-13 $12.255 < \lambda < 13.245$

9-14 $.50 < p < .54$

9-15 **(a)** 3.2%; **(b)** $\gamma = .78$

9-17 $n = 2{,}500$

9-18 $\hat{p} = .567$

9-21 $-5.17 < \eta_x - \eta_y < -2.83$

9-22 $87.7 < \eta < 92.3, \qquad 3.44 < \sigma < 9.13$

9-23 $0.44 < \sigma < 0.276$

9-25 $.308 < r < .572$

9-26 Use the identity $\Sigma[(\mathbf{x}_i + \mathbf{y}_i) - (\bar{\mathbf{x}} + \bar{\mathbf{y}})]^2 = \Sigma(\mathbf{x}_i - \bar{\mathbf{x}})^2 + \Sigma(\mathbf{y}_i - \bar{\mathbf{y}})^2 + 2(n - 1)\hat{\boldsymbol{\mu}}_{11}$.

9-27 **(a)** Note that $\Sigma(a - zb_i)^2 \ge 0$ for all z; **(b)** Set $x_i - \bar{x} = a_i$, $y_i - \bar{y} = b_i$.

9-28 $P\{\mathbf{y}_9 \le x_{.5} < \mathbf{y}_{11}\} = .5598 \simeq .5587$

9-30 Note that

$$G\left(\frac{0.5}{\sqrt{n/4}}\right) - G\left(\frac{-0.5}{\sqrt{n/4}}\right) \simeq \sqrt{\frac{2}{\pi n}}$$

9-31 $\delta = .1$, $c = .02$; $n = 3745$

9-32 $\hat{c} = 1.29$

9-33 $\hat{c} = 1/(\bar{x} - x_0)$

9-36 Use the identities

$$\frac{\partial^2 f}{\partial \theta^2} = \frac{\partial^2 L}{\partial \theta^2} f + \left(\frac{\partial L}{\partial \theta}\right)^2 f$$

$$0 = \int_{-\infty}^{\infty} \frac{\partial f}{\partial \theta} dx = \int_{-\infty}^{\infty} \frac{\partial^2 f}{\partial \theta^2} dx$$

9-37 **(a)** $I = 1/\theta^2$; **(b)** Use Problem 7-24.

9-38 Note that the statistic $\mathbf{w} = \mathbf{y} - \alpha\mathbf{z}$ is unbiased with $\sigma_w^2 = \sigma_y^2 - 2\alpha\mu_{yz} + \alpha^2\sigma_z^2$.

9-43 The function $J(y, \theta)$ in (9-95) is known, and

$$f_y(y, \theta) = f_y(y, \theta_0) \frac{J(y, \theta)}{J(y, \theta_0)}$$

9-44 Show that $f(y, z) = c^2 e^{-cy}$ for $0 < z < y$.

9-45 Use Problem 9-38.

CHAPTER 10

10-1 **(a)** $\bar{x} > 8.58$, $\beta = .32$;
(b) $n = 129$, $\bar{x} > 8.41$

10-3 $t_{.05}(63) = -1.67$, $t_{.005}(63) = -2.62$, $q = -1.6$; accept H_0

10-5 $t_{.975}(40) \simeq 2$, $\hat{\sigma}_{\bar{w}} = 0.336$, $q = 1.19 < 2$; accept H_0

10-6 Test $\eta_y = 0$ against $\eta_y \neq 0$ where

$$\bar{\mathbf{x}} = \frac{1}{n_k} \sum_{i=1}^{n_k} \bar{\mathbf{x}}_{k_i} \qquad \sigma_y^2 = \sigma^2 \sum_{k=1}^{r} \frac{c_k^2}{n_k}$$

10-7 $k_1 = 24$, $k_2 = 40$; reject H_0

10-9 $n\lambda_0 = 110$, $k_1 = 110 - 1.645\sqrt{110} = 93 > q = 90$; reject the hypothesis that $\lambda = 5$

10-10 Under hypothesis H_1, the RV $\mathbf{q}\sigma_0^2/\sigma^2$ is $\chi^2(n)$ and

$$P\{\mathbf{q} \le c|H_1\} = P\left\{\frac{\sigma_0^2}{\sigma^2}\mathbf{q} < \frac{\sigma_0^2}{\sigma^2}c|H_1\right\}$$

10-12 $c = .136 > q = .1$; accept H_0

10-13 $k_1 = 31 < k = 36 < k_2 = 47$; accept H_0'; no decision about H_0

10-14 **(a)** $G(2 - 0.5\sqrt{n}) - G(-2 - 0.5\sqrt{n}) = \beta(30.1) = 0.1$; find n. Accept H_0 iff $30 - 0.4/\sqrt{n} < x < 30 + 0.4/\sqrt{n}$;
(b) $1 - \beta(30.1) = 0.9$

10-15 **(a)** UCL = 93 Ω; **(b)** $p = .79$

10-16 $q = 9.76 < \chi^2_{.95}(5) = 11$; accept H_0

10-17 $q = 2.36 < \chi^2_{.95}(2) = 5.99$; yes

10-18 $q = 17.76 < \chi^2_{.95}(9) = 16.92$; reject H_0

10-20 Note that $\sum p_i = 1$, $\sum \frac{\partial p_i}{\partial \theta_j} = 0$.

10-25 Reject H_0 if $x > 1.384$; $\beta = .40$

10-26 **(a)** $f_r(r, \theta_0)\, dr = P\{r < \mathbf{r} \le r + dr|H_0\} = f(X, \theta_0)\, dV$, $f_r(r, \theta_1)\, dr = P\{r < \mathbf{r} \le r + dr|H_1\} = f(X, \theta_1)\, dV$;

(b) $$f_r(r, \theta) = \frac{\theta^2 x}{r(\theta_1 - \theta_0)} e^{-\theta x}$$

$$\frac{\theta_0^2}{\theta_1^2} e^{-(\theta_0 - \theta_1)x} = r$$

10-27 Note that the RVs $2\theta\mathbf{x}$ and $2\theta\mathbf{q}$ are $\chi^2(4)$ and $\chi^2(4n)$ respectively. Show that

$$\alpha = P\{\mathbf{q} < c|H_0\} = P\{2\theta_0\mathbf{q} < 2\theta_0 c\}$$
$$= \int_0^{2\theta_0} \chi^2(4n, x)\, dx$$
$$\beta(\theta) = P\{\mathbf{q} > c|H_1\} = P\{2\theta\mathbf{q} > 2\theta c|H_1\}$$
$$= \int_{2\theta c}^{\infty} \chi^2(4n, x)\, dx$$

10-28 Use the lognormal approximation (Problem 7-20) for the density of the product $\mathbf{x}_1 \cdots \mathbf{x}_n$.

10-32 Show that $\theta_{m0} = \theta_0$, $\theta_m = \bar{x}$, $w = 2n[\theta_0 - \bar{x} + \bar{x}\ln(\bar{x}/\theta_0)]$.

10-33 Show that the N RVs $\mathbf{x}_j^i$ are i.i.d. with joint density $f(X) \sim \exp\left\{-\frac{1}{2V}Q\right\}$ where $Q = Q_1 + Q_2$ as in Problem 10-22.

CHAPTER 11

11-3 Use (11-26) with $w_1 = 1$, $w_2 = x$, $w_3 = y$, $y = z$.

11-6 **(a)** Maximize the sum

$$\Sigma(x_i - \alpha)^2 + \Sigma(y_i - \beta)^2 + \Sigma(z_i - \gamma)^2$$

subject to the constraint $\alpha + \beta + \gamma = \pi$ and show that

$$\hat{\alpha} - \bar{x} = \hat{\beta} - \bar{y} = \hat{\gamma} - \bar{z} = \frac{\pi - (\bar{x} + \bar{y} + \bar{z})}{3}$$

(b) Maximize the sum

$$\frac{1}{\sigma_x^2}\sum (x_i - \alpha)^2 + \frac{1}{\sigma_y^2}\sum (y_i - \beta)^2 + \frac{1}{\sigma_z^2}\sum (z_i - \gamma)^2$$

subject to the same constraint and show that

$$\frac{\hat{\alpha} - \bar{x}}{\sigma_x^2} = \frac{\hat{\beta} - \bar{y}}{\sigma_y^2} = \frac{\hat{\gamma} - \bar{z}}{\sigma_z^2} = \frac{\pi - (\bar{x} + \bar{y} + \bar{z})}{3}$$

11-8 Use Lagrange multipliers.

(a) Note that

$$I = \sum \alpha_i^2 - \lambda\left(\sum \alpha_i - 1\right) - \mu \sum \alpha_i x_i$$

$$\frac{\partial I}{\partial \alpha_i} = 2\alpha_i - \lambda - \mu x_i = 0$$

for any λ and μ. Solve the $n + 2$ equations $2\alpha_i = \lambda + \mu x_i$, $\Sigma\alpha_i = 1$, $\Sigma\alpha_i x_i = 0$ for the $n + 2$ unknowns α_i, λ, and μ;

(b) Proceeding similarly, solve the system $2\beta_i = \lambda + \mu x_i$, $\Sigma\beta_i = 0$, $\Sigma\beta_i x_i = 1$.

11-9 **(a)** Accept H_0 iff [see (11-52)]

$$|\hat{b}| < \sigma_{\hat{b}} z_{1-\alpha/2}$$

where

$$\sigma_{\hat{b}}^2 = \frac{1}{\Sigma(x_i - \bar{x})^2}\sigma^2$$

(b) Accept H_0 iff [see (11-57)]

$$|\hat{b}| < \sigma_{\hat{b}} t_{1-\alpha/2}(n-2)$$

where

$$\sigma_{\hat{b}}^2 = \frac{1}{\Sigma(x_i - \bar{x})^2}\,\frac{\Sigma[y_i - (\hat{\alpha} + \hat{b}x_i)]^2}{n-2}$$

11-12 Note that

(a) $\text{Cov}(\bar{\mathbf{y}}, \hat{\mathbf{b}}) = \dfrac{\sigma^2}{n}\sum \beta_i = 0$

(b) $$\frac{1}{\sigma^2}\sum(\hat{\boldsymbol{\eta}}_i - \eta_i)^2 = \left[\frac{\bar{\mathbf{y}} - (a + b\bar{x})}{\sigma/\sqrt{n}}\right]^2 + \left(\frac{\hat{\mathbf{b}} - b}{\sigma_{\hat{b}}}\right)^2$$

(c) $\Sigma\boldsymbol{\nu}_i^2 = \Sigma\boldsymbol{\epsilon}_i^2 - \Sigma(\hat{\boldsymbol{\eta}}_i - \eta_i)^2$

11-13 Maximize the sum $\Sigma w_i[y_i - (a + bx_i)]^2$.

11-15 Using the independence of the RVs $\mathbf{y} - a\mathbf{x}$ and $\mathbf{x}$, show that $E\{\mathbf{y} - a\mathbf{x}|\mathbf{x}\} = 0$, $E\{(\mathbf{y} - a\mathbf{x})^2\} = Q$, and $E\{\mathbf{y} - a\mathbf{x}|\mathbf{x}\} = E\{\mathbf{y}|\mathbf{x}\} - aE\{\mathbf{x}|\mathbf{x}\} = E\{\mathbf{y}|\mathbf{x}\} - a\mathbf{x}$.

11-16 Note that the RVs $\mathbf{y} - \hat{\mathbf{y}}$ and $\mathbf{x}_i$ are orthogonal; hence, they are independent, and $E\{\mathbf{y} - \hat{\mathbf{y}}|\mathbf{x}_i, \ldots, \mathbf{x}_n\} = E\{\mathbf{y} - \hat{\mathbf{y}}\} = 0$.

CHAPTER 12

12-1 $p_1 = p_3 = p_5 = .2$, $p_2 = p_4 = p_0 = .4/3$

12-2 $p_1 = .42$, $p_2 = .252$, $p_3 = .151$, $p_4 = .09$, $p_5 = .054$, $p_6 = .034$

12-3 $f(x) = 0.425e^{-x}$ for $-1 < x < 1$

12-4 $f(x) = 0.212e^{-(x-1)}$ for $1 < x < 5$

12-5 $f(x) = \dfrac{1.465}{\sqrt{2\pi}}e^{(x-9)^2/2}$

12-6 $f(x) = Ze^{-\lambda\cos x}$ for $|x| < \pi$ and $f(x) = 0$ for $|x| > \pi$

12-7 $p_k = .25 \times .75^k$, $0 \le k \le 30$

12-8 $p_1 = p_2 = .064$, $p_3 = p_4 = .138$, $p_5 = p_6 = .298$

12-9 **(a)** $H(\mathbf{y}) = H(\mathbf{x})$;
(b) $H(\mathbf{y}) = H(\mathbf{x}) + \ln 3$

12-10 Show that

$$E\left\{\ln\frac{c(\mathbf{x}, \mathbf{y})}{f(\mathbf{x}, \mathbf{y})}\right\} \le E\left\{\frac{c(\mathbf{x}, \mathbf{y})}{f(\mathbf{x}, \mathbf{y})} - 1\right\} = 0$$

12-12 $$f(x, y) = \frac{1}{2\pi\sqrt{7}}\exp\left\{-\frac{2}{7}\left(x^2 - \frac{3}{2}xy + y^2\right)\right\}$$

12-13 Show that

$$E\left\{\frac{\mathbf{z}^2}{\sigma_z^2} - 2r\frac{\mathbf{zw}}{\sigma_z\sigma_w} + \frac{\mathbf{w}^2}{\sigma_\varphi^2}\right\} = 2(1 - r^2)$$

12-14 $H(\mathbf{x}, \mathbf{y}) = 3.734$

12-17 **(a)** $H(A) = 0.655$;
(b) $n_t \simeq 2.79 \times 10^{28}$, $n_a \simeq 5.36 \times 10^{47}$

Index

D

E

F

G

H

I

J

K

L

M

N

O

P

Q

R

S

T

U

V

W

Z